过渡金属催化的
药物合成反应

Drug Synthesis
Reactions Catalyzed
by Transition Metals

赵继全　等 编著

化学工业出版社

·北京·

内容简介

《过渡金属催化的药物合成反应》主要介绍了过渡金属催化偶联反应的反应机理、基本反应和在药物合成上的应用实例。本书共分9章：第1章赫克（Heck）反应，第2章铃木（Suzuki）反应，第3章根岸（Negishi）反应，第4章熊田（Kumada-Corriu）偶联反应，第5章薗头（Sonogoshira）反应，第6章施蒂勒（Stille）反应，第7章桧山（Hiyama）偶联反应，第8章Tsuji-Trost反应，以及第9章串联反应。

本书可用作制药工程、药物化学和化学专业高年级本科生以及研究生的教学用书，也可供相关领域科研人员参考。

图书在版编目（CIP）数据

过渡金属催化的药物合成反应 / 赵继全等编著.
北京：化学工业出版社，2024.10. -- ISBN 978-7-122-46125-4

Ⅰ. TQ426.8

中国国家版本馆 CIP 数据核字第 2024N75D34 号

责任编辑：马泽林　杜进祥　　　文字编辑：张瑞霞
责任校对：田睿涵　　　　　　　装帧设计：刘丽华

出版发行：化学工业出版社
　　　　　（北京市东城区青年湖南街13号　邮政编码100011）
印　　刷：北京云浩印刷有限责任公司
装　　订：三河市振勇印装有限公司
710mm×1000mm　1/16　印张21¼　字数404千字
2024年9月北京第1版第1次印刷

购书咨询：010-64518888　　　　售后服务：010-64518899
网　　址：http://www.cip.com.cn
凡购买本书，如有缺损质量问题，本社销售中心负责调换。

定　价：129.00元　　　　　　　　版权所有　违者必究

前言

药物是用来治疗、预防、诊断疾病的化学物质,多数为有机化合物。历史上药物是通过萃取从药用植物中获得的,而今则主要通过药物合成制备和生产。药物合成依赖于有机合成反应,其中碳-碳键的构建最为重要,应用最普遍。长期以来,碳-碳键的构建多借助于传统的有机反应,如傅-克(Friedel-Crafts)反应、维尔斯迈尔-哈克(Vilsmeier-Haack)反应等芳环上的亲电取代反应,普林斯(Prins)反应、曼尼希(Mannich)反应等酸催化的加成和缩合反应,迈克尔(Michael)加成、雷夫马斯基(Reformasky)反应等负离子为亲核试剂的加成反应,以及卤代烃与稳定碳负离子如碱金属氰化物和炔钠等的亲核取代反应。在药物合成上,这些反应在未来仍将起不可或缺的作用。如今,药物合成面临不断提高的现实要求,如更高的区域与立体选择性、反应过程的高效性、生态效益与经济效益的有机结合等。为满足这些要求,在药物合成中必须挖掘、利用新的有机合成反应。

20 世纪 70 年代以来,一系列过渡金属特别是钯催化构建碳-碳键的偶联反应不断被发现并得到迅速发展,成为有机化学研究领域的里程碑之一,为有机和药物分子的高效、洁净和多方位合成奠定了基础。由于此类反应的重要性及对有机合成领域的重大贡献,瑞典皇家科学院将 2010 年诺贝尔化学奖授予此类反应中的赫克(Heck)反应、铃木(Suzuki)反应和根岸(Negishi)反应的发现和开拓者 Richard F. Heck、Akira Suzuki 和 Ei-ichi Negishi,以表彰他们在"有机合成中钯催化的交叉偶联反应"优异的工作。尽管这些偶联反应也可以在其他过渡金属催化剂催化下进行,但钯催化剂具有更高的选择性和非常温和的反应条件,因此应用最为广泛。此外,根据反应特点引入特定的配体可使非活性的底物和多反应途径的底物顺利且高选择性地进行反应,极大地拓展了底物范围;而且,手性配体的引入可实现部分反应的不对称化,得到光学活性的目标产物。

相较于传统的构建碳-碳键的反应,过渡金属尤其是钯催化的偶联反应具有如下优点:通过单步催化转化有时可以替代多步化学计量反应,这种捷径反应可有效简化反应工艺,降低目标产物;反应通常在温和的反应条件下进行,无需官能团的保护-去保护步骤,使药物合成中梦寐以求的利用高度官能团化的反应片段偶联的收敛合成策略得以实现；反应通常在低温下进行并很少需要高活性试剂，提高了反应的选择性并减少废弃物生成；此外，通过精心设计反应路线，还可以实现此类催化反应之间或与其他反应之间的串联，完成复杂药物分子的简捷合成。

正是由于这些催化偶联反应具有上述优点，其在药物合成上的应用越来越普遍，地位也越来越重要。大力发展医药事业，始终把人民群众的健康放在首位，坚持人民至上、生命至上，这是中国共产党执政为民理念的最好诠释。我国科研人员在过渡金属催化的偶联反应研究领域也取得了一系列重大的研究成果，在本书中有相应的介绍。

然而，国内尚无相关书籍专门系统介绍此类过渡金属催化的偶联反应及其在药物合成上的应用。为此，希望本书可以弥补此类反应在国内现有书籍中的缺失。由于相关反应的研究成果庞大，无法做到面面俱到，编者只能介绍一些成熟的反应和应用成果。为了拓展读者知识面，更加深入地理解相关知识内涵，每一反应和实例均著有参考文献。相关知识按反应类型列章介绍，每章内容包括反应机理、基本反应和药物合成应用实例。希望本书能够促进学生更新、拓展有机反应理论，强化新有机反应在药物合成中的应用。

本书由赵继全负责统稿，全书由赵继全（绪论、第一、二章）、张月成（第三、四章）、张宏宇（第五、六、七章）和韩亚平（第八、九章）共同撰写。

本书编写得到了河北工业大学化工学院领导的支持与鼓励，课题组研究生也提供了有益的帮助，在此表示感谢。

编者在书稿撰写过程中力求精益求精，但由于编者水平有限，书中仍可能存在许多不足，恳请广大读者不吝赐教。

赵继全于河北工业大学
2024 年 3 月

目录

绪论 / 1

0.1 概述 ·· 1
0.2 Heck 偶联反应 ·· 3
0.3 Suzuki 偶联反应 ·· 4
0.4 Negishi 偶联反应 ··· 5
0.5 Kumada-Corriu 偶联反应 ·· 6
0.6 Sonogashira 偶联反应 ··· 8
0.7 Stille 偶联反应 ·· 9
0.8 Hiyama 偶联反应 ··· 10
0.9 Tsuji-Trost 偶联反应 ·· 12
0.10 结语 ·· 13
参考文献 ·· 14

第 1 章
Heck 反应及其在药物合成上的应用 / 17

1.1 Heck 反应及其反应机理 ·· 17
1.2 基本的 Heck 反应 ··· 18
1.2.1 分子间 Heck 反应 ·· 18
1.2.2 分子内 Heck 反应 ·· 21
1.2.3 Heck 串联（tandem）反应 ··· 29
1.2.4 不对称 Heck 反应 ·· 33
1.3 Heck 反应在药物合成上的应用实例 ·· 40

1.3.1 分子间反应的应用 ---- 40
1.3.2 分子内反应的应用 ---- 46
参考文献 ---- 49

第 2 章
Suzuki 反应及其在药物合成上的应用 / 54

2.1 Suzuki 反应及其机理 ---- 54
  2.1.1 Suzuki 反应 ---- 54
  2.1.2 反应机理 ---- 55
2.2 基本的 Suzuki 反应 ---- 55
  2.2.1 分子间 Suzuki 反应 ---- 55
  2.2.2 分子内 Suzuki 反应 ---- 71
  2.2.3 不对称 Suzuki 反应 ---- 73
2.3 Suzuki 反应在药物合成上的应用实例 ---- 76
  2.3.1 分子间反应的应用 ---- 76
  2.3.2 分子内反应的应用 ---- 80
参考文献 ---- 83

第 3 章
Negishi 反应及其在药物合成上的应用 / 89

3.1 Negishi 反应及其机理 ---- 89
  3.1.1 Negishi 反应 ---- 89
  3.1.2 反应机理 ---- 90
3.2 基本 Negishi 反应 ---- 90
  3.2.1 $C(sp^2)$-$C(sp^2)$ 键的构建 ---- 91
  3.2.2 $C(sp)$-$C(sp^2)$ 键的构建 ---- 93
  3.2.3 $C(sp^3)$-$C(sp^2)$ 键的构建 ---- 96
  3.2.4 $C(sp^3)$-$C(sp^3)$ 键的构建 ---- 99
  3.2.5 不对称 Negishi 反应 ---- 101
3.3 Negishi 反应在药物合成上的应用实例 ---- 103
参考文献 ---- 112

第4章
Kumada-Corriu 偶联反应及其在药物合成上的应用 / 116

- 4.1 Kumada-Corriu 偶联反应及其反应机理 ············ 116
  - 4.1.1 Kumada-Corriu 偶联反应 ············ 116
  - 4.1.2 反应机理 ············ 117
- 4.2 基本的 Kudama-Corriu 偶联反应 ············ 118
  - 4.2.1 格氏试剂与乙烯基的偶联 ············ 118
  - 4.2.2 格氏试剂与芳基或杂芳基的偶联 ············ 120
  - 4.2.3 格式试剂与烷基的偶联 ············ 125
  - 4.2.4 特殊的 Kumada-Corriu 偶联反应 ············ 127
  - 4.2.5 不对称 Kumada-Corriu 偶联反应 ············ 129
- 4.3 Kumada-Corriu 偶联反应在药物合成上的应用实例 ············ 132
- 参考文献 ············ 139

第5章
Sonogashira 反应及其在药物合成上的应用 / 143

- 5.1 Sonogashira 反应及机理 ············ 143
  - 5.1.1 Sonogashira 反应 ············ 143
  - 5.1.2 反应机理 ············ 144
- 5.2 基本的 Sonogashira 反应 ············ 145
  - 5.2.1 经典催化体系下的反应 ············ 146
  - 5.2.2 单纯钯催化的 Sonogashira 反应 ············ 152
  - 5.2.3 非贵金属催化的 Sonogashira 反应 ············ 155
- 5.3 Sonogashira 反应在药物合成上的应用实例 ············ 156
- 参考文献 ············ 162

第6章
Stille 反应及其在药物合成上的应用 / 166

- 6.1 Stille 反应及其机理 ············ 166
  - 6.1.1 Stille 反应 ············ 166
  - 6.1.2 反应机理 ············ 167

6.2	基本的 Stille 反应	168
6.2.1	基本的分子间 Stille 反应	168
6.2.2	分子内 Stille 反应	178
6.3	Stille 反应在药物合成上的应用实例	180
参考文献		188

第 7 章
Hiyama 偶联反应及其在药物合成上的应用 / 192

7.1	Hiyama 偶联反应及其反应机理	192
7.1.1	Hiyama 偶联反应	192
7.1.2	反应机理	193
7.2	基本的 Hiyama 偶联反应	193
7.2.1	分子间 Hiyama 偶联反应	193
7.2.2	分子内 Hiyama 反应	209
7.3	Hiyama 反应在药物合成上的应用实例	210
参考文献		217

第 8 章
Tsuji-Trost 反应及其在药物合成上的应用 / 220

8.1	Tsuji-Trost 反应及其机理	220
8.1.1	Tsuji-Trost 反应	220
8.1.2	反应机理	222
8.2	基本的 Tsuji-Trost 反应	223
8.2.1	分子间 Tsuji-Trost 反应	223
8.2.2	分子内 Tsuji-Trost 反应	240
8.3	Tsuji-Trost 反应在药物和天然产物合成上的应用实例	251
参考文献		257

第 9 章
过渡金属催化串联反应及其在药物合成上的应用 / 262

| 9.1 | Heck/Tsuji-Trost 串联反应及机理 | 262 |

9.1.1	Heck/Tsuji-Trost 串联反应	262
9.1.2	反应机理	263
9.1.3	基本的 Heck/Tsuji-Trost 反应	264
9.1.4	Heck/Tsuji-Trost 反应在药物和天然产物合成上的应用	278
9.2	Heck/Heck 串联反应	281
9.2.1	Heck/Heck 串联反应及机理	281
9.2.2	基本的 Heck/Heck 反应	281
9.2.3	Heck/Heck 串联反应在药物和天然产物合成上的应用	288
9.3	Heck/Sonogashira 串联反应	292
9.3.1	Heck/Sonogashira 串联反应及机理	292
9.3.2	基本的 Heck/Sonogashira 反应	294
9.3.3	Heck/Sonogashira 串联反应在药物和天然产物合成上的应用	299
9.4	Heck/Suzuki 串联反应	301
9.4.1	Heck/Suzuki 串联反应及反应机理	301
9.4.2	基本的 Heck/Suzuki 反应	302
9.4.3	Heck/Suzuki 串联反应在药物和天然产物合成上的应用	313
9.5	Heck/C-H 官能团化串联反应	314
9.5.1	Heck/C-H 官能团化串联反应及机理	314
9.5.2	基本的 Heck/C-H 官能团化串联反应	316
9.5.3	Heck/C-H 官能团化串联反应在药物和天然产物合成上的应用	322
参考文献		324

绪论

0.1 概述

药物是用来治疗、预防、诊断疾病的化学物质。现代治疗药剂主要是有机药物，数量庞大，用于治疗人类已发现的各类疾病，保障人们的身体健康。作为有机药物的有机化合物来源广泛，其中的一部分属于天然产物，提取于动植物，而大部分则通过药物合成获得。随着人类活动的不断增加，自然环境的变迁，新的疾病还将不断产生，人类必须不断进行药物合成研究以获取新的药物。碳-碳键的生成是构筑有机分子骨架、延长碳链、引入官能团的主要手段。生成碳-碳键的反应众多，但从大的方面分类主要有经典的酸或碱催化的有机反应、金属或元素有机试剂参与的化学计量反应以及过渡金属催化的偶联反应。长期以来，碳-碳键的构建多借助于传统的有机反应，如狄尔斯-阿尔德(Diels-Alder)反应、傅-克（Friedel-Crafts）反应、维尔斯迈尔-哈克(Vilsmeier-Haack)反应等芳环上的亲电取代反应，普林斯（Prins）反应、曼尼希（Mannich）反应等酸催化的加成和缩合反应，迈克尔（Michael）加成反应、雷夫马斯基（Reformasky）反应等负离子为亲核试剂的加成反应，以及卤代烃与稳定碳负离子如碱金属氰化物和炔钠等的亲核取代反应[1]。在药物合成上，这些反应未来仍将起不可或缺的作用。如今，药物合成面临不断提高的现实要求，如更高的区域与立体选择性、反应过程的高效性、生态效益与经济效益的有机结合等。为满足这些要求，在药物合成中必须挖掘、利用新的有机反应。

因此，过渡金属催化的偶联反应应运而生。20世纪60年代中期至70年代初利用钯催化构建碳-碳键的开拓性研究引发了有机和药物合成的革命性发展。受氧作为亲核试剂（水或氢氧根离子）与烯烃-钯配合物反应形成碳-氧键，即著名的Wacker反应的启发，Tsuji试图在Wacker反应条件下用碳亲核试剂替代氧亲核试剂以构建碳-碳键，结果成功实现了π-烯丙基氯化钯与丙二酸酯（或乙酰乙酸酯）偶联生成烯丙基化丙二酸酯（或烯丙基化乙酰乙酸酯）[2]。几年后，Heck[3]和Mizoroki等[4]分别报道了芳基卤化物与烯烃在钯催化下的偶联反应。芳基卤化物与烯烃的偶联反应可被看作金属有机催化在有机合成发展和应用中的里程碑。该反应的发现激发了化学家的研究热情，一些钯催化的偶联反应在 20 世纪 70 年代先后被发现，其中之一是芳基卤化物与炔烃的偶联，即 Sonogashira 偶联反应，通常在催化量的钯和铜共同存在下进行[5]。与烯烃和炔烃作为偶联伙伴的 Heck 和 Sonogashira 偶联反应不同，Negishi[6]和 Murahashi[7]分别将芳基锌和芳基镁试剂作为偶联伙伴用于钯催化的偶联反应得到联二芳基化合物，如今分别被称作 Negishi 偶联和 Kumada 偶联反应。随后，Suzuki 和 Miyaura[8]发展了钯催化的芳基硼酸和酯与芳基卤化物的偶联反应，用于合成对称和不对称的联二芳基化合物，如今被称作 Suzuki 偶联反应或 Suzuki-Miyaura 反应。与此类似，Stille[9]和 Hiyama[10,11]发现了钯催化的芳基锡和芳基硅烷分别与卤代芳基化合物合成联二芳基化合物的反应，分别称作 Stille 偶联和 Hiyama 偶联反应。这些反应在有机合成、药物合成以及有机材料合成上得到广泛应用，使得有机合成发生了革命性的发展。为此，瑞典皇家科学院将 2010 年诺贝尔化学奖授予此类反应中的 Heck 偶联反应、Suzuki 偶联反应和 Negishi 偶联反应的发现和开拓者 R. F. Heck、A.Suzuki 和 E-i. Negishi，以表彰他们在"有机合成中钯催化的交叉偶联反应"方面的卓越贡献。此后，许多有机合成化学家在此领域进行了大量工作，取得了大量突破性的成果，相继发展了更多过渡金属催化的偶联反应，偶联反应底物更加广泛，催化剂或催化体系更加多元化。就历史地位和在药物合成上的应用程度而言，以下八个偶联反应更加重要，分别是 Heck 偶联反应[12]、Suzuki 偶联反应[13]、Negishi 偶联反应[14]、Hiyama 偶联反应[15]、Kumada-Corriu 偶联反应[16]、Sonogashira 偶联反应[17]、Stille 偶联反应[9]以及 Tsuji-Trost 偶联反应[18]。其中，前七个偶联反应具有更多的共同点，都属于芳基卤化物或其类似物与亲核试剂的偶联反应，只是金属有机亲核试剂不同，这可由图 0-1 直观地看出。

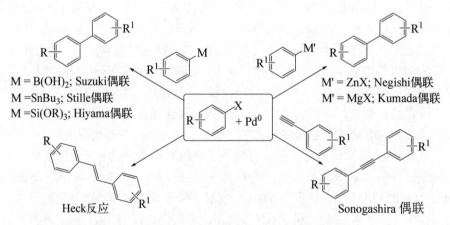

图 0-1　重要的钯催化构建碳-碳键的偶联反应

尽管这些过渡金属催化的偶联反应（包括 Tsuji-Trost 偶联反应）的底物或亲核试剂存在差异，但从反应机理的角度来看基本上遵循图 0-2 所示的由氧化加成、金属转移和还原消除三个基元步骤构成的催化循环进行。

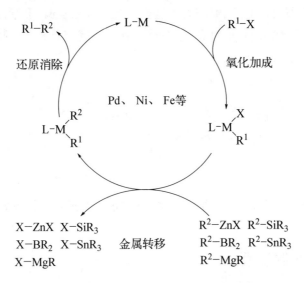

图 0-2 过渡金属催化的偶联反应的基本催化循环

经过几十年的发展，如今这些过渡金属催化的偶联反应已成为药物合成的可靠和不可或缺的手段，也为复杂性不断增加的药物分子设计、合成提供新的思路。为了使读者对这些过渡金属催化的偶联反应有一个初级认识，下面对它们进行简要概述。本书正文将对相关偶联反应的反应机理、基本反应和应用实例进行更详细的阐述。

0.2 Heck 偶联反应

在 Heck 偶联反应中，（杂）芳基、烯基和苄基卤化物与各种烯烃类化合物在钯催化剂催化下进行偶联反应生成相应的取代烯烃，见反应式（1）。

$$R^1\text{-}X \ + \ \diagup\!\!\diagdown R^2 \xrightarrow{\text{Pd, 碱}} R^1\diagup\!\!\diagdown\!\!\diagup R^2 \tag{1}$$

X=卤素、三氟甲磺酸酯基等；
R^1, R^2=芳基、杂芳基、烯基、苄基

Heck 反应通常以很高的立体和区域选择性顺利进行。自从 20 世纪 70 年代早

期发现并经几十年的发展,无论是催化剂的种类还是底物的范围都有大的拓展。目前,Heck 反应在有机合成特别是药物合成中占有重要地位,使合成界受益匪浅,一些采用传统方法难以合成的复杂药物分子得以顺利合成。图 0-3 给出了利用 Heck 反应作为关键合成步骤合成的两个临床药物孟鲁斯特(montelukast)和紫杉醇(taxol)[19]。临床上孟鲁斯特用于治疗过敏性鼻炎和哮喘,而紫杉醇对各种肿瘤特别是肺癌有一定的疗效。

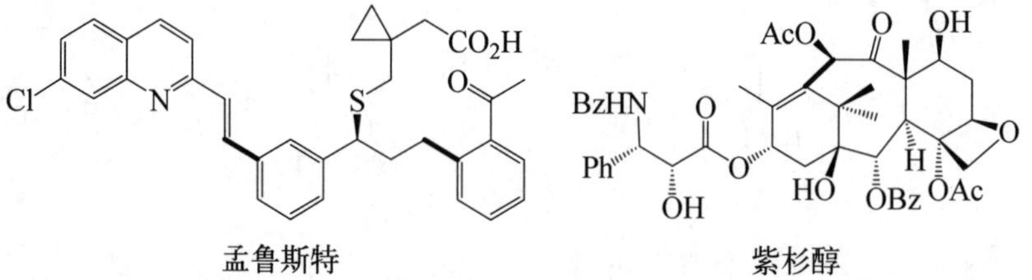

图 0-3 利用 Heck 反应合成的两个代表性的药物分子(加粗键经 Heck 反应生成)

0.3 Suzuki 偶联反应

Suzuki 偶联反应也称作 Suzuki-Miyaura 反应,是有机硼试剂和有机卤化物或三氟甲磺酸酯在钯催化下发生的交叉偶联反应,可由反应式(2)表示。

$$R^1-X + R_2B-R^2 \xrightarrow{Pd, 碱} R^1-R^2 \tag{2}$$

X=卤素、三氟甲磺酸酯基等;
R^1, R^2=芳基、杂芳基、烯基、烷基

Suzuki 偶联反应最早报道于 1979 年[20, 21],目前已成为构建芳基-芳基键的最有效方法。Suzuki 反应的反应条件温和、官能团耐受性广泛并且所需硼酸容易获得,使得该偶联反应更适合于高度官能团化的有机分子的合成。需要注意的是,该反应对氧气特别敏感,反应过程需彻底脱除空气并在惰性气体保护下进行,以避免自偶联和脱硼副产物的生成。Suzuki 偶联反应常用于新药物分子的合成和医药工业生产,钯是主要的催化剂,也有镍作为催化剂的实例,但并不常见。图 0-4 是利用 Suzuki 偶联反应合成的一些药物分子的结构式,彰显 Suzuki 反应在药物合成与生产中的重要性[22-24]。氯沙坦(losartan)是治疗高血压药物,SB-245570 用于治疗抑郁症,而 OSU 6162 是中枢神经疾病治疗药物。

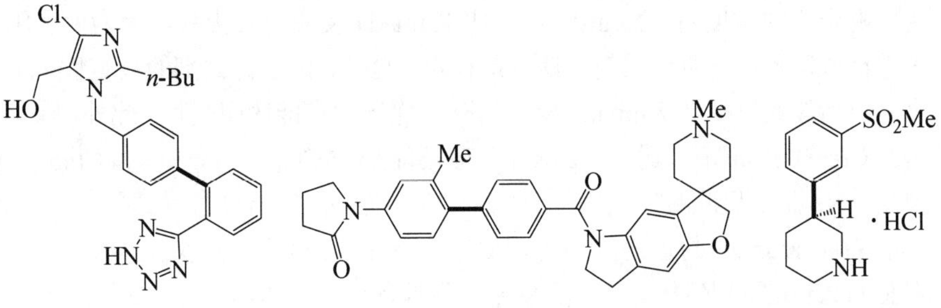

图 0-4 利用 Suzuki 偶联反应合成的三个代表性的药物分子

（氯沙坦　　SB-245570　　OSU 6162）

0.4 Negishi 偶联反应

Negishi 偶联反应建立在 Negishi 对铝、镁、锌和锆等金属有机化合物与芳基卤化物在钯或镍催化剂催化下进行偶联反应研究的基础上。1976 年，Negishi[25]首次报道了有机铝试剂与芳基卤化物在镍催化剂催化下的交叉偶联反应。然而，在相同条件下合成共轭二烯却发生立体专一性明显下降的问题，而以钯催化剂替代镍催化剂，使问题迎刃而解。不久，Negishi[6]用锌试剂与芳基卤化物进行交叉偶联反应，取得很好的结果，如反应式（3）～式（4）。

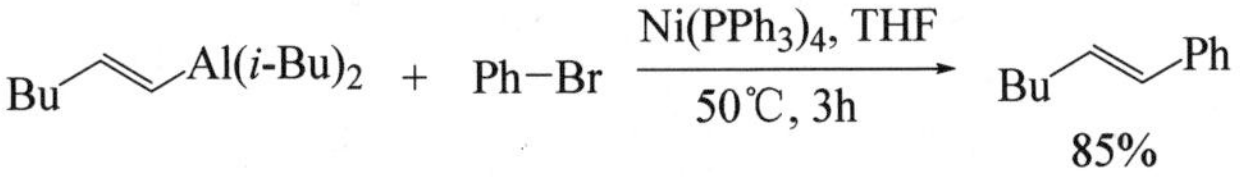

$$Bu-CH=CH-Al(i\text{-}Bu)_2 + Ph\text{-}Br \xrightarrow{Ni(PPh_3)_4,\ THF}_{50℃,\ 3h} Bu-CH=CH-Ph \quad 85\% \quad (3)$$

$$Ph\text{-}ZnCl + O_2N\text{-}C_6H_4\text{-}I \xrightarrow{PdCl_2(PPh_3)_2,\ (i\text{-}Bu)_2AlH}_{THF,\ rt,\ 1\sim 2h} O_2N\text{-}C_6H_4\text{-}Ph \quad 74\% \quad (4)$$

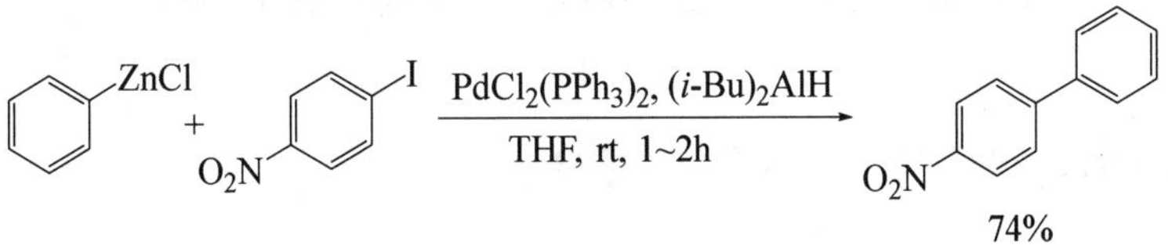

在上述工作的基础上，Negishi 等又进一步研究并最终形成了如反应式（5）所示的 Negishi 偶联反应，即各种芳基卤化物或三氟甲磺酸酯等与有机锌试剂在钯或镍催化剂催化下进行的交叉偶联反应。

$$R^1\text{-}X + R^2\text{-}ZnX' \xrightarrow{Pd\ 或\ Ni} R^1\text{-}R^2 \quad (5)$$
$$X=\text{卤素、三氟甲磺酸酯基等};\ X'=\text{卤素};$$
$$R^1, R^2=\text{芳基、杂芳基、烯基、烷基}$$

从有机合成的角度看，Negishi 反应比 Kumada 反应更具优势，因为后者用有机镁试剂进行偶联反应，而锌试剂比镁试剂稳定。显然，由于锌试剂的稳定性，Negishi 反应比利用活泼镁试剂的 Kumada 反应具有更优异的官能团耐受性。此外，除了能形成经典的 $C(sp^2)$-$C(sp^2)$键，通过 Negishi 反应还能形成 $C(sp^3)$-$C(sp^2)$键和 $C(sp^3)$-$C(sp^3)$键。因此，Negishi 反应被广泛用于天然产物和药物分子的合成，如图 0-5 所示的β-胡萝卜素（β-carotene）与盘皮海绵内酯（discodermolide）[19]、达芙文（differin）[26]、依折麦布（ezetimibe）[27]。其中盘皮海绵内酯具有抗肿瘤作用，达芙文用于治疗轻度至中度痤疮，依折麦布则能抑制胆固醇的吸收,从而降低血液中的胆固醇含量。

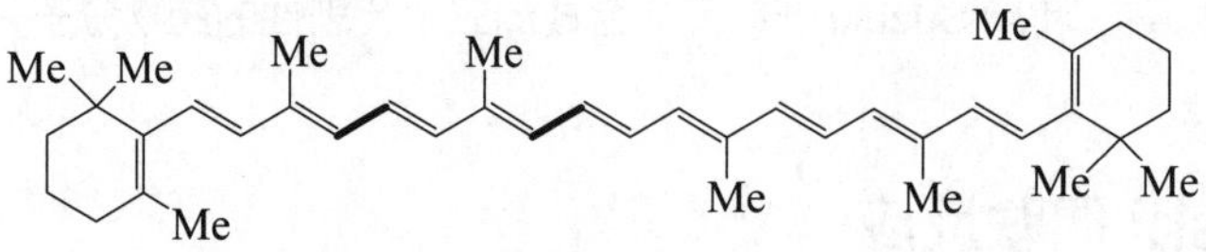

β-胡萝卜素

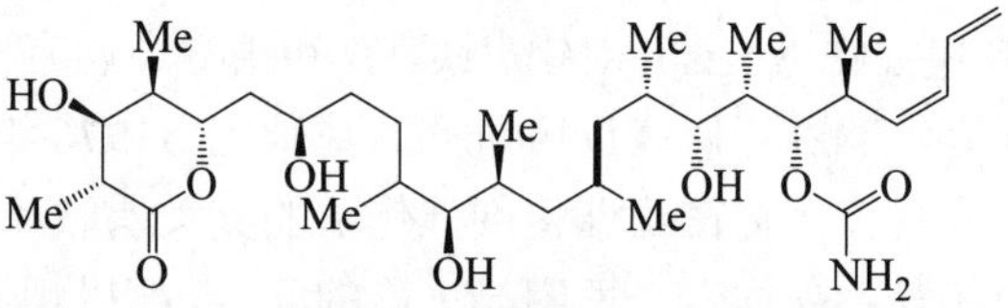

盘皮海绵内酯

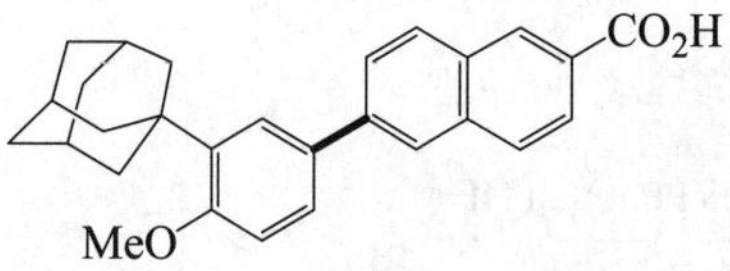

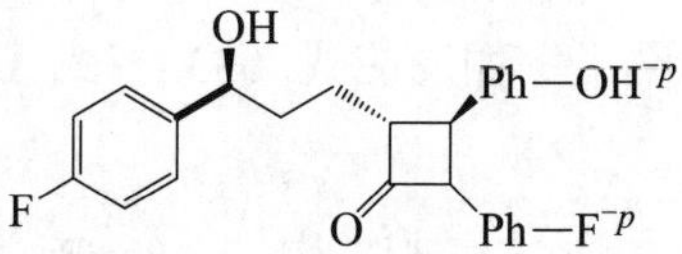

达芙文　　　　依折麦布

图 0-5　利用 Negishi 偶联反应合成的天然产物和药物分子实例

0.5　Kumada-Corriu 偶联反应

如反应式（6）所示，Kumada-Corriu 偶联反应是在过渡金属如钯、镍、铜或铁催化下格氏试剂与芳基或乙烯基卤代物或类卤代物间的交叉偶联反应，也简称 Kumada 偶联反应。

$$R^1\text{–}X + R^2\text{–}MgX' \xrightarrow{M} R^1\text{–}R^2 + MgXX' \quad (6)$$

M=Ni, Pd, Cu, Fe等；R^1, R^2=芳基、烯基、烷基；
X=卤素、三氟甲磺酸酯基等；X'=卤素

该反应发展于 1972 年 Kumada[28]和 Corriu[29]独立发现的分别由镍和钯催化剂催化的烯基或芳基卤化物与芳基或烷基格氏试剂间的交叉偶联反应,反应式如式（7）和式（8）所示。后来,又发展了包括铁、铜、钴和锰等金属催化剂催化的 Kumada-Corriu 偶联反应[26]。

$$\text{Ph-Cl} + \text{EtMgBr} \xrightarrow[\text{0℃, 10min, 回流, 20h}]{\text{NiCl}_2(\text{dppe}), \text{Et}_2\text{O}} \text{Ph-Et} \quad (7)$$
$$98\%$$

$$\text{Ph}\diagup=\diagdown\text{Br} + \text{MeMgBr} \xrightarrow[\text{rt, 3h}]{\text{Pd(PPh}_3)_4, \text{C}_6\text{H}_6} \text{Ph}\diagup=\diagdown\text{Me} \quad (8)$$

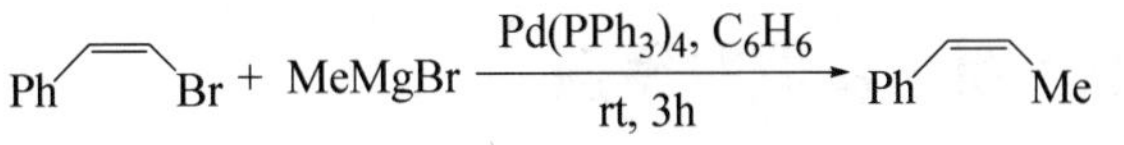

dppe：1,2-双(二苯膦)乙烷 rt—室温

Kumada-Corriu 偶联反应中的亲核试剂是格氏试剂,廉价易得,而且反应活性高,反应在温和的条件下就能顺利进行。然而,任何事情都有其两面性。正是因为格氏试剂的高活性特征,其能够与许多亲电基团发生反应,使得 Kumada-Corriu 偶联反应的官能团耐受性差,反应底物范围狭窄。因此,Kumada-Corriu 偶联反应特别不适合于高度官能团化的化合物的合成。为了解决这一问题,化学家试图以低活性的亲核试剂如有机锌、有机硼和有机锡试剂来替代格氏试剂,因而又促成了 Negishi、Suzuki 和 Stille 偶联反应的发现与发展。

尽管 Kumada-Corriu 偶联反应存在官能团耐受性低的缺点,但对于一些偶联步骤中不含活性官能团的药物分子的合成,该反应仍是一种上佳的选择。图 0-6 是几

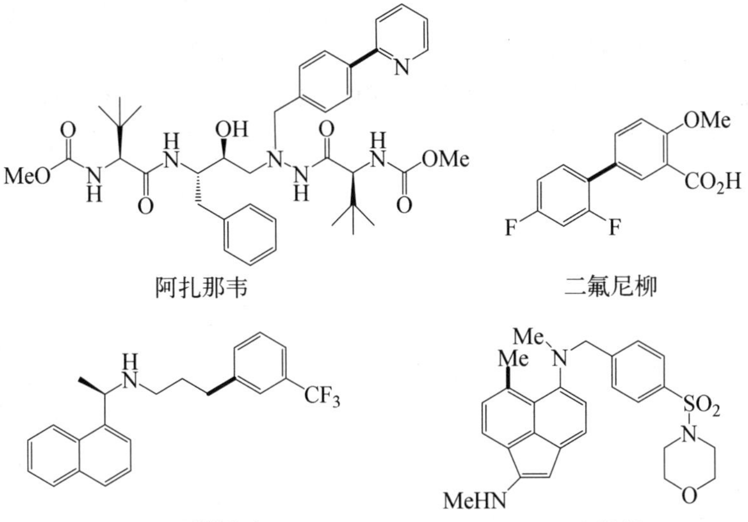

阿扎那韦

二氟尼柳

西那卡塞

AG341

图 0-6　利用 Kumada-Corriu 偶联反应合成的药物分子实例

个利用 Kumada-Corriu 偶联反应成功合成的药物分子的实例[26, 30, 31]。阿扎那韦（atazanavir）是蛋白酶抑制剂（PI）类的抗逆转录病毒药物，用于治疗人类免疫缺陷病毒（HIV）的感染。二氟尼柳（diflunisal）是一种非甾体抗炎镇痛药，临床上用于治疗风湿性关节炎、类风湿关节炎、骨关节炎、扭伤、劳损和镇痛。西那卡塞（cinacalcet）用于治疗进行透析的慢性肾病患者的继发性甲状旁腺功能亢进症。AG341 是胸苷酸合成酶的有效抑制剂，具有抗癌活性。

0.6 Sonogashira 偶联反应

经典的 Sonogashira 偶联反应是指在有机碱和铜盐存在下，钯催化剂催化的端炔与芳基卤化物交叉偶联反应，可由反应式（9）表示。标准的反应条件是钯配合物 $Pd(PPh_3)_4$ 为催化剂、碘化亚铜 CuI 为助催化剂、三乙胺作碱，溶剂可根据反应的要求进行跳变[32]。

$$R^1\text{-}X \ + \ H\text{≡≡}R^2 \xrightarrow{\text{Pd, Cu, 碱}} R^1\text{≡≡}R^2 \quad (9)$$
$$X=卤素、三氟甲磺酸酯基等；$$
$$R^1=芳基、杂芳基、烯基$$

1975 年三个研究团队各自独立完成了在钯催化剂催化下端炔的芳基化，见反应式（10）。Heck [33]和 Cassar[34]采用无铜存在的 Heck 反应条件进行反应，而 Sonogashira[5]则采用在钯、铜共同存在下进行反应。恰恰是后者最后发展成向芳基或乙烯基引入端炔的经典反应，即 Sonogashira 偶联反应。该反应具有工艺简单、高效、官能团相容性好和产率高等优点。虽然后来也有单独钯配合物和单独铜配合物为催化剂的偶联反应，但都不如钯-铜共同催化的反应条件温和、底物范围广泛。

$$\text{Ar-Br} \ + \ R\text{≡≡}H \xrightarrow[\text{或 } Pd^0/CuI/PPh_3, \text{ 碱}]{Pd(OAc)_2/PPh_3} R\text{≡≡Ar} \quad (10)$$

如今，钯-铜共催化的 Sonogashira 偶联反应已无可争辩地成为通过端炔与 sp^2 杂化碳原子（如芳基、杂芳基和烯基卤化物）偶联制备共轭芳-炔和共轭烯-炔的最有效的方法。虽然炔基在药物分子中并不常见，但由 Sonogashira 偶联反应生成的炔基可以进一步转化为烯基、烷基、碳环、杂芳基和芳基结构片段，因而该反应在药物合成中具有重要的地位。图 0-7 所示的实例彰显 Sonogashira 偶联反应在药物合成中的重要性[26,31]。由于炔烃单元的转化，有些药物分子并不显示

Sonogashira 偶联反应的结构单元。芬留顿（fenleuton）是一种 5-脂氧合酶抑制剂，用于骨关节炎及相关炎症疾病的治疗。他扎罗汀（tazarotene）主要用作抗皮肤角化异常药，用于治疗银屑病、痤疮，并用于角化异常性疾病、毛囊皮脂腺疾病、皮肤癌前期病变。PHA-529311 对 HCMV、HSV-1 和 VZV 聚合酶表现出强大的抑制作用，对疱疹病毒 DNA 聚合酶的选择性优于人类聚合酶。恩尿嘧啶（eniluracil）是一种口服有效的二氢嘧啶脱氢酶 (DPD) 抑制剂，被用作抗癌药物 5-氟尿嘧啶（5-Fu）的增效剂。

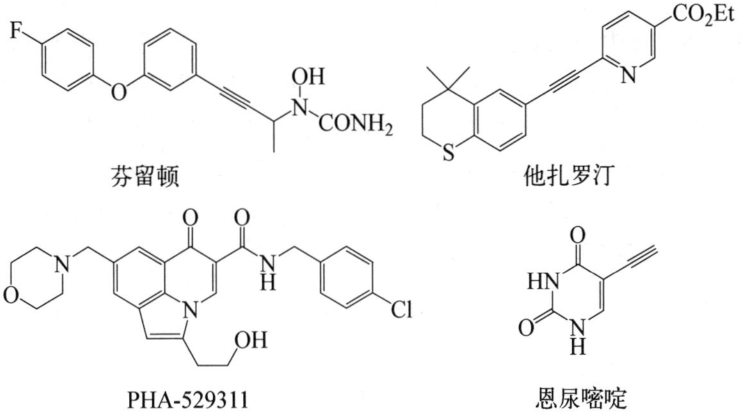

图 0-7 利用 Sonogashira 偶联反应合成的药物分子实例

0.7 Stille 偶联反应

有机锡与亲电试剂在钯催化剂催化下的交叉偶联反应通常称作 Stille 偶联反应，可用反应式（11）表示。

$$R^1\text{-}X + R_3Sn\text{-}R^2 \xrightarrow{Pd} R^1\text{-}R^2 + R_3SnX \tag{11}$$

X=卤素、三氟甲磺酸酯基等；
R^2=芳基、杂芳基、烯基、烷基；
R^1=芳基、杂芳基、烯基；R=烷基

实际上，Stille 并非第一个报道此类反应的化学家，在其发表第一篇相关反应[35]之前，已有相关反应的报道[36, 37]。但是，自从 1978 年起，其持续的合成及反应机理研究对该领域的巨大贡献使该反应称作 Stille 反应实至名归。如果不是因为 Stille 在 59 岁时因飞机失事遇难，他很可能会因"在钯催化的交叉偶联反应方面的工作"而分享 2010 年诺贝尔化学奖。Stille 反应的突出优点是反应条件温和，使其

能与各种官能团相容,因而经常被用于复杂分子的合成;而且,有机锡对水和氧气相对来说不敏感,使其能够耐受更加苛刻的条件。此外,有机锡来源广泛,容易制备。然而,对于该反应的担心除了用昂贵的钯作为催化剂以外,潜在的锡污染也是一种不能忽视的问题。也许,单纯从化学的角度看这一问题似乎被夸大了,因为反应中更常用的三丁基锡衍生物的毒性要远远低于三乙基和三甲基锡衍生物的毒性,但有机锡的毒性的确不能被忽视,特别是在药物生产中。对于毒理学家而言,锡含量的上限是 20ppm(1ppm=1×10^{-6})[38]。也正是如此,相较于其他偶联反应,Stille 反应在药物合成上的应用相对较少,而多用于其他反应不容易实现的复杂天然产物的合成。图 0-8 是利用 Stille 偶联反应作为关键步骤合成的药物[39]和复杂天然产物分子[40]的几个实例。ITA 是一种 VEGFR 激酶抑制剂,具有良好的抗肿瘤活性。actinoallolide A 是一种独特的大环内酯,最初分离自泰国的一种辣椒根,有显著的抗锥虫活性,而且无毒。ripostain B 来源于纤维堆囊菌(*Sorangium cellulosum*)培养液的上清液,对一些革兰氏阳性 RNA 聚合酶有抑制作用。(−)-leiodematolide 是一种分离自海洋海绵 *Leiodermatium* 的大环内酯,对纳摩尔范围内的几种癌症细胞具有杀伤活性。

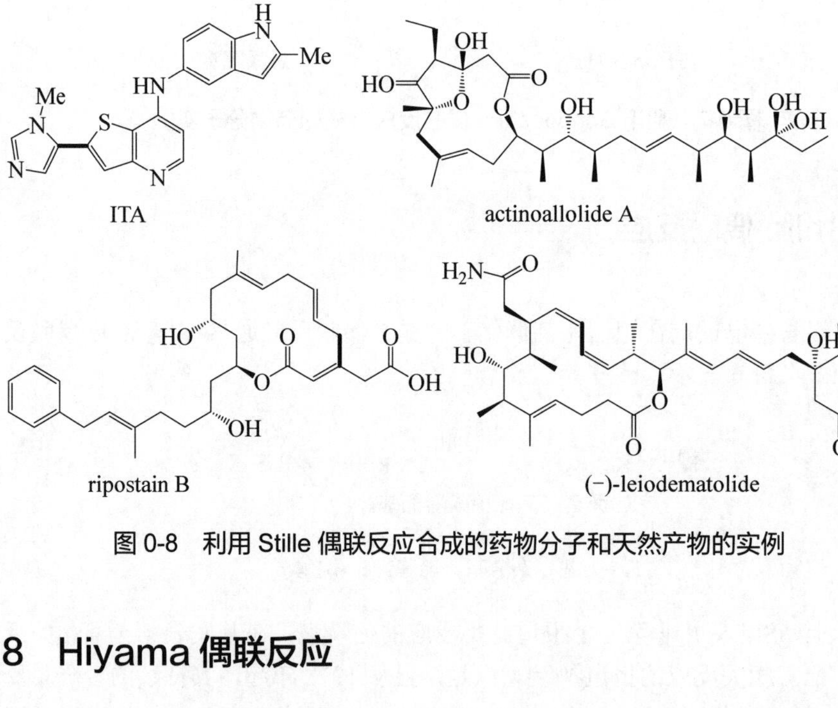

图 0-8　利用 Stille 偶联反应合成的药物分子和天然产物的实例

0.8　Hiyama 偶联反应

如前所述,1979 年 Suzuki 和 Miyaura [21, 41]首次报道了 Suzuki 偶联反应。Suzuki 反应的优点,使其得到迅速发展并在药物合成上得到广泛应用。尽管

Suzuki 反应已得到广泛应用，但 Suzuki 反应也面临着一些挑战，包括质子脱硼基化（protodeborylation）、交叉偶联中选择性降低、起始反应试剂的自偶联以及膦配体上的芳基嵌入偶联产物等[42]。因此，有必要发展新的过渡金属催化的偶联反应。

十年后，Hiyama[10]开拓性地发现在氟离子活化下有机硅烷能够与芳基、烯基和烯丙基卤化物或三氟甲磺酸酯在过渡金属催化剂催化下发生交叉偶联反应。通过创新硅烷试剂和改进反应条件，使 Hiyama 偶联反应的底物范围进一步拓宽。反应式（12）基本反映了 Hiyama 偶联反应的发展现状。

$$R-SiR^1R^2R^3 \ + \ R'-X \ \xrightarrow{M, F^-} \ R-R' \qquad (12)$$

R=芳基,烯基,炔基； R'=芳基,烯基,炔基,烷基；
R^1, R^2, R^3=Cl, F, OH, OM,烷基,烷氧基；
X=Cl, Br, I, OTf 等； M = Pd, Ni, Rh, Cu 等

Hiyama 偶联反应相较于传统的利用金属有机试剂的偶联反应具有更广泛的底物范围、更温和的反应条件、更专一的区域和立体选择性、更低的毒性和更高的化学稳定性。显然，Hiyama 偶联反应更适合于药物分子的合成。图 0-9 是利用 Hiyama 偶联反应合成的药物分子的实例[43]。软骨藻酸 H（isodomoic acid H）是一种非蛋白氨基酸，具有强烈的神经毒性，N-乙酰基异亮氨酸甲酯（N-acetyl colchinol-O-methyl ester）是一种抗肿瘤药物，(+)-阜孢假丝菌素 D [(+)-papulacandin D] 具有高效的抗真菌活性。

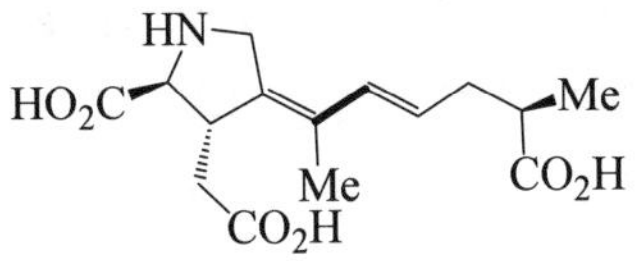

软骨藻酸H

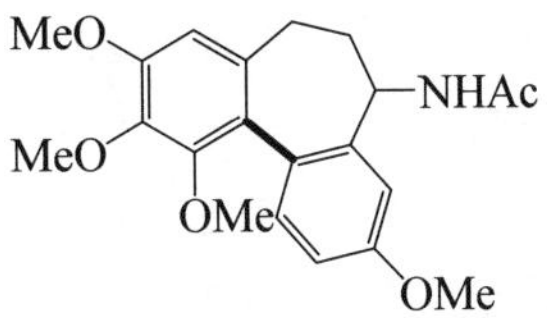

N-乙酰基异亮氨酸甲酯

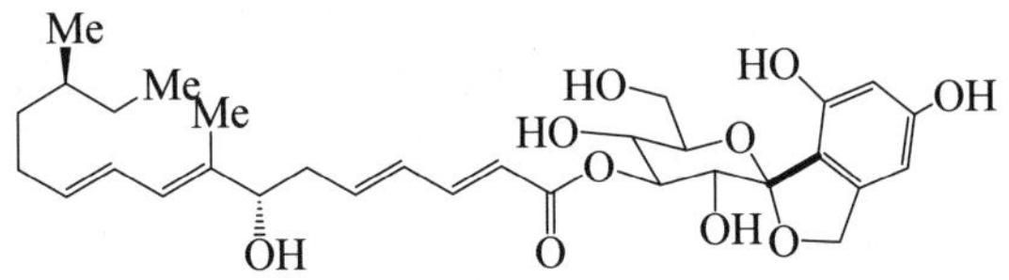

(+)-阜孢假丝菌素D

图 0-9　利用 Hiyama 偶联反应合成的天然产物和药物分子实例

0.9 Tsuji-Trost 偶联反应

Tsuji-Trost 偶联反应是指含烯丙基化合物在钯催化剂催化下烯丙位离去基团被碳亲核试剂取代的反应，反应经由π-烯丙基钯中间体进行，可用反应式（13）表示。式中亲核试剂 Nu^- 的种类很多，往往由含活泼氢的化合物在碱存在下原位生成，而且随着催化体系不断改进，范围越来越广泛。详细内容将在正文第 8 章介绍。

$$R^1\diagup\!\!\!\diagup X \xrightarrow{Pd^0,\text{碱}} \underset{X}{R^1\diagup\!\!\!\diagup\!\!\!\diagup\!\!\!\diagup}_{Pd(\text{II})} \xrightarrow{Nu^-} R^1\diagup\!\!\!\diagup Nu \qquad(13)$$

X=OCOR, OCO$_2$R, OCONHR, OP(O)(OR)$_2$, OPh,
Cl, NO$_2$, SO$_2$Ph, NR$_3$X, SR$_2$X, OH

Tsuji-Trost 偶联反应源自π-烯丙基氯化钯与丙二酸二乙酯、乙酰乙酸乙酯等软亲核试剂反应生成烯丙基丙二酸二乙酯、烯丙基乙酰乙酸乙酯的化学计量反应，由 Tsuji 等于 1965 年发现[2]。研究发现π-烯丙基钯具有亲电性，与亲核试剂反应后生成 Pd^0，预示反应能够以催化反应进行的可能性。经过几年不懈的探索，1970 年经由π-烯丙基钯中间体进行的亲核试剂的催化烯丙基化反应最终被发现[44, 45]，从而诞生了 Tsuji-Trost 偶联人名反应，也是第一个过渡金属催化的人名反应[46, 47]。除了软亲核试剂底物的烯丙基化，π-烯丙基钯中间体与主族金属有机化合物类的硬碳亲核试剂的交叉偶联反应也能进行。反应通过π-烯丙基钯中间体与 Mg、Zn、B、Al、Si、Sn 和 Hg 等金属有机化合物间的金属转移以及随后的还原消除完成。大量研究表明，相较于无催化剂催化的烯丙基化反应，钯催化的 Tsuji-Trost 偶联反应显示更高的区域和立体选择性。如今，Tsuji-Trost 偶联反应在有机和药物合成，特别是在环状结构单元的构建中得到广泛应用。图 0-10 是利用 Tsuji-Trost 偶联反应作为关键步骤合成的药物和天然产物分子的一些实例[26, 48–50]。Kujounin A$_1$ 是从葱属植物分离出的一个天然产物，具有许多药物活性，包括抗菌活性、抑制巨噬细胞活化和抑制发炎等。喹红霉素（cethromycin）是一种酮内酯类抗生素，对社区获得性肺炎（CAP）的治疗和预防暴露后吸入性炭疽均有一定的疗效。环桑色烯（cyclomorusin）是从桑科木材中提取的一种天然化合物，具有抗癌和杀菌活性。carinatine A 是一种从长须草中分离得到的四环生物碱，具有抗肿瘤和乙酰胆碱酯酶（AChE）抑制活性。

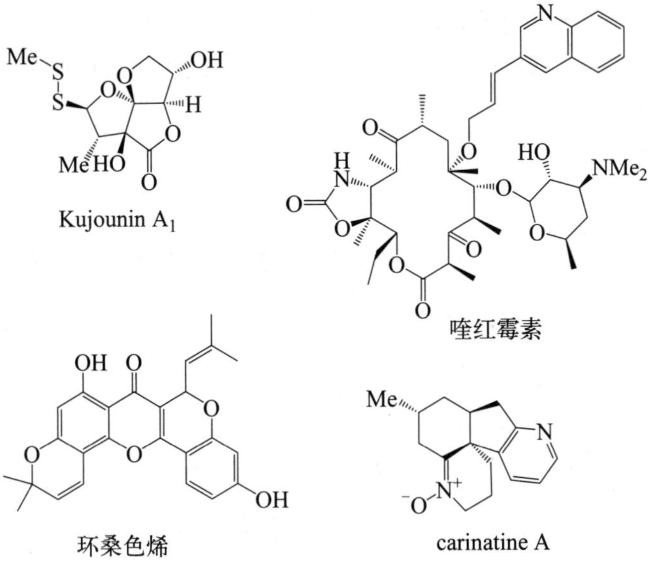

图 0-10 利用 Tsuji-Trost 偶联反应合成的药物分子和天然产物的实例

0.10 结语

受 20 世纪 60 年代 Tsuji 发现的π-烯丙基氯化钯与含活泼亚甲基亲核试剂间化学计量偶联反应的启发,经过以三名诺贝尔奖得主为代表的众多化学家的不懈努力,过渡金属特别是钯催化构建碳-碳键的偶联反应得到巨大的发展,目前已成为极其重要和可以信赖的有机反应,在复杂药物和生物活性天然产物合成中得到广泛应用。坦率而言,以上每一种偶联反应都有自身的优缺点,根据目标产物的特点选用相适的反应往往会取得令人满意的结果。特别需要强调的是,这些经典的偶联反应从机理层面上看往往都经历π-过渡金属配合物中间体与σ-过渡金属配合物中间体之间的生成以及二者之间的转化,因此,通过精准设计可以实现不同偶联反应的串联,使反应得到进一步拓展和延伸,合成应用更加广泛,这将在本书正文第 9 章得以体现。

相较于非过渡金属催化的有机反应,过渡金属催化的构建碳-碳键的偶联反应在药物合成上显示以下优势:一个精心选择的过渡金属催化的偶联反应有时可以替代多步化学计量反应,这种合成捷径可以显著提高合成效率,降低生产成本;过渡金属催化的构建碳-碳键的反应通常在温和的条件下就能成功实现,无须对反应物官能团进行保护和脱保护,因而特别适合于两个高度官能团化合成片段的交叉偶联;催化偶联反应通常在低温下进行,很少用到高活性试剂,使得反应选择性大大提高,同时减少了废弃物的生成,这符合习近平总书记生态文明思想,践行节约集约、绿色低碳发展。应该看到，过渡金属催化的偶联反应的研究并未停滞，更适合药物合成的新催化反应体系不断被发现，将为不断创新的复杂结构的药物分子的合成提供更加简捷高效的途径。

参考文献

[1] Carey F A, Sundberg R J. Advanced organic chemistry: part B: reactions and synthesis (fifth edition) [M]. 5th ed. New York: Springer Science & Business Media, 2007.

[2] Tsuji J, Takahashi H, Morikawa M. Organic syntheses by means of noble metal compounds XVII. Reaction of π-allylpalladium chloride with nucleophiles[J]. Tetrahedron Letters, 1965, 6(49): 4387-4388.

[3] Heck R F. Acylation, methylation, and carboxyalkylation of olefins by Group VIII metal derivatives[J]. Journal of the American Chemical Society, 1968, 90(20): 5518-5526.

[4] Mizoroki T, Mori K, Ozaki A. Arylation of olefin with aryl iodide catalyzed by palladium[J]. Bulletin of the Chemical Society of Japan, 1971, 44(2): 581-581.

[5] Sonogashira K, Tohda Y, Hagihara N. A convenient synthesis of acetylenes: catalytic substitutions of acetylenic hydrogen with bromoalkenes, iodoarenes and bromopyridines[J]. Tetrahedron Letters, 1975, 16(50): 4467-4470.

[6] Negishi E, King A O, Okukado N. Selective carbon-carbon bond formation via transition metal catalysis. 3. A highly selective synthesis of unsymmetrical biaryls and diarylmethanes by the nickel-or palladium-catalyzed reaction of aryl-and benzylzinc derivatives with aryl halides[J]. The Journal of Organic Chemistry, 1977, 42(10): 1821-1823.

[7] Yamamura M, Moritani I, Murahashi S I. The reaction of σ-vinylpalladium complexes with alkyllithiums. Stereospecific syntheses of olefins from vinyl halides and alkyllithiums[J]. Journal of Organometallic Chemistry, 1975, 91(2): C39-C42.

[8] Miyaura N, Yanagi T, Suzuki A. The palladium-catalyzed cross-coupling reaction of phenylboronic acid with haloarenes in the presence of bases[J]. Synthetic Communications, 1981, 11(7): 513-519.

[9] Stille J K. The palladium-catalyzed cross-coupling reactions of organotin reagents with organic electrophiles [new synthetic methods (58)][J]. Angewandte Chemie International Edition, 1986, 25(6): 508-524.

[10] Hatanaka Y, Hiyama T. Cross-coupling of organosilanes with organic halides mediated by a palladium catalyst and tris (diethylamino) sulfonium difluorotrimethylsilicate[J]. The Journal of Organic Chemistry, 1988, 53(4): 918-920.

[11] Hatanaka Y, Hiyama T. Alkenylfluorosilanes as widely applicable substrates for the palladium-catalyzed coupling of alkenylsilane/fluoride reagents with alkenyl iodides[J]. The Journal of Organic Chemistry, 1989, 54(2): 268-270.

[12] Cabri W, Candiani I. Recent developments and new perspectives in the Heck reaction[J]. Accounts of Chemical Research, 1995, 28(1): 2-7.

[13] Suzuki A. Recent advances in the cross-coupling reactions of organoboron derivatives with organic electrophiles, 1995—1998[J]. Journal of Organometallic Chemistry, 1999, 576(1-2): 147-168.

[14] Knochel P, Singer R D. Preparation and reactions of polyfunctional organozinc reagents in organic synthesis[J]. Chemical Reviews, 1993, 93(6): 2117-2188.

[15] Hiyama T, Hatanaka Y. Palladium-catalyzed cross-coupling reaction of organometalloids through

activation with fluoride ion[J]. Pure and Applied Chemistry, 1994, 66(7): 1471-1478.

[16] Banno T, Hayakawa Y, Umeno M. Some applications of the Grignard cross-coupling reaction in the industrial field[J]. Journal of Organometallic Chemistry, 2002, 653(1-2): 288-291.

[17] Sonogashira K. Metal-catalyzed cross-coupling reactions[M]. New York: Wiley-VCH, 1998: 203-229.

[18] Trost B M, Van Vranken D L. Asymmetric transition metal-catalyzed allylic alkylations[J]. Chemical Reviews, 1996, 96(1): 395-422.

[19] Wu X F, Anbarasan P, Neumann H, et al. From noble metal to Nobel prize: palladium-catalyzed coupling reactions as key methods in organic synthesis[J]. Angewandte Chemie International Edition, 2010, 49(48): 9047-9050.

[20] Miyaura N, Yamada K, Suzuki A. A new stereospecific cross-coupling by the palladium-catalyzed reaction of 1-alkenylboranes with 1-alkenyl or 1-alkynyl halides[J]. Tetrahedron Letters, 1979, 20(36): 3437-3440.

[21] Miyaura N, Suzuki A. Stereoselective synthesis of arylated (*E*)-alkenes by the reaction of alk-1-enylboranes with aryl halides in the presence of palladium catalyst[J]. Journal of the Chemical Society, Chemical Communications, 1979 (19): 866-867.

[22] Larsen R D, King A O, Chen C Y, et al. Efficient synthesis of losartan, a nonpeptide angiotensin II receptor antagonist[J]. The Journal of Organic Chemistry, 1994, 59(21): 6391-6394.

[23] Ennis D S, McManus J, Wood-Kaczmar W, et al. Multikilogram-scale synthesis of a biphenyl carboxylic acid derivative using a Pd/C-mediated Suzuki coupling approach[J]. Organic Process Research & Development, 1999, 3(4): 248-252.

[24] Lipton M F, Mauragis M A, Maloney M T, et al. The synthesis of OSU 6162: efficient, large-scale implementation of a Suzuki coupling[J]. Organic Process Research & Development, 2003, 7(3): 385-392.

[25] Negishi E, Baba S. Novel stereoselective alkenyl-aryl coupling via nickel-catalysed reaction of alkenylanes with aryl halides[J]. Journal of the Chemical Society, Chemical Communications, 1976 (15): 596b-597b.

[26] Magano J, Dunetz J R. Large-scale applications of transition metal-catalyzed couplings for the synthesis of pharmaceuticals[J]. Chemical Reviews, 2011, 111(3): 2177-2250.

[27] Sasikala C, Reddy Padi P, Sunkara V, et al. An improved and scalable process for the synthesis of ezetimibe: an antihypercholesterolemia drug[J]. Organic Process Research & Development, 2009, 13(5): 907-910.

[28] Tamao K, Sumitani K, Kumada M. Selective carbon-carbon bond formation by cross-coupling of Grignard reagents with organic halides. Catalysis by nickel-phosphine complexes[J]. Journal of the American Chemical Society, 1972, 94(12): 4374-4376.

[29] Corriu R J P, Masse J P. Activation of Grignard reagents by transition-metal complexes. A new and simple synthesis of trans-stilbenes and polyphenyls[J]. Journal of the Chemical Society, Chemical Communications, 1972 (3): 144a-144a.

[30] Marzoni G, Varney M D. An improved large-scale synthesis of benz [*cd*] indol-2 (1*H*)-one and 5-methylbenz[*cd*]indol-2 (1*H*)-one[J]. Organic Process Research & Development, 1997, 1(1): 81-84.

[31] King A O, Yasuda N. Palladium-catalyzed cross-coupling reactions in the synthesis of pharmaceuticals. In: organometallics in process chemistry. [J]. Topics in Organometallic Chemistry, 2004, 6: 205-245.

[32] Chinchilla R, Nájera C. The Sonogashira reaction: a booming methodology in synthetic organic chemistry[J]. Chemical Reviews, 2007, 107(3): 874-922.

[33] Dieck H A, Heck F R. Palladium catalyzed synthesis of aryl, heterocyclic and vinylic acetylene

derivatives[J]. Journal of Organometallic Chemistry, 1975, 93(2): 259-263.

[34] Cassar L. Synthesis of aryl-and vinyl-substituted acetylene derivatives by the use of nickel and palladium complexes[J]. Journal of Organometallic Chemistry, 1975, 93(2): 253-257.

[35] Milstein D, Stille J K. A general, selective, and facile method for ketone synthesis from acid chlorides and organotin compounds catalyzed by palladium[J]. Journal of the American Chemical Society, 1978, 100(11): 3636-3638.

[36] Azarian D, Dua S S, Eaborn C, et al. Reactions of organic halides with R_3 MMR$_3$ compounds (M= Si, Ge, Sn) in the presence of tetrakis (triarylphosphine) palladium[J]. Journal of Organometallic Chemistry, 1976, 117(3): C55-C57.

[37] Kosugi M, Sasazawa K, Shimizu Y, et al. Reactions of allyltin compounds Iii. Allylation of aromatic halides with allyltributyltin in the presence of tetrakis (triphenylphosphine) palladium (O)[J]. Chemistry Letters, 1977, 6(3): 301-302.

[38] Cordovilla C, Bartolomé C, Martinez-Ilarduya J M, et al. The Stille reaction, 38 years later[J]. Acs Catalysis, 2015, 5(5): 3040-3053.

[39] Ragan J A, Raggon J W, Hill P D, et al. Cross-coupling methods for the large-scale preparation of an imidazole-thienopyridine: synthesis of [2-(3-methyl-3*H*-imidazol-4-yl)-thieno[3,2-*b*]pyridin-7-yl]-(2-methyl-1*H*-indol-5-yl)-amine[J]. Organic Process Research & Development, 2003, 7(5): 676-683.

[40] Heravi M M, Mohammadkhani L. Recent applications of Stille reaction in total synthesis of natural products: an update[J]. Journal of Organometallic Chemistry, 2018, 869: 106-200.

[41] Miyaura N, Yamada K, Suzuki A. A new stereospecific cross-coupling by the palladium-catalyzed reaction of 1-alkenylboranes with 1-alkenyl or 1-alkynyl halides[J]. Tetrahedron Letters, 1979, 20(36): 3437-3440.

[42] Miyaura N, Suzuki A. Palladium-catalyzed cross-coupling reactions of organoboron compounds[J]. Chemical Reviews, 1995, 95(7): 2457-2483.

[43] Sore H F, Galloway W R J D, Spring D R. Palladium-catalysed cross-coupling of organosilicon reagents[J]. Chemical Society Reviews, 2012, 41(5): 1845-1866.

[44] Hata G, Takahashi K, Miyake A. Palladium-catalysed exchange of allylic groups of ethers and esters with active-hydrogen compounds [J]. Journal of the Chemical Society D: Chemical Communications, 1970. 21: 1392-1393.

[45] Atkins K E, Walker W E, Manyik R M. Palladium catalyzed transfer of allylic groups[J]. Tetrahedron Letters, 1970, 11(43): 3821-3824.

[46] Trost B M. Transition metal templates for selectivity in organic synthesis[J]. Pure and Applied Chemistry, 1981, 53(12): 2357-2370.

[47] Tsuji J. Catalytic reactions via π-allylpalladium complexes[J]. Pure and Applied Chemistry, 1982, 54(1): 197-206.

[48] Burtea A, Rychnovsky S D. Biosynthesis-inspired approach to Kujounin A2 using a stereoselective Tsuji-Trost alkylation[J]. Organic Letters, 2018, 20(18): 5849-5852.

[49] Smith R J, Bower R L, Ferguson S A, et al. The synthesis of (±)-oxyisocylointegrin[J]. European Journal of Organic Chemistry, 2019(7): 1571-1573.

[50] Meng L. Total synthesis of (−)-Carinatine A and (+)-Lycopladine A[J]. The Journal of Organic Chemistry, 2016, 81(17): 7784-7789.

第 1 章
Heck 反应及其在药物合成上的应用

1.1　Heck 反应及其反应机理

Heck 反应是指在碱存在下钯催化剂催化烯烃的芳基化和烯基化反应，由 Mizoroki[1]和 Heck[2]于 20 世纪 60 年代末独立发现。其基本反应如式（1）所示。

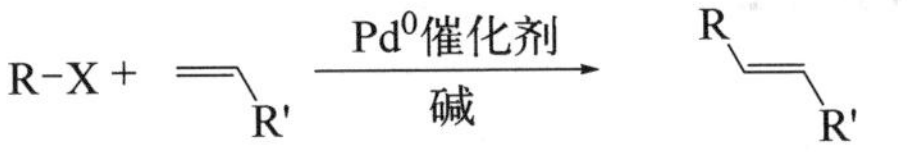

$$\text{R-X} + \underset{\text{R'}}{=\!=} \xrightarrow[\text{碱}]{\text{Pd}^0\text{催化剂}} \underset{\text{R'}}{\overset{\text{R}}{=\!=}} \quad (1)$$

R,R'=芳基、乙烯基、苄基；X = I、Br、OTf 等

钯催化的 Heck 反应是一个复杂的过程。由于催化体系组成的变化，基元反应也有所不同，因而很难给出所有 Heck 反应详尽的机理，但基于 $Pd^0/Pd(II)$ 氧化-还原循环的基本催化机理被广泛认可[3]。该催化机理由图 1-1 所示的 4 个独立的基本反应和几个关键的中间体构成。第一步，由催化剂前体原位生成的配位不饱和的 14 电子的富电子活性物种 Pd^0L_2(**A**)，具有明显的亲核性和空的配位位，能与亲电底物 RX 发生氧化加成，生成反式的加合物 $RPdL_2X$(**B**)，其中 R（芳基或烯基）与 Pd 通过 σ 键结合。实际上反式构型的 **A** 由热力学不稳定的顺式异构体异构化生成。第二步，**B** 中的一个配体 L 或 X 被取代，进而与烯烃配位生成 π-配合物中间体（**C**），该中间体可能是电中性或是电正性的。第三步，可能通过协同四元过渡态进行顺式插入生成不稳定的 σ-烷基钯配合物（**D**），反应的区域选择性通常由立体和电子效应共同决定。第四步，如果一个处于顺式位置的 β-氢原子能够被接近，这个氢原子可以通过内旋转调节与金属的方向使 β-氢原子与 Pd 原子顺式共平面，在钯原子存在空的配位位的情况下发生 β-氢消除，生成 Heck 偶联产物和钯氢化物 $HPdXL_2$。最后，体系中的碱从钯氢化物 $HPdXL_2$ 上夺氢再生活性物种 Pd^0L_2(**A**)，实现催化循环。显然，碱在 Heck 反应中是必需的。

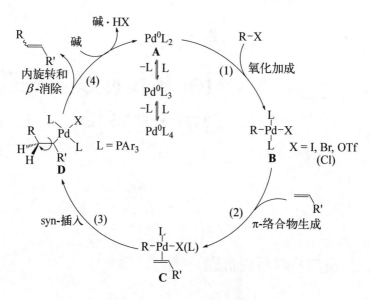

图 1-1　Heck 反应的催化机理

1.2　基本的 Heck 反应

1.2.1　分子间 Heck 反应

许多官能团都与 Heck 反应相容。传统的 Heck 反应通常需要芳基或乙烯基碘化物或溴化物作为亲电底物［反应式（1）～式（8）］，后来发现三氟甲磺酸酯［反应式（9）］、酰氯［反应式（10）、式（11）］、磺酰氯［反应式（12）］、酸酐［反应式（13）］以及芳基氯［反应式（14）］等多种化合物也能作为亲电底物。相较于其他芳基化物，芳基氯作为亲电底物具有廉价易得的优势。就烯烃而言，带有吸电子基的端烯烃在钯催化剂催化下很容易反应［反应式（1）～式（8），反应式（10）～式（13），反应式（22）、式（23）］，但带有电子中性和给电子取代基的端烯烃则不容易反应。随着对催化体系的不断改进，一些带有给电子基的烯烃也可进行反应［反应式（9），反应式（18）、式（19）］[4,5]。如果烯烃的邻位含有羟基或者是烯醚结构，反应过程中会发生碳碳双键的迁移，最终生成偶联重排产物［反应式（15）～式（17），式（20）］。钯催化剂可以是简单的钯盐，如 $Pd(OAc)_2$ 或 $PdCl_2$，引入膦配体如三苯基膦等可以促进反应的进行［反应式（7）、式（18）、式（19）］，尤其是在芳基氯作为亲电底物时效果更加明显[6]；而且，利用三(邻甲苯基)膦 $P(o\text{-tolyl})_3$ 作膦配体还可对催化剂物种起稳定化作用，延长催化剂寿命［反应式（4）、式（5）、式（8）］。后来发现其他廉价的过渡金属可以替代钯催化 Heck 反应进行[7][反应式（23）]。反应中的碱可以是无机碱，如碳酸钾、碳酸氢钠等[反应式（6）、式（10）、式（12）、式（21）、式（22）]，也可以是有机碱，如三乙胺等[反应式（1）～式（3）、式（7）～式（9）、式（16）、式（17）]。反应既可以在有机溶剂中进行，也可以在水中或水-有机混合溶剂中进行[反应式（14）]，为使反应顺利进行，有时需要向反应体系加入相转移催化剂或其他助剂[反应式（10），式（12）～式（15）、式（20）～式（22）]。以下反应反映了传统 Heck 反应的底物结构、催化反应体系的一些信息[8]。

$$\text{2-I-C}_6\text{H}_4\text{NH}_2 + \text{CH}_2=\text{CHCO}_2\text{Me} \xrightarrow[100℃, 80h]{\text{Pd(OAc)}_2, \text{Et}_3\text{N}} \text{2-NH}_2\text{-C}_6\text{H}_4\text{-CH=CH-CO}_2\text{Me} \quad 72\% \tag{2}$$

$$\text{2-I-C}_6\text{H}_4\text{Br} + \text{CH}_2=\text{CHCO}_2\text{H} \xrightarrow[\substack{100℃, 1h,\\ \text{MeCN}}]{\text{Pd(OAc)}_2, \text{Et}_3\text{N}} \text{2-Br-C}_6\text{H}_4\text{-CH=CH-CO}_2\text{H} \quad 82\% \tag{3}$$

$$\text{PhBr} + \text{CH}_2=\text{CHCONH}_2 \xrightarrow[\text{MeCN}]{\substack{\text{Pd(OAc)}_2, \text{P}(o\text{-tol})_3 \\ \text{Et}_3\text{N}, 100℃, 1h}} \text{Ph-CH=CH-CONH}_2 \quad 70\% \tag{4}$$

$$\text{4-OHC-C}_6\text{H}_4\text{Br} + \text{CH}_2=\text{CHCN} \xrightarrow[\text{MeCN}]{\substack{\text{Pd(OAc)}_2, \text{P}(o\text{-tol})_3 \\ \text{NaOAc}, 130℃, 24h}} \text{4-OHC-C}_6\text{H}_4\text{-CH=CH-CN} \quad 79\% \tag{5}$$

$$\text{PhI} + \text{CH}_2=\text{CHCHO} \xrightarrow[\text{DMF}]{\substack{\text{Pd(OAc)}_2, n\text{-Bu}_4\text{NCl} \\ \text{NaHCO}_3, 20℃, 60h}} \text{Ph-CH=CH-CHO} \quad 90\% \tag{6}$$

$$\text{2-I-C}_6\text{H}_4\text{Br} + \text{CH}_2=\text{CHSO}_2\text{NH}_2 \xrightarrow[\text{DMF}]{\substack{\text{Pd(OAc)}_2, \text{PPh}_3, \text{Et}_3\text{N} \\ 140℃, 24h}} \text{2-Br-C}_6\text{H}_4\text{-CH=CH-SO}_2\text{NH}_2 \quad 61\% \tag{7}$$

$$\text{4-Me}_2\text{N-C}_6\text{H}_4\text{Br} + \text{CH}_2=\text{CHP(O)(OEt)}_2 \xrightarrow[\text{MeCN}]{\substack{\text{Pd(OAc)}_2, \text{P}(o\text{-tol})_3, \\ \text{Et}_3\text{N}, 100℃, 6h}} \text{4-Me}_2\text{N-C}_6\text{H}_4\text{-CH=CH-P(O)(OEt)}_2 \quad 65\% \tag{8}$$

NMP—N-甲基-2-吡咯烷酮

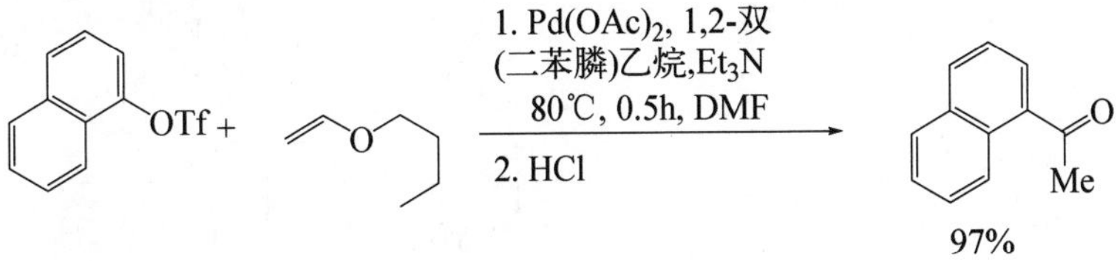

1.2.2 分子内 Heck 反应

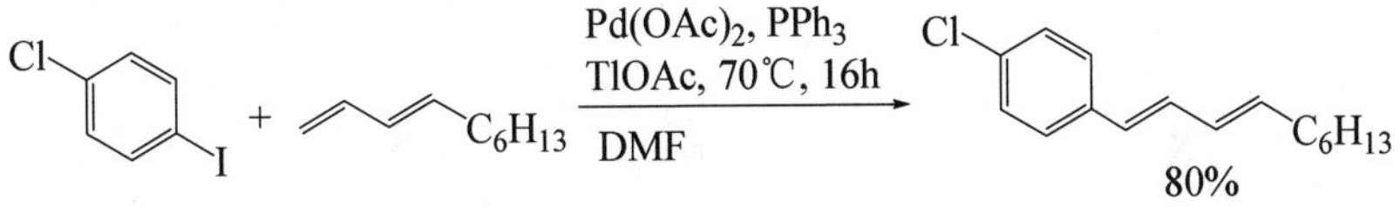

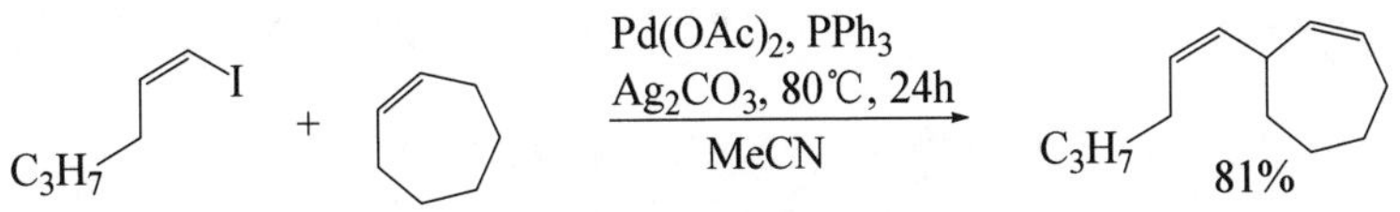

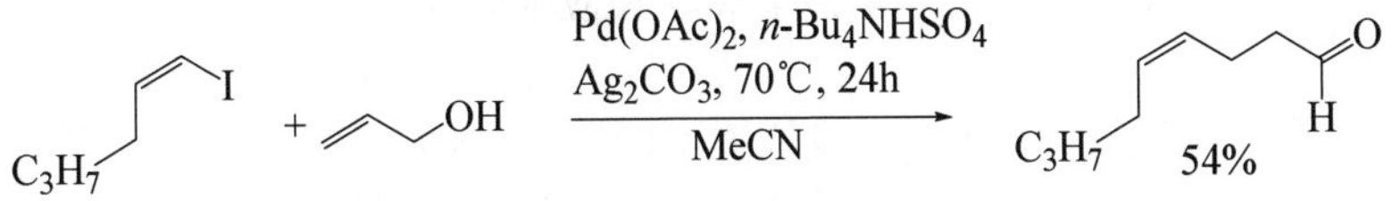

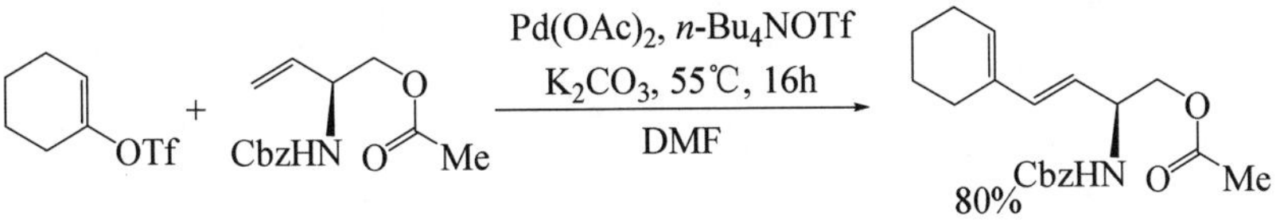

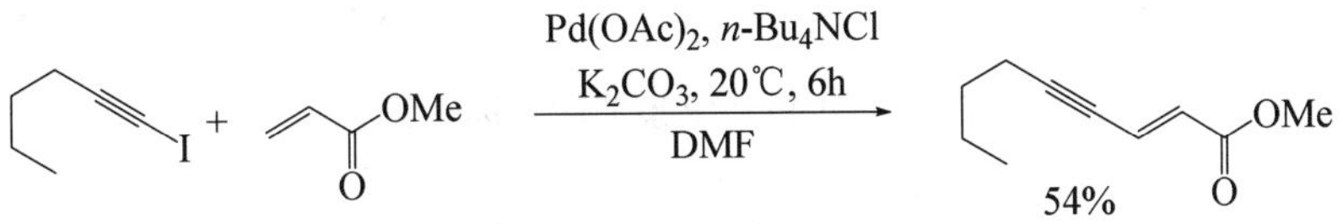

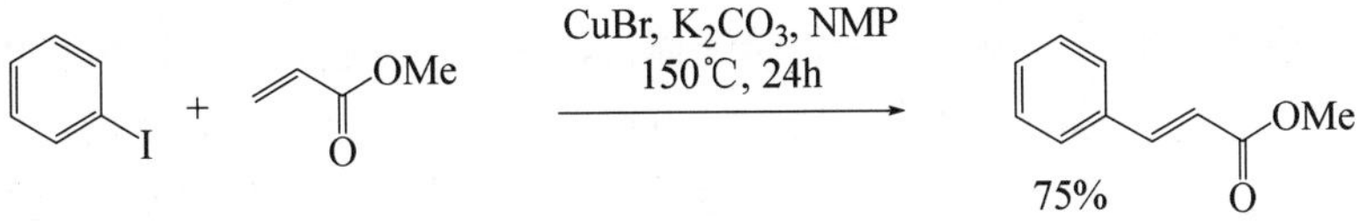

1.2.2.1 基本原理

当芳基卤代物或三氟甲磺酸芳基酯结构单元与不饱和基团如乙烯基共处于同一分子结构,并且二者之间的位置适宜时,在 Heck 反应条件下就可以发生分子内 Heck 反应,从而生成各种结构不同、大小各异的环状化合物;有时分子内 Heck 反应还伴随各种串联反应,最终生成重排反应产物[9]。分子内 Heck 反应可简单地由图 1-2 来表示。

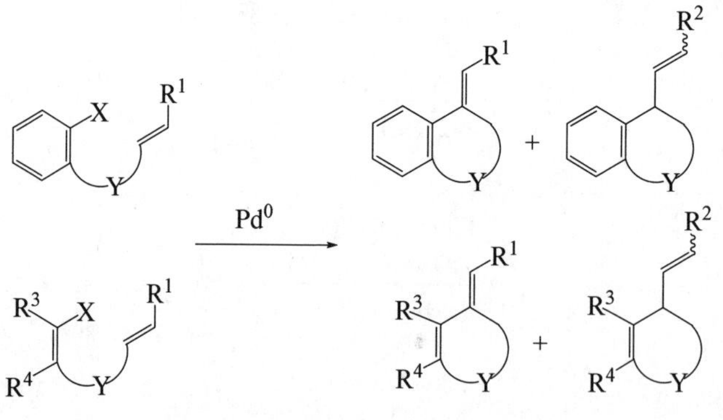

X=I、Br、三氟甲磺酰基等；Y=C、N、O等

图 1-2　分子内 Heck 反应示意图

在关环过程中，根据新σ键的生成位置不同分为 *exo*-关环和 *endo*-关环，图 1-3 是 *exo*-和 *endo*-关环的示意图。一般而言 *exo*-关环更常见，而 *endo*-关环则较少。以下反应如无特别提醒，皆为 *exo*-关环反应。

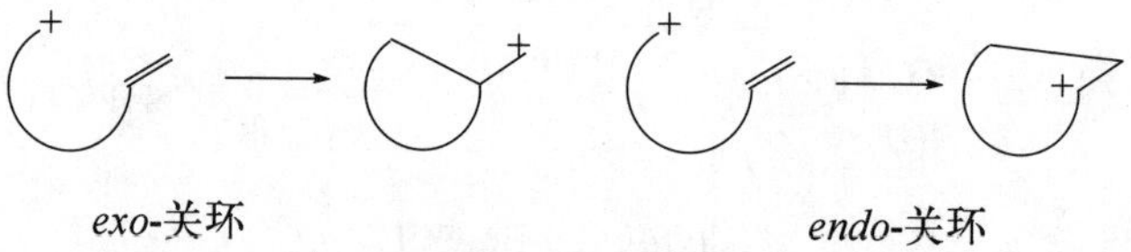

exo-关环　　　　　　　　　　*endo*-关环

图 1-3　分子内 Heck 反应的关环方式

分子内 Heck 反应的底物通常是芳基、乙烯基卤化物或全氟代磺酸酯。碘化物、溴化物和三氟甲磺酸酯最常见。若引入膦配体，如三叔丁基膦，氯代苯也可以顺利地进行反应[10]。不饱和基团包括烯、炔、联烯和芳基都可进行反应。连接链可以是全碳的碳链也可以是含杂原子的碳链，长度可以变化，杂原子可以是氮、氧等。分子内 Heck 反应可以用来构建四元环到二十七元环的环状化合物，而且具有广泛的官能团耐受性。通过分子内 Heck 反应生成一个新的烯烃，在反应条件下会发生重排反应。

1.2.2.2　四元环的合成

由于四元环存在较大的张力，一般不容易形成，但通过底物的精准设计也可以由分子内 Heck 反应构建。如反应式(24)所示结构的全氟代丁基磺酸酯（NfOR）在标准的 Heck 反应条件下发生环合反应以中等的收率生成二烯产物，其中一个双键处于环外[11]。

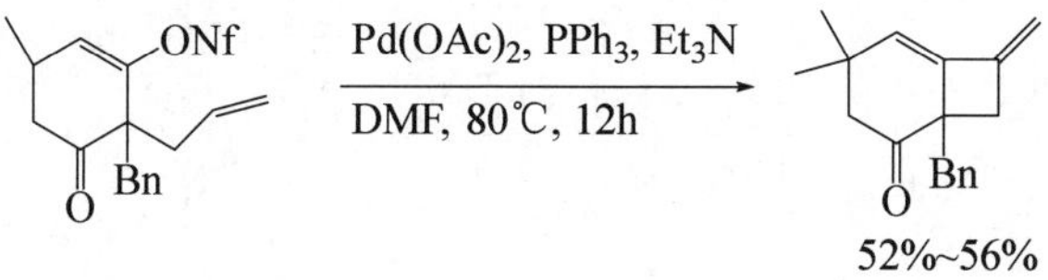

$$\text{(24)} \quad 52\%\sim56\%$$

1.2.2.3 五元环的合成

五元环最容易通过分子内 Heck 反应构建，特别适合于含对酸敏感官能团的底物，因为 Heck 反应通常在碱性条件下进行，无须酸的存在。因此，利用分子内 Heck 反应可合成吲哚、苯并呋喃、苯并噻吩、氧化吲哚、苯并呋喃酮以及其他各类杂环化合物。

首先以五元碳环化合物的合成为例。同时含有邻碘苯或碘代乙烯和丙烯酸酯结构单元的底物分子，若连接两个结构单元间碳链的长度适宜很容易进行分子内 Heck 反应形成五元碳环[12]。

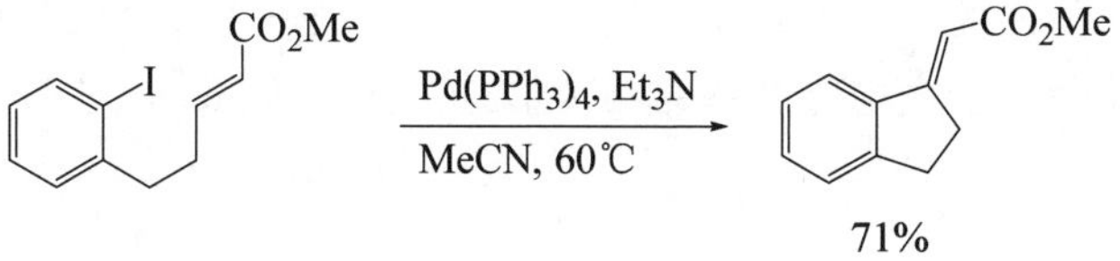

$$\text{(25)} \quad 71\%$$

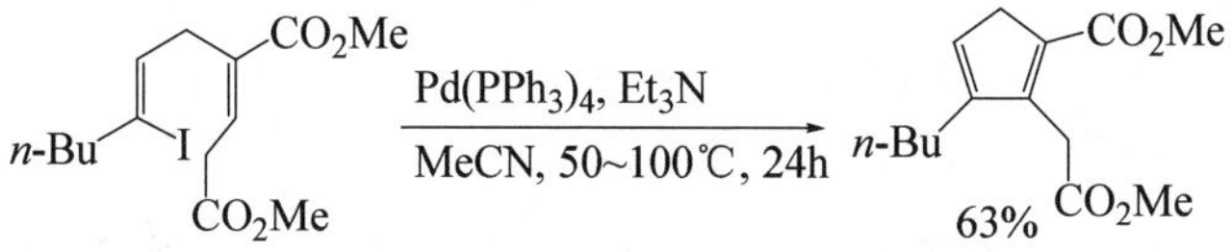

$$\text{(26)} \quad 63\%$$

值得注意的是，环合反应往往生成顺式的稠环产物，如反应式（27）中碘代乙烯化合物的环合产物中，氢原子与氰基处于环的同侧[13]。加入 Ag_2CO_3 的作用是阻止烯烃的异构化。

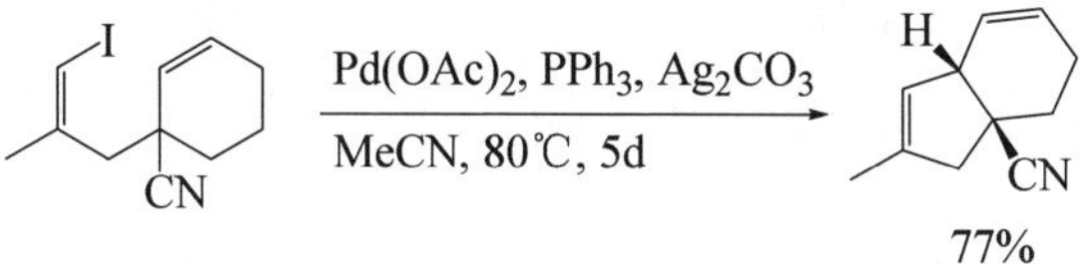

$$\text{(27)} \quad 77\%$$

由于利用氨基的亲核性很容易得到烯丙胺类衍生物，因而可以通过分子内 Heck 反应构建各类含吡咯环的化合物。

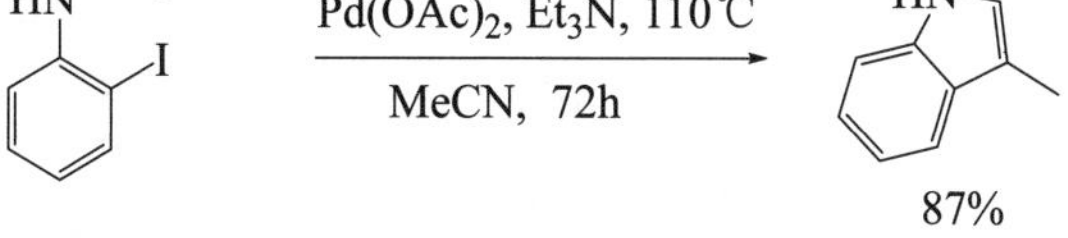

$$\text{(28)} \quad 87\%$$

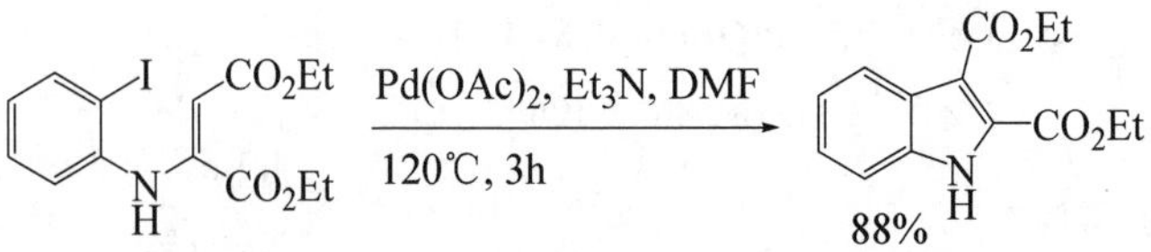

$$\text{(29)} \quad 88\%$$

当乙烯基与卤代芳基通过酰胺键连接时,通过分子内 Heck 反应形成吡咯烷酮结构,同时非对映选择性地构建季碳中心,使得不对称分子内 Heck 反应成为可能[14]。

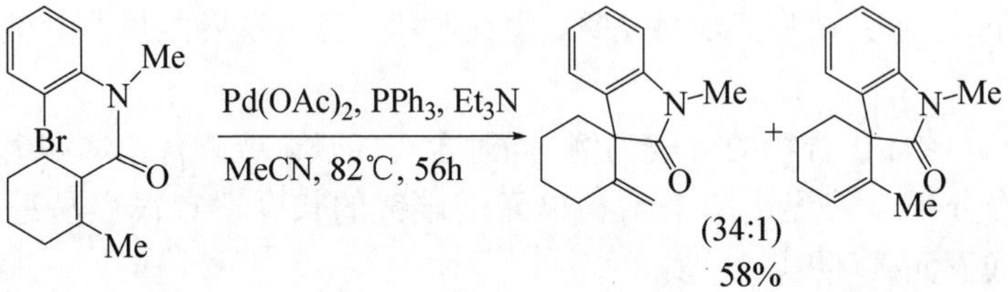

$$\text{(30)} \quad (34:1) \ 58\%$$

卤代烯或苯基与乙烯基通过氧原子连接的化合物则可用以合成呋喃衍生物。以下是最简单的含氧杂原子底物烯丙基邻碘苯基醚的分子内 Heck 反应,产物为 3-甲基苯并呋喃[15]。

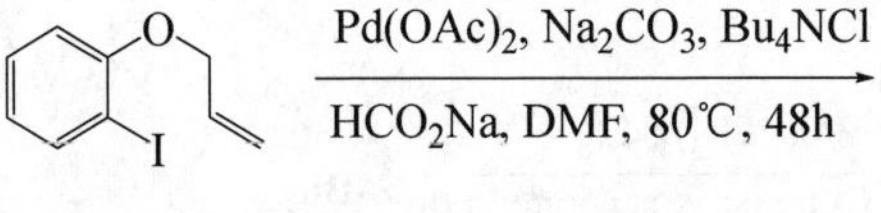

$$\text{(31)} \quad 47\%$$

含有两个羟基的溴乙烯类底物在 Heck 反应条件下环合生成稠环化合物,与碳环化反应(27)类似,产物为顺式结构。反应先通过 β-氢消除生成烯醇,再经互变异构生成酮。这两个例子也说明 Heck 反应广泛的官能团耐受性[9]。

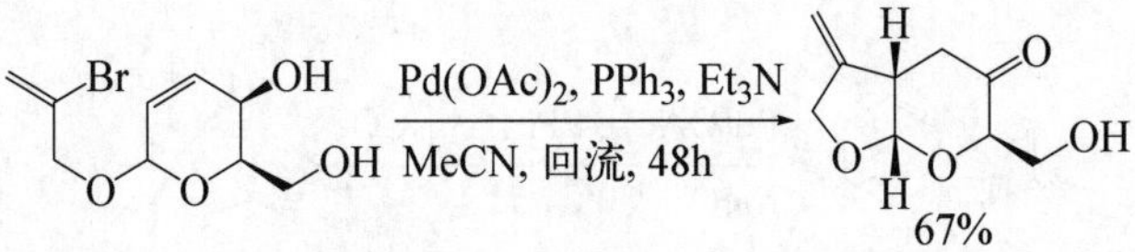

$$\text{(32)} \quad 67\%$$

β-氢消除是 Heck 反应的关键步骤,如果底物结构中没有β-氢,需要外加氢源反应才能进行,但生成的是氢化芳基化产物,例如反应式(33)生成的三环稠合体系[16]。该反应的氢源是预先由三乙胺和甲酸生成的盐。

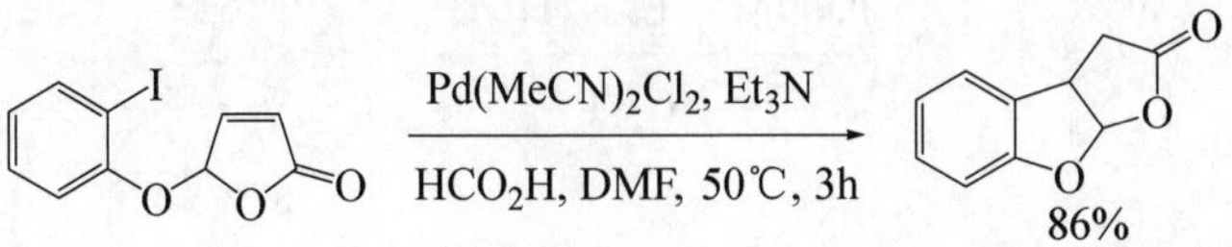

$$\text{(33)} \quad 86\%$$

尽管利用分子内 Heck 反应很容易构建五元碳环和氮、氧五元杂环,但含硫五元环化合物则鲜有合成,可能原因是硫与催化剂钯强配位作用使催化剂失活所致。式(34)是极少利用分子内 Heck 反应合成五元含硫杂环的实例,烯丙基碘代苯基硫醚在较高反应温度下生成 3-甲基苯并噻吩[17]。

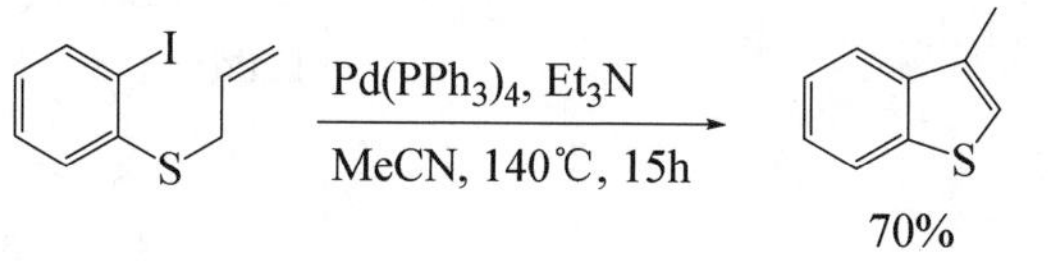

$$\text{反应式 (34)}$$

1.2.2.4 六元环的合成

利用分子内 Heck 反应合成六元环也很容易,通常生成环外烯烃。对于碳链底物,与五元环情形类似,只要碳链长度合适就能容易地进行分子内 Heck 反应,生成六元碳环[18]。

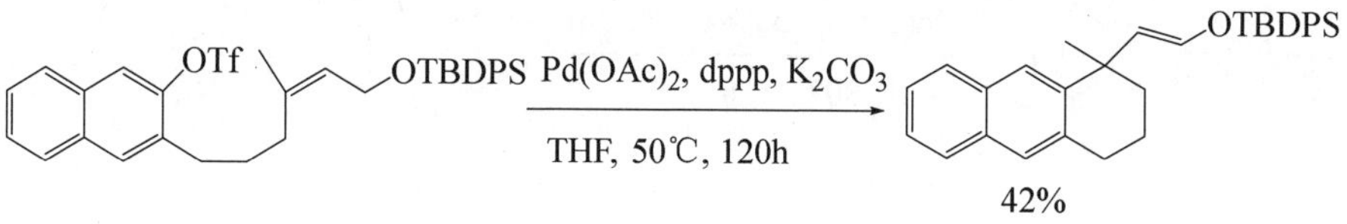

$$\text{反应式 (35)}$$

如果涉及顺反异构,则生成顺式的稠环结构,如下面芳基溴化物在 Herrmann-Better 催化剂即反式二-μ-双[2-(二邻甲苯基膦)苄基]乙酸二钯(Ⅱ)(HBC)存在下转化为四环稠合化合物,其中两个氢原子处于环的同侧[19]。

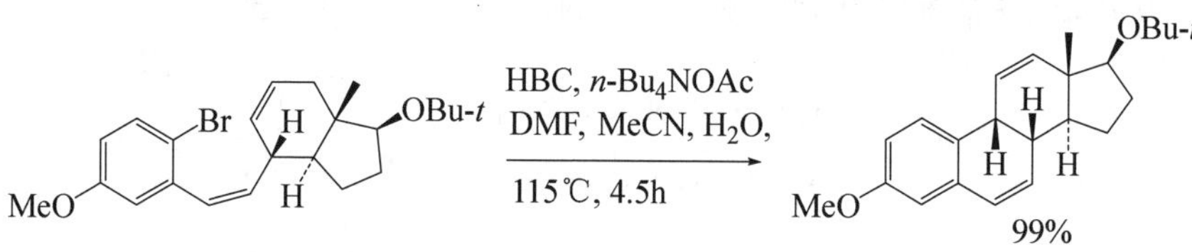

$$\text{反应式 (36)}$$

反应式(37)是一个构建六元碳环的特殊例子,乍一看好像是通过 endo-关环实现的,但仔细分析产物的构型却并非如此。实际上该反应仍通过 exo-关环进行,只不过是在反应关环后发生重排,中间形成三元碳环,然后三元碳环开环重排,生成构型更稳定的双烯结构[20]。该反应表明在分子内 Heck 反应过程中会存在重排问题。

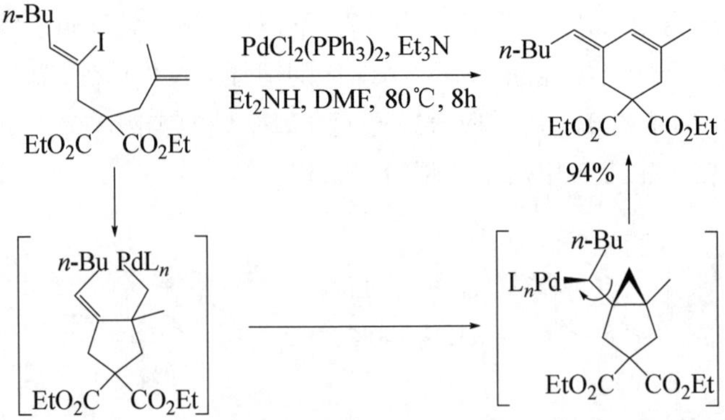

$$\text{(37)}$$

分子内 Heck 反应生成六元氮杂环的实例不胜枚举,这里重点介绍生成喹啉、异喹啉及其部分加氢产物的反应。芳基溴或碘与乙烯基若处于含氮碳链的适当位置,在 Heck 催化剂存在下可以反应成环生成取代的喹啉和异喹啉,典型的反应如下[14]。

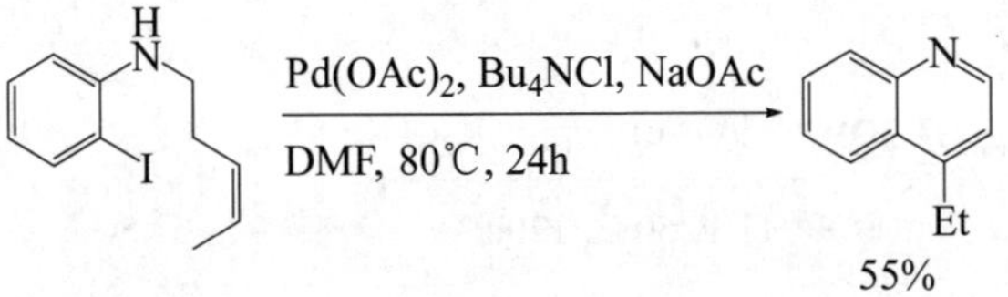

$$\text{(38)}$$

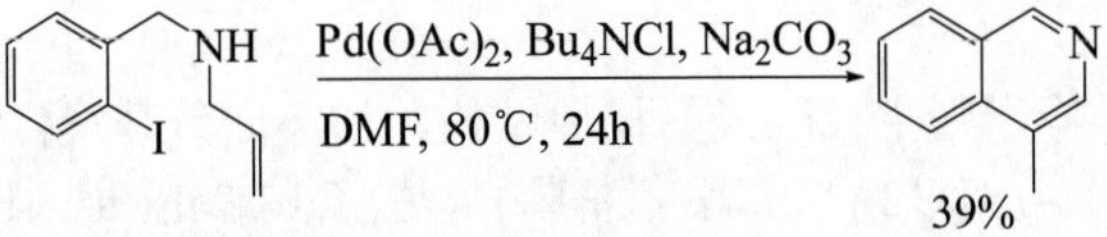

$$\text{(39)}$$

如果乙烯基本身是环的一部分,那么反应可生成螺环、稠环或者桥环化合物。反应式(40)就是生成桥环化合物的一个实例[21]。

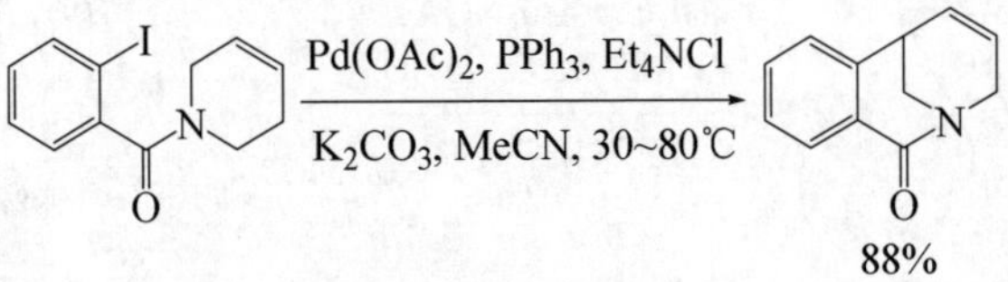

$$\text{(40)}$$

另一个实例是以磺酰胺基为桥链的溴代苯与乙烯基的分子内反应,以较高的收率生成三环桥环体系[22]。

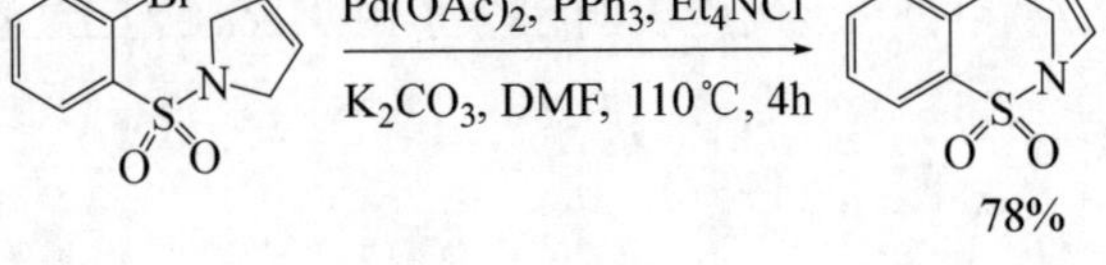

$$\text{(41)}$$

相较于生成六元氮杂环的分子内 Heck 反应，生成六元氧杂环的反应要少得多，下式是生成六元杂环的一反应实例。底物中溴代苯基和丙烯酸乙酯通过缩醛结构相连，即使如此，该底物在标准 Heck 反应条件下，也以较高的收率转化为三环稠环化合物。反应之所以在高温下进行，是因为烯烃在高温下才能异构化为热力学稳定的 *Z*-式异构体。缩醛结构在反应后得到保持，进一步说明 Heck 反应可以耐受各种敏感的官能团[23]。

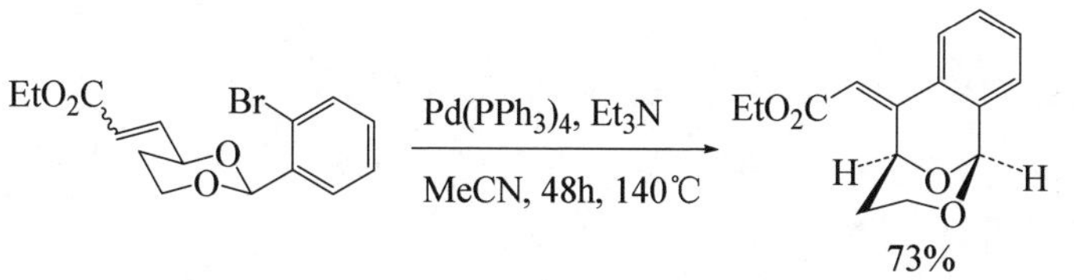

EtO_2C
Br
O
O
$Pd(PPh_3)_4$, Et_3N
MeCN, 48h, 140℃
EtO_2C
H
O
H
O
73%

(42)

芳基碘与乙烯基砜环合生成六元环醚，反应体系中加入银盐促进了反应进行，提高了目标产物的收率[24]。

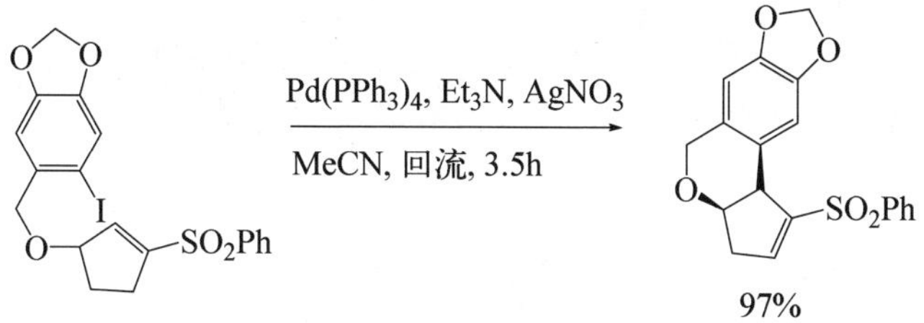

O
O
I
O
SO_2Ph
$Pd(PPh_3)_4$, Et_3N, $AgNO_3$
MeCN, 回流, 3.5h
O
O
O
SO_2Ph
97%

(43)

1.2.2.5 七元环的合成

分子内 Heck 反应也可用来构建七元环，尽管不像构建五元和六元环一样容易。由于底物难以制备，合成七元碳环的实例较少，但只要有与分子内 Heck 反应相匹配的底物，反应则相对容易，甚至能构建张力大的环。反应式（44）就是一个构建七元碳环的实例。由于生成的七元环中含有桥头双键（anti-bredt alkene）、偕二甲基取代的桥碳，以及五个 sp^2 杂化中心，因而存在很高的张力[25]。

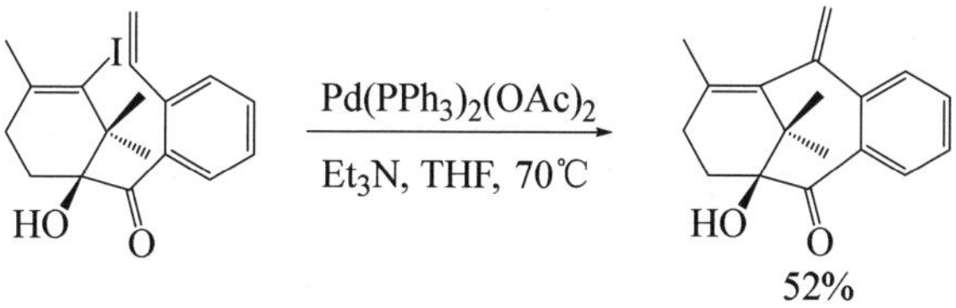

I
HO
O
$Pd(PPh_3)_2(OAc)_2$
Et_3N, THF, 70℃
HO
O
52%

(44)

同样，含氮七元环也可以通过分子内 Heck 反应构建，而且更容易生成顺式构型的稠环产物[26]。

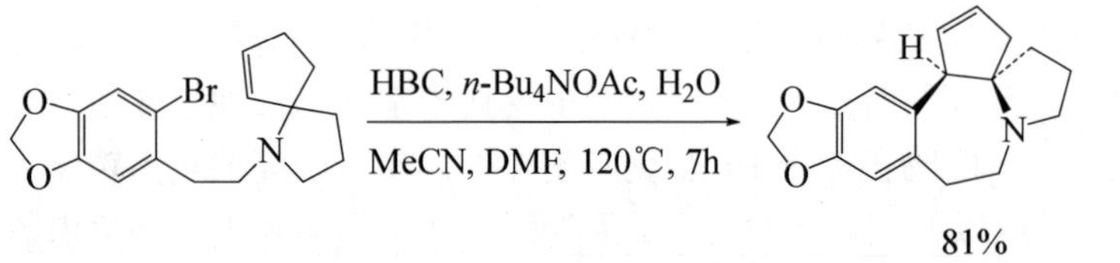

$$\text{(45)} \quad 81\%$$

结构合适的底物也可以通过 *endo*-关环合成七元氮杂环，反应在无水条件下进行[27]。

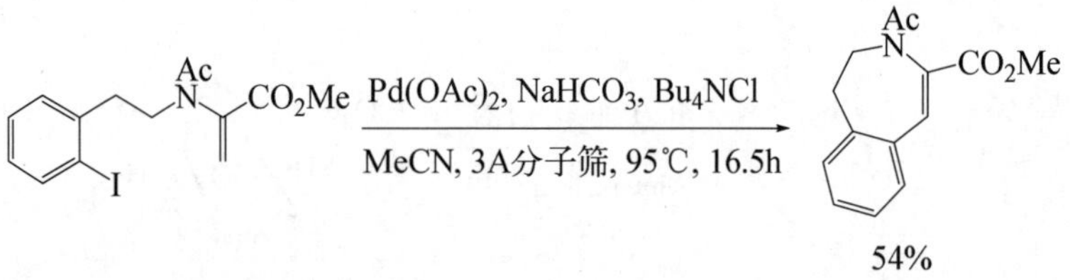

$$\text{(46)} \quad 54\%$$

七元含氧杂环也不例外，也可以由适当的底物容易地合成，例如碘代苯甲基环己烯醚环合生成稠合三环化合物[28]。

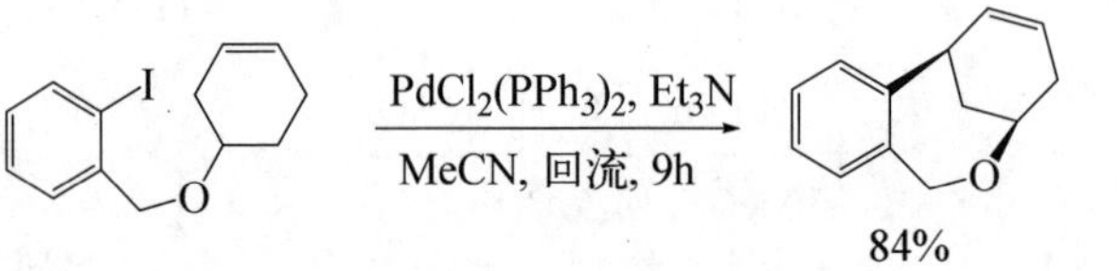

$$\text{(47)} \quad 84\%$$

1.2.2.6 八元和大环的合成

对于八元环，通常由丙二烯基替代乙烯基与卤代芳基反应合成，丙二烯基比乙烯基更活泼，反应发生在 sp 杂化的中间碳上[29, 30]。

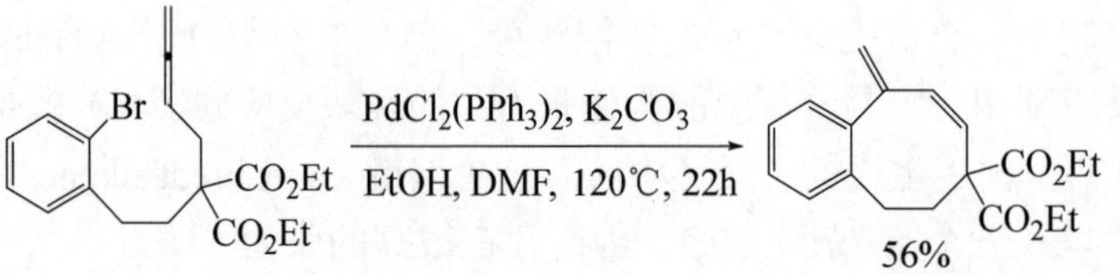

$$\text{(48)} \quad 56\%$$

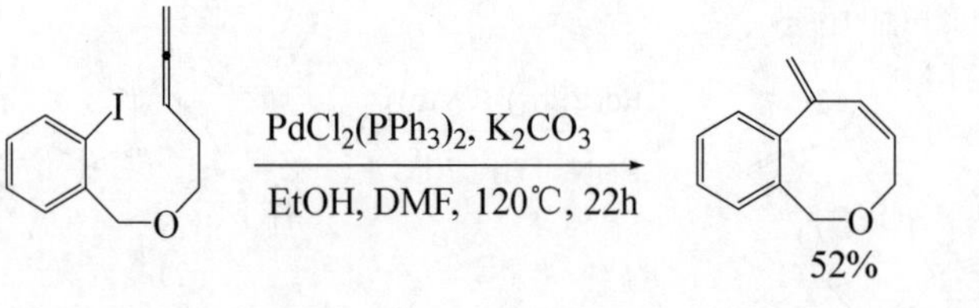

$$\text{(49)} \quad 52\%$$

大环也可以通过分子内 Heck 反应合成。例如，16 元内酯以 55%的收率由结构中同时含碘代烯和乙烯基的底物合成。为了阻止反应过程中底物的二聚和低聚，需要加入化学计量的钯配合物,而且需要将底物缓慢地加入反应液中[31]。利用丙烯酸酯基提供的乙烯基与碘代苯基进行 *endo*-关环反应可以得到 22 元环的肉桂酸酯[32]。

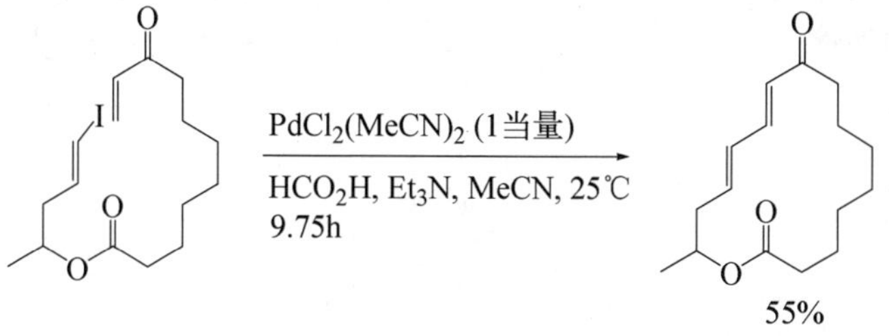

$$\text{(50)}$$

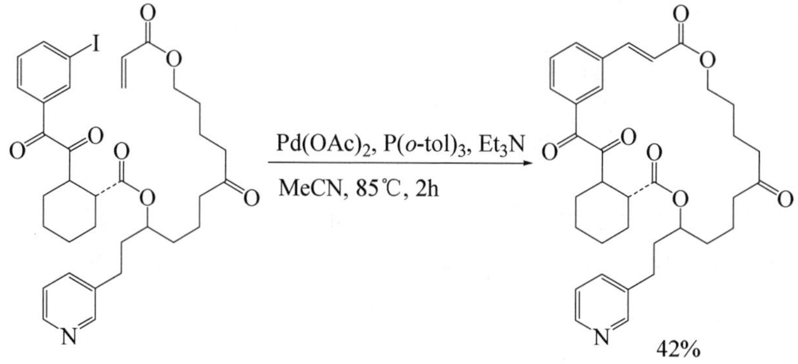

$$\text{(51)}$$

1.2.3　Heck 串联(tandem)反应

在图 1-1 所示的反应机理中,σ-烷基钯配合物一旦生成,如果没有顺式的 β-氢原子存在,则不能发生 β-氢消除生成烯烃。然而,σ-烷基钯配合物比较稳定,具有足够长的寿命,因此当有其他亲核试剂存在时可以与亲核试剂偶联,并引起串联反应,为合成丰富多彩的有机分子提供可行的途径。

1.2.3.1　分子间串联反应

在反应式(52)所示的反应中,底物中碘代苯与其中的乙烯基在钯催化剂催化下先生成苯并四氢吡喃烷基σ-钯配合物,该配合物与外加的四苯硼酸钠反应生成苯并四氢吡喃烷基与苯基偶联的产物[33]。

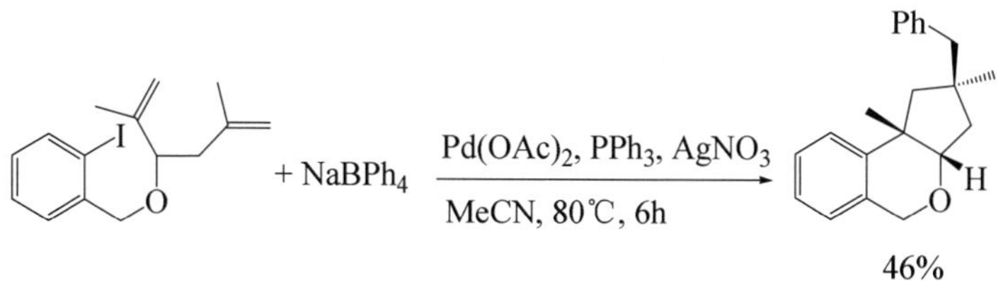

$$\text{(52)}$$

邻碘代苯-4-羟-环戊烯醚在钯催化下生成σ-烷基钯配合物，由于羟基的存在无法生成β-氢与钯碘顺式的钯中间体，因而与 1-辛烯-3-酮发生偶联反应生成相应的偶联产物苯并前列环素衍生物[34]。值得注意的是，由于底物中原有手性中心的诱导，又生成两个新的手性中心。

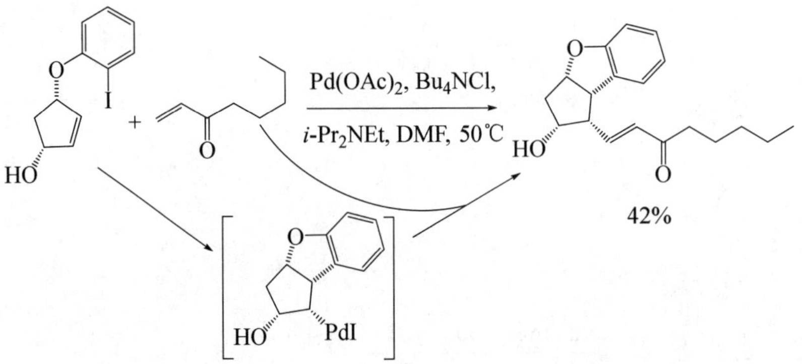

降冰片烯与手性的碘乙烯衍生物在标准的 Heck 催化剂催化下偶联生成稳定的σ-钯配合物，用氰化钾捕捉该中间体使氰基引入降冰片烷骨架生成氰基与乙烯衍生物基团处于顺式、exo-构型的产物，并且反应是对映选择性的[35]。

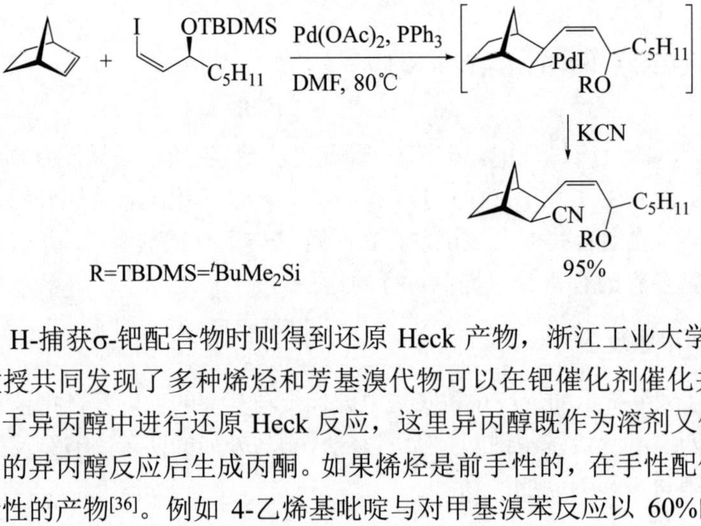

当用 H-捕获σ-钯配合物时则得到还原 Heck 产物，浙江工业大学靳立群教授、胡信全教授共同发现了多种烯烃和芳基溴代物可以在钯催化剂催化并在氢氧化钾存在下，于异丙醇中进行还原 Heck 反应，这里异丙醇既作为溶剂又作为氢源，起氢源作用的异丙醇反应后生成丙酮。如果烯烃是前手性的，在手性配体存在下则得到光学活性的产物[36]。例如 4-乙烯基吡啶与对甲基溴苯反应以 60%的收率得到相应的还原偶联产物。

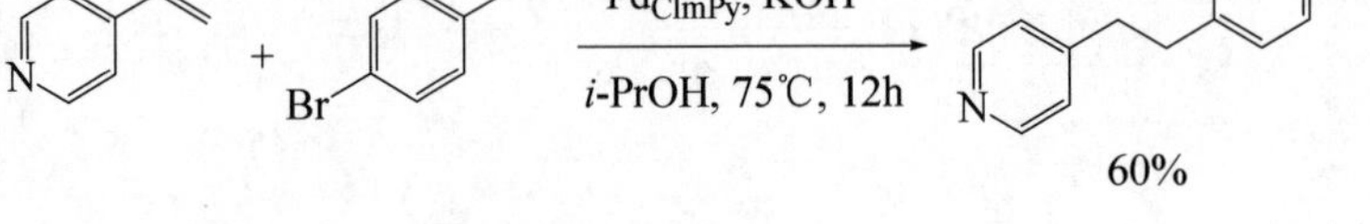

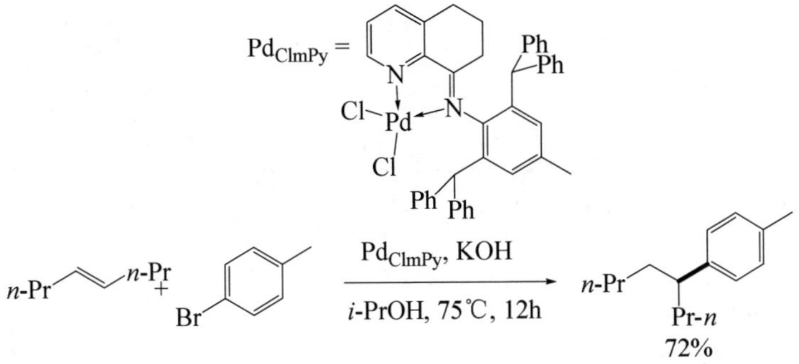

$$\text{Pd}_{\text{ClmPy}} = \text{[结构式]}$$

$$n\text{-Pr}\diagup\!\!\!\diagdown\!\!n\text{-Pr} + \text{4-MeC}_6\text{H}_4\text{Br} \xrightarrow[i\text{-PrOH, 75℃, 12h}]{\text{Pd}_{\text{ClmPy}}, \text{KOH}} \text{产物} \quad 72\%$$ (56)

当以醇作溶剂在一氧化碳气氛中进行 Heck 反应时，如果σ-钯烷基配合物的β-氢顺式消除受阻，σ-钯烷基配合物则与 CO 加成生成σ-酰基钯配合物，然后醇解生成酯，TlOAc 促进反应的进行[37]。此类反应可称作 Heck 羰基化反应。

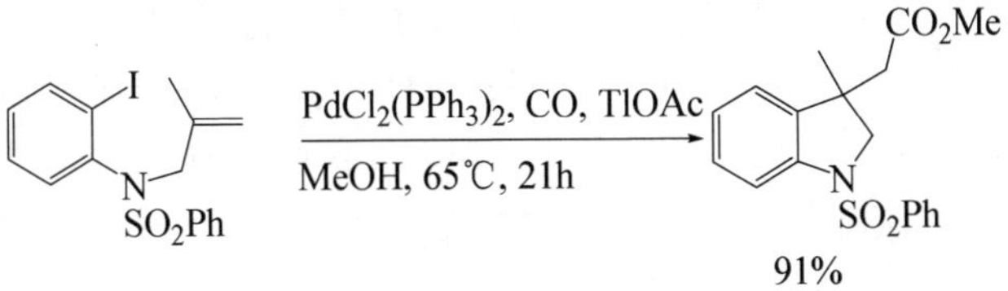

$$\xrightarrow[\text{MeOH, 65℃, 21h}]{\text{PdCl}_2(\text{PPh}_3)_2, \text{CO, TlOAc}} \quad 91\%$$ (57)

用其他亲核试剂如有机还原剂、胺替代醇反应也能进行，生成相应的酰基与亲核试剂的偶联产物，如在二苯基甲基硅烷存在下可生成醛[38,39]。

用胺作为弱亲核试剂，则生成酰胺，反应如下[40]。

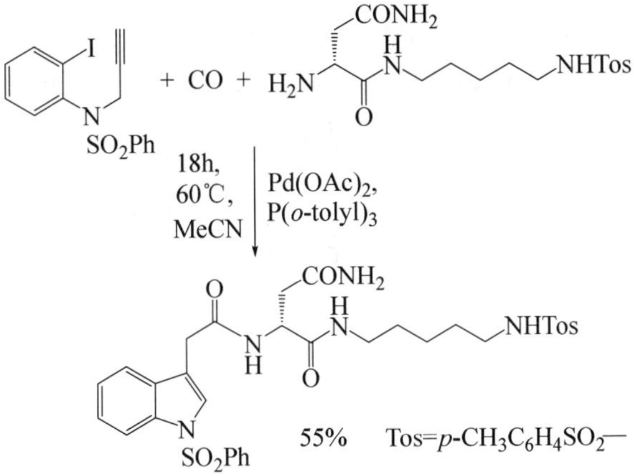

$$\xrightarrow[\text{MeCN}]{\substack{18\text{h},\\60℃,}} \xrightarrow{\text{Pd(OAc)}_2,\ \text{P}(o\text{-tolyl})_3}$$ (58)

55%   Tos=p-CH$_3$C$_6$H$_4$SO$_2$—

丙二烯在 Heck 串联反应中起重要角色。例如，在邻碘苯基炔丙基醚进行分子内 Heck 反应中先生成 5-exo 乙烯基钯加合物，加合物与丙二烯反应生成π-烯丙基中间体，该中间体被(S)-(+)-2-吡咯烷甲醇捕获生成二烯胺[41]。

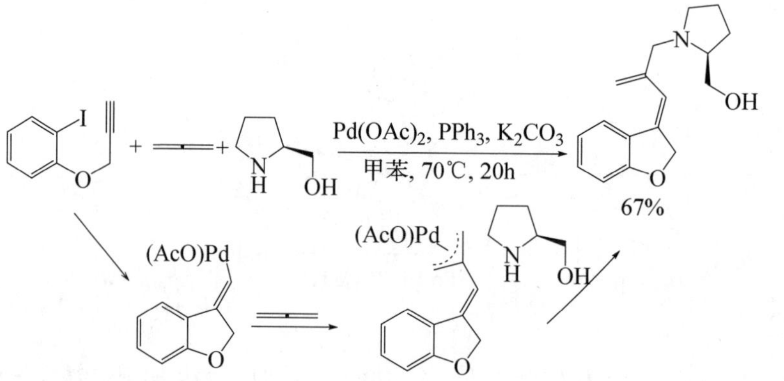

(59)

1.2.3.2 分子内串联反应

Heck 分子内串联反应更容易发生,通过多重插入,可以合成多环骨架化合物。在反应式(60)中,氮原子同时含有乙烯基和烯丁基的邻碘苯甲酰胺在 Heck 催化剂催化下先由乙烯基与邻碘苯基进行 5-exo 关环生成新戊基σ-钯中间体,该中间体不能发生β-氢消除,因而与烯丁基立刻发生 6-exo 偶联高收率生成三环稠合化合物,同时构建一个季碳中心[42]。

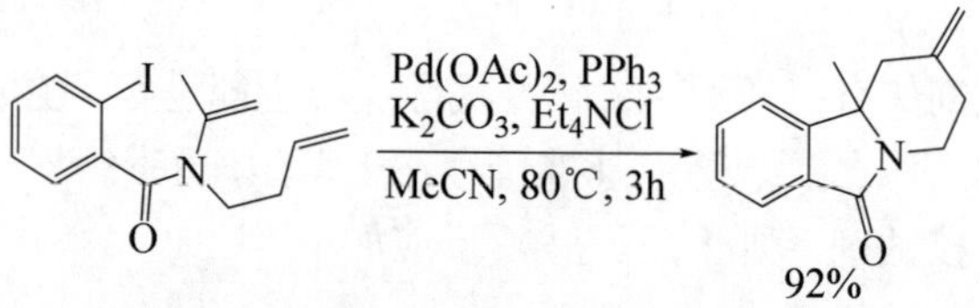

(60)

与上面反应类似,通过分子内 Heck 串联反应最后生成的σ-烷基钯中间体与距离适当的芳环偶联,完成 Heck/C-H 官能团化(Heck/C-H functionalization)反应,高收率得到茚满与氧化吲哚螺合的化合物[43]。

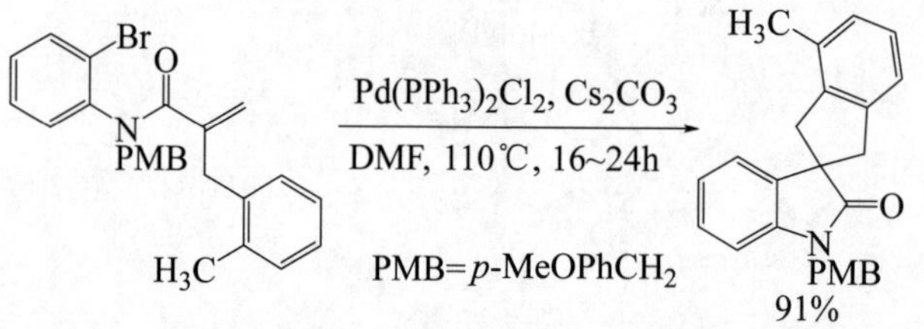

(61)

利用分子内 Heck 串联反应中的多次插入,合成三元环与其他环稠合的化合物。如下结构的乙烯衍生物的三氟甲磺酸酯在标准的 Heck 反应条件下经 5-exo 关环生成新戊基σ-钯配合物中间体,该中间体再进行 3-exo 关环、β-氢消除非对映选择性地生成含三元环的三环稠合化合物[44]。

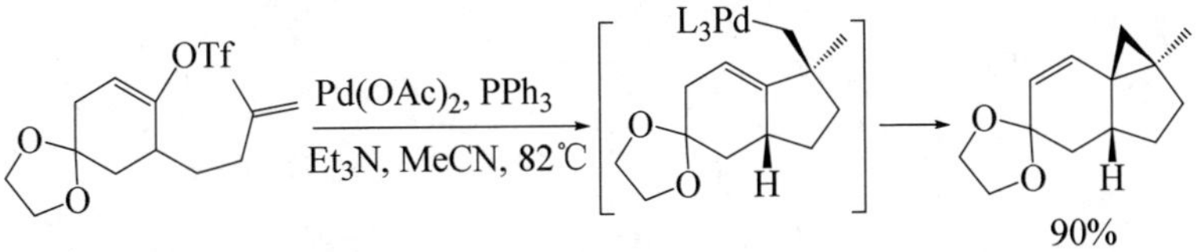

$$\text{(62)}$$

炔烃是用于分子内 Heck 串联反应构建稠合多环化合物有效的反应单元。例如，下面含有炔键和烯键的乙烯基碘代物在标准 Heck 反应条件下先由碘乙烯单元与炔键发生 6-exo 合环，再经两次 5-exo 插入关环形成两个五元环，高收率地得到三环稠合的化合物。这种"拉链(zipper)"类型的多环化，高效地生成棱角状多环稠合的网络化合物[45]。

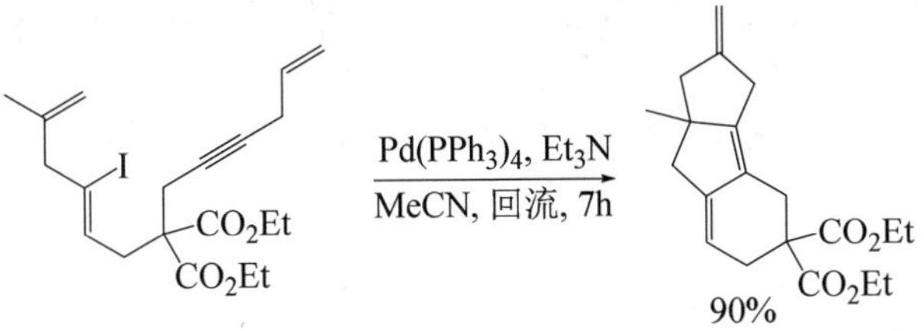

$$\text{(63)}$$

1.2.4 不对称 Heck 反应

跟大多数过渡金属催化的有机反应类似，当构成催化体系的膦配体为适当结构的手性膦配体时，可进行不对称 Heck 反应。相较于其他过渡金属催化的不对称有机反应，高对映选择性地进行不对称 Heck 反应要难得多，这是由不对称 Heck 反应的机理决定的。不对称 Heck 反应的机理比较复杂，文献多有介绍[46,47]，这里不再涉及。总之，手性膦配体是实现高收率、高对映选择性不对称 Heck 反应的关键。一个好的手性膦配体应满足不对称 Heck 反应区域选择性(regioselectivity)和对映选择性（enantioselectivity）的共同要求。一般而言，分子内不对称反应比分子间不对称反应要容易进行。

1.2.4.1 分子间不对称 Heck 反应

分子间不对称 Heck 反应不容易实现，而且反应底物范围较小。乙烯底物多为杂环烯，而其偶联伙伴则为芳基或烯基三氟甲磺酸酯。首次报道的分子间不对称 Heck 反应是 2,3-二氢呋喃在 Pd/(R)-BINAP 催化下与芳基三氟甲磺酸酯之间的偶联反应[48]。反应产物以 (R)-2-芳基-2,3-二氢呋喃为主，并伴有 (S)-2-芳基-2,5-二氢呋喃的生成。当芳基为苯基时，主、副产物的分离收率分别为 71%和 7%，对映选择性分别为 93%和 67%。

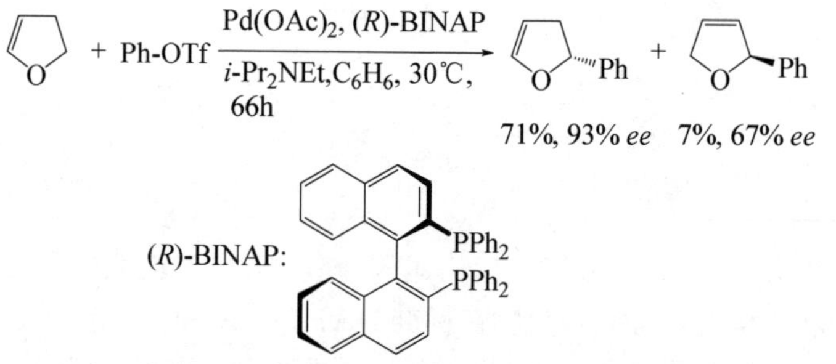

$$\text{(二氢呋喃)} + \text{Ph-OTf} \xrightarrow[\substack{i\text{-Pr}_2\text{NEt, C}_6\text{H}_6, 30℃, \\ 66\text{h}}]{\text{Pd(OAc)}_2, (R)\text{-BINAP}} \text{(2-Ph-2,5-二氢呋喃)} + \text{(2-Ph-2,3-二氢呋喃)} \quad (64)$$

71%, 93% ee   7%, 67% ee

(R)-BINAP: [结构式]

值得注意的是该不对称 Heck 反应的主副产物的构型相反,这可能是由图 1-4 所示的 **A** 和 **B** 参与的动力学拆分决定的。首先,配位的烯烃(二氢呋喃)插入 Pd-Ar 键分别生成两个 σ-配合物异构体,然后分别进行 β-氢消除生成氢-烯钯配合物 **A** 及其非对映异构体 **B**。**A** 的构型更有利于接下来的烯烃插入和 β-氢消除过程从而生成主产物,而 **B** 则很容易释放出烯烃生成构型相反的副产物[47]。

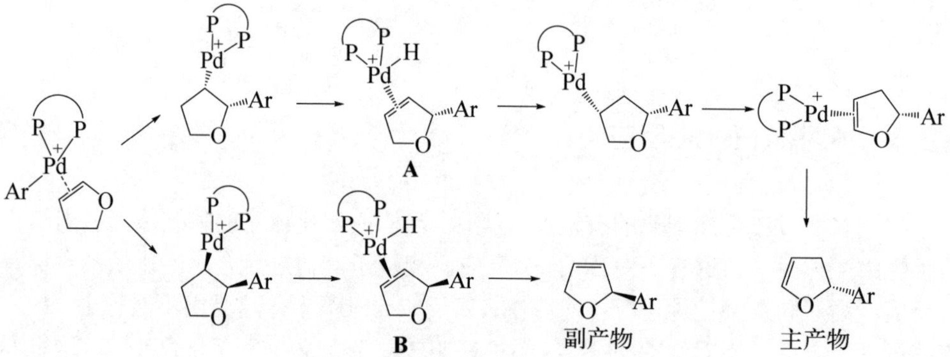

图 1-4 2,3-二氢呋喃不对称芳基化反应的途径

2,3-二氢呋喃与三氟甲磺酸烯酯之间也可进行不对称 Heck 反应。例如,2,3-二氢呋喃与 2-乙氧羰基环己烯三氟甲磺酸酯反应以高达 96%的对映选择性生成不对称偶联产物,没有区域异构体的生成[49]。

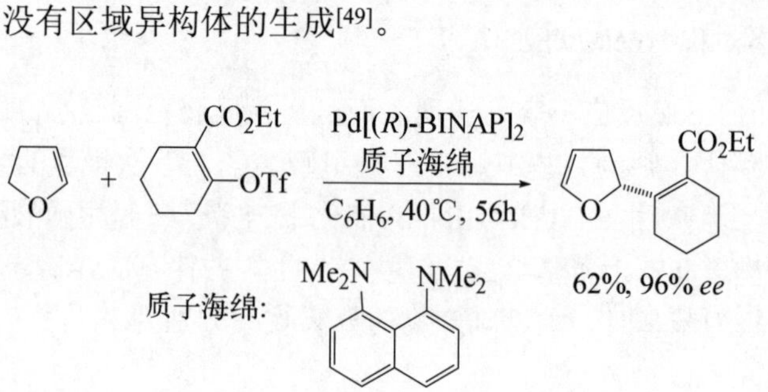

$$\text{(二氢呋喃)} + \text{(2-CO}_2\text{Et-环己烯基-OTf)} \xrightarrow[\substack{\text{质子海绵} \\ \text{C}_6\text{H}_6, 40℃, 56\text{h}}]{\text{Pd}[(R)\text{-BINAP}]_2} \text{产物} \quad (65)$$

62%, 96% ee

质子海绵: [1,8-双(二甲氨基)萘结构]

4H-1,3-二噁英也可以中等的收率和对映选择性实现不对称芳基化[50]。不对称芳基化产物很容易水解生成相应的 1,3-二醇，该 1,3-二醇是 Sharpless 合成(R)-氟西汀（fluoxetine）的中间体[51]。

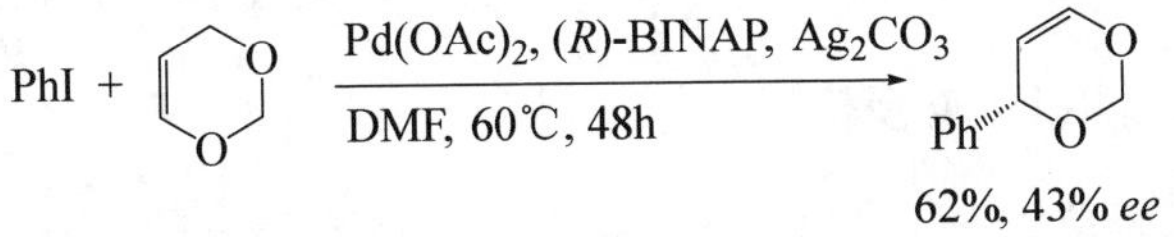

$$\text{PhI} + \underset{}{\text{[4H-1,3-二噁英]}} \xrightarrow[\text{DMF, 60℃, 48h}]{\text{Pd(OAc)}_2, (R)\text{-BINAP, Ag}_2\text{CO}_3} \underset{62\%,\ 43\%\ ee}{\text{Ph-[产物]}} \quad (66)$$

$$\xrightarrow{\ \ } \text{Ph-CH(OH)-CH}_2\text{-CH}_2\text{OH} \xrightarrow{\ \ } (R)\text{-氟西汀}$$

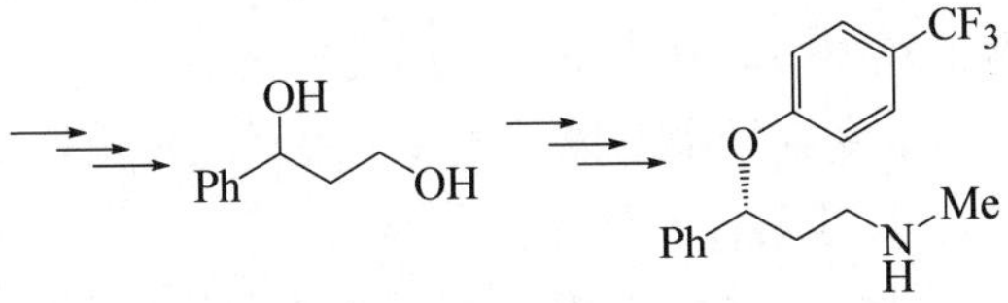

类似地，4,7-二氢-1,3-二噁英与三氟甲磺酸苯酯在(S)-BINAP 诱导下发生不对称 Heck 反应，以良好的收率和对映选择性生成相应的偶联产物[52]。值得注意的是，加入分子筛可以同时提高反应的产率和对映选择性。同样的反应，当手性配体改为一种手性 P,N-配体时，反应的对映体过量值可提高到 92%[53]，表明手性配体在不对称 Heck 反应中的重要性。

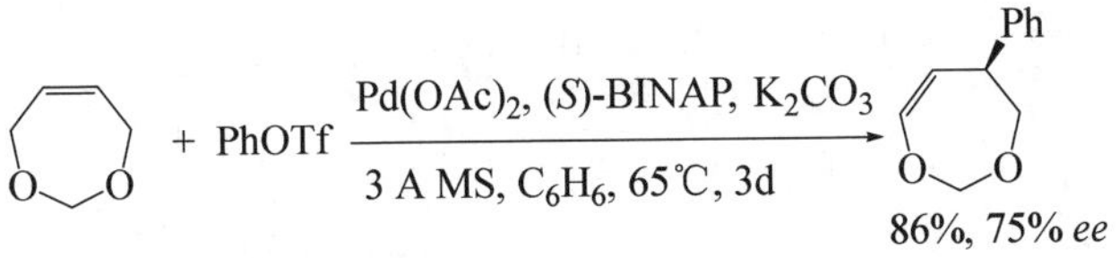

$$\text{[底物]} + \text{PhOTf} \xrightarrow[\text{3 Å MS, C}_6\text{H}_6,\ 65℃,\ 3d]{\text{Pd(OAc)}_2,\ (S)\text{-BINAP, K}_2\text{CO}_3} \underset{86\%,\ 75\%\ ee}{\text{[产物]}} \quad (67)$$

$$\text{[底物]} + \text{PhOTf} \xrightarrow[\text{THF, 70℃, 7d}]{\text{Pd(dba)}_2,\ \text{P,N-L},\ i\text{-Pr}_2\text{NEt}} \underset{70\%,\ 92\%\ ee}{\text{[产物]}} \quad (68)$$

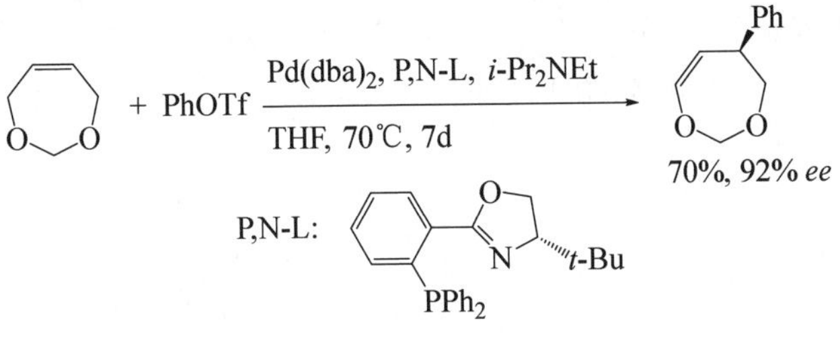

P,N-L: [2-(4-t-Bu-oxazolinyl)phenyl]diphenylphosphine

在手性钯催化体系催化下一些氮杂环化合物也能与三氟甲磺酸烯酯发生不对称 Heck 反应。例如，2,3-二氢吡咯-N-甲酸甲酯与 2-乙氧羰基环己烯三氟甲磺酸酯反应，与以 2,3-二氢呋喃为底物时相近的区域和对映选择性生成相应的不对称偶联产物[49]。

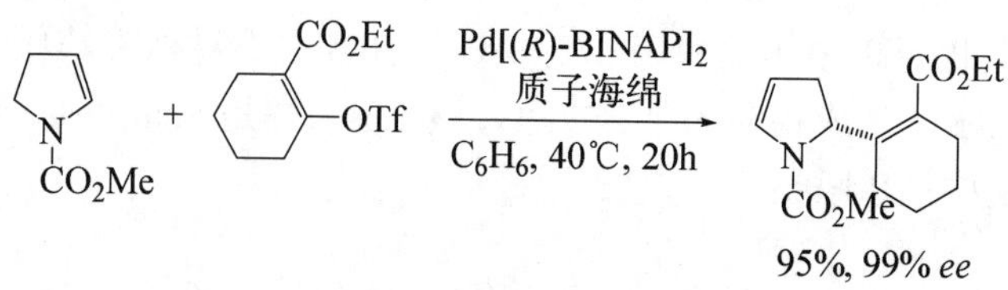

$$\text{(69)}$$

反应式 (69): 95%, 99% ee

1.2.4.2 分子内不对称 Heck 反应

相较于分子间反应，分子内不对称 Heck 反应的底物范围要宽泛得多。根据底物的结构可生成稠环和螺环化合物，而根据连接两个反应基团之间碳链的长短可生成不同大小的环；根据潜手性中心的结构不同可形成叔碳手性中心和季碳手性中心。同时，反应的对映选择性除了受底物中乙烯基上的离去基团影响外，还与手性配体、钯催化剂前体、碱、溶剂及反应条件密切相关。

（1）叔碳中心的形成   如下结构的碘代烯最早作为分子内不对称 Heck 反应的底物，在优化的反应条件下只以较低的对映选择性生成顺式十氢萘衍生物，而当底物中的离去基团碘替代为三氟甲磺酸根时，反应的对映选择性达到 91%[54]。

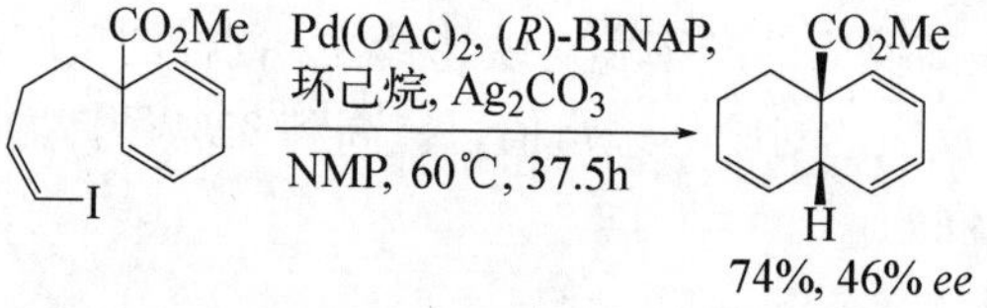

$$\text{(70)}$$

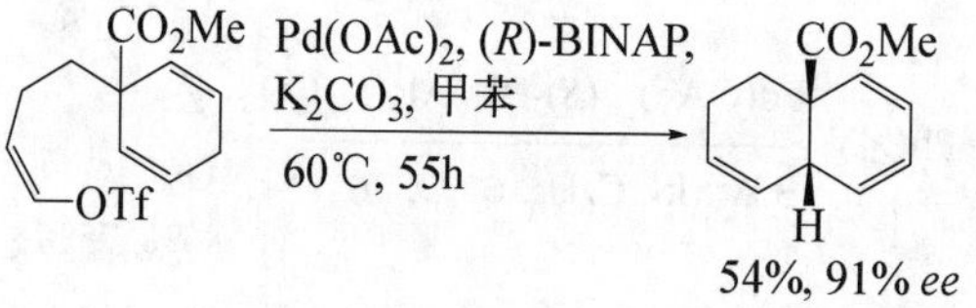

74%, 46% ee

$$\text{(71)}$$

54%, 91% ee

在类似的催化体系下，还可以不对称合成 5,6-稠环体系的氢化茚满类化合物，2-位取代的碘代物和三氟甲磺酸烯酯作为底物均可，磷酸银是最适宜的银盐促进剂[55]。

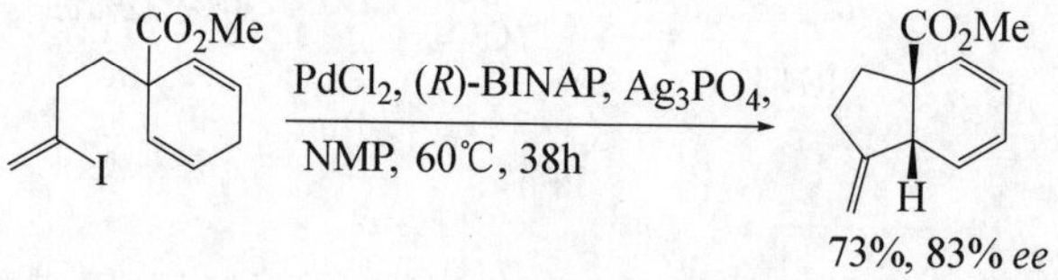

$$\text{(72)}$$

73%, 83% ee

通过底物的精准设计以及催化体系的调整，利用分子内不对称 Heck 反应还可以合成 5,5-稠环体系，该体系是许多天然产物的骨架结构。值得注意的是，当底物为碘代烯类化合物时，合成 5,6-稠环体系所需的银盐在这里因引起底物的分解反而起反作用，而四丁基醋酸铵则必不可少，因为醋酸根起亲核试剂作用[56]。

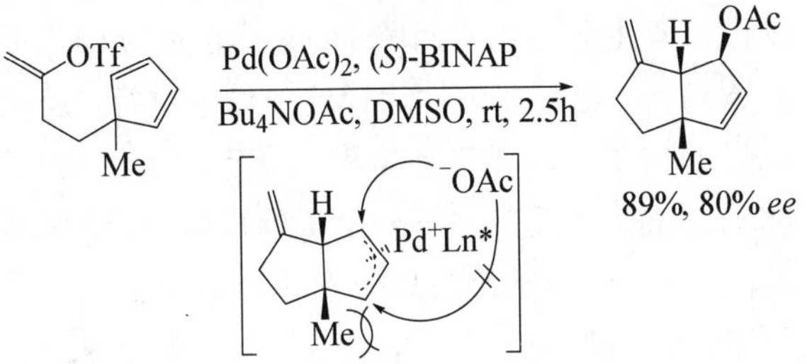

（2）季碳中心的形成　利用分子内不对称 Heck 反应构建季碳中心可为含螺环结构的天然产物的合成提供一条便捷的途径。第一个用于分子内不对称 Heck 反应的底物是一个含环状烯烃的底物 N-甲基-N-（碘代苯基）四氢苯甲酰胺-4-酮缩乙二醇，该底物的特殊结构可避免反应过程中碳碳双键的重排。值得注意的是，催化体系中有磷酸银存在时生成 S-构型的含吡咯烷酮环的螺环产物，而无银盐但有叔胺 1,2,2,6,6-五甲基哌啶（PMP）存在时则生成 R-构型的产物，二者互为对映异构体[57]。原因是有银盐存在时反应按离子中间体路径进行，而叔胺 PMP 存在时则按中性中间体途径进行。如图 1-5 所示，在离子中间体和中性中间体中，乙烯基与中心钯离子配位的构型正好相反，因而生成构型正好相反的环合产物[47]。

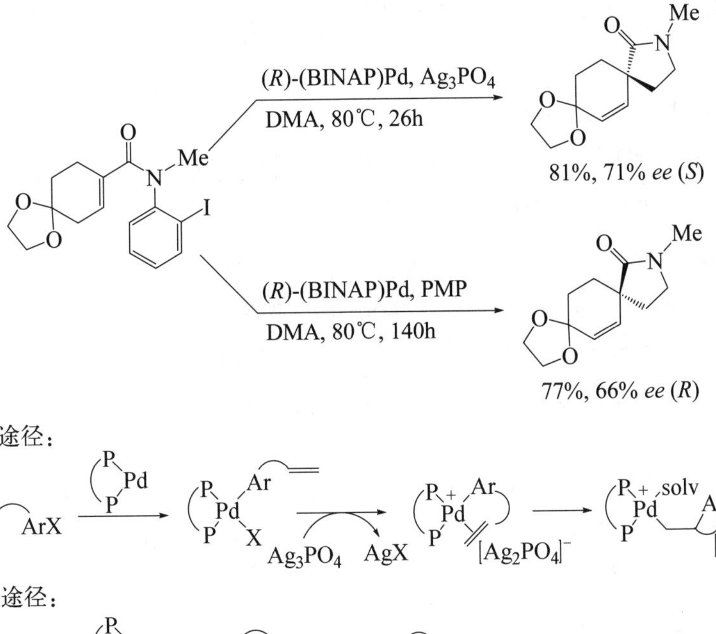

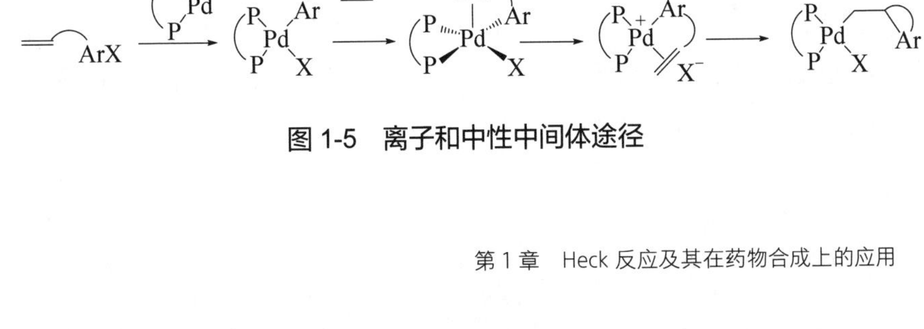

图 1-5　离子和中性中间体途径

非环状烯烃底物也可发生分子内不对称 Heck 环合反应构建季碳中心，反应既可在离子途径的条件下进行，也可在中性途径的条件下进行；反应的对映选择性和产物的构型与底物烯烃的构型、取代基的体积以及反应条件相关。需要注意的是，在离子途径条件下和中性条件下反应并不一定得到构型相反的产物。例如，下面反应式是含不同取代基的（Z）-构型碘代烯烃分别在离子途径（银盐存在）和中性途径（PMP 存在）条件下的反应结果。当 α-取代基（R^1）是甲基时，无论是在离子条件还是中性条件下反应，反应的对映选择性都为中等。当 α-取代基是体积大的苯基和叔丁基时，在离子条件下得到较好的对映选择性，而中性条件下反应的对映选择性仍然较低。末端取代基似乎对立体选择性无太大影响，如体积大的硅氧基与小的氢原子时的反应结果相近。特别需要注意的是 **1c**［式（75）］在两种条件下得到的产物都为（R）构型，与其他底物反应的结果不同。这一结果也说明产物的构型并不完全与中性或离子条件相对应。

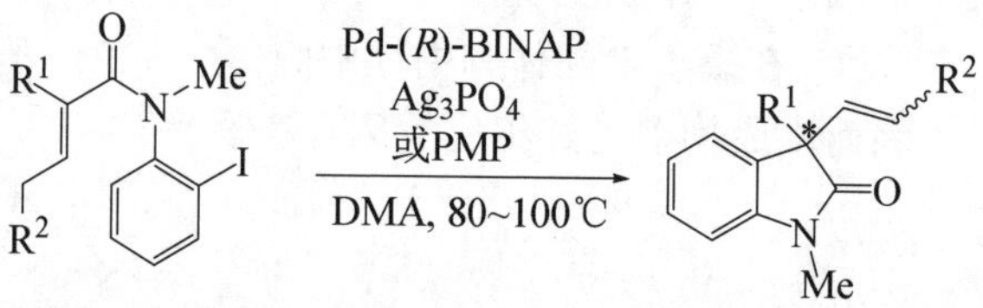

$$\text{(75)}$$

1a: R^1=Me; R^2=OTBDMS　　OTBDMS: 叔丁基二甲硅氧基

1b: R^1=Ph; R^2=OTIPS　　　OTIPS: 三异丙基硅氧基

1c: R^1=t-Bu; R^2=OTIPS

PMP条件：　　　　　　　　Ag_3PO_4条件：

1a: 58%, 38% ee (R)　　　**1a**: 80%, 45% ee (S)

1b: 74%, 35% ee (R)　　　**1b**: 93%, 73% ee (S)

1c: 90%, 27% ee (R)　　　**1c**: 41%, 72% ee (R)

（E）-构型碘代烯烃在中性条件下（PMP）的分子内不对称 Heck 环合反应与其（Z）-式异构体的反应有所不同。如反应式（76）所示，多数情况下（E）-构型碘代烯烃在中性条件下以较高的对映选择性生成环合产物，有的反应的对映选择性甚至高于 90%（**2a** 和 **2c**）。当溶剂由 DMA 改为 THF 时，**2c** 以良好的收率和高达 97%的对映选择性生成环合产物。大体积 α-取代的底物在中性和离子两个条件下分别以低的和中等对映选择性生成环合产物（**2b**）。与（Z）-异构体的情况相反，所有（E）-构型的底物在两种条件下都得到相同构型的产物。

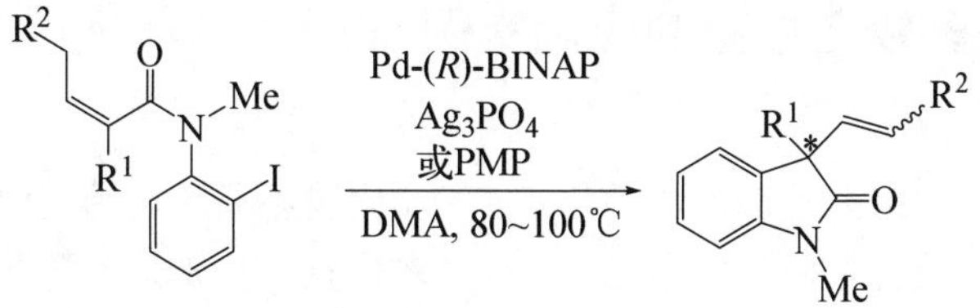

$$\text{(76)}$$

2a: R^1=CH$_2$CH(OMe)$_2$; R^2=OTBDMS
2b: R^1= Ph; R^2= OTIPS
2c: R^1=Me; R^2= OTBDMS

OTBDMS: 叔丁基二甲硅氧基
OTIPS: 三异丙基硅氧基

PMP条件：
2a: 93%, 91% ee (S)
2b: < 5%, 19% ee (R)
2c: 80%, 92% ee (R)

Ag$_3$PO$_4$条件：
2b: 86%, 65% ee (R)

含烯基取代基的萘的三氟甲磺酸酯进行分子内不对称 Heck 反应形成季碳中心，同时高对映选择性地生成三环稠合的化合物，该化合物是合成药物分子哈里醌 (halenaquinone)的关键中间体[58]。

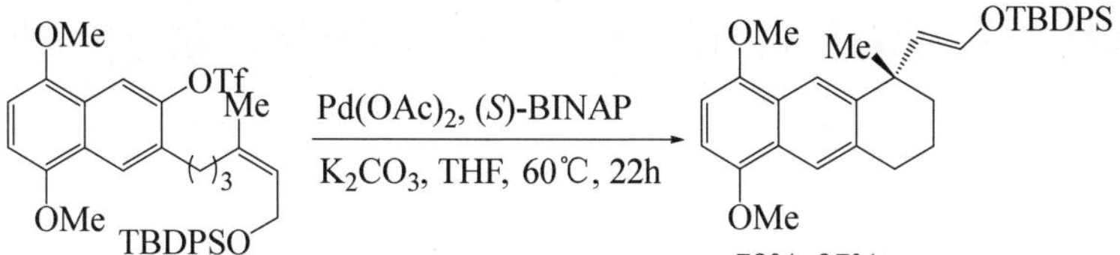

$$\text{(77)}$$

78%, 87% ee

OTBDPS: 叔丁基二丙基硅氧基

季碳中心的构建也可通过分子内不对称 Heck 串联反应得以实现。以下是芳基碘代化合物经由不对称分子内 Heck 串联环合-氢捕获过程以中等收率和高的对映选择性生成苯并四氢呋喃衍生物的实例[59]。反应结果与碘代苯单元上甲酯基与碘原子的相对位置有关。当甲酯基处于碘原子的对位时目标产物的收率较低但对映选择性较高，而甲酯基处于碘原子的间位则提高了目标产物的收率但对映选择性有所降低。

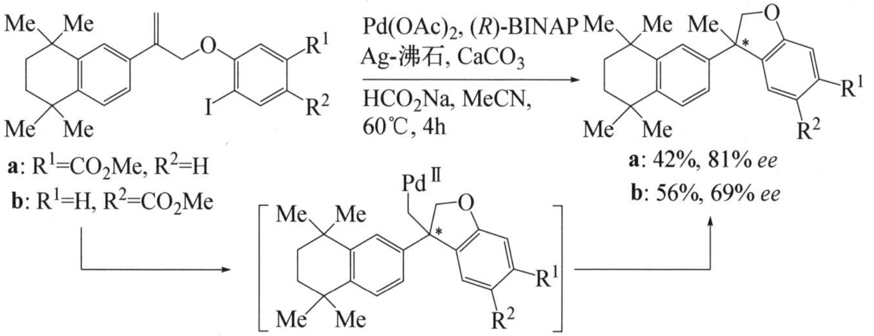

$$\text{(78)}$$

a: R^1=CO$_2$Me, R^2=H
b: R^1=H, R^2=CO$_2$Me

a: 42%, 81% ee
b: 56%, 69% ee

1.3 Heck 反应在药物合成上的应用实例

由上述基本反应可知,Heck 反应底物适用性广泛,底物不局限于活化的烯烃,简单烯烃也可进行反应。此外,Heck 反应还具有诸多优点,特别是广泛的官能团耐受性,酮羰基、酯基、酰氨基、醚键或各种杂环都不影响 Heck 反应的进行,可用于合成多种多样的多官能团化合物。因此,Heck 反应被广泛用于合成结构多样性的各类药物分子。以下是 Heck 反应在药物合成应用中的几个实例。为了描述方便,把 Heck 反应生成的产物命名为 H-x,x 为整数,代表序号。

1.3.1 分子间反应的应用

阿昔替尼(axitinib)是一种其他系统治疗无效的晚期肾癌治疗的临床用药。其合成的关键一步是芳硫基苯并吡唑碘代物与 2-乙烯基吡啶间的 Heck 反应。反应以 Pd(OAc)$_2$ 和 P(o-tolyl)$_3$ 为催化剂前体,LiBr 为助剂,1,8-二(二甲氨基)萘为碱,N-甲基吡咯烷酮(NMP)为溶剂在 110℃反应 28h,以中等收率生成产物 H-1,即阿昔替尼[60]。

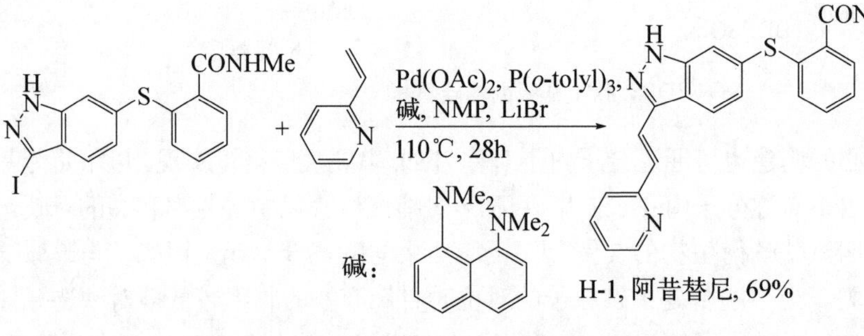

反应式(80)最终的目标化合物是一种有效的血栓素受体拮抗剂,用于治疗哮喘、不稳定型心绞痛、深静脉血栓形成和冠状动脉粥样硬化。利用芳基碘与芳基溴在 Heck 反应中的活性差别,3-溴-5-碘苯基(4'-氟苯基)甲酮在 Pd(OAc)$_2$ 为催化剂,三乙胺为碱,乙腈为溶剂的催化反应体系下化学选择性地先在碘代碳上与丙烯酸甲酯进行第一次 Heck 反应生成偶联产物 H-2;然后在相同的催化剂存在下,用二异丙胺替代三乙胺,二甲苯替代乙腈,H-2 的溴代碳与 N-乙烯基邻苯二甲酰亚胺进行第二次 Heck 反应生成 H-2'。H-2'经后续反应得到目标化合物[61]。

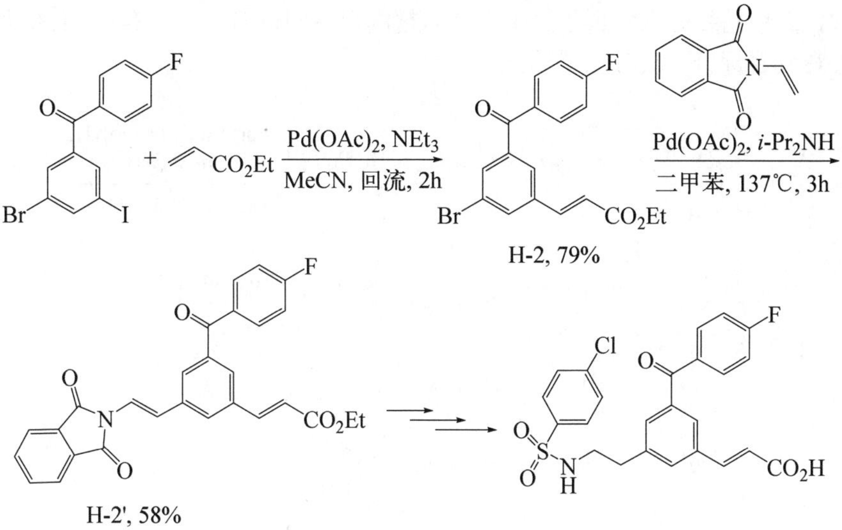

下面的实例反应式（81）的产物是一种口服活性$\alpha_{2/3}$-选择性 $GABA_A$ 激动剂（$GABA_A$ agonist），用于治疗焦虑、抽搐和认知障碍的一种候选药物。该药物可通过前体化合物咪唑三嗪衍生物与芳基溴之间的 Heck 偶联反应合成。反应以价格相对低廉以及性质稳定的 $Pd(OAc)_2/PPh_3$ 为催化体系，KOAc 为碱，DMAc 为溶剂，在 130℃反应 4h 顺利完成，以 86%的收率得到目标产物。反应的主要杂质是前体咪唑三嗪衍生物的二聚体以及另一前体芳基溴的脱溴产物，杂质产率均低于 3%[62]。

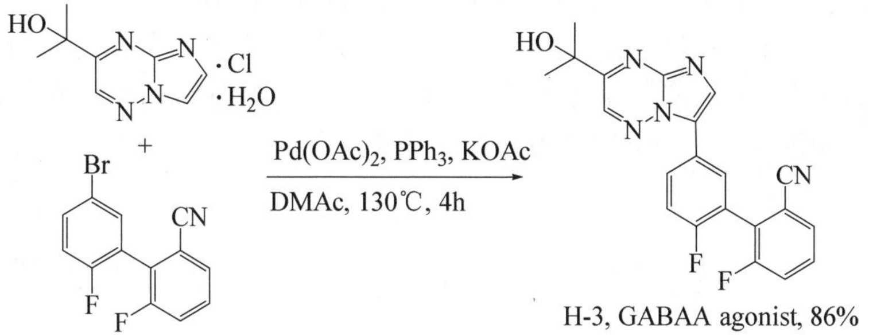

如下所示结构的 2,4,9-三取代-2-苯并䓬-3-酮是一个第二代$\alpha_v\beta_3$ 受体拮抗剂（$\alpha_v\beta_3$ receptor antagonist），用于治疗骨质疏松症。利用 Heck 反应能够合成该药物。在芳基溴与依康酸进行 Heck 偶联反应前，将芳基溴上的醛羰基生成二甲基缩醛来保护醛羰基是必要的，否则，偶联产物在 Heck 反应条件下发生分子内羟醛缩合导致生成萘二羧酸。一旦生成缩醛的反应完成，反应液直接加入依康酸、$Pd(OAc)_2$/P(o-tolyl)$_3$、NEt_3 和 n-Bu_4NBr 的乙腈溶液，然后在回流条件下反应 10h。由于依康酸是偕二取代的烯烃，其 Heck 反应比一般烯烃的反应难度要大。Heck 反应完成后用稀

盐酸处理，二甲基缩醛水解再生醛基，然后精制得中间体 H-5[63]。H-5 经后续多步反应最终得到目标分子 $\alpha_v\beta_3$ 受体拮抗剂。

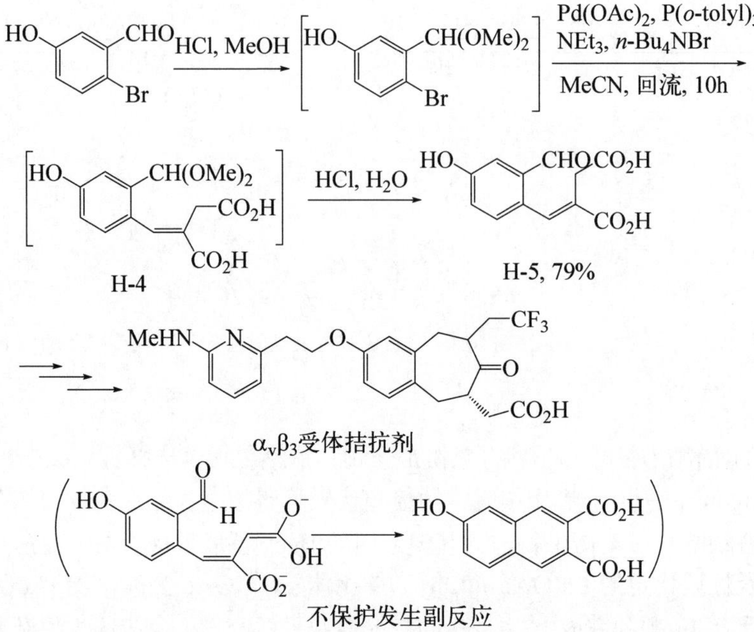

$$(82)$$

利匹韦林（rilpivirine）是一种非核苷逆转录酶抑制剂。其合成的关键中间体（H-6）由 4-碘-2,6-二甲基苯胺在钯催化剂催化下与丙烯腈进行 Heck 偶联反应合成。由于均相催化剂存在钯残留问题，用 Pd/C 作为催化剂进行反应。偶联产物 H-6 生成后用 6mol/L 的 HCl 异丙醇溶液处理成盐，起到精制作用，得到 E/Z 为 98∶2 H-6 的盐酸盐 H-6·HCl[64]。H-6·HCl 再经多步转化生成利匹韦林。

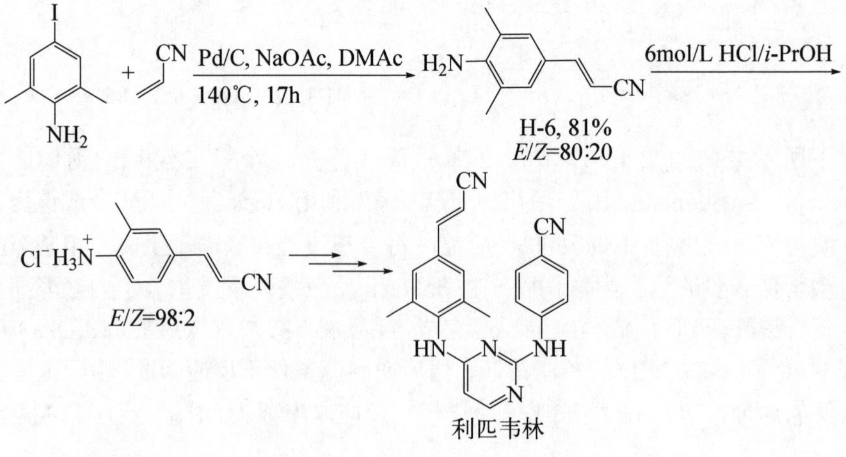

$$(83)$$

下面反应的目标产物是一类强效 NMDA 甘氨酸拮抗剂,用于治疗中风和神经退行性疾病如阿尔茨海默病和亨廷顿舞蹈症。其结构特征是三环稠合体系,可通过两次 Heck 反应构建。第一次 Heck 反应是 6-氯-4-碘吲哚-2-甲酸甲酯与烯丙醇在相转移催化剂 BnEt₃NCl 和碱 NaHCO₃ 存在及 Pd(OAc)₂ 催化下于 50℃反应 4h,在进行 Heck 反应的同时烯丙醇异构化为丙醛,以 90%的收率得到关键中间体 H-7;H-7 经多步反应转化为另一关键中间体 **a**,该中间体同时在吲哚环的 C_4 位含有 2-戊烯羧酸乙酯基以及在 C_3 位含有碘。第二个 Heck 反应是分子内反应,在标准的 Heck 反应条件〔Pd(OAc)₂、PPh₃、NEt₃〕下反应进展很慢,用 Pd(PPh₃)₄ 替代 Pd(OAc)₂ 和 PPh₃,并向反应体系加入适量水和 Ag₃PO₄,无须 NEt₃,反应顺利进行,以 81%的收率生成环合产物 H-7'。H-7'经多步反应生成目标产物[65]。

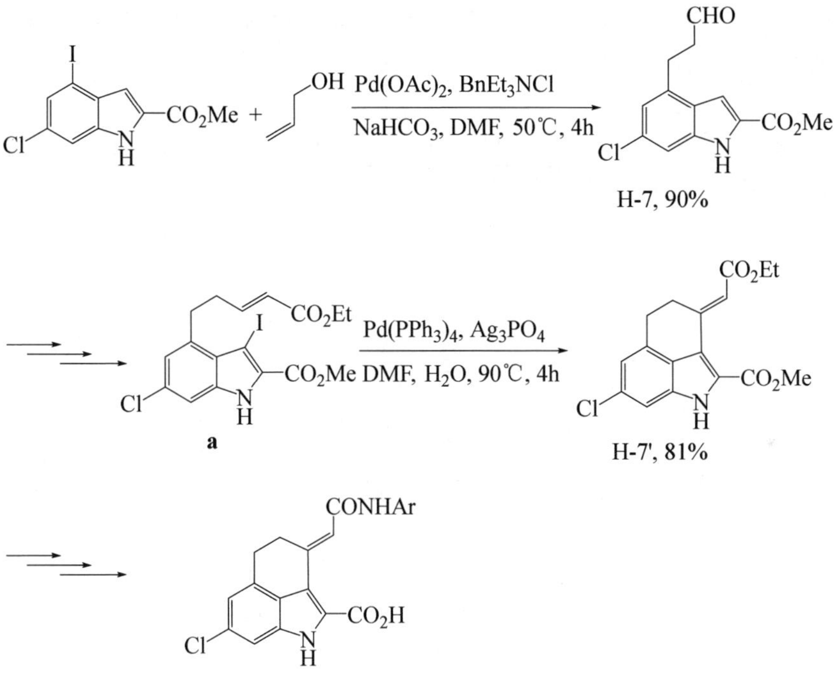

反应式(85)中的目标产物是一种 $R_v\beta_3$ 整合素拮抗剂,可用于治疗骨质疏松症。其合成的关键步骤之一是 5-溴-2-甲氧基吡啶与丙烯酸叔丁酯之间的 Heck 偶联反应。反应在 Pd(OAc)₂ / P(o-tolyl)₃ 催化下进行,几乎以定量的收率生成中间体 H-8。以丙烯酸叔丁酯为 Heck 反应的偶联伙伴是因为叔丁基可以阻止酯基中碳氧双键与有机钯中间体的加成反应。为避免反应过热,需要将丙烯酸烯丙酯在反应温度下缓慢滴加到反应混合物中[66]。

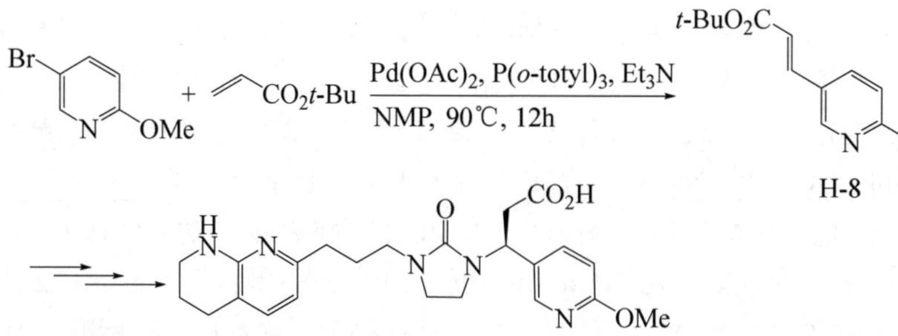

$$\text{(85)}$$

赛红霉素（cethromycin）是抗大环内酯类耐药呼吸道病原体的高效抗生素，其喹啉侧链中间体 H-9 由 3-溴喹啉与丙烯酸乙酯间的 Heck 反应得到，然后连接到母核结构上[67]。Heck 反应以 Pd(OAc)$_2$ 为催化剂、四丁基溴化铵（n-Bu$_4$N$^+$Br$^-$）为相转移催化剂、NaHCO$_3$ 为碱，无须膦配体，在 DMF 中顺利进行。之所以反应能够顺利进行，是因为溶剂 DMF 起配体作用，而且 n-Bu$_4$N$^+$Br$^-$ 对催化剂也起一定的稳定化作用。

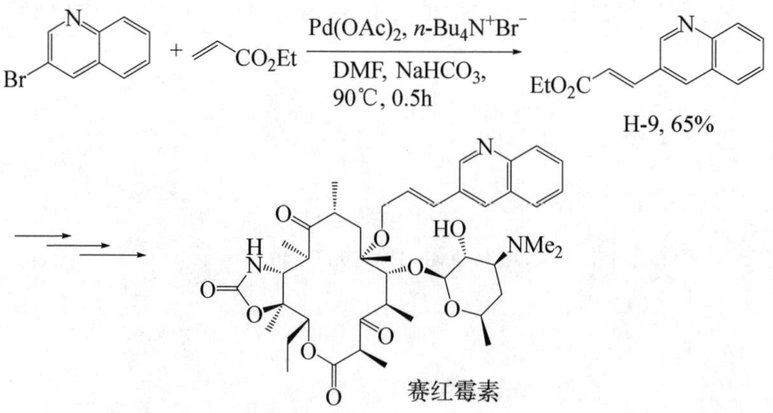

$$\text{(86)}$$

Heck 反应可用于合成多种甾体类药物，雌二醇（estradiol）是其中之一。在合成雌二醇过程中，首先四氢茚满衍生物 **b** 与 2-（β-溴乙烯基）-4-甲氧基溴苯（**c**）在醋酸钯催化下进行分子间 Heck 反应高收率生成中间体 H-8，碳碳键由 **c** 的乙烯基溴基团选择性地从角甲基的背面进攻 **b** 的 C-4 位形成。反应的第二步是分子内 Heck 反应，反应在特殊的催化剂（cat.d）催化下进行，以近乎定量的收率生成反常的 B 环和 C 环 cis-式趋向的单一非对映异构体 H-8'。H-8' 首先在 Wilkinson 催化剂催化下 B 环中碳碳双键加氢，紧接着 C 环中的双键原位异构化至 C$_9$～C$_{11}$ 之间，然后以 Pd/C 为催化剂、1,4-环己二烯为氢源进行氢转移加氢还原 C$_9$～C$_{11}$ 双键，最后再用 BBr$_3$ 同时脱除叔丁基和甲基，以三步 69%的总收率得到雌二醇[68]。

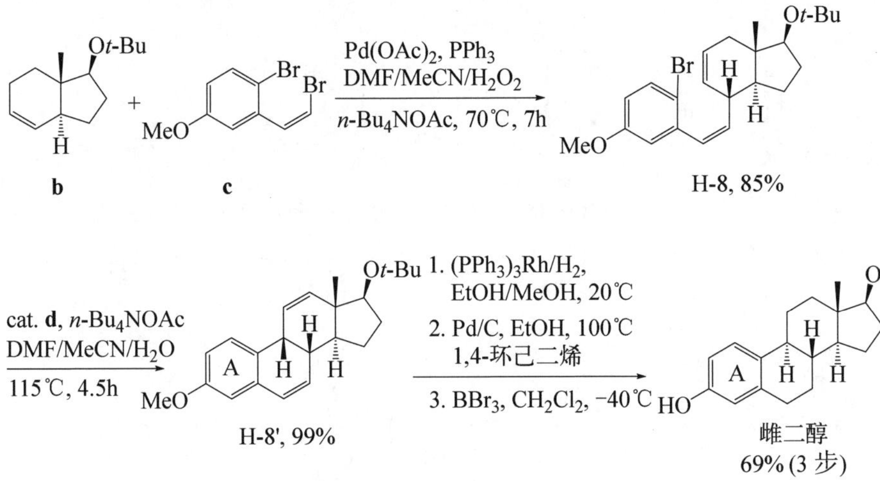

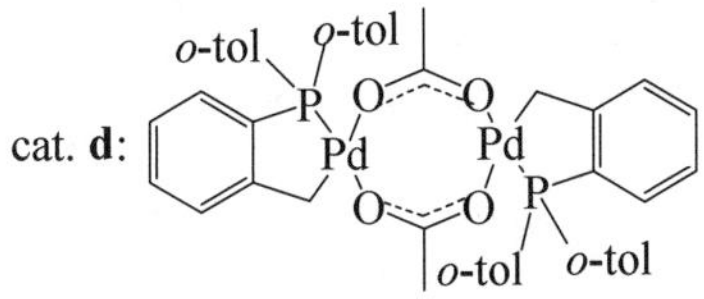

木香内酯（micheliolide）是 5/7/5 三环结构的愈创木烷型倍半萜内酯化合物之一，是临床上根除急性髓系白血病干细胞的首选药物。其内酯环上固有的端碳烯基团使其能够通过分子间 Heck 反应引入多种芳基，获得众多的治疗急性髓系白血病的备选药物分子。如下反应路线，木香内酯与三氟甲磺酸三乙基硅酯（TESOTf）反应使木香内酯上的羟基三乙基硅醚化，羟基得到保护，硅醚化的木香内酯在标准的 Heck 反应条件下与芳基碘偶联得到 E-构型的主产物 H-9（收率 18%～78%）和 Z-构型的副产物 H-9'（收率 9%～21%）。用三氟乙酸脱三乙基硅保护基得到相应的 E-木香内酯-Ar 和 Z-木香内酯-Ar[69]。

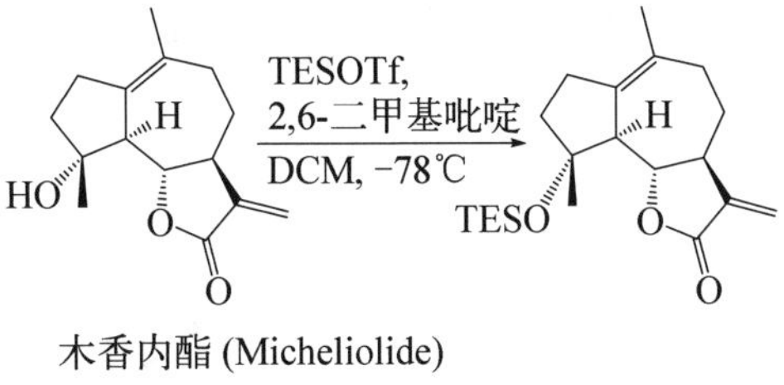

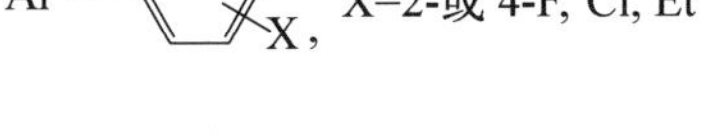

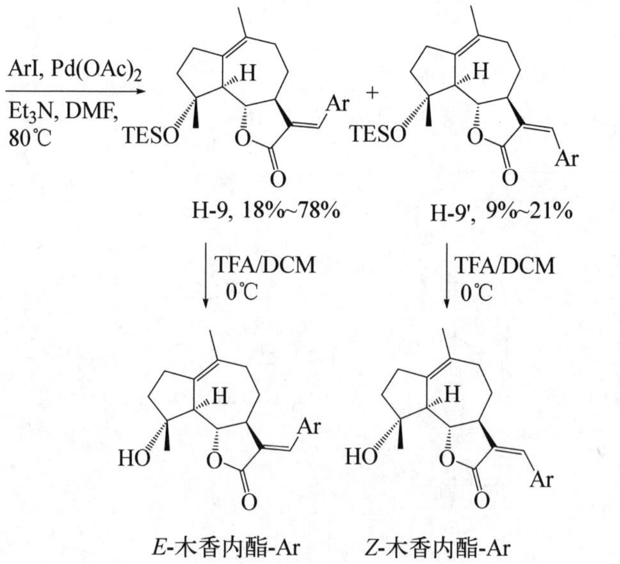

$$\xrightarrow[\substack{\text{Et}_3\text{N, DMF,}\\80℃}]{\text{ArI, Pd(OAc)}_2}$$

H-9, 18%~78%   H-9', 9%~21%

↓ TFA/DCM 0℃   ↓ TFA/DCM 0℃

E-木香内酯-Ar   Z-木香内酯-Ar

(88)

1.3.2 分子内反应的应用

前述强效 NMDA 甘氨酸拮抗剂和雌二醇合成中的第二个 Heck 反应均为分子内 Heck 反应。实际上，分子内 Heck 反应在药物和生物活性化合物合成上具有更广泛的应用。

利可拉明（lycoramine）属加兰他敏型生物碱（galanthamine-type alkaloids），其具有重要的生物活性，其结构特征是七元环上含有螺季碳中心并与取代的苯基稠合。含螺季碳中心的螺-[2,3-二氢-8-甲氧基-4H-1-苯并吡喃-4,1'-环己-2'-烯]-2,4'-二酮是合成利可拉明的关键中间体，通过分子内 Heck 反应合成 H-10 比其他方法更简捷[70]。如下反应，Birch 还原 4-甲氧基苯乙酸生成 γ-羧甲基-β,γ-环己烯酮（e），e 与 2-碘-6-甲氧基苯酚在缩合剂 1-乙基-（3-二甲基氨基丙基）碳二亚胺盐酸盐（EDCl）存在下缩合生成酯 f，f 在钯催化下进行分子内 Heck 反应生成 H-10，由 H-10 按文献方法经三步反应生成利可拉明[71]。

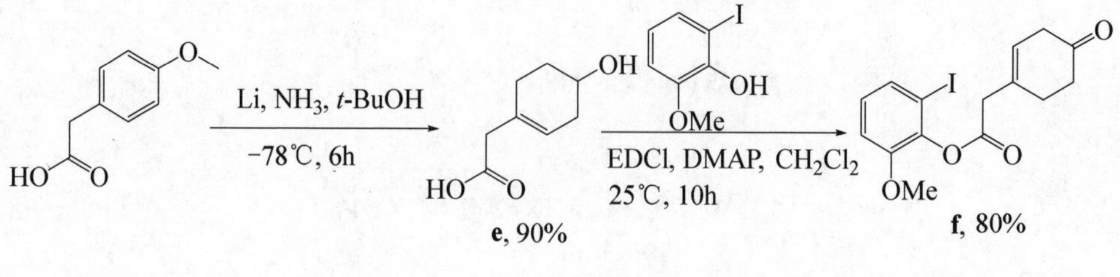

$$\xrightarrow[-78℃, 6h]{\text{Li, NH}_3, t\text{-BuOH}} \text{e, 90%} \xrightarrow[\substack{25℃, 10h}]{\text{EDCl, DMAP, CH}_2\text{Cl}_2} \text{f, 80%}$$

H-10, 50%　　利可拉明

(89)

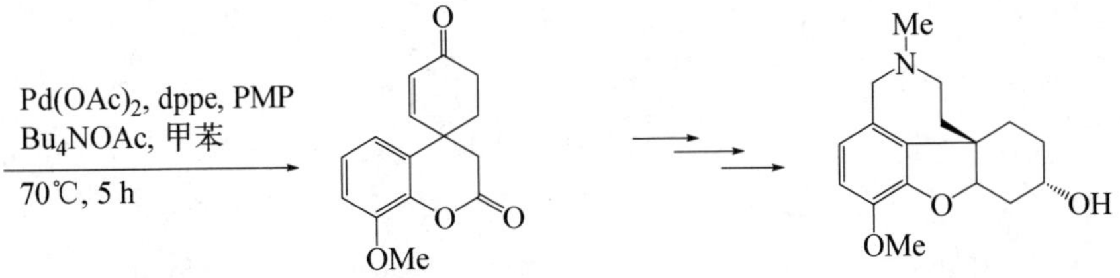

Heck 反应还可用于合成更复杂结构的天然产物，包括生物碱、萜和半萜、芳香天然产物、聚酮等[72]。南开大学张炜程教授与陈悦教授团队合作，成功实现高活性抗癌大环天然产物 nannocystin A 的全合成。细胞毒素肽 nannocystin A 是发现于黏细菌 *Nannocystis* sp.中的一个 21 元大环骨架化合物，由三肽和带有环氧酰胺基团的聚酮组成。其可由其前体在标准的 Heck 反应条件下通过分子内偶联反应合成[73]。

(90)

H-11, nannocystin A, 58%

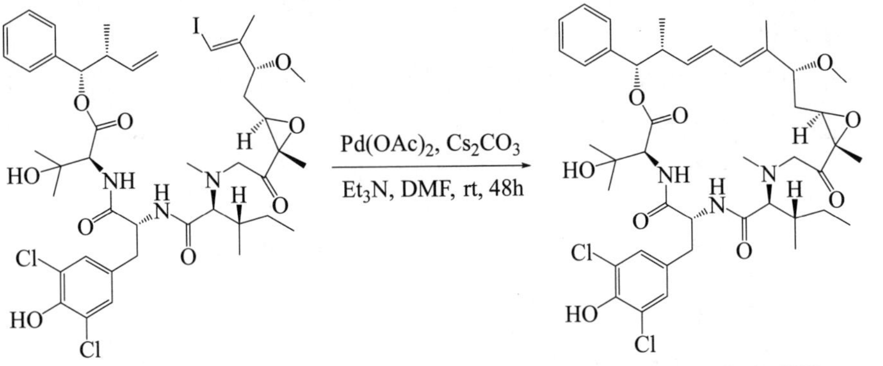

分子内不对称 Heck 反应被用于生物活性物质和天然产物的合成，通常是合成中关键一步。例如，利用分子内不对称 Heck 反应，成功合成了(+)-taiwaniaquinol D 和 (–)-taiwaniaquinone D [74, 75]。作为关键一步，*E*/*Z* 为 2∶8 的三氟甲磺酸酯 **g** 在 $Pd(OAc)_2$ 与(*R*)-(–)-[(5,6),(5',6')-双(乙烯二氧)联苯-2,2'-基]二苯基磷 [(*R*)-Synphos] 原位生成的手性催化剂催化下环合，不经分离在 Wilkinson 催化剂催化下加氢，以 72%的收率和 98%的对映选择性生成环合产物 H-12，然后再经后续多步反应生成 (+)-taiwaniaquinol D 和 (–)-taiwaniaquinone D [76]。

g　　H-12,72%, 98% *ee*

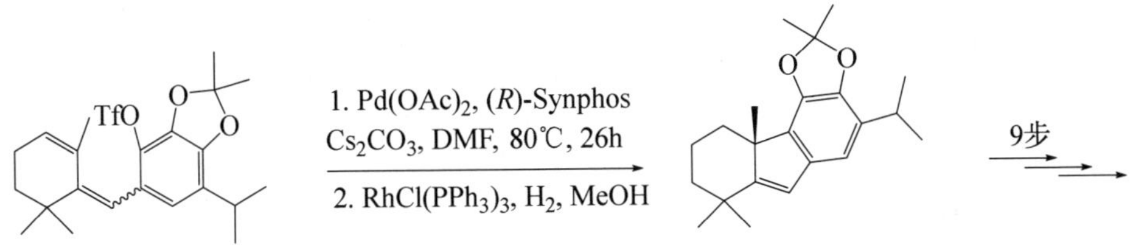

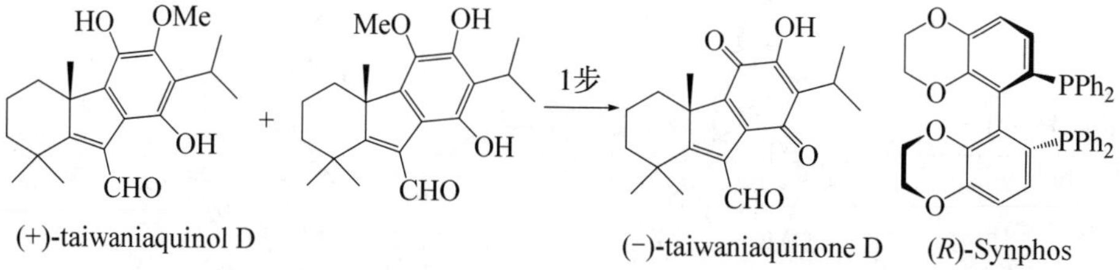

(+)-taiwaniaquinol D + [structure] →(1步) (−)-taiwaniaquinone D   (R)-Synphos (91)

分子内不对称 Heck 环合反应成功实现了聚吡咯烷吲哚生物碱类化合物 quadrigemine C 和其异构体 psycholeine 的合成,是药物和天然产物合成上的一个里程碑。合成从中间体 **h** 开始,**h** 由 Pd(OAc)$_2$/ (R)-tol-BINAP 催化并在 PMP 存在下可控地进行分子内不对称双 Heck 反应使 **h** 去对称化,生成 C1-对称的双吲哚酮 H-13。关键的不对称环合反应完成后,只需氢化脱氨基保护基 Ts 环合、溶解金属还原两步反应完成 quadrigemine C 的全合成。quadrigemine C 在酸催化下异构化则生成 psycholeine。十二元环的生物碱 quadrigemine C 和 psycholeine 的不对称全合成为不对称 Heck 环化策略在全合成中的作用提供了一个有强力说服力的例证。首先,**h** 环合生成 H-13 说明 Heck 反应显著的官能团耐受性;其次,表明分子内 Heck 反应生成拥挤的全碳季碳中心强大的能力[77]。

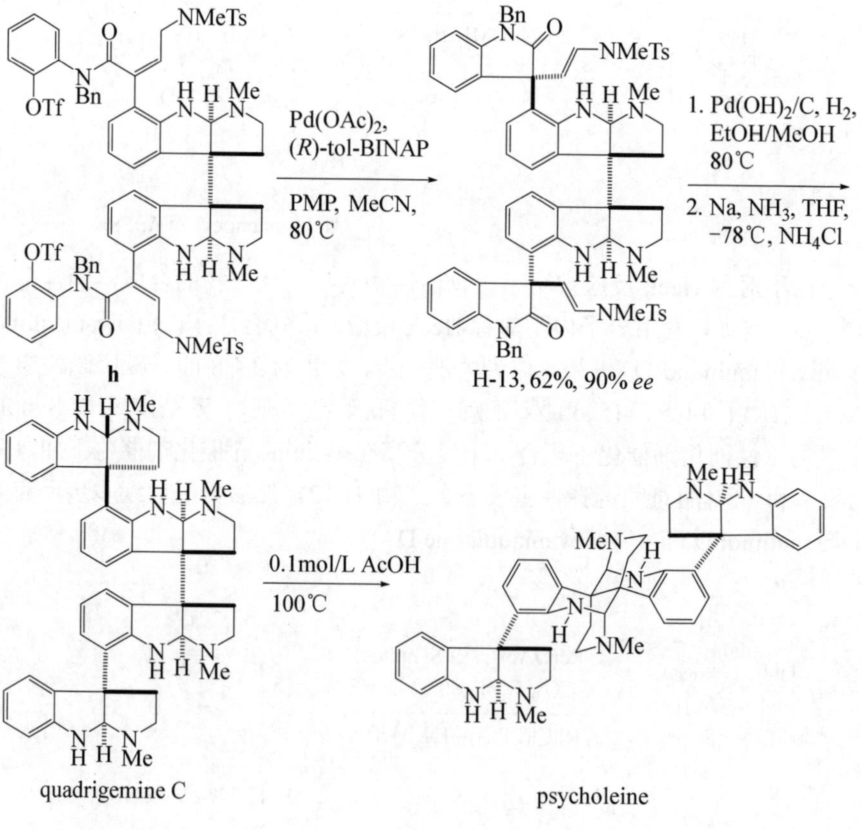

h →[Pd(OAc)$_2$, (R)-tol-BINAP, PMP, MeCN, 80℃] H-13, 62%, 90% ee →[1. Pd(OH)$_2$/C, H$_2$, EtOH/MeOH, 80℃; 2. Na, NH$_3$, THF, −78℃, NH$_4$Cl] quadrigemine C →[0.1mol/L AcOH, 100℃] psycholeine (92)

参考文献

[1] Mizoroki T, Mori K, Ozaki A. Arylation of olefin with aryl iodide catalyzed by palladium[J]. Bulletin of the Chemical Society of Japan, 1971, 44(2): 581.

[2] Heck R F, Nolley Jr J P. Palladium-catalyzed vinylic hydrogen substitution reactions with aryl, benzyl, and styryl halides[J]. The Journal of Organic Chemistry, 1972, 37(14): 2320-2322.

[3] Crisp G T. Variations on a theme-recent developments on the mechanism of the Heck reaction and their implications for synthesis[J]. Chemical Society Reviews, 1998, 27(6): 427-436.

[4] Larhed M, Andersson C M, Hallberg A. Chelation-controlled, palladium-catalyzed arylation of enol ethers with aryl triflates. Ligand control of selection for α-or β-arylation of [2-(dimethylamino) ethoxy] ethene[J]. Tetrahedron, 1994, 50(2): 285-304.

[5] Cabri W, Candiani I, Bedeschi A, et al. 1, 10-Phenanthroline derivatives: a new ligand class in the Heck reaction. Mechanistic aspects[J]. The Journal of Organic Chemistry, 1993, 58(26): 7421-7426.

[6] Lee D, Tather A, Hossain S, et al. An efficient and general method for the Heck and Buchwald-Hartwig coupling reactions of aryl-chlorides[J]. Organic Letters, 2011, 13(20):5540-5543.

[7] Peng Y, Chen J, Ding J, et al. Ligand-free copper-catalyzed arylation of Olefins by the Mizoroki-Heck reaction[J]. Synthesis, 2011, 2011(2): 213-216.

[8] Larhed M, Hallberg A. Intermolecular heck reaction: scope, mechanism, and other fundamental aspects of the Intermolecular Heck reaction. in: handbook of organopalladium chemistry for organic synthesis[M]. New York: John Wiley & Sons, Inc, 2002.

[9] Link J T. The intramolecular Heck reaction[J]. Organic Reactions, 2004, 60: 157-534.

[10] Littke A F, Fu G C. Heck reactions in the presence of $P(t-Bu)_3$: expanded scope and milder reaction conditions for the coupling of aryl chlorides[J]. The Journal of Organic Chemistry, 1999, 64(1): 10-11.

[11] Bräse S. Synthesis of bis(enolnonaflates) and their 4-*exo*-trig-cyclizations by intramolecular Heck reactions[J]. Synlett, 1999(10): 1654-1656.

[12] O'Connor B, Zhang Y, Negishi E, et al. Palladium-catalyzed cyclization of alkenyl and aryl halides containing α, β-unsaturated carbonyl groups via intramolecular carbopalladation[J]. Tetrahedron Letters, 1988, 29(32): 3903-3906.

[13] Larock R C, Song H, Baker B E, et al. Synthesis of bicyclic and polycyclic alkenes via palladium-catalyzed intramolecular arylation and vinylation[J]. Tetrahedron Letters, 1988, 29(24): 2919-2922.

[14] Gibson S E, Middleton R J. The intramolecular Heck reaction[J]. Contemporary Organic Synthesis, 1996, 3(6): 447-471.

[15] Larock R C, Stinn D E. Synthesis of benzofurans via palladium-promoted cyclization of ortho-substituted aryl allyl ethers[J]. Tetrahedron Letters, 1988, 29(37): 4687-4690.

[16] Wolff S, Hoffmann H M R. Aflatoxins revisited: convergent synthesis of the ABC-moiety[J]. Synthesis, 1988(10): 760-763.

[17] Arnau N, Moreno-Mañas M, Pleixats R. Preparation of benzo[*b*]thiophenes by Pd (0)-catalyzed intramolecular cyclization of allyl (and propargyl) *o*-iodophenyl sulfides[J]. Tetrahedron, 1993, 49(47): 11019-11028.

[18] Shibasaki M, Kojima A, Shimizu S. Catalytic asymmetric synthesis of natural products with heterocyclic rings[J]. Journal of Heterocyclic Chemistry, 1998, 35(5): 1057-1064.

[19] Tietze L F, Nöbel T, Spescha M. Stereoselective synthesis of steroids with the Heck reaction[J]. Angewandte Chemie International Edition, 1996, 35(19): 2259-2261.

[20] Owczarczyk Z, Lamaty F, Vawter E J, et al. Apparent *endo*-mode cyclic carbopalladation with inversion

of alkene configuration via *exo*-mode cyclization-cyclopropanation rearrangement[J]. Journal of the American Chemical Society, 1992, 114(25): 10091-10092.

[21] Grigg R, Sridharan V, Stevenson P, et al. Palladium (Ⅱ) catalysed construction of tetrasubstituted carbon centres, and spiro and bridged-ring compounds from enamides of 2-Iodobenzoic acids[J]. Journal of the Chemical Society, Chemical Communications, 1986(23): 1697-1699.

[22] Grigg R, Sridharan V, York M. Sequential and cascade olefin metathesis-intramolecular Heck reaction[J]. Tetrahedron Letters, 1998, 39(23): 4139-4142.

[23] Wünsch B, Dieckmann H, Höfner G. Stereoselective synthesis of 2-aminoethyl substituted tricycles with NMDA receptor affinity[J]. Tetrahedron: Asymmetry, 1995, 6(7): 1527-1530.

[24] Jin Z, Fuchs P L. Regiospecific silver(Ⅰ) promoted, palladium[O]-catalyzed intramolecular addition of aryl iodides to vinyl sulfones[J]. Tetrahedron Letters, 1993, 34(33): 5205-5208.

[25] Masters J J, Jung D K, Bornmann W G, et al. A concise synthesis of a highly functionalized C-aryl taxol analog by an intramolecular Heck olefination reaction[J]. Tetrahedron Letters, 1993, 34(45): 7253-7256.

[26] Tietze L F, Schirok H. Enantioselective highly efficient synthesis of (−)-cephalotaxine using two palladium-catalyzed transformations[J]. Journal of the American Chemical Society, 1999, 121(44): 10264-10269.

[27] Gibson S E, Middleton R J. Synthesis of 7-, 8-and 9-membered rings via *endo* Heck cyclisations of amino acid derived substrates [J]. Journal of the Chemical Society, Chemical Communications, 1995(17):1743-1744.

[28] Negishi E, Nguyen T, O'Connor B, et al. Synthesis of benzofurans, tetrahydrobenzopyrans, and related cyclic ethers via cyclic carbopalladation[J]. Heterocycles (Sendai), 1989, 28(1): 55-58.

[29] Ma S, Negishi E. Facile formation of seven-and eight-membered cycloalkenes via catalytic and cyclic carbopalladation of allenes[J]. The Journal of Organic Chemistry, 1994, 59(17): 4730-4732.

[30] Ma S, Negishi E. Palladium-catalyzed cyclization of. omega.-haloallenes: A new general route to common, medium, and large ring compounds via cyclic carbopalladation[J]. Journal of the American Chemical Society, 1995, 117(23): 6345-6357.

[31] Ziegler F E, Chakraborty U R, Weisenfeld R B. A palladium-catalyzed carbon-carbon bond formation of conjugated dienones: a macrocyclic dienone lactone model for the carbomycins[J]. Tetrahedron, 1981, 37(23): 4035-4040.

[32] Stocks M J, Harrison R P, Teague S J. Macrocyclic ring closures employing the intramolecular Heck reaction[J]. Tetrahedron Letters, 1995, 36(36): 6555-6558.

[33] Grigg R, Dorrity M J, Malone J F, et al. Palladium-catalysed polycyclisation-anion capture processes[J]. Tetrahedron Letters, 1990, 31(9): 1343-1346.

[34] Larock R C, Lee N H. Efficient free-radical and palladium-catalyzed tandem alkene insertions: a new approach to benzoprostacyclins[J]. The Journal of Organic Chemistry, 1991, 56(22): 6253-6254.

[35] De Meijere A, Meyer F E. Fine feathers make fine birds: the Heck reaction in modern garb[J]. Angewandte Chemie International Edition, 1995, 33(23-24): 2379-2411.

[36] Jin L, Qian J, Sun N, et al. Pd-Catalyzed reductive Heck reaction of olefins with aryl bromides for Csp^2-Csp^3 bond formation[J]. Chemical Communications, 2018, 54(45): 5752-5755.

[37] Grigg R, Kennewell P, Teasdale A J. Palladium catalysed cascade cyclisation-carbonylation processes. Rate enhancement by Tl(Ⅰ) salts[J]. Tetrahedron Letters, 1992, 33(50): 7789-7792.

[38] Grigg R, Sridharan V. Spirocycles via palladium catalysed cascade cyclisation-carbonylation-anion capture processes[J]. Tetrahedron Letters, 1993, 34(46): 7471-7474.

[39] Brown S, Clarkson S, Grigg R, et al. Palladium-catalysed cyclisation-carboformylation. Molecular queuing cascades[J]. Journal of the Chemical Society, Chemical Communications, 1995 (11): 1135-1136.

[40] Grigg R, Sridharan V. Heterocycles via Pd catalysed molecular queuing processes. Relay switches and the maximisation of molecular complexity[J]. Pure and Applied Chemistry, 1998, 70(5): 1047-1057.

[41] Grigg R, Savic V. Palladium catalysed termolecular queuing cascades. Facile cyclisation-anion capture routes to heterocyclic dienes via allene insertion processes[J]. Tetrahedron Letters, 1996, 37(36): 6565-6568.

[42] Burns B, Grigg R, Sridharan V, et al. Regiospecific palladium catalysed tandem cyclisation-anion capture processes. Stereospecific group transfer from organozinc and organoboron reagents[J]. Tetrahedron Letters, 1989, 30(9): 1135-1138.

[43] Ruck R T, Huffman M A, Kim M M, et al. Palladium-catalyzed tandem Heck reaction/C-H functionalization-preparation of spiro-indane-oxindoles[J]. Angewandte Chemie International Edition, 2008, 47(25): 4711-4714.

[44] Overman L E, Abelman M M, Kucera D J, et al. Palladium-catalyzed, polyene cyclizations[J]. Pure and Applied Chemistry, 1992, 64(12): 1813-1819.

[45] Negishi E. Zipper-mode cascade carbometallation for construction of polycyclic structures[J]. Pure and Applied Chemistry, 1992, 64(3): 323-334.

[46] Shibasaki M, Vogl E M, Ohshima T. Asymmetric heck reaction[J]. Advanced Synthesis & Catalysis, 2004, 346(13-15): 1533-1552.

[47] Mc Cartney D, Guiry P J. The asymmetric Heck and related reactions[J]. Chemical Society Reviews, 2011, 40(10): 5122-5150.

[48] Ozawa F, Kubo A, Hayashi T. Catalytic asymmetric arylation of 2, 3-dihydrofuran with aryl triflates[J]. Journal of the American Chemical Society, 1991, 113(4): 1417-1419.

[49] Ozawa F, Kobatake Y, Hayashi T. Palladium-catalyzed asymmetric alkenylation of cyclic olefins[J]. Tetrahedron Letters, 1993, 34(15): 2505-2508.

[50] Sakamoto T, Kondo Y, Yamanaka H. The palladium-catalyzed arylation of 4H-1, 3-dioxin[J]. Tetrahedron Letters, 1992, 33(45): 6845-6848.

[51] Gao Y, Sharpless K B. Asymmetric synthesis of both enantiomers of tomoxetine and fluoxetine. Selective reduction of 2, 3-epoxycinnamyl alcohol with Red-Al[J]. The Journal of Organic Chemistry, 1988, 53(17): 4081-4084.

[52] Koga Y, Sodeoka M, Shibasaki M. Palladium-catalyzed asymmetric arylation of 4, 7-dihydro-1, 3-dioxepin. Catalytic asymmetric synthesis of γ-butyrolactone derivatives[J]. Tetrahedron Letters, 1994, 35(8): 1227-1230.

[53] Loiseleur O, Hayashi M, Schmees N, et al. Enantioselective Heck reactions catalyzed by chiral phosphinooxazoline-palladium complexes[J]. Synthesis, 1997(11): 1338-1345.

[54] Shibasaki M, Vogl E M, Ohshima T. Asymmetric Heck reaction[J]. Advanced Synthesis & Catalysis, 2004, 346(13-15): 1533-1552.

[55] Sato Y, Honda T, Shibasaki M. A catalytic asymmetric synthesis of hydrindans[J]. Tetrahedron Letters, 1992, 33(18): 2593-2596.

[56] Kagechika K, Shibasaki M. Asymmetric Heck reaction: a catalytic asymmetric synthesis of the key intermediate for .DELTA. 9(12)-capnellene-3.beta., 8.beta., 10.alpha.-triol and .DELTA. 9(12)-capnellene-3.beta., 8.beta., 10.alpha., 14-tetrol[J]. The Journal of Organic Chemistry, 1991, 56(13):

4093-4094.

[57] Ashimori A, Overman L E. Catalytic asymmetric synthesis of quarternary carbon centers. Palladium-catalyzed formation of either enantiomer of spirooxindoles and related spirocyclics using a single enantiomer of a chiral diphosphine ligand[J]. The Journal of Organic Chemistry, 1992, 57(17): 4571-4572.

[58] Maddaford S P, Andersen N G, Cristofoli W A, et al. Total synthesis of (+)-xestoquinone using an asymmetric palladium-catalyzed polyene cyclization[J]. Journal of the American Chemical Society, 1996, 118(44): 10766-10773.

[59] Diaz P, Gendre F, Stella L, et al. New synthetic retinoids obtained by palladium-catalyzed tandem cyclisation-hydride capture process[J]. Tetrahedron, 1998, 54(18): 4579-4590.

[60] Flahive E J, Ewanicki B L, Sach N W, et al. Development of an effective palladium removal process for VEGF oncology candidate AG13736 and a simple, efficient screening technique for scavenger reagent identification[J]. Organic Process Research & Development, 2008, 12(4): 637-645.

[61] Waite D C, Mason C P. A scalable synthesis of the thromboxane receptor antagonist 3-{3-[2-(4-chlorobenzenesulfonamido) ethyl]-5-(4-fluorobenzyl) phenyl} propionic acid via a regioselective Heck cross-coupling strategy[J]. Organic Process Research & Development, 1998, 2(2): 116-120.

[62] Gauthier D R, Limanto J, Devine P N, et al. Palladium-catalyzed regioselective arylation of imidazo [1, 2-b][1, 2, 4] triazine: synthesis of an α2/3-selective GABA agonist[J]. The Journal of Organic Chemistry, 2005, 70(15): 5938-5945.

[63] Wallace M D, McGuire M A, Yu M S, et al. Multi-kiloscale enantioselective synthesis of a vitronectin receptor antagonist[J]. Organic Process Research & Development, 2004, 8(5): 738-743.

[64] Schils D, Stappers F, Solberghe G, et al. Ligandless Heck coupling between a halogenated aniline and acrylonitrile catalyzed by Pd/C: development and optimization of an industrial-scale Heck process for the production of a pharmaceutical intermediate[J]. Organic Process Research & Development, 2008, 12(3): 530-536.

[65] Katayama S, Ae N, Nagata R. Synthesis of tricyclic indole-2-caboxylic acids as potent NMDA-glycine antagonists[J]. The Journal of Organic Chemistry, 2001, 66(10): 3474-3483.

[66] Yasuda N, Hsiao Y, Jensen M S, et al. An efficient synthesis of an $\alpha_v\beta_3$ antagonist[J]. The Journal of Organic Chemistry, 2004, 69(6): 1959-1966.

[67] Plata D J, Leanna M R, Rasmussen M, et al. The synthesis of ketolide antibiotic ABT-773 (cethromycin)[J]. Tetrahedron, 2004, 60(45): 10171-10180.

[68] Tietze L F, Düfert A. Multiple Pd-catalyzed reactions in the synthesis of natural products, drugs, and materials[J]. Pure and Applied Chemistry, 2010, 82(7): 1375-1392.

[69] Ding Y H, Fan H X, Long J, et al. The application of Heck reaction in the synthesis of guaianolide sesquiterpene lactones derivatives selectively inhibiting resistant acute leukemic cells[J]. Bioorganic & Medicinal Chemistry Letters, 2013, 23(22): 6087-6092.

[70] Gras E, Guillou C, Thal C. A formal synthesis of (±)-lycoramine via an intramolecular Heck reaction[J]. Tetrahedron Letters, 1999, 40(52): 9243-9244.

[71] Ishizaki M, Ozaki K, Kanematsu A, et al. Synthetic approaches toward spiro [2, 3-dihydro-4H-1-benzopyran-4, 1'-cyclohexan]-2-one derivatives via radical reactions: total synthesis of (±)-lycoramine[J]. The Journal of Organic Chemistry, 1993, 58(15): 3877-3885.

[72] Paul D, Das S, Saha S, et al. Intramolecular Heck reaction in total synthesis of natural products: an

update[J]. European Journal of Organic Chemistry, 2021(14): 2057-2076.

[73] Yang Z, Xu X, Yang C H, et al. Total synthesis of nannocystin A[J]. Organic Letters, 2016, 18(21): 5768-5770.

[74] Ozeki M, Satake M, Toizume T, et al. First asymmetric total synthesis of (+)-taiwaniaquinol D and (−)-taiwaniaquinone D by using intramolecular Heck reaction[J]. Tetrahedron, 2013, 69(19): 3841-3846.

[75] Node M, Ozeki M, Planas L, et al. Efficient asymmetric synthesis of abeo-abietane-type diterpenoids by using the intramolecular Heck reaction[J]. The Journal of Organic Chemistry, 2010, 75(1): 190-196.

[76] Xie J Q, Liang R X, Jia Y X. Recent advances of catalytic enantioselective Heck reactions and reductive-Heck reactions[J]. Chinese Journal of Chemistry, 2021, 39(3): 710-728.

[77] Dounay A B, Overman L E. The asymmetric intramolecular Heck reaction in natural product total synthesis[J]. Chemical Reviews, 2003, 103(8): 2945-2964.

第 2 章
Suzuki 反应及其在药物合成上的应用

2.1 Suzuki 反应及其机理

2.1.1 Suzuki 反应

Suzuki 反应又称为 Suzuki-Miyaura 反应，由 Suzuki 等于 1979 年首次报道[1]。Suzuki 反应是指在碱性条件下芳基或乙烯基硼酸与芳基或乙烯基卤化物或三氟甲磺酸酯在钯催化剂催化下的交叉偶联反应。传统的 Suzuki 反应可由图 2-1 代表。

$$R^1\text{-}BY_2 + R^2\text{-}X \xrightarrow[\text{碱}]{[Pd^0L_n]} R^1\text{-}R^2$$

R^1=烷基、炔基、芳基、乙烯基
R^2=烷基、炔基、芳基、苄基、乙烯基
X=Br、Cl、I、OAc、OTf、OP(=O)(OR)$_2$

图 2-1 传统的 Suzuki 反应

即使经过多年的研究已取得了很大的进展，但 Suzuki 反应中的硼化物主要还是有机硼酸、硼酸酯和三氟硼酸钾，而偶联反应的亲电试剂除了卤代和三氟甲磺酸酯外，种类繁多的其他亲电试剂也被成功地引入 Suzuki 反应，如重氮盐、酰基等等[2-4]。钯催化剂通常为零价的钯配合物，其中配体对催化剂的性能有显著的影响，当配体具有手性时可实现不对称 Suzuki 反应[5]。为了降低催化剂在有机和药物合成中所占的成本，目前已发展了负载钯多相催化剂[6,7]和一些非贵重金属催化剂[8]。

Suzuki 反应是有机和药物合成上最有用的交叉偶联反应之一，相较于其他类似的反应有诸多优点。Suzuki 反应不需要难以制备的金属有机或元素有机化合物，所需试剂廉价易得；有机硼酸容易制备、无毒；芳基硼酸特别稳定，反应能够在温和

条件下在各种溶剂包括水中顺利进行，而且几乎耐受任何官能团，底物范围广泛；此外，含硼副产物相较于其他反应中金属有机试剂生成的副产物更容易去除，特别是在大规模生产中优势更加明显。以上优势使得 Suzuki 反应在药物和天然产物合成的碳碳键构建中有着广泛的应用。

2.1.2　反应机理

尽管不同催化体系下的 Suzuki 反应的机理会有所差异，但钯催化的 Suzuki 反应的机理相对简单，并被广泛认可。催化循环如图 2-2 所示[9]。首先，卤化物 2 氧化加成到 Pd⁰ 催化剂 1，生成有机钯物种 3；3 与碱叔丁醇钠进行复分解反应生成中间体 4；4 与由碱和有机硼酸 5 生成的配合物 6 进行转移金属反应生成有机钯中间体 8；8 发生还原消除反应生成目标偶联产物，并再生催化剂 1，完成催化循环。碱在反应中起三个作用：生成钯配合物 $ArPd(OR)L_2$，生成三烷基硼酸酯，以及通过烷氧化物与钯配合物反应加速还原消除步骤的进行[10]。

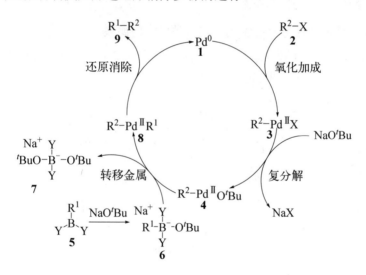

图 2-2　钯催化的 Suzuki 反应机理

2.2　基本的 Suzuki 反应

2.2.1　分子间 Suzuki 反应

尽管 Suzuki 反应容易进行，但与任何有机反应一样，Suzuki 反应也受众多因素的影响，但基本影响因素主要有底物的结构与种类、催化剂、碱的种类、溶剂。

因此，在进行 Suzuki 反应时需根据偶联反应亲电和亲核底物的结构特征调整影响反应的因素，使反应更高效、更实用。

就底物而言，亲电底物的反应活性顺序为：$ArN_2^+X^-$ >> ArI > ArBr > ArCl > ArOTf ⩾ ArOTs, ArOMe，似乎与离去基团的难易相关。而作为亲核底物的硼化合物的反应难易则与硼化物中有机基团的空间位阻相关，尤其是烷基硼酸酯为底物的 Suzuki 烷基化反应，因此需根据有机基团的空间位阻改变硼试剂的结构[11]。总之，Suzuki 反应底物范围广泛，偶联伙伴间的不同组合可以得到更多的偶联产物。有机硼化合物中有机骨架的结构不同，反应的难易程度各异，有时需要改变硼酸（酯）以适用反应的进行。同样作为亲电试剂的卤化物或其类似物也要根据偶联基团的结构加以调整，使反应顺利进行。

以下是一些基本的 Suzuki 反应，这些反应实例表明不同的底物需要不同的催化剂、碱和反应条件；同时也一定程度上反映出 Suzuki 反应催化剂的研究进展。

2.2.1.1 芳基硼化物的交叉偶联

（1）芳基硼化合物与芳基卤代物或其类似物的偶联

反应式（1）～式（3）表明，在相同反应条件下碘代物的反应显然比溴代物要容易，延长反应时间可改进反应的结果[12]。

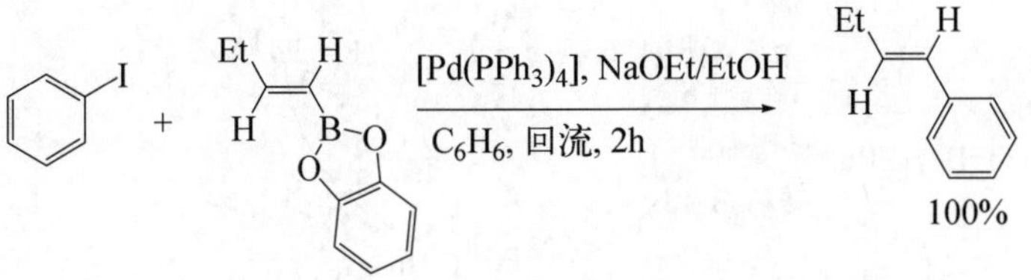

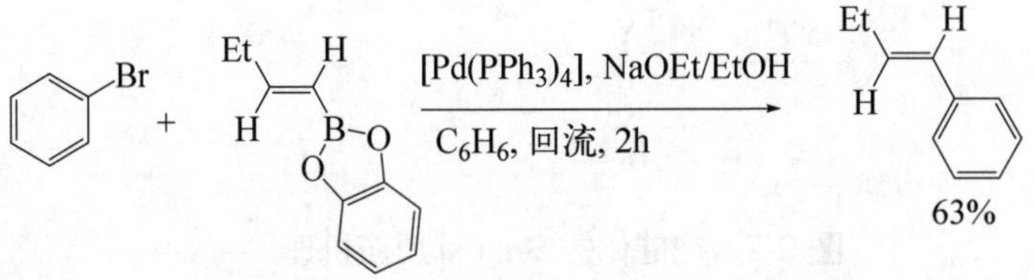

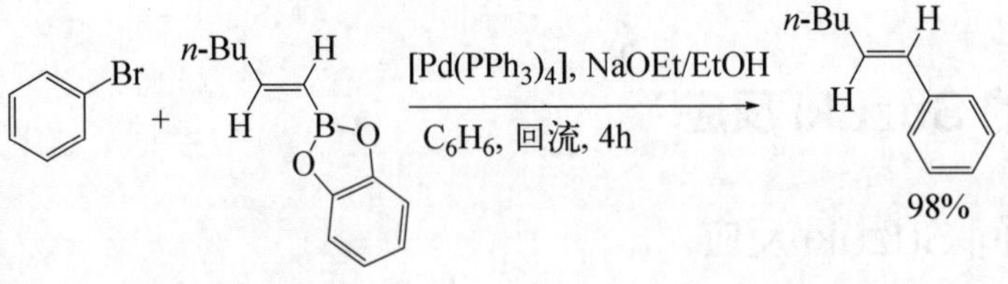

有机硼酸作为底物显然比硼酸酯更具优势，因为反应的副产物为无机硼酸，很容易与偶联产物分离。例如，均三甲苯基硼酸与碘代苯在更温和的条件下发生偶联，

高收率生成目标产物[13]。

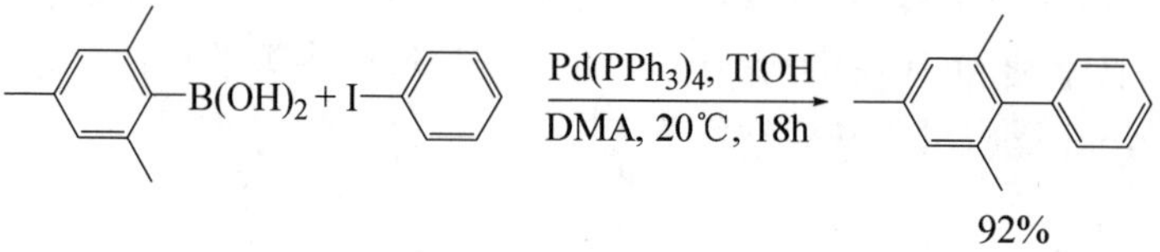

$$\text{(2,4,6-三甲基苯基)B(OH)}_2 + \text{I-C}_6\text{H}_5 \xrightarrow[\text{DMA, 20°C, 18h}]{\text{Pd(PPh}_3)_4,\ \text{TlOH}} \text{联芳烃}\quad 92\% \tag{4}$$

利用芳基硼酸还可对各种杂环化合物芳基化，无论是富电子的吡咯环还是缺电子的吡啶环。如 3,5-二甲基-4-溴吡咯甲酸乙酯与苯基硼酸在 DMF 中回流，几乎定量地生成 3,5-二甲基-4-苯基吡咯甲酸乙酯[14]。

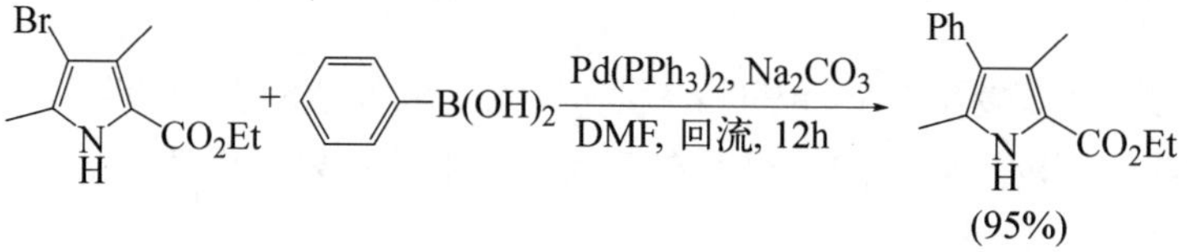

$$\text{溴代吡咯酯} + \text{PhB(OH)}_2 \xrightarrow[\text{DMF, 回流, 12h}]{\text{Pd(PPh}_3)_2,\ \text{Na}_2\text{CO}_3} \text{产物}\quad (95\%) \tag{5}$$

即使空间位阻特别大的芳基硼酸都能与卤代吡啶进行 Suzuki 反应，碱的强弱直接影响反应的进行。反应式（6）和表 2-1 中的结果充分说明碱的强弱在 Suzuki 反应中的作用，同时耦合反应底物上的取代基也与反应密切相关[15]。总的来说，碱的强度越强，反应越容易进行，例如叔丁醇钾作碱时，所有的底物都在较短的时间内以良好的收率生成偶联产物；卤代吡啶的空间位阻越大，反应越不容易进行；溴代吡啶比氯代吡啶更容易进行反应。

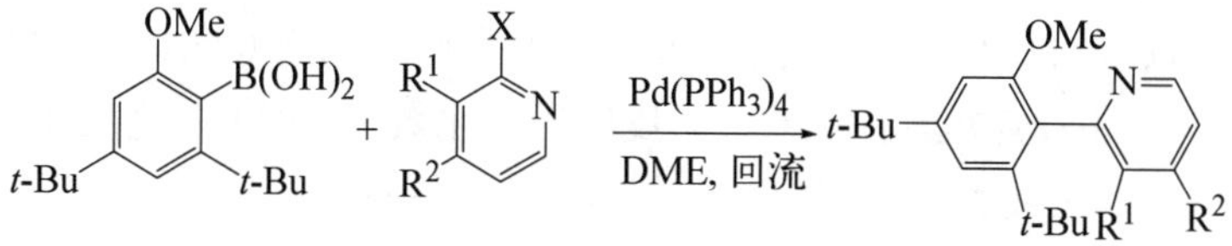

$$\text{芳基硼酸} + \text{卤代吡啶} \xrightarrow[\text{DME, 回流}]{\text{Pd(PPh}_3)_4} \text{联芳基产物} \tag{6}$$

1a: $R^1=R^2=H$, $X=Br$ **2a**: $R^1=R^2=H$
1b: $R^1=Me$, $R^2=H$, $X=Cl$ **2b**: $R^1=Me$, $R^2=H$
1c: $R^1-R^2=(CH_2)_2$, $X=Cl$ **2c**: $R^1-R^2=(CH_2)_2$

表 2-1　碱对芳基硼酸与卤代吡啶交叉偶联反应的影响

碱	2a (收率/%)/(时间/h)	2b (收率/%)/(时间/h)	2c (收率/%)/(时间/h)
Na_2CO_3	26/90	0/90	0/90
NaOH	40/140	22/24	44/26
NaOEt	74/4	0/12	45/26
KOt-Bu	86/4	83/16	77/10

当不同反应活性的卤原子处于同一分子中时,可以利用它们的不同反应活性进行选择性偶联。例如,4-甲氧基苯硼酸与 8-溴-2-碘邻菲咯啉进行区域选择性缩合生成 8-溴-2-(4-甲氧苯基)-1,10-邻菲咯啉[16]。由于产物结构中溴的存在,可以进一步通过 Heck 或 Suzuki 反应引入其他结构单元。

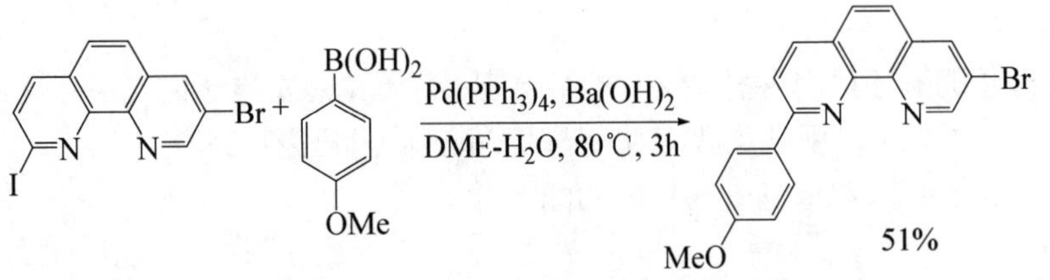

$$\text{(7)}$$

Pd(PPh$_3$)$_4$, Ba(OH)$_2$
DME-H$_2$O, 80℃, 3h
51%

杂芳基三氟甲磺酸酯与杂芳基硼酸顺利反应生成复杂的多芳杂芳环化合物,表明 Suzuki 反应广泛的底物适用性以及在药物合成中的重要应用性[17]。

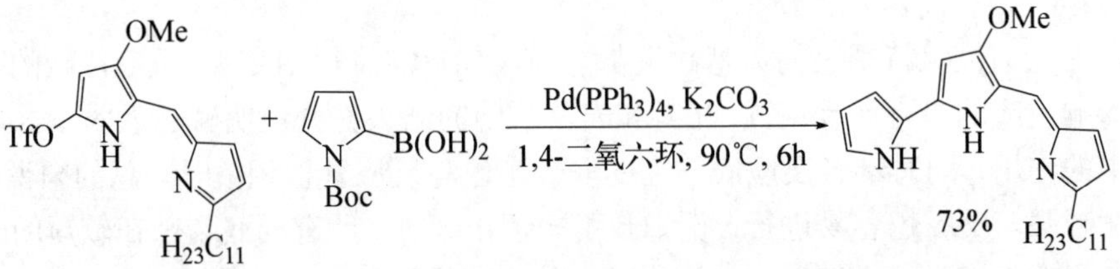

$$\text{(8)}$$

Pd(PPh$_3$)$_4$, K$_2$CO$_3$
1,4-二氧六环, 90℃, 6h
73%

配体在钯催化的 Suzuki 反应中起非常重要的作用,适宜的配体修饰钯催化剂不但可以促进反应的顺利进行,而且可以提高反应的转化数,使得 Suzuki 反应在规模化生产中得以应用。例如,4-溴代苯乙酮与苯基硼酸在只有 0.001%(摩尔分数)钯催化剂 cat.1 催化下反应就能顺利进行,反应的转化数(TON)高达 75000[18]。如此高的转化数可降低工业生产中催化剂在产品中所占的成本份额。

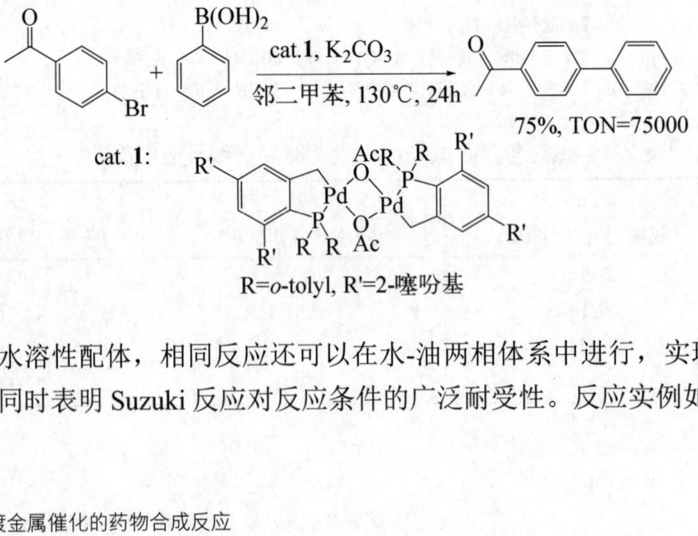

$$\text{(9)}$$

cat.1, K$_2$CO$_3$
邻二甲苯, 130℃, 24h
75%, TON=75000

cat. 1:
R=o-tolyl, R'=2-噻吩基

利用水溶性配体,相同反应还可以在水-油两相体系中进行,实现催化剂的循环使用,同时表明 Suzuki 反应对反应条件的广泛耐受性。反应实例如下[19]。

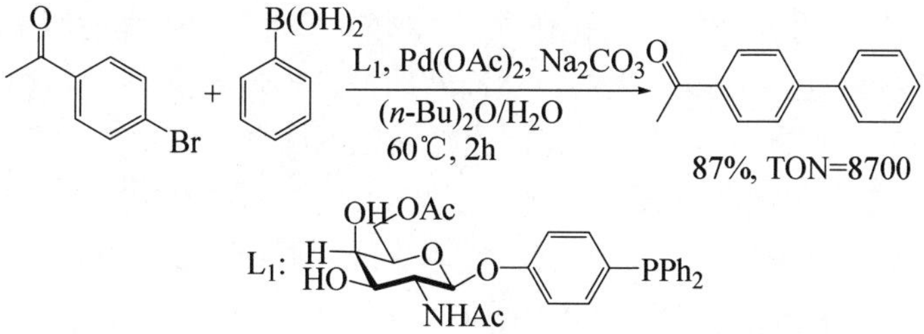

$$\text{(reaction 10)} \quad 87\%, \text{TON}=8700 \tag{10}$$

L₁: (glycoside-linked-C₆H₄-PPh₂ ligand structure)

适宜配体与钯催化剂的结合，使得 Suzuki 反应可以在很弱的碱性条件下进行，因而对强碱敏感的底物也能用于有机和药物分子的合成，实例如下[20]。

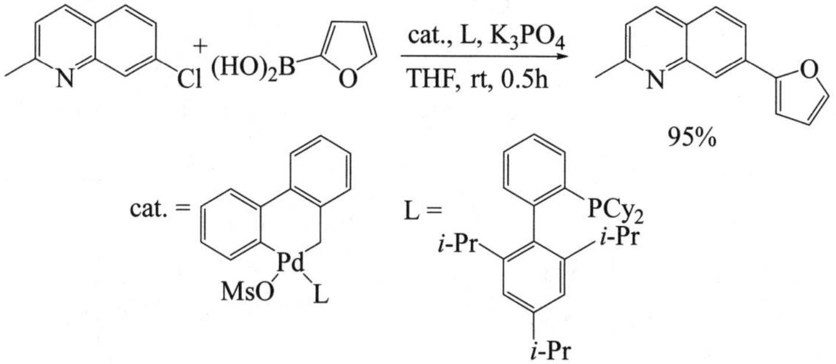

$$\text{(reaction 11)} \quad 95\% \tag{11}$$

cat. = (palladacycle with MsO and L), L = (2′-dicyclohexylphosphino-2,4,6-triisopropylbiphenyl, XPhos)

相较于钯配合均相催化剂，固体钯催化剂具有容易与产物分离的优点，因此，各种载体负载的钯纳米粒子（NPPs）作为 Suzuki 反应的催化剂成新的宠儿[21]。例如，以氧化石墨烯负载的钯纳米粒子为催化剂，芳基溴与芳基硼酸在乙醇水溶液中室温下就能顺利反应生成偶联产物，当碘苯与苯硼酸为底物时，转化频率（TOF）高达 237000[22]。芳基卤代物底物的反应活性受苯环上取代基的影响较小，而芳基硼酸底物上的取代基对反应有较大的影响，吸电子基降低反应的活性。

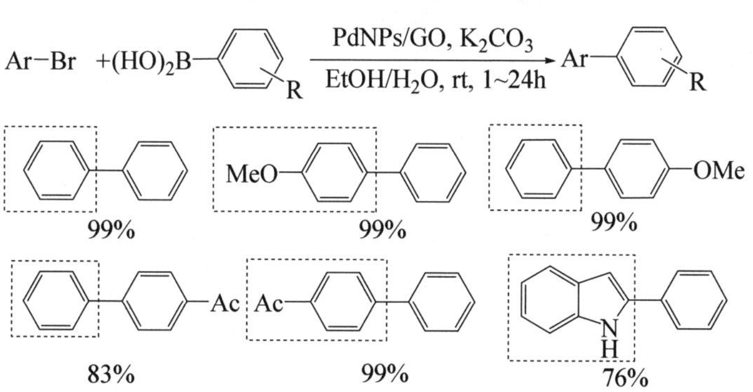

$$\text{Ar-Br} + (\text{HO})_2\text{B-C}_6\text{H}_4\text{-R} \xrightarrow[\text{EtOH/H}_2\text{O, rt, 1~24h}]{\text{PdNPs/GO, K}_2\text{CO}_3} \text{Ar-C}_6\text{H}_4\text{-R} \tag{12}$$

产物: 联苯 99%；MeO-联苯 99%；联苯-OMe 99%；联苯-Ac 83%；Ac-联苯 99%；2-苯基吲哚 76%

由于钯催化剂价格昂贵，人们试图用廉价的非贵重过渡金属替代钯进行 Suzuki 反应。如图 2-2 所示，之所以钯能够催化 Suzuki 反应顺利进行，是因为钯存在氧化还原对 Pd⁰/Pd²⁺，而镍也存在类似的氧化还原对 Ni⁰/Ni²⁺，而且由于镍的原子半径小，更容易进行亲核取代反应。因此，人们自然而然地把目光转向镍，并发现一些镍配合物是 Suzuki 反应良好的催化剂，在镍催化剂催化下许多卤代物或其类似物都能顺利地与芳基硼化物发生偶联反应[23]。例如，以 NiCl$_2$(dppp)为催化剂，2-氯苯腈与 4-甲基苯硼酸在无水磷酸钾存在下以优异的收率生成偶联产物 2-氰基-4'-甲基联苯，该产物是合成多个抗高血压药物的关键中间体。

$$\text{2-ClC}_6\text{H}_4\text{CN} + (HO)_2B\text{-C}_6\text{H}_4\text{-Me} \xrightarrow[\text{1,4-二氧六环, 100℃}]{\text{NiCl}_2(\text{dppp}), K_3PO_4} \text{2-CN-4'-Me-biphenyl (95\%)} \quad (13)$$

重氮盐可替代芳基卤代物和三氟甲磺酸酯进行 Suzuki 反应。相较于芳基卤代物和三氟甲磺酸酯类底物，重氮盐由于其独特的结构和更高的反应活性通常能以优异的收率得到交叉偶联产物，重氮盐的阴离子一般为四氟硼酸根；各种取代的芳基重氮盐与各种取代的芳基硼酸都能进行反应，尽管偶联产物的收率与反应双边底物的取代基有一定的关系，但没有明显的规律[24]。以下实例反映了重氮盐作为 Suzuki 反应底物的广泛性，以及反应的高效性[25]。

$$\text{O}_2\text{N-C}_6\text{H}_3(\text{N}_2\text{BF}_4)\text{-R}^1 + (HO)_2B\text{-C}_6\text{H}_4\text{-R} \xrightarrow[\text{rt, 1~3h}]{\text{Pd(OAc)}_2, \text{1,4-二氧六环}} \text{NO}_2\text{-biphenyl-R}^1, R^2 \quad (14)$$

R¹	R²	收率/%
H	2-OBn	85
4-OMe	4-CO$_2$Me	81
4-OMe	3,4,5-OMe	83
4,5-Cl	4-OMe	78
4,5-Me	4-Br	91

除了有机硼酸酯和有机硼酸，有机三氟硼酸盐（RBF$_3$K）是 Suzuki 反应中另一种多用途底物，几乎与任何卤代或三氟甲磺酸芳基酯进行交叉偶联反应。有机三氟硼酸盐具有稳定、适宜的物理性质、可规模化使用以及操作简单等固有的优点，还具有原子经济性、廉价的优势。与有机硼酸不同，有机三氟硼酸盐结构稳定，能耐受各种苛刻的反应条件，使其在复杂分子修饰与合成中得到青睐。以下是有机三氟硼酸盐作为底物的一些反应，各种底物包括四氟硼酸重氮盐、杂环卤代物、三氟甲磺酸芳基酯与杂环三氟硼酸盐都能顺利地进行反应，而杂环底物的 Suzuki 反应通常是挑战性的，这些实例充分说明有机三氟硼酸盐作为 Suzuki 反应偶联伙伴的广泛的底物适用性[26]。

一个更特殊的例子是芳基三氟硼酸盐与二碘代芳基化合物之间的双芳基交叉偶联反应,而相应的芳基硼酸片呐醇酯则不能发生该反应[27]。

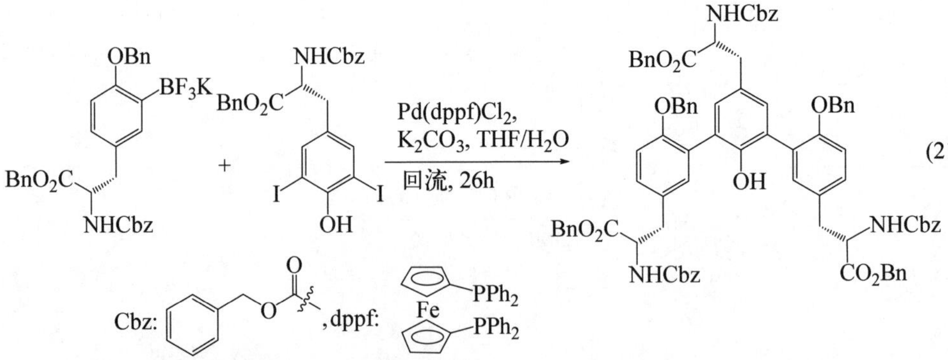

（2）芳基硼化物与其他卤代物的偶联

芳基硼化物与卤代乙烯或三氟甲磺酸乙烯酯类化合物甚至烷基化合物在钯催化下也能发生交叉偶联反应。溴代环戊烯与芳基硼酸在钯催化下很容易地缩合得到交叉偶联产物[28]。

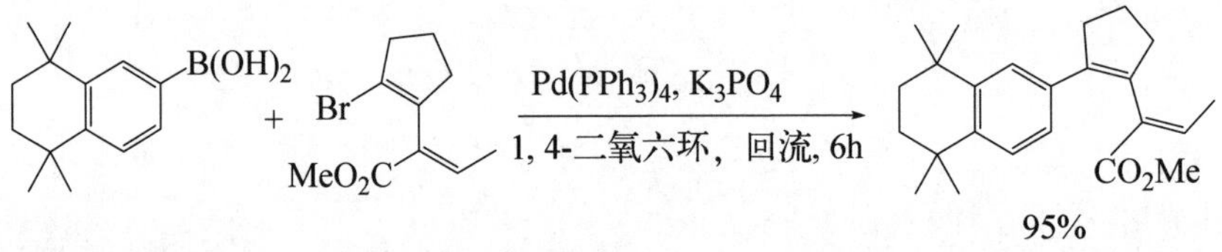

共轭的卤代烯烃也能与卤代硼酸进行 Suzuki 反应，如 4-氯代香豆素与芳基硼酸缩合得到相应的交叉偶联产物[29]。

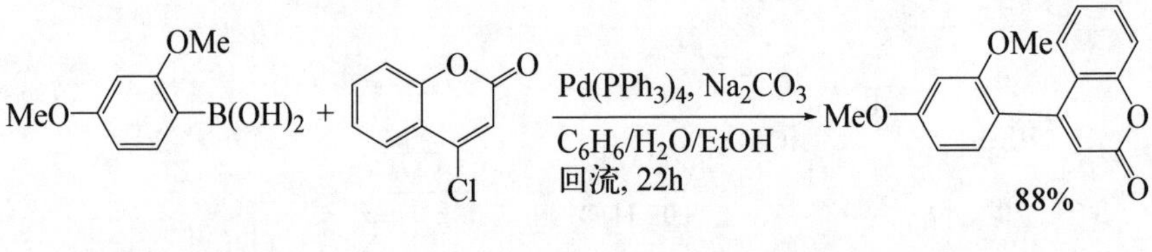

烯丙基卤代物也能与芳基硼酸偶联，反应在没有膦配体存在下进行[30]。

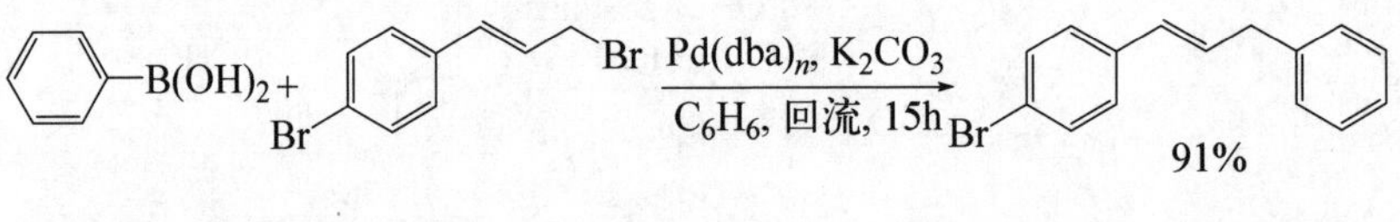

由羧酸合成芳基酮在合成天然产物上有着广泛的应用，最常用的反应当属 Friedel-Crafts 酰基化。然而，Friedel-Crafts 酰基化反应存在区域选择性和与一些官能团不相容的缺点限制了其在某些结构化合物合成上的应用。芳基硼酸与酰氯在钯催化下交叉偶联在一定程度上克服了上述问题。氯化钯作为催化剂在无配体存在的含水或无水丙酮中催化芳基硼酸与酰氯交叉反应几乎定量地生成交叉偶联产物[31]。

$$\text{R}\!-\!\!\!\overset{}{\underset{}{\bigcirc}}\!\!-\!\text{B(OH)}_2 + \text{PhCOCl} \xrightarrow[\text{CH}_3\text{COCH}_3/\text{H}_2\text{O, rt}]{\text{PdCl}_2,\ \text{Na}_2\text{CO}_3} \text{R-C}_6\text{H}_4\text{-CO-Ph} \quad (25)$$

R= 2-Me, 3-Me, 4-Me, 3-NO$_2$
约100%

类似地，2-氰基芳基硼酸酯在弱碱 K$_3$PO$_4$ 和少量水存在下于甲苯中反应生成交叉偶联产物，K$_3$PO$_4$ 的弱碱性降低酰氯的水解速率[32]。

$$\text{(2-CN-C}_6\text{H}_4\text{)B(OCH}_2\text{CH}_2\text{O)} + \text{PhCOCl} \xrightarrow[\text{H}_2\text{O, 甲苯, 回流, 4h}]{\text{Pd(PPh}_3)_2\text{Cl}_2,\ \text{K}_3\text{PO}_4} \text{2-CN-C}_6\text{H}_4\text{-CO-Ph} \quad (26)$$

70%

与酰氯类似结构的氯甲酰乙酯或氯甲酰二丁酰胺也能与芳基硼酸交叉偶联生成芳甲酸酯或芳甲酰胺，反应在钯催化剂和氧化亚铜协同催化下在温和条件下进行[33]。

$$\text{R}\!-\!\!\!\overset{}{\underset{}{\bigcirc}}\!\!-\!\text{B(OH)}_2 + \begin{array}{c}\text{ClCOOEt} \\ \text{ClCON(Bu)}_2\end{array} \xrightarrow[\text{K}_3\text{PO}_4\cdot3\text{H}_2\text{O, 甲苯, 80℃, 24h}]{\text{Pd(PPh}_3)_4,\ \text{Cu}_2\text{O}} \begin{array}{c}\text{R-C}_6\text{H}_4\text{-COOEt}\quad 21\%\sim78\% \\ \text{R-C}_6\text{H}_4\text{-CON(Bu)}_2\quad 60\%\sim93\% \end{array} \quad (27)$$

R = H, 3-Me, 4-Me, 4-Cl, 2-OAc, 4-OAc, 4-CF$_3$

酸酐替代酰氯也可以顺利地进行反应，而且对试剂级别要求不高，无须特别处理既可直接使用，并且官能团耐受性更广[34]。

$$\text{Ar-B(OH)}_2 + (\text{RCO})_2\text{O} \xrightarrow[\text{THF, 回流, 12h}]{\text{Pd(OAc)}_2,\ \text{P(}p\text{-MeOPh)}_3} \text{R-CO-Ar} \quad (28)$$

Ar=Ph, p-MeOPh, o-tol, 2-呋喃基, 3-噻吩基

R=n-C$_5$H$_{11}$, CH$_3$, 烯丙基, Ph, p-MeOPh

71%~98%

甚至羧酸在一些活化剂存在下也可以直接与芳基硼酸交叉偶联生成酮，图 2-3 是部分活化剂的结构[35]。例如，当以 PyClU 为活化剂时，各种取代的苯甲酸与苯基硼烷交叉缩合生成二芳基甲酮，具有广泛的官能团耐受性，PyClU 的作用是与羧酸原位生成酰氯[36]。

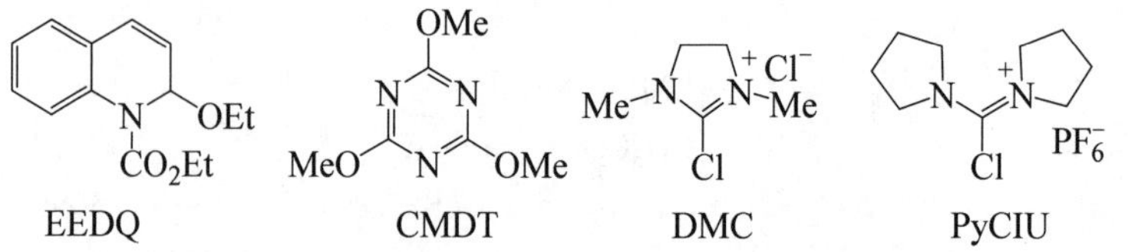

EEDQ　　CMDT　　DMC　　PyClU

图 2-3　部分活化剂的结构与名称缩写

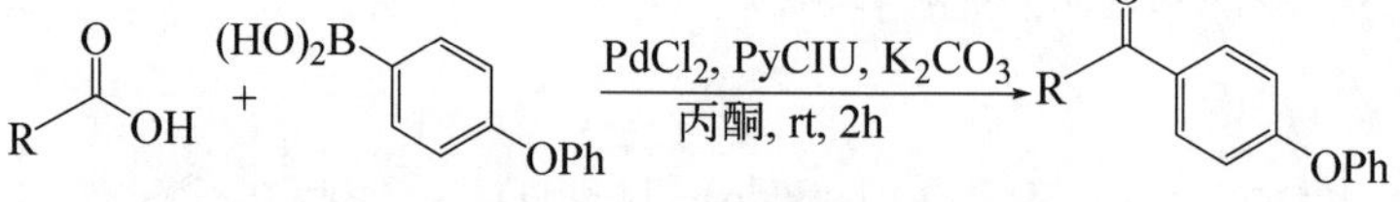

$$R-COOH + (HO)_2B-C_6H_4-OPh \xrightarrow[\text{丙酮, rt, 2h}]{\text{PdCl}_2,\ \text{PyClU},\ \text{K}_2\text{CO}_3} R-CO-C_6H_4-OPh \quad (29)$$

R=2-MePh, 3-MePh, 4-MePh, 2-ClPh, 3-ClPh, 4-ClPh,
3-FPh, 4-FPh, 3-CNPh, 4-CNPh, 4-CF$_3$Ph, 4-CF$_3$OPh
4-NO$_2$Ph, 3-MeOPh, 4-NHBocPh, 2-呋喃, 5-Me-2-噻吩, 6-MeO-3-吡啶

38%~92%

2.2.1.2　炔基硼化物的交叉偶联

相较于其他硼化物，炔基硼烷具有更强的路易斯（Lewis）酸性，使得炔基硼烷在典型的 Suzuki 反应条件下容易发生水解，因而不能直接进行 Suzuki 交叉偶联反应。令人欣慰的是，炔基锂可与 B-甲氧基-9-硼双环[3,3,1]-壬烷（9-甲氧基-9-BBN）反应生成稳定配合物，而该配合物能与溴代物在钯催化下顺利发生 Suzuki 反应生成各种交叉偶联产物[37]。如下反应，R 可以是芳基、烷基和硅烷基，R'可以是芳基和烯基；当 R 和 R′分别为三甲基硅烷基（SiMe$_3$）和 α-苯乙烯基（CH$_2$=CC$_6$H$_5$）时，交叉偶联产物的收率高达 88%。

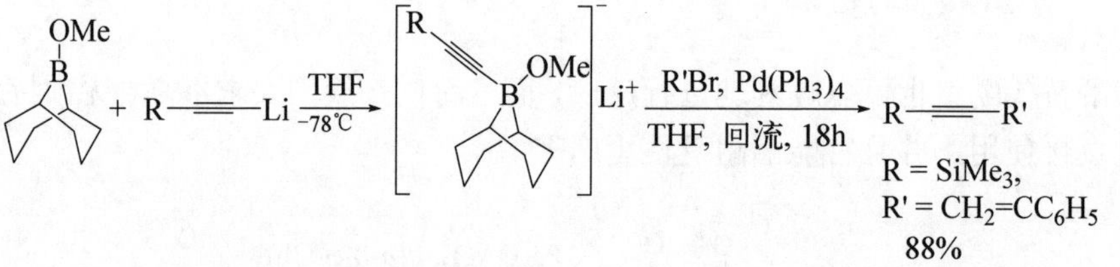

$$\text{9-OMe-9-BBN} + R-\equiv-\text{Li} \xrightarrow[-78℃]{\text{THF}} [R-\equiv-B(\text{OMe})(\text{9-BBN})]^-\text{Li}^+ \xrightarrow[\text{THF, 回流, 18h}]{R'\text{Br, Pd(PPh}_3)_4} R-\equiv-R' \quad (30)$$

R = SiMe$_3$, R' = CH$_2$=CC$_6$H$_5$
88%

用硼酸三异丙酯替代 9-甲氧基-9-BBN 生成炔基锂-硼酸三异丙酯加合物，也能实现炔基与溴代芳基或烯基的 Suzuki 交叉偶联[38]。

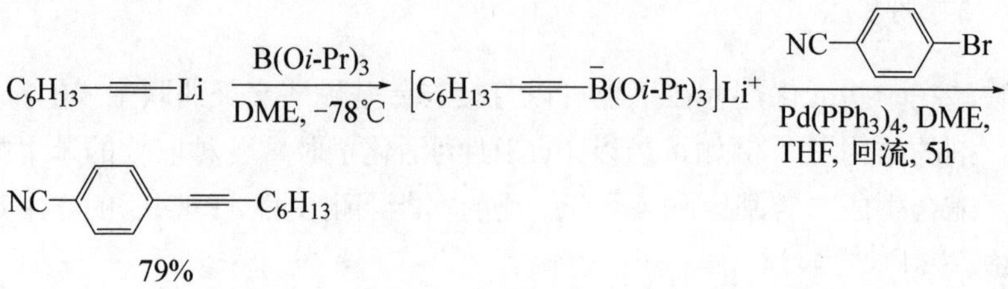

$$C_6H_{13}-\equiv-\text{Li} \xrightarrow[\text{DME, }-78℃]{B(Oi\text{-Pr})_3} [C_6H_{13}-\equiv-\bar{B}(Oi\text{-Pr})_3]\text{Li}^+ \xrightarrow[\text{THF, 回流, 5h}]{\text{NC-C}_6\text{H}_4\text{-Br, Pd(PPh}_3)_4,\ \text{DME,}}$$

$$\text{NC-C}_6\text{H}_4-\equiv-C_6H_{13} \quad (31)$$

79%

2.2.1.3 乙烯基硼化物的交叉偶联

（1）与芳基卤代物或其类似物的偶联

乙烯基硼化物能与芳基卤代物或芳基三氟甲磺酸酯顺利地进行交叉偶联反应，而且属区域选择性反应，芳基化只发生在烯烃的端碳上。如 1-己烯硼酸邻苯二酚酯与碘苯反应定量生成 1-苯基己烯而不生成 2-苯基己烯[39]。

$$\text{Bu-CH=CH-B(cat)} + \text{PhI} \xrightarrow[\text{C}_6\text{H}_6, \text{回流, 2h}]{\text{Pd(PPh}_3)_4, \text{NaOH}} \text{Bu-CH=CH-Ph} \ (100\%) + \text{Bu-C(=CH}_2)\text{-Ph} \ (0\%) \quad (32)$$

烯基硼酸片呐醇酯也是 Suzuki 反应的很好底物，可以合成结构更复杂的化合物，而且反应生成的片呐醇硼酸酯副产物通常为晶状固体，很容易与交叉偶联产物分离[40]。

$$\quad (33) \quad 97\%$$

烯基硼酸也是很好的底物，与三氟甲磺酸吡啶衍生物酯顺利反应，利用该反应可合成维生素 A[41]。

$$\quad (34) \quad 92\%$$

（2）与烯基卤代物或其类似物的偶联

乙烯基硼酸酯可以与卤代烯烃偶联，卤代烯烃的构型在交叉偶联反应中保持不变，常用来合成二烯、三烯甚至多烯[42]。首先看两个合成二烯的实例，反应产物的两个碳碳双键的构型与两个反应底物的构型一致[43,44]。

$$\text{C}_4\text{H}_9\text{-CH=CH-B(OCH}_2\text{CH}_2\text{O)} + \text{Br-CH=CH-Ph} \xrightarrow[\text{C}_6\text{H}_6, \text{回流, 2h}]{\text{Pd(PPh}_3)_4/\text{NaOEt}} \text{C}_4\text{H}_9\text{-CH=CH-CH=CH-Ph} \quad (35)$$

$$\text{C}_4\text{H}_9\text{-CH=CH-B(O}i\text{-Pr})_2 + \text{Br-CH=CH-C}_6\text{H}_{13} \xrightarrow[\text{C}_6\text{H}_6, \text{回流, 2h}]{\text{Pd(PPh}_3)_4/\text{NaOEt}} \text{C}_4\text{H}_9\text{-CH=CH-CH=CH-C}_6\text{H}_{13} \quad (36)$$

再如，Z-构型的溴代二烯与烯基硼酸酯交叉偶联得到 Z-构型的三烯[45]。

$$\text{Ph} \diagup \diagup \text{Br} + \underset{\text{苯并二氧硼烷-异丁烯基}}{\text{BO}} \xrightarrow[\text{C}_6\text{H}_6, \text{回流}, 10\text{min}]{\text{Pd(PPh}_3)_4, \text{EtONa}} \text{Ph} \diagup\diagup\diagup\diagup \quad 85\% \qquad (37)$$

基于乙烯基硼酸与乙烯基卤代物 Suzuki 偶联反应优异的化学、区域和立体选择性, 该反应用于合成维生素 A, 氢氧化铊作为碱促进反应的进行[46]。

$$(38)$$

维生素A 83%

烯基硼酸酯甚至可以与碘代环丙烷交叉偶联, 似乎实现了 $C(sp^2)$-$C(sp^3)$ 键的构建, 但环丙烷环上碳近似 sp^2 杂化的实质是反应得以进行的原因[47]。例如, 2-环丙烷基乙烯硼酸酯与碘代环丙烷在钯催化下在 DMF-H$_2$O 混合溶剂中于 90℃反应 20 h 几乎定量地生成反式的 1,2-环丙基乙烯衍生物, 反应加入相转移催化剂四丁基氯化铵以提高无机碱碳酸钾在有机溶剂中的溶解度[48]。

$$(39) \quad 95\%$$

2.2.1.4 烷基硼化物的交叉偶联

(1) 与芳基卤代物或其类似物的交叉偶联

芳基卤代物或芳基三氟甲磺酸酯与烷基硼酸、烷基硼烷、烷基硼酸酯等烷基化试剂进行 Suzuki 交叉偶联反应, 实现芳环的烷基化或 $C(sp^3)$-$C(sp^2)$ 的构建。碳原子数较少的烷基化反应可以采用以上任何烷基化试剂, 但烷基硼酸作为烷基化试剂更经济。以下是短链烷基化反应的一些实例, 反应以 PdCl$_2$(dppf)为催化剂, Ag$_2$O 作为添加剂促进反应的进行, 底物上的取代基对反应影响不大[49]。

$$\text{R} \diagdown \text{B(OH)}_2 + \text{Br} \diagdown \diagup \text{R}' \xrightarrow[\text{THF, Ag}_2\text{O, 80℃,}]{\text{PdCl}_2(\text{dppf}), \text{K}_2\text{CO}_3} \text{R} \diagdown \diagup \text{R}' \qquad (40)$$

R	R'	
n-Pr	OMe	80%
n-Pr	COMe	84%
n-Pr	CO$_2$Et	82%
SiMe$_3$	COMe	80%
CH$_2$CH$_2$CH=CH$_2$	CO$_2$Et	77%

当碳链较长时则以烷基硼烷或烷基硼酸酯为烷基化试剂，因为长链烷基硼烷或烷基硼酸酯可以通过烯烃的加成或卤代烷的取代容易地制备。图 2-4 是一些烷基硼烷或烷基硼酸酯的合成途径，其中烷基-9-硼双环[3,3,1]-壬烷（烷基-9-BBN）由于 9-BBN 的低空间位阻使交叉偶联反应更加容易[50]。

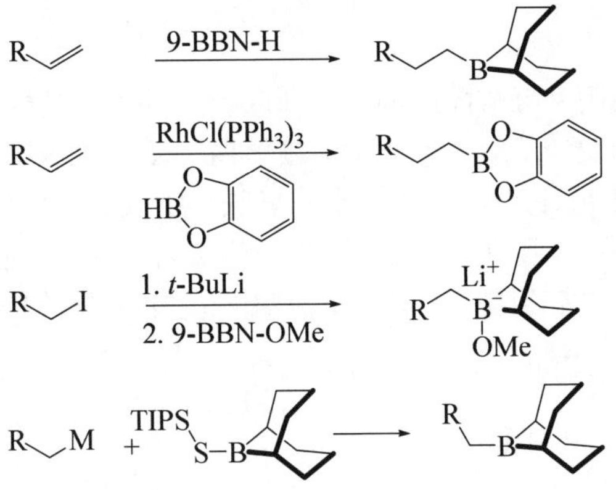

图 2-4 部分烷基硼烷或烷基硼酸酯的合成途径

长链硼烷或酯类化合物都能与芳基卤代物或芳基三氟甲磺酸酯进行交叉偶联，在相同或相近的反应条件下以烷基-B-9-BBN 作为烷基化试剂通常获得更高的收率[50]。

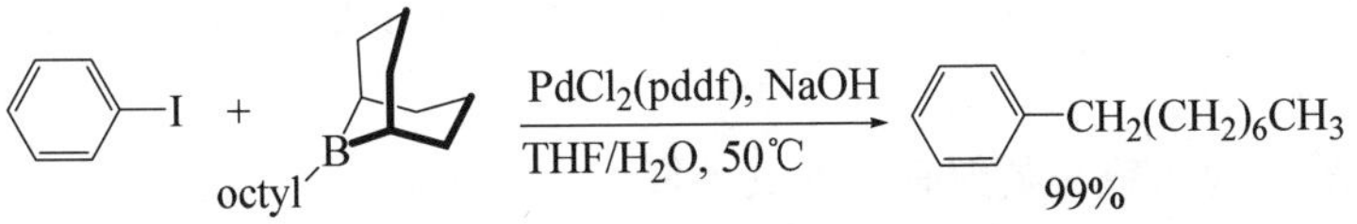

$$\text{Ph-I} + \text{octyl-B(9-BBN)} \xrightarrow[\text{THF/H}_2\text{O, 50℃}]{\text{PdCl}_2(\text{pddf}), \text{NaOH}} \text{Ph-CH}_2(\text{CH}_2)_6\text{CH}_3 \quad 99\% \tag{41}$$

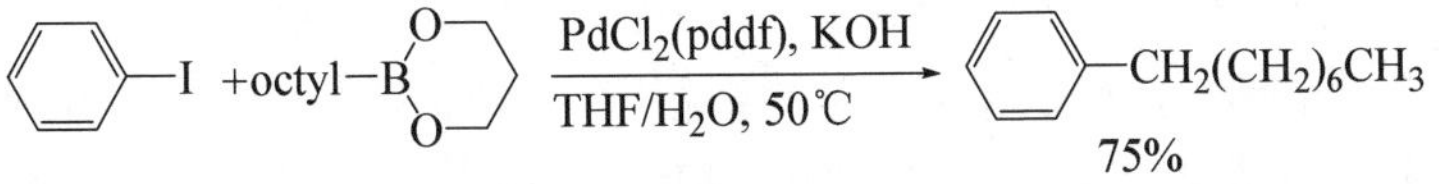

$$\text{Ph-I} + \text{octyl-B(OCH}_2\text{CH}_2\text{CH}_2\text{O)} \xrightarrow[\text{THF/H}_2\text{O, 50℃}]{\text{PdCl}_2(\text{pddf}), \text{KOH}} \text{Ph-CH}_2(\text{CH}_2)_6\text{CH}_3 \quad 75\% \tag{42}$$

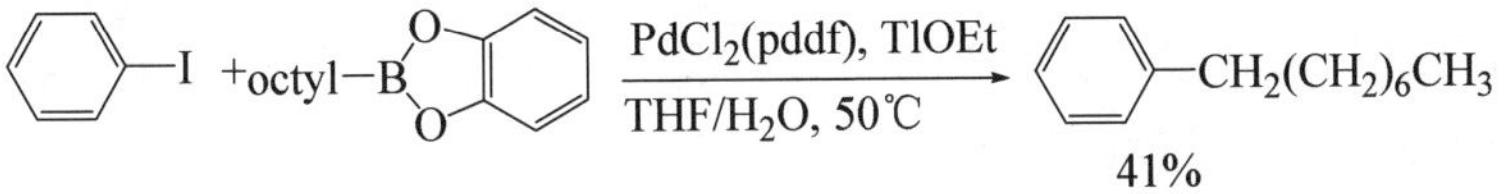

$$\text{Ph-I} + \text{octyl-B(catechol)} \xrightarrow[\text{THF/H}_2\text{O, 50℃}]{\text{PdCl}_2(\text{pddf}), \text{TlOEt}} \text{Ph-CH}_2(\text{CH}_2)_6\text{CH}_3 \quad 41\% \tag{43}$$

需要注意的是，配体对反应有很大的影响，通常氯代芳基化合物不发生烷基化 Suzuki 反应，但通过适当的配体调节，反应可以顺利进行，如下反应所示[51,52]。

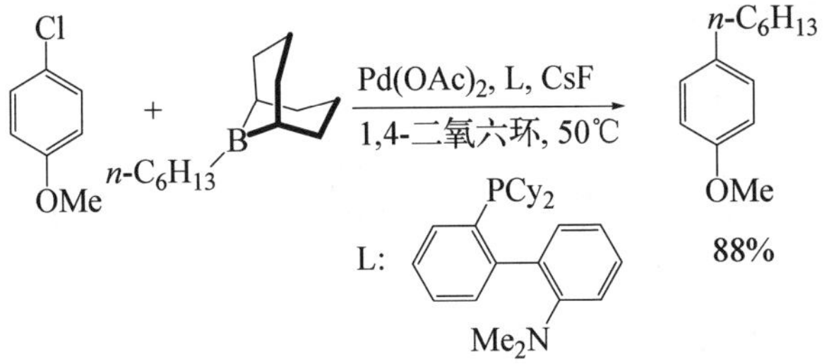

$$\text{4-MeO-C}_6\text{H}_4\text{-Cl} + n\text{-C}_6\text{H}_{13}\text{-B(9-BBN)} \xrightarrow[\text{1,4-二氧六环, 50℃}]{\text{Pd(OAc)}_2, \text{L}, \text{CsF}} \text{4-MeO-C}_6\text{H}_4\text{-}n\text{-C}_6\text{H}_{13} \quad 88\% \tag{44}$$

L: 2-PCy$_2$-2'-NMe$_2$-联苯

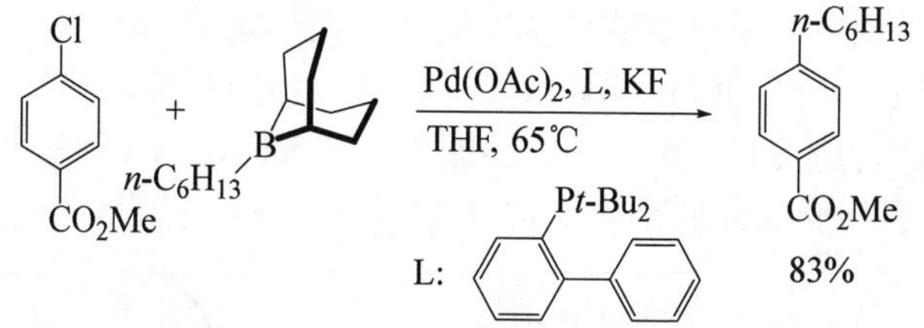

$$\text{(45)}$$

芳基卤代物与烯丙基硼酸酯间的交叉偶联也属于芳环的烷基化反应,可用来合成 3-芳基丙烯类化合物,反应的选择性依赖于耦合反应的条件。例如,烯丙基片呐醇硼酸酯与碘代苯在 Pd(PPh$_3$)$_4$ 催化下于四氢呋喃中反应,以甲醇钠作碱只以较低的收率得到目标产物;当用 CsF 替代甲醇钠后烯丙基片呐醇硼酸酯则能与多种取代的芳基碘反应,以中等到良好的收率得到相应的交叉偶联产物[53]。

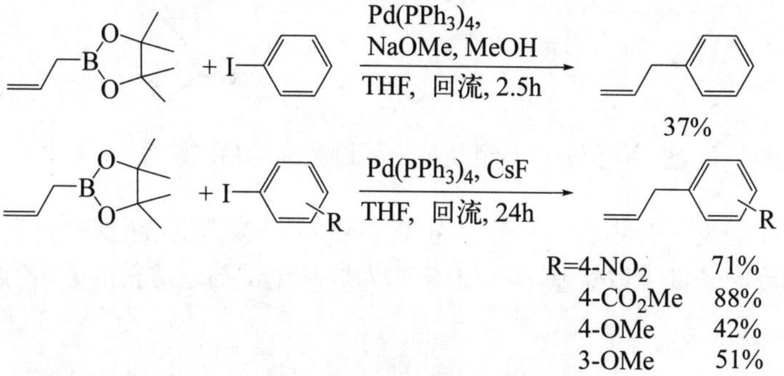

$$\text{(46)}$$

$$\text{(47)}$$

R=4-NO$_2$	71%
4-CO$_2$Me	88%
4-OMe	42%
3-OMe	51%

(2) 与烯基卤代物或其类似物的交叉偶联

烷基硼化物可以与包括简单烯烃和 α,β-不饱和羰基化合物在内的各种烯基卤代物或三氟甲磺酸酯进行交叉偶联反应得到相应的烯烃烷基化产物。烷基-9-BBN 是最常用的烷基化试剂,可由烯烃与 9-BBN-H 原位加成制得。以下两个反应是简单的卤代烯烃的烷基化反应,可以看到卤代端烯烃和内烯烃都能进行烷基化,而且反应后碳碳双键的构型得到保持[50]。

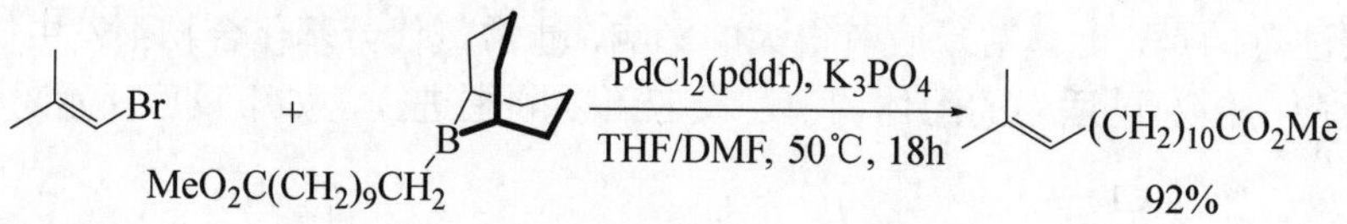

$$\text{(48)}$$

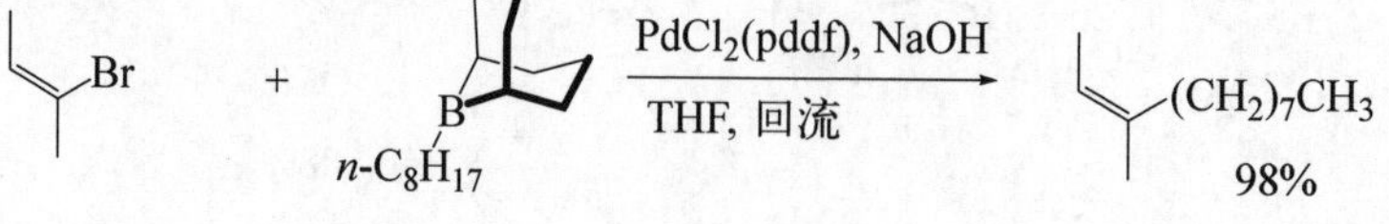

$$\text{(49)}$$

另外,反应兼容各种官能团,特别是底物中硫原子的存在不影响钯催化剂的催化性能，预示 Suzuki 烯烃烷基化反应在药物合成中的广泛应用性[50]。

$$\text{PhS} \diagdown \text{Br} + \text{烯烃硼烷} \xrightarrow[\text{THF, 回流}]{\text{PdCl}_2(\text{pddf}), \text{NaOH}} \text{产物} \quad 90\% \tag{50}$$

卤代 α,β-不饱和羰基化合物，无论卤素处于 α 位还是 β 位，均能与烷基硼化物进行交叉偶联反应。5,5-二甲基-3-溴环己烯酮和 2-甲基-3-溴丙烯酸甲酯在相同的温和反应条件下与烷基硼酸邻苯二酚酯反应以较高的收率生成相应的烷基化产物[53]。

$$\text{邻苯二酚硼酸酯-}(CH_2)_4COCH_3 + \text{5,5-二甲基-3-溴环己烯酮} \xrightarrow[\text{THF, 50℃, 16h}]{\text{PdCl}_2(\text{dppf}), \text{Tl}_2\text{CO}_3} CH_3CO(CH_2)_4\text{-环己烯酮} \quad 66\% \tag{51}$$

$$\text{邻苯二酚硼酸酯-}(CH_2)_4COCH_3 + \text{2-甲基-3-溴丙烯酸甲酯} \xrightarrow[\text{THF, 50℃, 16h}]{\text{PdCl}_2(\text{dppf}), \text{Tl}_2\text{CO}_3} CH_3CO(CH_2)_4\text{-CO}_2Me \quad 68\% \tag{52}$$

α-碘代环戊烯酮衍生物与酯烷基-9-BBN 交叉偶联得到 α-酯烷基取代的环戊烯酮衍生物，是合成前列腺素的关键一步[54]。

$$\text{TBDMSO-环戊烯酮-I} + MeO_2C(CH_2)_6\text{-B}(9\text{-BBN}) \xrightarrow[\text{Cs}_2\text{CO}_3, \text{DMF/THF/H}_2\text{O, rt, 1.5h}]{\text{PdCl}_2(\text{pddf}), \text{Ph}_3\text{As}} \text{TBDMSO-环戊烯酮-}(CH_2)_6CO_2Me \quad 80\% \tag{53}$$

烷基硼烷与碘代烯烃在一氧化碳存在下发生羰基化交叉偶联生成 α,β-不饱和酮。反应以 $Pd(PPh_3)_4$ 或 $PdCl_2(PPh_3)_2$ 为催化剂，磷酸钾为碱，在 1～3 大气压的一氧化碳气氛下顺利进行[55]。

$$I\diagdown C_4H_9 + CH_3(CH_2)_7\text{-B}(9\text{-BBN}) \xrightarrow[\text{K}_3\text{PO}_4, C_6H_6, \text{rt, 5h}]{0.1\text{MPa CO}, Pd(PPh_3)_4} CH_3(CH_2)_7\text{-CO-}C_4H_9 \quad 99\% \tag{54}$$

$$I\diagdown C_5H_{11}\text{-TBDMSO} + \text{烯烃硼烷} \xrightarrow[1,4\text{-二氧六环, rt, 5h}]{0.3\text{MPa CO}, Pd(PPh_3)_4, K_3PO_4} \text{产物} \quad 53\% \tag{55}$$

（3）与烷基的偶联

烷基卤代物不容易与烷基硼化物发生 Suzuki 交叉偶联反应，因为它们与钯氧化加成反应较慢，而且加成后生成的烷基钯中间体更容易发生消除。但在适宜的反应条件，如以 K_3PO_4 作碱，$Pd(PPh_3)_4$ 为催化剂，烷基碘与烷基-9-BBN 之间的交叉偶联反应也可以进行，以中等到良好的收率得到相应的偶联产物。下面的反应是 9-BBN-H 先与烯烃反应生成烷基-9-BBN，进而与碘代烷交叉偶联反应的结果[50]。由于 9-BBN-H 更容易与端烯烃加成生成相应的烷基-9-BBN，偶联反应发生在端烯烃一端，而非发生在内烯烃一端。

$$R^1HC=CH_2 \xrightarrow[THF]{9\text{-BBN-H}} 9\text{-R-9-BBN} \xrightarrow[K_3PO_4, 1,4\text{-二氧六环} \atop 60℃]{R^2-I, Pd(PPh_3)_4} R-R^2 \qquad (56)$$

$R^1HC=CH_2$	R^2	$R-R^2$	
$CH_2=CH(CH_2)_8CO_2Me$	Me	$CH_3(CH_2)_{10}CO_2Me$	71%
$CH_2=CH(CH_2)_5CH_3$	$CH_3(CH_2)_5$	$CH_3(CH_2)_{12}CH_3$	64%
$CH_2=CH(CH_2)_8CO_2Me$	Me_3CCH_2	$Me_3C(CH_2)_{11}CO_2Me$	45%
(结构式)	$NC(CH_2)_3$	(结构式) $NC(CH_2)_6$	61%
(结构式)	$MeO_2C(CH_2)_3$	(结构式) $MeO_2C(CH_2)_5$	57%

与卤代烯烃交叉偶联类似，卤代烷与烷基硼烷在一氧化碳气氛中也可发生羰基化交叉偶联得到酮，但反应需要额外光照才能进行[56]。

$$C_6H_{13}I + CH_3(CH_2)_7\text{-B(9-BBN)} \xrightarrow[K_3PO_4, C_6H_6, rt, h\nu\ 24h]{0.1MPa\ CO, Pd(PPh_3)_4} C_6H_{13}\text{-CO-}C_8H_{17} \qquad (57)$$
67%

$$\text{(双环碘化物)} + \text{(9-BBN-(CH}_2)_3\text{CMe}_2\text{CO}_2\text{Me)} \xrightarrow[K_3PO_4, C_6H_6, rt, h\nu\ 24h]{0.1MPa\ CO, Pd(PPh_3)_4} \text{(酮产物-CO}_2\text{Me)} \qquad (58)$$

2.2.2 分子内 Suzuki 反应

2.2.2.1 五元和六元环的构建

同一分子结构中同时存在 Suzuki 交叉偶联伙伴基团时，若两个基团间距合适，在适宜的 Suzuki 催化反应体系下可发生分子内交叉偶联反应生成各种环状化合物。最先发现烷基硼化物与卤代烯或芳基间的分子内 Suzuki 反应可以构筑五元和六元环。如下反应，同时含有端烯和卤代芳基、卤代乙烯基的化合物先与 9-BBN-H 进行加成反应，将端烯烃转化为烷基-9-BBN，进而在 Suzuki 催化反应体系下卤代芳基或卤代乙烯基团与烷基-9-BBN 基团进行分子内交叉偶联反应生成五元和六元环状化合物[57]。

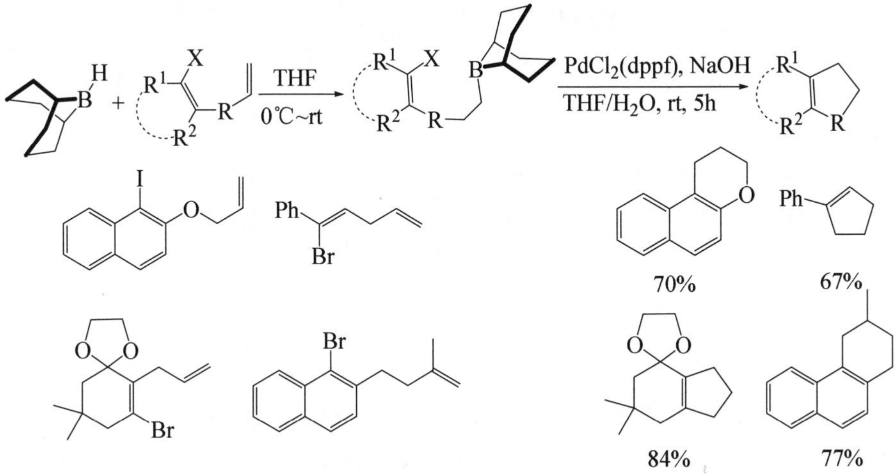

上面反应底物中的卤代乙烯单元的碳-碳双键处于卤素原子和端烯烃的中间，因而交叉偶联反应得到环内烯烃。如果底物中卤代乙烯的碳-碳双键处于卤素原子和端烯烃之外，并且卤代碳原子与端烯烃之间连接链的碳原子数是 2 或 3，分子内交叉偶联反应也能进行，此时生成环外烯烃产物。

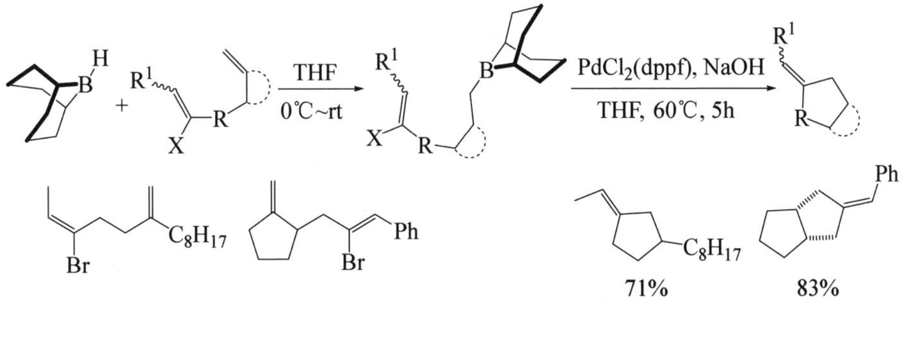

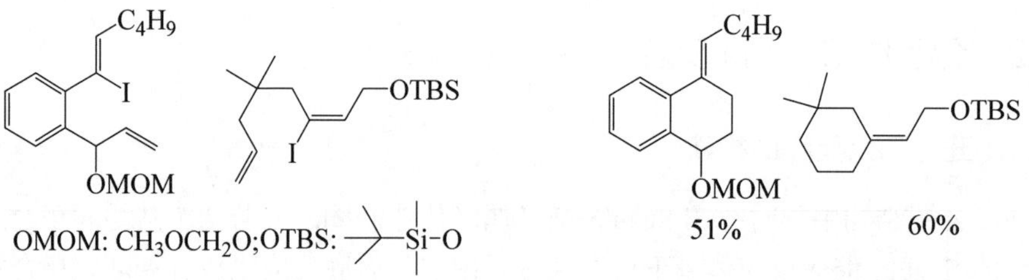

OMOM: CH₃OCH₂O; OTBS: (structure)

随后发现含有端烯基团的芳基三氟甲磺酸酯和烯基三氟甲磺酸酯与 9-BBN-H 进行硼氢化反应生成 B-烷基化合物,也是分子内 Suzuki 反应构筑五元和六元环的很好底物[58]。

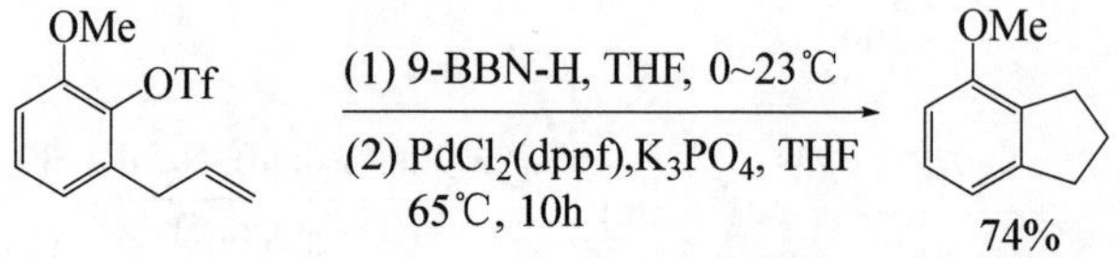

$$\text{(structure)} \xrightarrow[\text{(2) PdCl}_2\text{(dppf), K}_3\text{PO}_4\text{, THF, 65℃, 10h}]{\text{(1) 9-BBN-H, THF, 0~23℃}} \text{(structure) 74\%} \quad (61)$$

$$\text{(structure)} \xrightarrow[\text{(2) PdCl}_2\text{(dppf), K}_3\text{PO}_4\text{, 1,4-二氧六环/THF, 85℃, 10h}]{\text{(1) 9-BBN-H, THF, 0~23℃}} \text{(structure) 76\%} \quad (62)$$

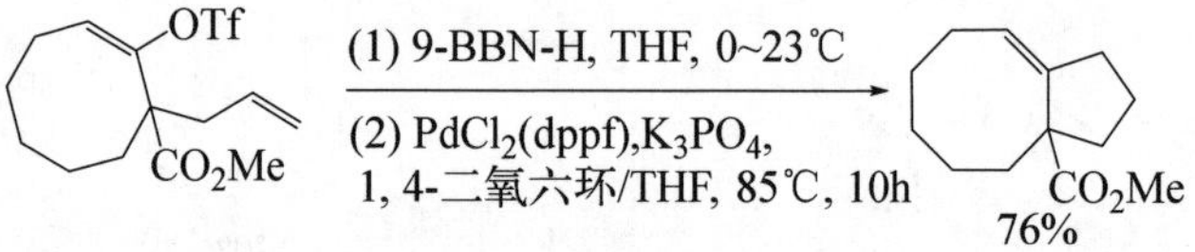

2.2.2.2 中环和大环的构建

对于简单结构的底物分子,分子内 Suzuki 交叉偶联形成多于六个碳的中环比较困难,主要原因是在相同的反应条件下更容易发生分子间 Suzuki 反应生成低聚物;即使用很稀的底物浓度来抑制分子间反应,也只能以很低的收率得到分子内反应的产物。随着偶联伙伴基团间连接链增长,分子内 Suzuki 反应逐渐变得容易,一些复杂结构的底物在适宜的反应条件下也能以中等甚至优异的收率生成环合产物。例如,将底物烯烃与 9-BBN-H 进行硼氢化反应得到烷基硼烷,先用 TlOEt 处理,然后将混合液缓慢滴加到催化反应体系以控制底物的浓度在 0.003mol/L,下面两个底物能以中等的收率得到环合产物,而且反应后原来碘代烯烃的构型得到保持[59]。

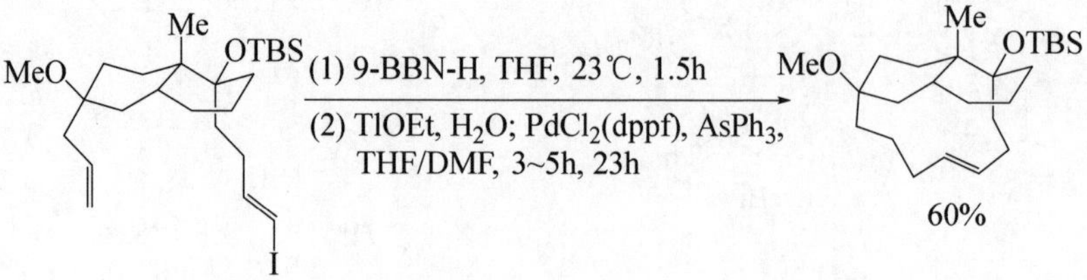

$$\text{(structure)} \xrightarrow[\text{(2) TlOEt, H}_2\text{O; PdCl}_2\text{(dppf), AsPh}_3\text{, THF/DMF, 3~5h, 23h}]{\text{(1) 9-BBN-H, THF, 23℃, 1.5h}} \text{(structure) 60\%} \quad (63)$$

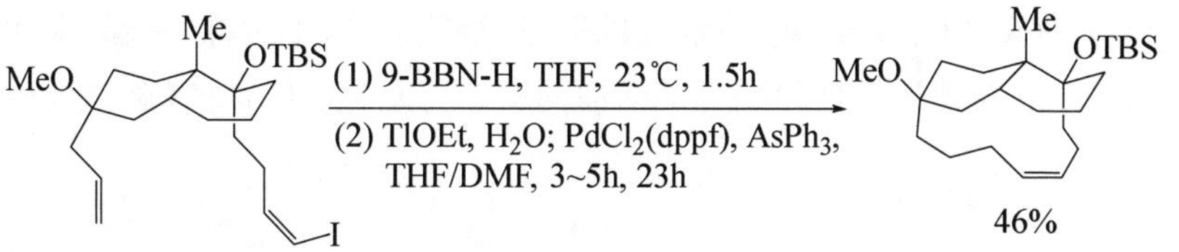

有时，通过提高催化剂的量并延长反应时间，可以改善分子内 Suzuki 烷基化反应，使目标产物的产率明显提高，甚至生成八元中环[60]。

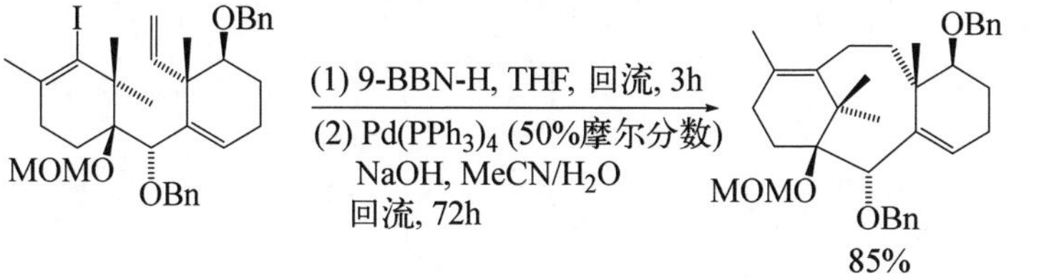

2.2.3 不对称 Suzuki 反应

Suzuki 反应的催化剂多数情况下为钯-膦配体配合物或者由钯盐与膦配体原位生成的配合物，可以预期手性配体可以诱导不对称 Suzuki 反应。由前述内容可知，Suzuki 反应涉及多种卤代物和有机硼化物偶联伙伴之间的交叉偶联，其中芳基与芳基之间的交叉偶联可形成联二芳基不对称化合物（若芳基相同则含 C_2 对称轴），而仲卤代烷与有机硼化物间的交叉偶联可形成手性碳。因此，在这些交叉偶联反应中如果有手性配体介入有可能实现不对称 Suzuki 反应。

2.2.3.1 双芳基的不对称偶联

不对称 Suzuki 反应并不容易实现，目前不对称 Suzuki 反应主要是邻位取代的芳基萘硼酸与其他邻位取代的芳基（包括芳基萘）卤化物间的交叉偶联。手性配体在不对称 Suzuki 反应中起关键作用，早期多采用手性膦配体，后来逐渐开发出氮配体或氮-膦混合配体[61]。以下按交叉偶联底物的不同介绍几个实例。各种邻位取代的α-溴代萘与邻位取代的α-萘硼酸在二茂铁基手性配体存在下可进行不对称交叉偶联反应，其中交叉偶联伙伴中β位取代基为甲基时结果最好，对映选择性达到 85%，生成的 α,α'-二甲基联二萘含 C_2 对称轴[62]。

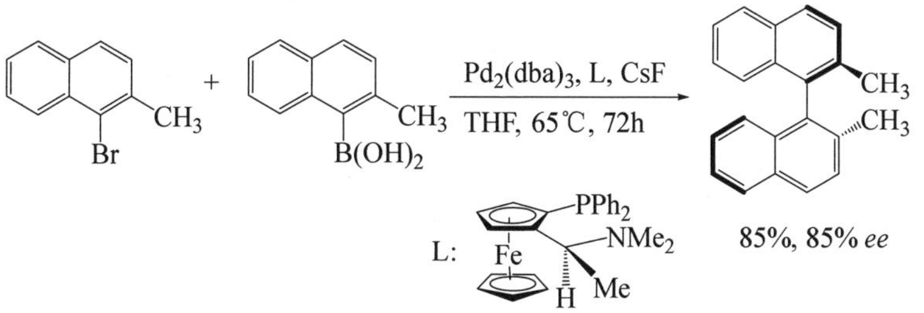

α-溴代-β-萘磷酸酯与各种邻位取代的芳基硼酸在 C_2-不对称单齿膦配体诱导下以良好到优异的对映选择性得到光学活性的联二芳基化合物,对映体过量值最高可达 92%[63]。

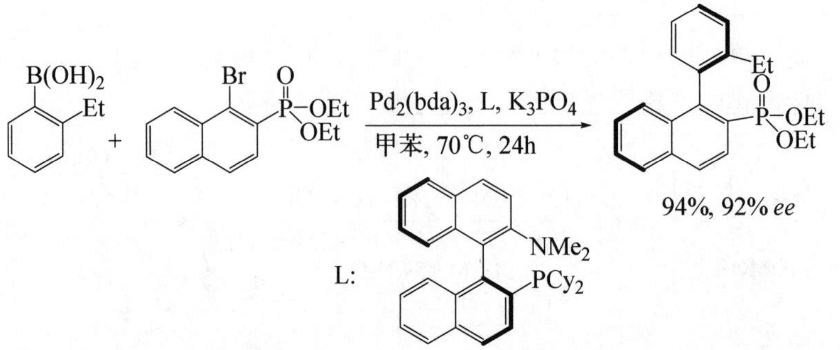

$$\text{(67)}$$

$$\text{94\%, 92\% } ee$$

邻位含大空间位阻酰胺基团的芳基卤代物与 β-甲基-α-萘硼酸在上述相同手性 N,P-手性配体存在下也以优异的对映体选择性生成光学活性的偶联产物[64]。

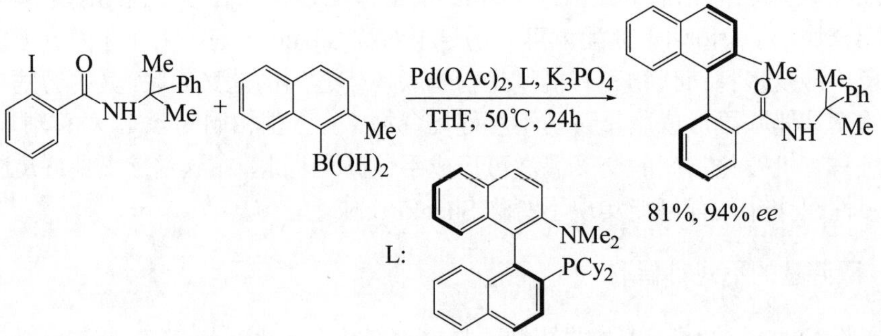

$$\text{(68)}$$

$$\text{81\%, 94\% } ee$$

萘硼酸与采用特殊磷酰基［双（2-氧代-3-噁唑基）膦酰基］（BOP）保护的萘酚在一种手性单齿膦配体诱导下以收率和对映体过量值均佳的结果得到交叉偶联产物[5]。

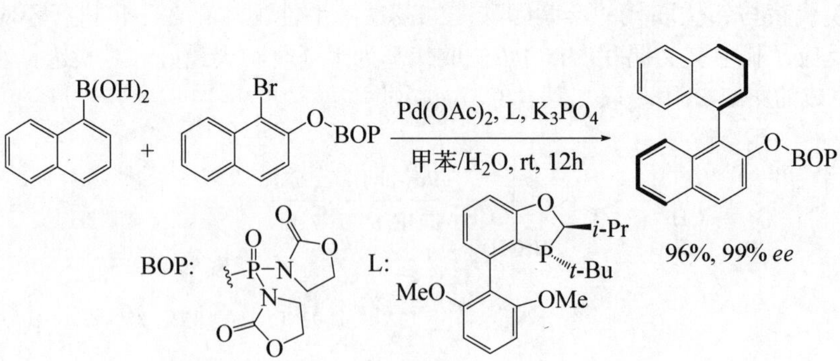

$$\text{(69)}$$

$$\text{96\%, 99\% } ee$$

甲酰基在有机合成中是一个多用途官能团。采用不对称 Suzuki 偶联合成含甲酰基的手性联双芳基化合物可为其他新的含各类官能团的手性双芳基化合物的合成奠定基础。其中，2-甲酰基硼酸与二烷氧基膦酰基溴代萘在联二萘单齿膦配体诱导和钯催化下以很高的收率和对映体过量值得到偶联产物[65]。

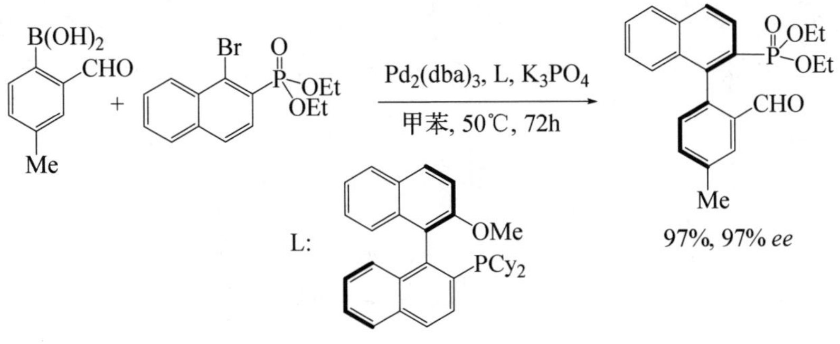

$$\text{(70)}$$

2.2.3.2 仲卤代烷与烷基或芳基的不对称偶联

仲卤代烷与烷基硼化物的交叉偶联形成手性碳，若催化剂是手性的，交叉偶联反应有可能是不对称的。然而，实现不对称 Suzuki 双烷基化反应并非容易。手性的钯催化体系往往在此反应中无手性诱导作用。最早发现，一些手性双氮配体与镍配合物构成的催化体系在仲烷基卤代物与烷基-9-BBN 的交叉偶联反应中显示优良的催化活性和对映选择性，但反应条件要求严格，反应实例如下。

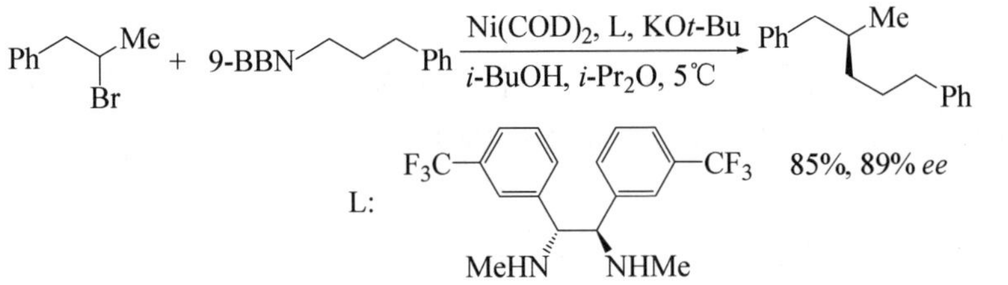

$$\text{(71)}$$

在与上面类似的催化反应体系下，2-氯丁酰胺与苯基-9-BBN 交叉偶联以优良的收率和对映选择性生成光学活性的 2-苯基酰胺[66]。

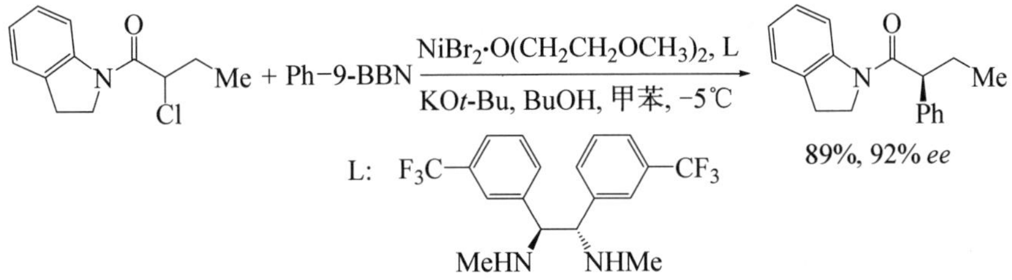

$$\text{(72)}$$

许多药物分子都含有光学活性的杂环骨架核心，因此，杂环底物的不对称交叉偶联反应在手性药物的合成中更有应用潜力。在杂环卤代物与杂环硼酸之间的不对称交叉偶联反应中，通过偶联伙伴间结构的变化与组合，可以衍生出众多光学活性的杂环化合物[67]。手性铑催化剂在烯丙基（含杂环）卤代物与杂环硼酸不对称交叉偶联反应中显示良好的催化性能。在下面两个反应中，手性双齿膦配体与铑催化剂前体原位生成手性铑催化剂催化氯代二氢吡喃和氯代四氢吡啶与芳基硼酸的不对称交叉偶联，以优异的对映选择性生成光学活性的杂环化合物。

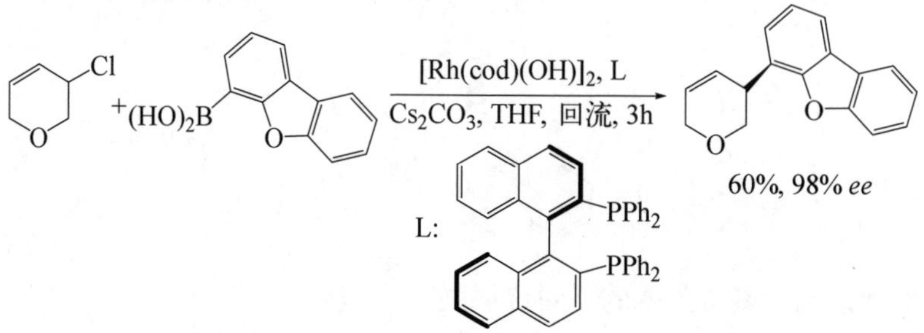

Suzuki 反应不像 Heck 反应那样可以直接构建季碳中心，因而不对称 Suzuki 反应也相对较少，但即使如此其在光学活性药物分子的合成中也有应用。

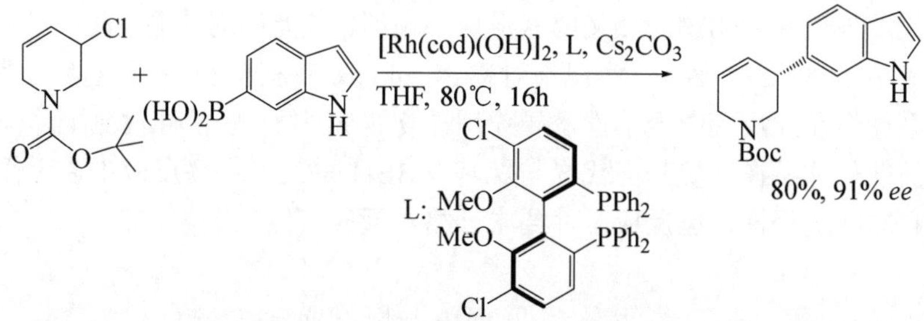

2.3 Suzuki 反应在药物合成上的应用实例

Suzuki 反应是构建碳-碳键最有效和最有用的偶联反应之一，具有反应条件温和、底物适用性广，以及目标产物选择性高的优点，因此在药物和生物活性分子的合成上具有广泛应用[68,69]。Suzuki 反应在药物分子和天然产物合成中作用不胜枚举，以下按交叉偶联伙伴的结构类型介绍几个利用 Suzuki 反应合成药物或生物活性物质的实例。为了便于描述，将 Suzuki 反应目标产物编号为 SA-x，x 为整数，代表序号。

2.3.1 分子间反应的应用

SA-1 是磷酸二酯酶-4 抑制剂，用于治疗慢性阻塞性肺病和哮喘，其合成的最后一步是中间体氯代物与芳基硼酸在 Suzuki 反应条件下的交叉偶联。直接分离得到的偶联产物的反-顺（trans/cis）异构体比例为 82∶12，用含水 10%的乙腈重结晶后可将反式异构体的比例提高到高于 99%[70]。

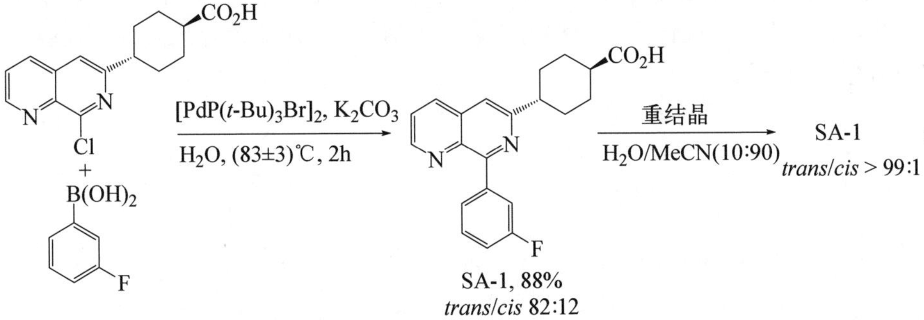

$$\text{反应式 (75)}$$

SA-1, 88%
trans/cis 82:12

重结晶 $H_2O/MeCN(10:90)$ → SA-1, trans/cis > 99:1

2-氨基-5-溴吡嗪与苯硼酸在钯催化剂催化下在甲苯-H_2O-DMF 三元混合溶剂中回流容易交叉偶联生成 SA-2，SA-2 经后续转化生成一个高选择性 NPY-5 受体拮抗剂，用于治疗肥胖病[71]。

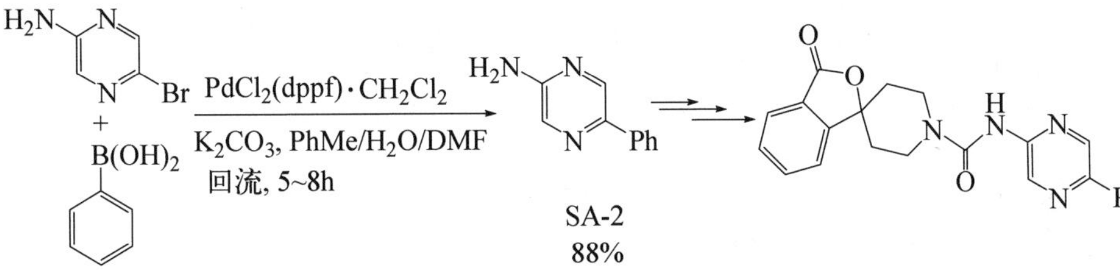

$$\text{反应式 (76)}$$

SA-2 88%

碘代苯衍生物与 4-羧基苯硼酸在钯黑催化下进行 Suzuki 反应以优异的收率得到交叉偶联产物 SA-3，反应中以钯黑替代醋酸钯解决了催化剂与产物的分离难题并提高了产品的纯度。SA-3 与甲胺在缩合剂羰基咪唑（CDI）作用下缩合生成相应的酰胺，是治疗帕金森综合征的候选药物[72]。

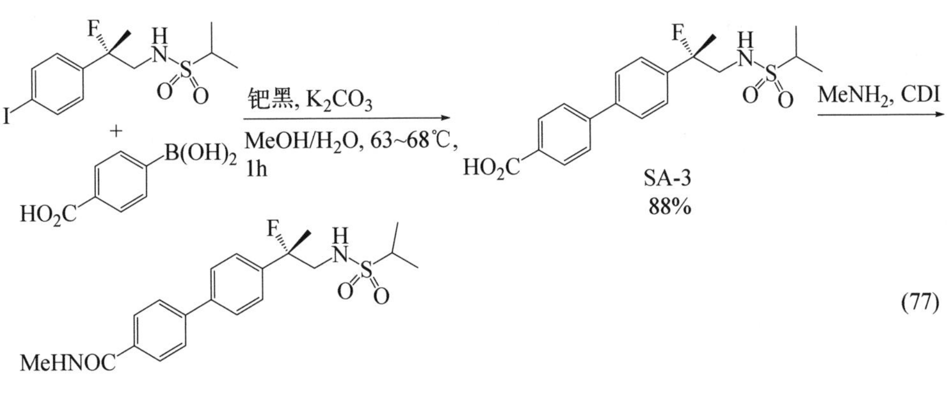

$$\text{反应式 (77)}$$

SA-3 88%

三氟甲磺酸酯替代卤代物也可以与芳基硼酸交叉偶联合成药物分子。5-溴-3-乙基胡椒环与丁基锂反应生成芳基锂,进而与硼酸三异丙酯进行硼化反应并在稀乙酸中水解生成 3-乙基胡椒环-5-硼酸,不经分离直接与芳基三氟甲磺酸酯在简单的 Suzuki 催化体系下交叉偶联生成 SA-4[73]。SA-4 再在氢氧化钾水溶液-丁醇体系中皂化水解转化成钾盐,该钾盐是一种内皮素受体拮抗剂,用于治疗原发性肺动脉高压和充血性心力衰竭。

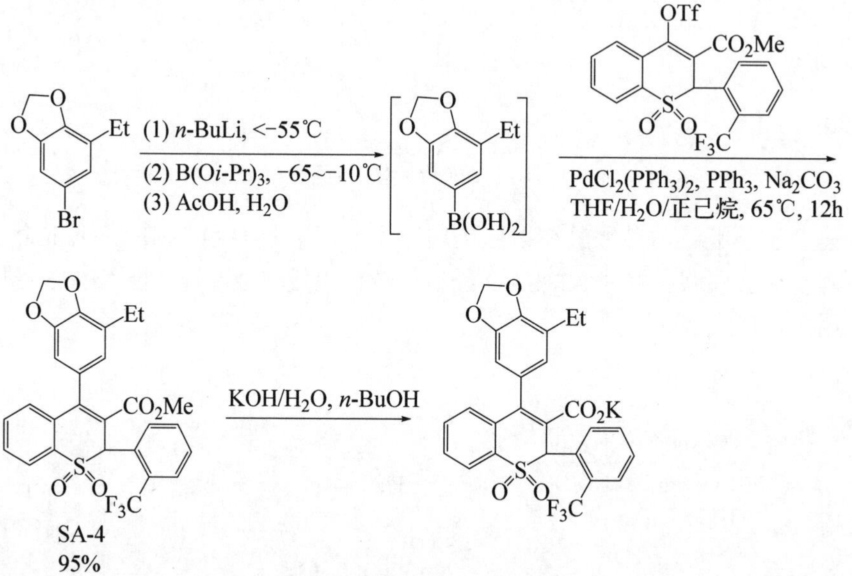

氯代吡啶衍生物与丙醛缩乙二醇-9-BBN 在标准的 Suzuki 催化反应体系下交叉偶联生成 SA-5,SA-5 不经分离进行缩醛水解生成杂芳基丙醛基中间体,两步总收率达 87%。该中间体经后续反应生成一种非肽 $R_v\beta_3$ 拮抗剂,是治疗骨质疏松症的候选药物[74]。

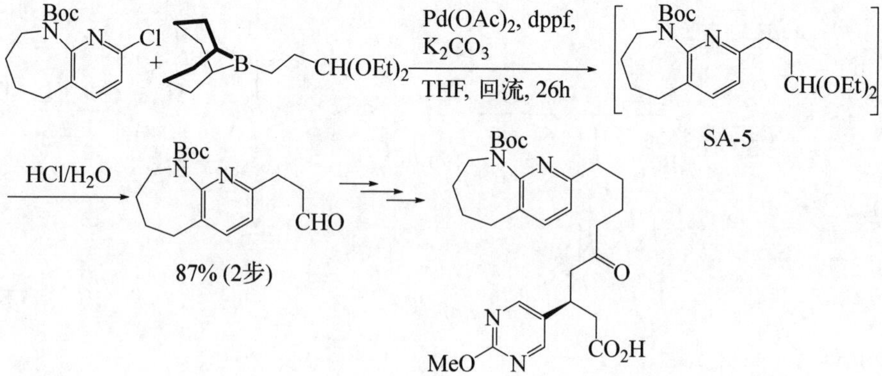

硼酸酯与（杂）芳基卤代物间的 Suzuki 反应在药物合成上也广泛应用。在一种潜在的用于治疗高血压、心脏肥大、心力衰竭和冠心病药物的血管紧张素-(1~7)-受体激动剂的合成中,芳基硼酸片呐醇酯与溴代噻吩衍生物间的交叉偶联生成 SA-6 的反应是整个药物合成的关键一步[75]。

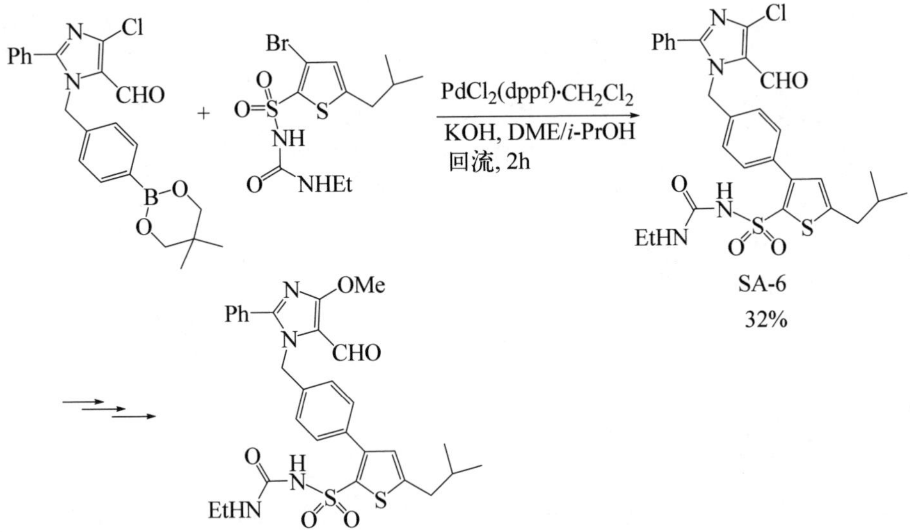

(80)

（+）-圆皮海绵内酯（(+)-discodermolide）是一种微管稳定剂（microtubule-stabilizing agent），属聚酮化合物,对双向淋巴细胞反应有抑制作用。在其合成过程中,碘代物中间体先在叔丁基锂作用下与 MeO-9-BBN 反应生成硼锂配合物,不经分离直接加入碘代乙烯衍生物,在 PdCl$_2$(dppf) 和 Cs$_2$CO$_3$ 催化体系下于室温反应以较好的收率得到交叉偶联产物 SA-7。SA-7 经后续多步反应生成（+）-圆皮海绵内酯[76]。

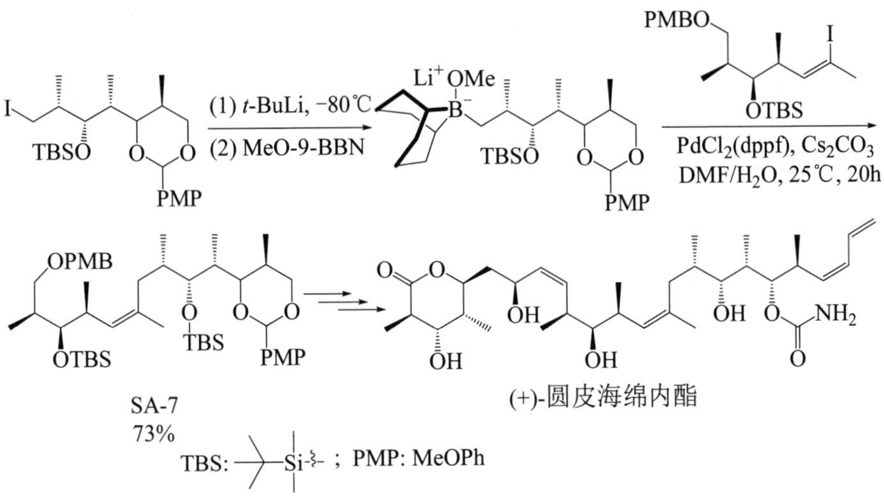

(81)

替米沙坦（telmisartan）是治疗原发性高血压的强血管紧张素Ⅱ受体拮抗剂,其可通过Suzuki反应合成。其前体1-甲基-2-溴苯并咪唑与含咪唑基的有机三氟硼酸钾盐在简单的Suzuki催化反应体系下并协以微波活化,高收率地得到替米沙坦[77]。该合成路线的优势还表现在两个偶联前体容易高选择性地合成。

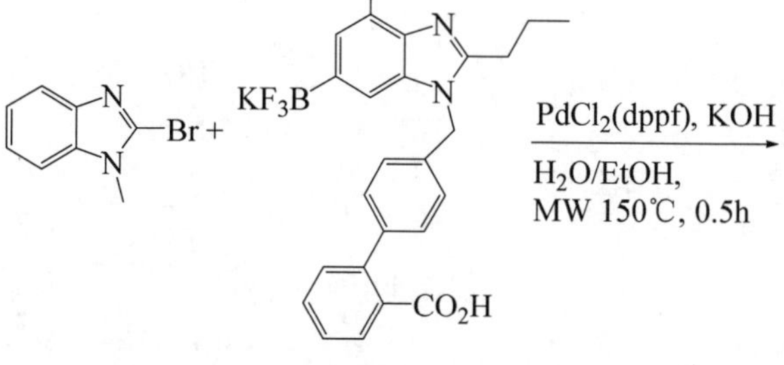

(82)

SA-8,替米沙坦
89%

2.3.2 分子内反应的应用

Ripostatins A 是一种从纤维素索氏黏杆菌中提取的天然产物,通过抑制细菌RNA聚合酶（RNAP）而具有抗菌活性,并以酮-醇式和半缩酮互变异构的形式存在。分子内Suzuki反应是全合成Ripostatins A的关键一步。首先含乙烯基的碘代化合物与9-BBN-H硼氢化生成烷基-9-BBN前体,同时含有碘代乙烯基和烷基硼烷基单元的前体在典型Suzuki催化反应体系下进行分子内交叉偶联生成SA-9,SA-9再转化为Ripostatins A[78]。

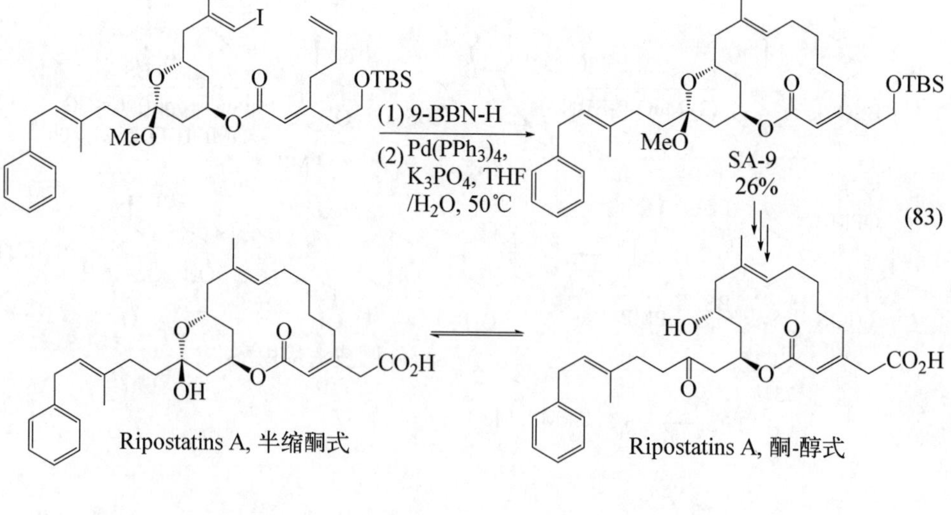

(83)

Ripostatins A,半缩酮式

Ripostatins A,酮-醇式

Nannocystin A 是一个 21 元杂环化合物，同时含有肽和聚酮骨架片段，它在低浓度下对多种人类癌细胞有强抑制增长作用。我国科学家北京大学深圳研究生院叶涛教授基于分子内 Suzuki 反应设计该分子的全合成方案。由 Boc 保护的 β-（3,5-二氯-4-羟基苯基）-α-氨基丙酸经多步合成带有 C5-C6 乙烯基碘和 C3-C4 乙烯硼酸酯结构单元的前体。该前体在催化量的 Pd(PPh₃)₄ 和 Ag₂O 存在下于含水的四氢呋喃中进行分子内 Suzuki 反应，高收率地得到 Nannocystin A[79]。

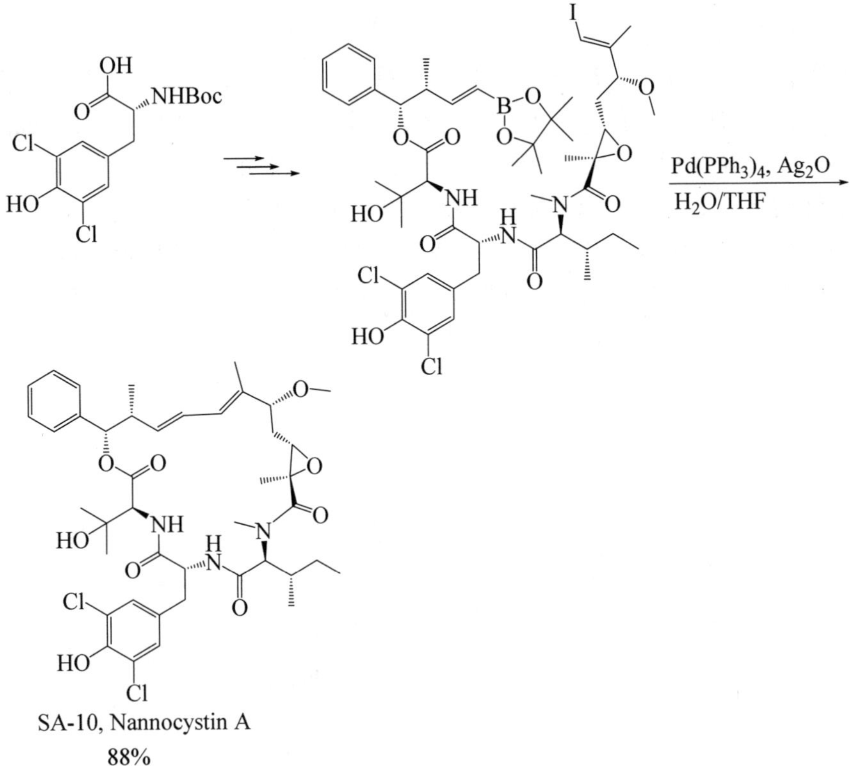

(84)

SA-10, Nannocystin A
88%

Arylomycins A₂ 属于六肽化合物，对多重耐药革兰氏阴性菌（如大肠杆菌）感染具有强效、广谱抗菌活性。其全合成的关键一步是钯催化的 Suzuki 分子内双芳基交叉偶联。反应在 PdCl₂(SPhos)₂ 催化下于甲苯-H₂O 混合溶剂中 90℃反应，收率高达 91%[80]。

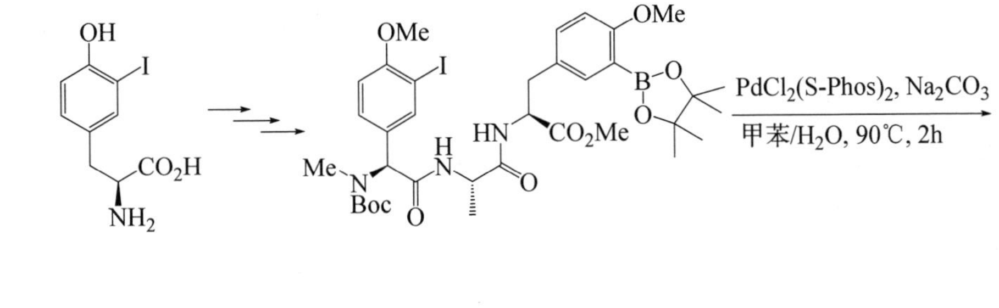

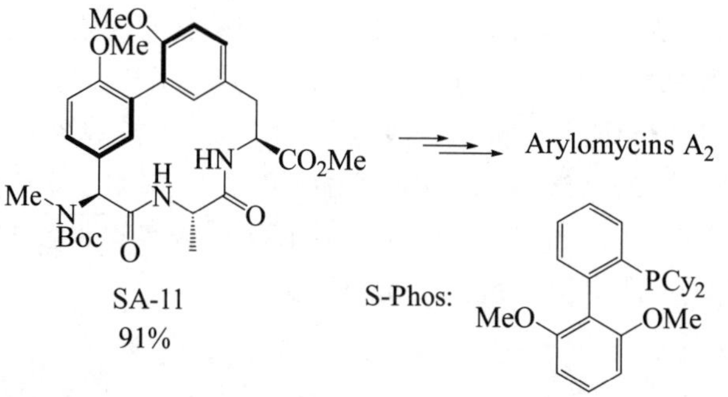

SA-11
91%

(85)

科鲁普钩枝藤碱 A（korupensamine A）是一种抗疟疾天然药物，具有轴手性的结构特征。中国科学院上海有机化学研究所汤文军研究员利用不对称 Suzuki 芳基交叉偶联成功实现了轴手性的构建，并结合其他反应成功合成了科鲁普钩枝藤碱 A。在轴手性构建一步，邻溴苯甲醛衍生物与 7-甲基-5-甲氧基-4-苄氧基-1-萘硼酸在 $Pd(OAc)_2$ 与手性配体原位生成的手性钯配合物催化下在甲苯-H_2O 混合溶剂中于 35℃反应 8h，以 96%的收率和 93%的对映体过量值生成光学活性的中间体 SA-12，SA-12 经后续多步转化生成科鲁普钩枝藤碱 A[81]。

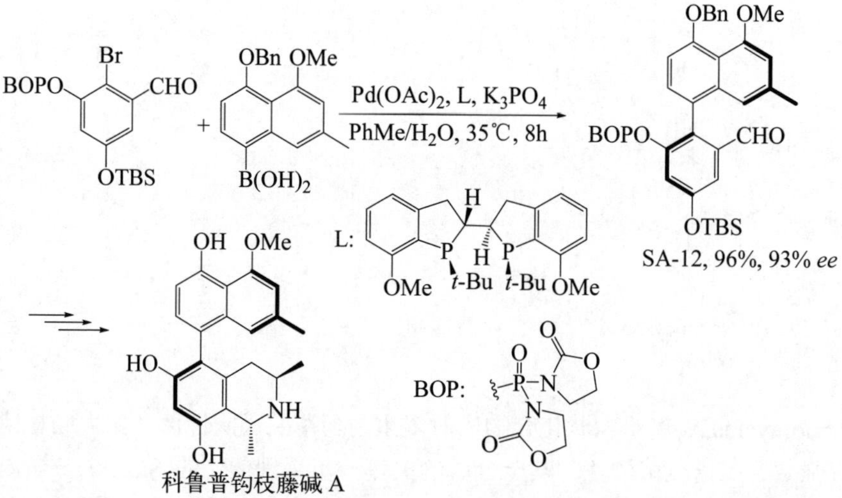

科鲁普钩枝藤碱 A

(86)

尼拉帕尼（niraparib）是一种聚 ADP 核糖聚合酶抑制剂，具有潜在的抗肿瘤活性，有望用于各种癌症，包括乳腺癌、卵巢癌、肺癌和前列腺癌的治疗。利用分子内不对称 Suzuki 反应为尼拉帕尼的合成提供了一条便捷的途径。首先将芳基溴转化为芳基硼酸片呐醇酯，然后该硼酸酯与 Boc-保护的 5-氯哌啶烯在 $[Rh(cod)(OH)]_2$/(*S*)-BINAP 催化下交叉偶联以 94%的收率和 98%的对映体过量值生成关键的光学活性中间体 SA-13，SA-13 在 Wilkinson 催化下几乎定量地将哌啶烯催化加氢转化为下一个关键中间体[82]；该中间体按文献方法进一步转化为尼拉帕尼[83]。

(87)

参考文献

[1] Miyaura N, Yamada K, Suzuki A. A new stereospecific cross-coupling by the palladium-catalyzed reaction of 1-alkenylboranes with 1-alkenyl or 1-alkynyl halides[J]. Tetrahedron Letters, 1979, 20(36): 3437-3440.

[2] Rosen B M, Quasdorf K W, Wilson D A, et al. Nickel-catalyzed cross-couplings involving carbon–oxygen bonds[J]. Chemical Reviews, 2011, 111(3): 1346-1416.

[3] Yu D G, Li B J, Shi Z J. Exploration of new C-O electrophiles in cross-coupling reactions[J]. Accounts of Chemical Rresearch, 2010, 43(12): 1486-1495.

[4] Rocard L, Hudhomme P. Recent developments in the Suzuki-Miyaura reaction using nitroarenes as electrophilic coupling reagents[J]. Catalysts, 2019, 9(3): 213.

[5] Xu G, Fu W, Liu G, et al. Efficient syntheses of korupensamines A, B and michellamine B by asymmetric Suzuki-Miyaura coupling reactions[J]. Journal of the American Chemical Society, 2014, 136(2): 570-573.

[6] Pourjavadi A, Motamedi A, Marvdashti Z, et al. Magnetic nanocomposite based on functionalized salep as a green support for immobilization of palladium nanoparticles: reusable heterogeneous catalyst for Suzuki coupling reactions[J]. Catalysis Communications, 2017, 97: 27-31.

[7] Mondal P, Bhanja P, Khatun R, et al. Palladium nanoparticles embedded on mesoporous TiO_2 material

(Pd@ MTiO$_2$) as an efficient heterogeneous catalyst for Suzuki-coupling reactions in water medium[J]. Journal of Colloid and Interface Science, 2017, 508: 378-386.

[8] Beletskaya I P, Alonso F, Tyurin V. The Suzuki-Miyaura reaction after the Nobel prize[J]. Coordination Chemistry Reviews, 2019, 385: 137-173.

[9] Gujral S S, Khatri S, Riyal P, et al. Suzuki cross coupling reaction-a review[J]. Indo Global Journal of Pharmaceutical Sciences, 2012, 2(4): 351-367.

[10] Amatore C, Jutand A, Le Duc G. Kinetic data for the transmetalation/reductive elimination in palladium-catalyzed Suzuki-Miyaura reactions: unexpected triple role of hydroxide ions used as base[J]. Chemistry: A European Journal, 2011, 17(8): 2492-2503.

[11] Chemler S R, Trauner D, Danishefsky S J. The B-alkyl Suzuki-Miyaura cross-coupling reaction: development, mechanistic study, and applications in natural product synthesis[J]. Angewandte Chemie International Edition, 2001, 40(24): 4544-4568.

[12] Miyaura N, Suzuki A. Stereoselective synthesis of arylated (*E*)-alkenes by the reaction of alk-1-enylboranes with aryl halides in the presence of palladium catalyst[J]. Journal of the Chemical Society, Chemical Communications, 1979 (19): 866-867.

[13] Suzuki A. Recent advances in the cross-coupling reactions of organoboron derivatives with organic electrophiles, 1995—1998[J]. Journal of Organometallic Chemistry, 1999, 576(1-2): 147-168.

[14] Chang C K, Bag N. Phenylpyrroles by suzuki cross coupling and a synthesis of type I tetramethyltetraphenylporphyrin[J]. The Journal of Organic Chemistry, 1995, 60(21): 7030-7032.

[15] Zhang H, Chan K S. Base effect on the cross-coupling of bulky arylboronic acid with halopyridines[J]. Tetrahedron Letters, 1996, 37(7): 1043-1044.

[16] Toyota S, Woods C R, Benaglia M, et al. Synthesis of unsymmetrical 2,8-and 2,9-dihalo-1,10-phenanthrolines and derivatives[J]. Tetrahedron Letters, 1998, 39(18): 2697-2700.

[17] DAlessio R, Rossi A. Short synthesis of undecylprodigiosine. A new route to 2,2'-bipyrrolyl-pyrromethene systems[J]. Synlett, 1996, (6): 513-514.

[18] Beller M, Fischer H, Herrmann W A, et al. Palladacycles as efficient catalysts for aryl coupling reactions[J]. Angewandte Chemie International Edition, 1995, 34(17): 1848-1849.

[19] Beller M, Krauter J G E, Zapf A. Carbohydrate-substituted triarylphosphanes: a new class of ligands for two-phase catalysis[J]. Angewandte Chemie International Edition, 1997, 36(7): 772-774.

[20] Bruno N C, Tudge M T, Buchwald S L. Design and preparation of new palladium precatalysts for C-C and C-N cross-coupling reactions[J]. Chemical Science, 2013, 4(3): 916-920.

[21] Beletskaya I P, Alonso F, Tyurin V. The Suzuki-Miyaura reaction after the Nobel prize[J]. Coordination Chemistry Reviews, 2019, 385: 137-173.

[22] Yamamoto S, Kinoshita H, Hashimoto H, et al. Facile preparation of Pd nanoparticles supported on single-layer graphene oxide and application for the Suzuki-Miyaura cross-coupling reaction[J]. Nanoscale, 2014, 6(12): 6501-6505.

[23] Han F S. Transition-metal-catalyzed Suzuki-Miyaura cross-coupling reactions: a remarkable advance from palladium to nickel catalysts[J]. Chemical Society Reviews, 2013, 42(12): 5270-5298.

[24] Bonin H, Fouquet E, Felpin F X. Aryl diazonium versus iodonium salts: preparation, applications and mechanisms for the Suzuki-Miyaura cross-coupling reaction[J]. Advanced Synthesis & Catalysis, 2011, 353(17): 3063-3084.

[25] Kuethe J T, Childers K G. Suzuki-Miyaura cross-coupling of 2-nitroarenediazonium tetrafluoroborates:

synthesis of unsymmetrical 2-nitrobiphenyls and highly functionalized carbazoles[J]. Advanced Synthesis & Catalysis, 2008, 350(10): 1577-1586.

[26] Molander G A, Canturk B. Organotrifluoroborates and monocoordinated palladium complexes as catalysts-a perfect combination for Suzuki-Miyaura coupling[J]. Angewandte Chemie International Edition, 2009, 48(49): 9240-9261.

[27] Skaff O, Jolliffe K A, Hutton C A. Synthesis of the side chain cross-linked tyrosine oligomers dityrosine, trityrosine, and pulcherosine[J]. The Journal of Organic Chemistry, 2005, 70(18): 7353-7363.

[28] Yue X, Qing F, Sun H, et al. A Suzuki coupling approach to double bonds locked analogues of strobilurin A[J]. Tetrahedron Letters, 1996, 37(45): 8213-8216.

[29] Boland G M, Dannelly D M X, Finet J P, et al. Synthesis of neoflavones by Suzuki arylation of 4-substituted coumarins[J]. Journal of the Chemical Society, Perkin Transactions 1, 1996, (21): 2591-2597.

[30] Moreno-Manas M, Pajuelo F, Pleixats R. Preparation of 1, 3-diarylpropenes by phosphine-free palladium(0)-catalyzed Suzuki-type coupling of allyl bromides with arylboronic acids[J]. The Journal of Organic Chemistry, 1995, 60(8): 2396-2397.

[31] Blangetti M, Rosso H, Prandi C, et al. Suzuki-Miyaura cross-coupling in acylation reactions, scope and recent developments[J]. Molecules, 2013, 18(1): 1188-1213.

[32] Urawa Y, Ogura K. A convenient method for preparing aromatic ketones from acyl chlorides and arylboronic acids via Suzuki-Miyaura type coupling reaction[J]. Tetrahedron Letters, 2003, 44(2): 271-273.

[33] Duan Y Z, Deng M Z. Palladium-catalyzed cross-coupling reaction of arylboronic acids with chloroformate or carbamoyl chloride[J]. Synlett, 2005, 2005, (02): 355-357.

[34] Gooßen L J, Ghosh K. Palladium-catalyzed synthesis of aryl ketones from boronic acids and carboxylic acids or anhydrides[J]. Angewandte Chemie International Edition, 2001, 40(18): 3458-3460.

[35] Buchspies J, Szostak M. Recent advances in acyl Suzuki cross-coupling[J]. Catalysts, 2019, 9(1): 53.

[36] Garcia-Barrantes P M, McGowan K, Ingram S W, et al. One pot synthesis of unsymmetrical ketones from carboxylic and boronic acids via PyClU-mediated acylative Suzuki coupling[J]. Tetrahedron Letters, 2017, 58(9): 898-901.

[37] Soderquist J A, Matos K, Rane A, et al. Alkynylboranes in the Suzuki-Miyaura coupling[J]. Tetrahedron Letters, 1995, 36(14): 2401-2402.

[38] Castanet A S, Colobert F, Schlama T. Suzuki-Miyaura coupling of alkynylboronic esters generated in situ from acetylenic derivatives[J]. Organic Letters, 2000, 2(23): 3559-3561.

[39] Suzuki A. Organoborane coupling reactions (Suzuki coupling)[J]. Proceedings of the Japan Academy, Series B, 2004, 80(8): 359-371.

[40] Kamei T, Itami K, Yoshida J. Catalytic carbometalation/cross-coupling sequence across alkynyl (2-pyridyl) silanes leading to a diversity-oriented synthesis of tamoxifen-type tetrasubstituted olefins[J]. Advanced Synthesis & Catalysis, 2004, 346(13-15): 1824-1835.

[41] Torrado A, Lopez S, Alvarez R, et al. General synthesis of retinoids and arotinoids via palladium-catalyzed cross-coupling of boronic acids with electrophiles[J]. Synthesis, 1995, (3): 285-293.

[42] Miyaura N, Suzuki A. Palladium-catalyzed cross-coupling reactions of organoboron compounds[J]. Chemical Reviews, 1995, 95(7): 2457-2483.

[43] Miyaura N, Suzuki A. Palladium-catalyzed reaction of 1-alkenylboronates with vinylic halides: (1Z, 3E)-1-phenyl-1, 3-octadiene[benzene, 1,3-octadienyl-, (Z, E)-][J]. Organic Syntheses, 2003, 68: 130-130.

[44] Miyaura N, Suginome H, Suzuki A. A stereospecific synthesis of conjugated (*E*, *Z*)-and (*Z*, *Z*)-alikadienes by a palladium-catalyzed cross-coupling reaction of 1-alkenylboranes with 1-alkenyl bromides[J]. Tetrahedron Letters, 1981, 22(2): 127-130.

[45] Uenishi J, Kawahama R, Yonemitsu O, et al. Palladium-catalyzed stereoselective hydrogenolysis of conjugated 1, 1-dibromo-1-alkenes to (*Z*)-1-bromo-1-alkenes. An application to stepwise and one-pot synthesis of enediynes and dienynes[J]. The Journal of Organic Chemistry, 1996, 61(17): 5716-5717.

[46] Torrado A, Iglesias B, López S, et al. The Suzuki reaction in stereocontrolled polyene synthesis: retinol (vitamin A), its 9-and/or 13-demethyl analogs, and related 9-demethyl-dihydroretinoids[J]. Tetrahedron, 1995, 51(8): 2435-2454.

[47] Wiberg K B. Bent bonds in organic compounds[J]. Accounts of Chemical Research, 1996, 29(5): 229-234.

[48] Charette A B, Giroux A. Palladium-catalyzed suzuki-type cross-couplings of iodocyclopropanes with boronic acids: synthesis of trans-1,2-dicyclopropyl alkenes[J]. The Journal of Organic Chemistry, 1996, 61(25): 8718-8719.

[49] Zou G, Reddy Y K, Falck J R. Ag (Ⅰ)-promoted Suzuki-Miyaura cross-couplings of *n*-alkylboronic acids[J]. Tetrahedron Letters, 2001, 42(41): 7213-7215.

[50] Chemler S R, Trauner D, Danishefsky S J. The *B*-alkyl Suzuki-Miyaura cross-coupling reaction: development, mechanistic study, and applications in natural product synthesis[J]. Angewandte Chemie International Edition, 2001, 40(24): 4544-4568.

[51] Old D W, Wolfe J P, Buchwald S L. A highly active catalyst for palladium-catalyzed cross-coupling reactions: room-temperature Suzuki couplings and amination of unactivated aryl chlorides[J]. Journal of the American Chemical Society, 1998, 120(37): 9722-9723.

[52] Wolfe J P, Singer R A, Yang B H, et al. Highly active palladium catalysts for Suzuki coupling reactions[J]. Journal of the American Chemical Society, 1999, 121(41): 9550-9561.

[53] Doucet H. Suzuki-Miyaura cross-coupling reactions of alkylboronic acid derivatives or alkyltrifluoroborates with aryl, alkenyl or alkyl halides and triflates[J]. European Journal of Organic Chemistry, 2008, (12): 2013-2030.

[54] Johnson C R, Braun M P. A two-step, three-component synthesis of PGE1: utilization of. alpha.-iodo enones in Pd(0)-catalyzed cross-couplings of organoboranes[J]. Journal of the American Chemical Society, 1993, 115(23): 11014-11015.

[55] Ishiyama T, Miyaura N, Suzuki A. Palladium-catalyzed carbonylative cross-coupling reaction of 1-halo-1-alkenes with 9-alkyl-9-BBN derivatives. A direct synthesis of α, β-unsaturated ketones[J]. Bulletin of the Chemical Society of Japan, 1991, 64(6): 1999-2001.

[56] Ishiyama T, Miyaura N, Suzuki A. Palladium-catalyzed carbonylative cross-coupling reaction of iodoalkanes with 9-alkyl-9-BBN derivatives. A direct and selective synthesis of ketones[J]. Tetrahedron Letters, 1991, 32(47): 6923-6926.

[57] Miyaura N, Ishiyama T, Sasaki H, et al. Palladium-catalyzed inter-and intramolecular cross-coupling reactions of *B*-alkyl-9-borabicyclo [3.3.1] nonane derivatives with 1-halo-1-alkenes or haloarenes. Syntheses of functionalized alkenes, arenes, and cycloalkenes via a hydroboration-coupling sequence[J]. Journal of the American Chemical Society, 1989, 111(1): 314-321.

[58] Ohe T, Miyaura N, Suzuki A. Palladium-catalyzed cross-coupling reaction of organoboron compounds with organic triflates[J]. The Journal of Organic Chemistry, 1993, 58(8): 2201-2208.

[59] Chemler S R, Danishefsky S J. Transannular macrocyclization via intramolecular B-alkyl Suzuki reaction[J]. Organic Letters, 2000, 2(17): 2695-2698.

[60] Kawada H, Iwamoto M, Utsugi M, et al. Synthetic studies on the taxane skeleton: construction of eight-membered carbocyclic rings by the intramolecular B-alkyl Suzuki-Miyaura cross-coupling reaction[J]. Organic Letters, 2004, 6(24): 4491-4494.

[61] Zhang D, Wang Q. Palladium catalyzed asymmetric Suzuki-Miyaura coupling reactions to axially chiral biaryl compounds: chiral ligands and recent advances[J]. Coordination Chemistry Reviews, 2015, 286: 1-16.

[62] Genov M, Almorín A, Espinet P. Efficient synthesis of chiral 1, 1'-binaphthalenes by the asymmetric Suzuki-Miyaura reaction: dramatic synthetic improvement by simple purification of naphthylboronic acids[J]. Chemistry: A European Journal, 2006, 12(36): 9346-9352.

[63] Yin J, Buchwald S L. A catalytic asymmetric Suzuki coupling for the synthesis of axially chiral biaryl compounds[J]. Journal of the American Chemical Society, 2000, 122(48): 12051-12052.

[64] Shen X, Jones G O, Watson D A, et al. Enantioselective synthesis of axially chiral biaryls by the Pd-catalyzed Suzuki-Miyaura reaction: substrate scope and quantum mechanical investigations[J]. Journal of the American Chemical Society, 2010, 132(32): 11278-11287.

[65] Zhou Y, Wang S, Wu W, et al. Enantioselective synthesis of axially chiral multifunctionalized biaryls via asymmetric Suzuki-Miyaura coupling[J]. Organic Letters, 2013, 15(21): 5508-5511.

[66] Lundin P M, Fu G C. Asymmetric Suzuki cross-couplings of activated secondary alkyl electrophiles: arylations of racemic α-chloroamides[J]. Journal of the American Chemical Society, 2010, 132(32): 11027-11029.

[67] Schäfer P, Palacin T, Sidera M, et al. Asymmetric Suzuki-Miyaura coupling of heterocycles via rhodium-catalysed allylic arylation of racemates[J]. Nature Communications, 2017, 8(1): 15762.

[68] Magano J, Dunetz J R. Large-scale applications of transition metal-catalyzed couplings for the synthesis of pharmaceuticals[J]. Chemical Reviews, 2011, 111(3): 2177-2250.

[69] Taheri Kal Koshvandi A, Heravi M M, Momeni T. Current applications of Suzuki-Miyaura coupling reaction in the total synthesis of natural products: an update[J]. Applied Organometallic Chemistry, 2018, 32(3): e4210.

[70] Jiang X, Lee G T, Villhauer E B, et al. A scalable synthesis of a 1, 7-naphthyridine derivative, a PDE-4 inhibitor[J]. Organic Process Research & Development, 2010, 14(4): 883-889.

[71] Itoh T, Kato S, Nonoyama N, et al. Efficient synthesis of a highly selective NPY-5 receptor antagonist: a drug candidate for the treatment of obesity[J]. Organic Process Research & Development, 2006, 10(4): 822-828.

[72] Magnus N A, Aikins J A, Cronin J S, et al. Diastereomeric salt resolution based synthesis of LY503430, an AMPA (α-amino-3-hydroxy-5-methyl-4-isoxazolepropionic acid) potentiator[J]. Organic Process Research & Development, 2005, 9(5): 621-628.

[73] Zhao D, Xu F, Chen C, et al. Efficient syntheses of 2-(3',5'-difluorophenyl)-3-(4'-methylsulfonylphenyl) cyclopent-2-enone, a potent COX-2 inhibitor[J]. Tetrahedron, 1999, 55(19): 6001-6018.

[74] Keen S P, Cowden C J, Bishop B C, et al. Practical asymmetric synthesis of a non-peptidic $\alpha_v\beta_3$ antagonist[J]. The Journal of Organic Chemistry, 2005, 70(5): 1771-1779.

[75] Derdau V, Oekonomopulos R, Schubert G. ^{14}C-Labeled and large-scale synthesis of the angiotensin-(1−7)-receptor agonist AVE 0991 by cross-coupling reactions[J]. The Journal of Organic Chemistry,

2003, 68(13): 5168-5173.

[76] Marshall J A, Johns B A. Total synthesis of (+)-discodermolide[J]. The Journal of Organic Chemistry, 1998, 63(22): 7885-7892.

[77] Martin A D, Siamaki A R, Belecki K, et al. A convergent approach to the total synthesis of telmisartan via a suzuki cross-coupling reaction between two functionalized benzimidazoles[J]. The Journal of Organic Chemistry, 2015, 80(3): 1915-1919.

[78] Tang W, Liu S, Degen D, et al. Synthesis and evaluation of novel analogues of ripostatins[J]. Chemistry: A European Journal, 2014, 20(38): 12310-12319.

[79] Liao L, Zhou J, Xu Z, et al. Concise total synthesis of nannocystin A[J]. Angewandte Chemie, 2016, 55(42): 13263-13266.

[80] Dufour J, Neuville L, Zhu J. Intramolecular Suzuki-Miyaura reaction for the total synthesis of signal peptidase inhibitors, arylomycins A_2 and B_2[J]. Chemistry: A European Journal, 2010, 16(34): 10523-10534.

[81] Xu G, Fu W, Liu G, et al. Efficient syntheses of korupensamines A, B and michellamine B by asymmetric Suzuki-Miyaura coupling reactions[J]. Journal of the American Chemical Society, 2014, 136(2): 570-573.

[82] Schäfer P, Palacin T, Sidera M, et al. Asymmetric Suzuki-Miyaura coupling of heterocycles via rhodium-catalysed allylic arylation of racemates[J]. Nature Communications, 2017, 8(1): 15762.

[83] Chung C K, Bulger P G, Kosjek B, et al. Process development of C-N cross-coupling and enantioselective biocatalytic reactions for the asymmetric synthesis of niraparib[J]. Organic Process Research & Development, 2014, 18(1): 215-227.

第 3 章
Negishi 反应及其在药物合成上的应用

3.1 Negishi 反应及其机理

3.1.1 Negishi 反应

Negishi 反应或偶联是指在过渡金属钯或镍催化下有机卤代物或三氟甲磺酸酯与有机锌试剂交叉偶联生成碳-碳键的反应。除了利用有机锌试剂直接进行反应外，很多金属有机化合物（如铝、锆等）在锌盐（如 $ZnBr_2$、$ZnCl_2$ 等）存在下与卤代烃的交叉偶联反应也归为 Negishi 反应范畴。狭义的 Negishi 反应可由图 3-1 表示。

$$R-X + R^1-ZnX' \xrightarrow{ML_n} R-R^1$$

X=Cl⁻、Br⁻、I⁻、$CF_3SO_3^-$、AcO⁻；R=烯基、芳基、烯丙基、炔基、炔丙基；
X'=Cl⁻、Br⁻、I⁻；R^1=芳基、烯丙基、烷基、苄基、高烯丙基、高炔丙基；
M=Pd、Ni；L = PPh_3、dppe、BINAP，等等

图 3-1 Negishi 反应及范围

在 Negishi 反应中，钯催化剂比镍催化剂具有更好的催化活性和官能团耐受性。Negishi 反应在全合成中用于复杂中间体间碳-碳键的构建，可使 sp^3、sp^2 和 sp 碳原子间发生偶联，这在其他钯催化的偶联反应中是不多见的。反应中的有机锌对水和氧气敏感，因此与其他偶联反应如 Suzuki 反应相比应用范围较窄。然而，有机锌与有机硼烷或硼酸酯相比活性更高，因而反应时间亦相应较短。Negishi 反应源于 1976 年 Negishi 与其同事利用钯、镍催化剂催化的有机铝试剂的交叉偶联反应。他们发现当用镍作催化剂时发生立体特异性的衰减，而钯作催化剂则没有这种现象发生[1]。因此，他们将注意力由有机铝试剂转向有机锌试剂，广泛开展了钯配合物催化的有机锌试剂的交叉偶联反应并发展出目前常用的反应条件[2]。为此，Negishi 与

Heck 和 Suzuki 一道获得了 2010 年诺贝尔化学奖，彰显他们所发展的钯催化交叉偶联反应在有机合成中的重要作用。

3.1.2 反应机理

无论是钯催化还是镍催化的 Negishi 交叉偶联反应，人们普遍认为由三个关键步骤构成，即氧化加成、金属转移以及还原消除。以钯催化的 Negishi 反应为例，其催化循环如图 3-2 所示：第一步氧化加成，是有机卤代物或三氟甲磺酸酯氧化加成到零价的钯物种；第二步，亲核试剂的亲核碳进攻钯配合物发生金属转移；第三步，与钯配位的两个有机基团从钯上消除形成碳-碳键实现偶联，并再生零价钯活性物种完成催化循环。关于有机卤代物或三氟甲磺酸酯 R—X 氧化加成反应的顺序一般为 I＞OTf＞Br＞＞Cl；就有机锌而言，有机锌卤化物和二有机锌都可作为起始原料，在金属转移过程中前者生成顺式加合物($\overset{R\ \ L}{\underset{R'\ \ L}{Pd}}$)导致下一步还原消除迅速，而后者生成反式加合物（$\overset{L}{\underset{L}{R-Pd-R'}}$）使下一步还原消除变慢[3]。

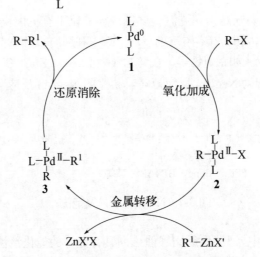

图 3-2 钯催化 Negishi 反应的催化循环

3.2 基本 Negishi 反应

Negishi 反应与交叉偶联底物的结构、催化剂中金属的种类和配体结构有关，因而所涉及的内容范围广泛。通过改变交叉偶联底物的结构并进行组合可以派生出多种类型的反应；而通过调变催化剂金属的种类以及与之相适应的反应条件可以最大限度地改进反应，而且通过手性配体还可诱导不对称的 Negishi 反应。以下将按不同结构特征的交叉偶联底物的组合介绍基本的 Negishi 的反应。

3.2.1　C(sp^2)-C(sp^2)键的构建

3.2.1.1　芳基-芳基偶联

芳基与芳基的偶联属于 Negishi 反应中 C(sp^2)-C(sp^2)键构建反应之一，可用于合成不同芳基偶联的化合物。例如，邻甲基苯基氯化锌与对溴硝基苯在 Pd(PPh$_3$)$_4$ 催化下于室温反应就能以良好的收率生成 2-甲基-4'-硝基联苯[2]。

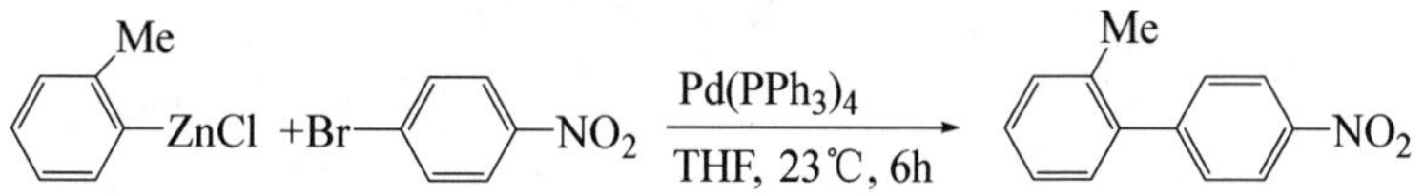

$$\text{(1)}$$

芳基锌化合物可以由格氏试剂与氯化锌原位生成，使反应容易实现，而且使用廉价的镍催化剂在合成上更有应用潜力[4]。

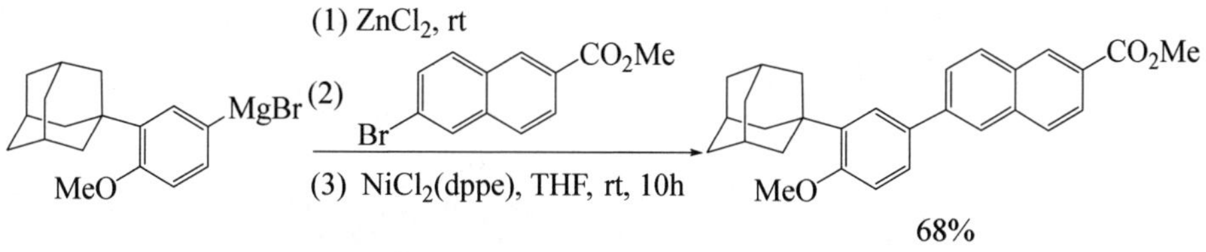

$$\text{(2)}$$

芳杂环之间也能通过镍催化的 Negishi 反应实现交叉偶联。例如 N-甲基吡咯基溴化锌与 3-溴吡啶在四氢呋喃与 N-乙基吡咯烷酮（NEP）混合溶剂中以 4-二甲氨基吡啶（DMAP）和亚磷酸二乙酯[(EtO)$_2$P(O)H]为配体的条件下顺利反应得到相应的交叉偶联反应产物[5]。

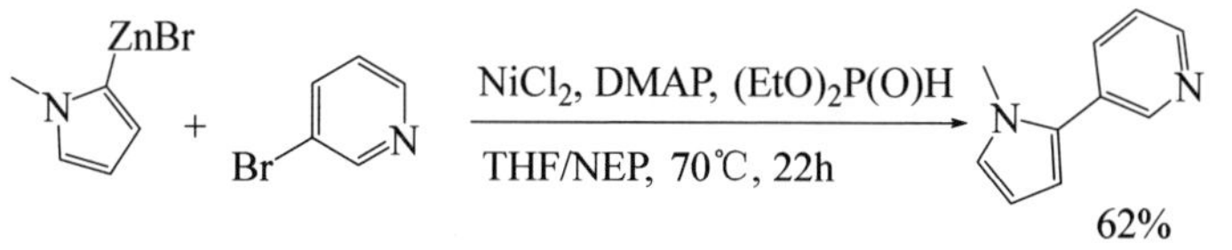

$$\text{(3)}$$

3.2.1.2　烯基-芳基和芳基-烯基偶联

烯基-芳基和芳基-烯基偶联也属于 C(sp^2)-C(sp^2)键构建反应。这两种交叉偶联反应均用于合成芳基取代的烯烃或苯乙烯衍生物，反应容易进行。其中正辛烯基氯化锌（烯基作为亲核试剂）与碘代苯在 Pd(PPh$_3$)$_4$ 催化下几乎定量地生成交叉偶联产物，而且烯烃的 E-构型在反应中得到保持[6]。

$$n\text{-}C_6H_{13}\text{-}CH=CH\text{-}ZnCl + IPh \xrightarrow[\text{THF, rt, 17min}]{\text{Pd(PPh}_3)_4} n\text{-}C_6H_{13}\text{-}CH=CH\text{-}Ph \quad \text{约100\%} \quad (4)$$

反过来，芳基卤化锌（芳基作为亲核试剂）与卤代烯烃同样能进行交叉偶联。如果卤代烯烃是 E-式构型，反应后烯烃的 E-构型也得到保持[5]。另外，相同条件下卤代芳烯比卤代脂肪烯更容易反应。

$$\text{MeO-C}_6\text{H}_4\text{-ZnBr} + n\text{-}C_6H_{13}\text{-CH=CH-Br} \xrightarrow[\text{25°C, 0.5h}]{\substack{\text{Ni(acac)}_2\text{, THF/NEP}\\ \text{(EtO)}_2\text{P(O)H, DMAP}}} \text{MeO-C}_6\text{H}_4\text{-CH=CH-}n\text{-}C_6H_{13} \quad 79\% \quad (5)$$

$$\text{MeO-C}_6\text{H}_4\text{-ZnBr} + \text{Ph-CH=CH-Br} \xrightarrow[\text{25°C, 0.25h}]{\substack{\text{Ni(acac)}_2\text{, THF/NEP}\\ \text{(EtO)}_2\text{P(O)H, DMAP}}} \text{MeO-C}_6\text{H}_4\text{-CH=CH-Ph} \quad 85\%$$

如果卤代烯是 Z-式构型，交叉偶联反应也能进行，但通常情况下 Z-式构型不能保持，产物烯烃的构型转化为 E-式[7]。

$$\text{Ph-CH=CH-Br} + \text{PhZnI} \xrightarrow[\text{THF, rt, 24h}]{\text{PdCl}_2\text{(Amphos)}_2} \text{Ph-CH=CH-Ph} \quad 94\% \quad (6)$$
$$E/Z=87/13$$

Amphos: (CH$_3$)$_2$N-C$_6$H$_4$-P(t-Bu)$_2$

在有机合成过程中是以芳基锌卤代物为亲核试剂还是以烯基锌卤代物为亲核试剂决定于具体反应中哪一个更容易制备。

3.2.1.3 烯基-烯基偶联

钯催化的 Negishi 烯基-烯基偶联属另一种构建 C(sp^2)-C(sp^2)键的反应，是合成共轭烯烃或低聚烯烃的最有效手段之一，反应在经典的 Negishi 反应条件下顺利进行。例如，正辛烯基氯化锌与 1-碘己烯在 Pd(PPh$_3$)$_4$ 催化下以优异的收率生成共轭二烯[6]。

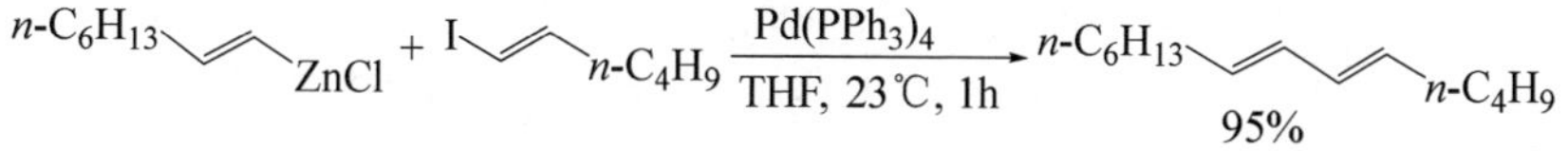

$$n\text{-}C_6H_{13}\diagup\!\!\!\diagup ZnCl + I\diagup\!\!\!\diagup n\text{-}C_4H_9 \xrightarrow[\text{THF, 23℃, 1h}]{Pd(PPh_3)_4} n\text{-}C_6H_{13}\diagup\!\!\!\diagup\!\!\!\diagup\!\!\!\diagup n\text{-}C_4H_9 \quad (7)$$
$$95\%$$

利用共轭二烯烃溴化锌与碘代烯反应生成共轭三烯，而且反应发生在内烯烃上，反应后两个交叉偶联底物烯烃的构型均得到保持[8]。

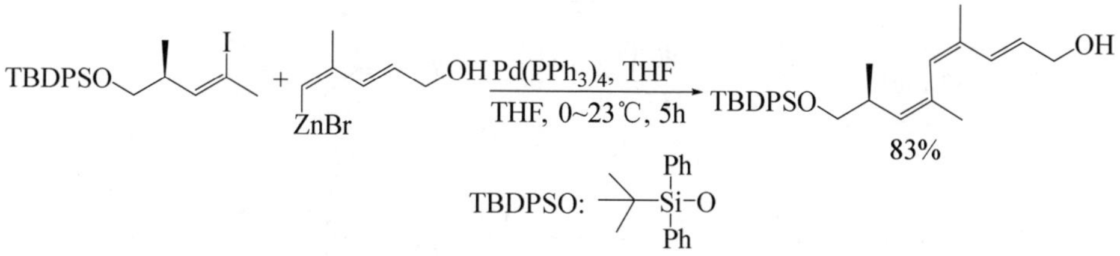

(8) 式反应图

与上面反应类似，还可以合成共轭体系更大的多烯化合物。在下面反应中，反应所需的烯基碘化锌由相应的碘代烯与丁基锂和溴化锌反应原位生成，反应生成四个羟基全保护的交叉偶联产物共轭五烯，然后不经分离直接用四丁基氟化铵(TBAF)选择性脱除端羟基上的 TBS 保护基，反应总收率达 61%，而且所有烯烃单元的构型得到保持[9]。

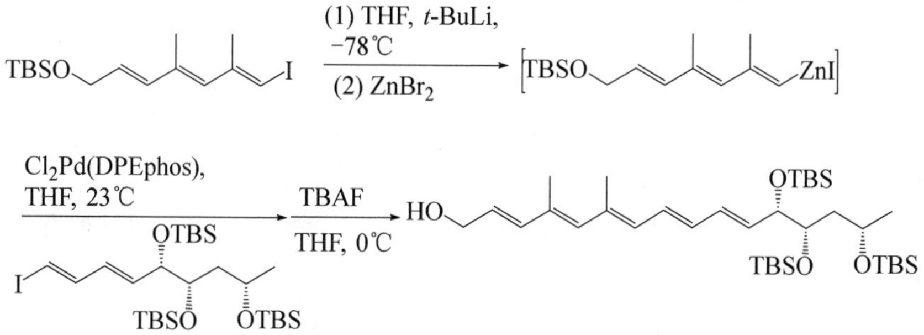

(9) 式反应图

3.2.2　C(sp)–C(sp²)键的构建

3.2.2.1　炔基-芳基偶联

炔基氯化锌或炔基溴化锌在标准的 Negighi 催化反应体系下能够与芳基碘 [反应式（10）～式（14）] 或芳基溴 [反应式（15）] 发生交叉偶联得到芳基炔，芳基可以是取代的苯基 [反应式（11）] 或杂芳基 [反应式（10），式（12）～式（15）][10,11]。富电子的杂芳基比缺电子的杂芳基更容易与炔基卤化锌发生反应 [反应式（12），式（14）]。

$$CH_3C\equiv CZnCl + \underset{S}{\text{[thiophene]}}-I \xrightarrow[\text{rt, 1h}]{Pd(PPh_3)_4, THF} \underset{S}{\text{[thiophene]}}-C\equiv CCH_3 \quad 92\% \qquad (10)$$

$$n\text{-}C_5H_{11}C\equiv CZnCl + NC\text{-}\underset{}{\text{[C}_6H_4\text{]}}-I \xrightarrow[\text{rt, 4h}]{Pd(PPh_3)_4, THF} NC\text{-}\underset{}{\text{[C}_6H_4\text{]}}-C\equiv CC_5H_{11}\text{-}n \quad 93\% \qquad (11)$$

$$HC\equiv CZnBr + \underset{Me}{\text{[N-Me-pyrrole]}}-I \xrightarrow[\text{23°C, 4h}]{Pd(PPh_3)_4, THF} HC\equiv C\text{-}\underset{Me}{\text{[N-Me-pyrrole]}} \quad 76\% \qquad (12)$$

$$HC\equiv CZnBr + \underset{O}{\text{[benzofuran]}}-I \xrightarrow[\text{23°C, 4h}]{Pd(PPh_3)_4, THF} HC\equiv C\text{-}\underset{O}{\text{[benzofuran]}} \quad 78\% \qquad (13)$$

$$EtO_2CC\equiv CZnCl + \underset{N}{\text{[pyridine]}}-I \xrightarrow[\text{己烷, 50°C, 2h}]{Pd(PPh_3)_4, THF} EtO_2CC\equiv C\text{-}\underset{N}{\text{[pyridine]}} \quad 38\% \qquad (14)$$

$$HC\equiv CZnBr + \underset{N}{\text{[isoquinoline]}}\text{-}Br \xrightarrow[\text{50~55°C, 24h}]{Pd(PPh_3)_4, THF/DMF} \underset{N}{\text{[isoquinoline]}}\text{-}C\equiv CH \quad 58\% \qquad (15)$$

3.2.2.2 炔基-烯基偶联

与芳基卤化物相比，卤代烯与炔基氯化锌的交叉偶联更容易，而且烯基的构型得到保持[11,12]。

$$PhC\equiv CZnCl + \underset{Cl}{\text{I-CH=CH}} \xrightarrow[\text{rt, 3h}]{Pd(PPh_3)_4, THF} Ph\text{-}C\equiv C\text{-CH=CH-Cl} \qquad (16)$$

$$n\text{-}C_5H_{11}C\equiv CZnCl + \underset{Bu\text{-}n}{\text{I-CH=CH}} \xrightarrow[\text{THF, 25°C}]{Pd(PPh_3)_4} n\text{-}C_5H_{11}C\equiv C\text{-CH=CH-}n\text{-Bu} \quad 82\% \qquad (17)$$

$$\underset{S}{\text{[thiophene]}}\text{-}C\equiv CZnCl + Br\text{-CH=CH-}SiMe_3 \xrightarrow[\text{0°C, 15h}]{Pd(PPh_3)_4, THF} \underset{S}{\text{[thiophene]}}\text{-}C\equiv C\text{-CH=CH-}SiMe_3 \quad 87\% \qquad (18)$$

炔基卤化锌与卤代烯的 Negighi 交叉偶联可以耐受多种官能团。例如，结构中同时含有环氧基、羰基的 α-碘代环己烯酮衍生物可以选择性地与炔基溴化锌进行

Negighi 反应生成相应的交叉偶联产物［反应式（19）］[13]；丁炔基氯化锌与 β-碘代丙烯酸甲酯缩合也得到相应的偶联产物［反应式（20）][14]；己炔基氯化锌甚至与含羧基的碘代烯发生交叉偶联生成烯炔酸，而且烯烃的构型得到保持［反应式（21）～式（22）][15]。

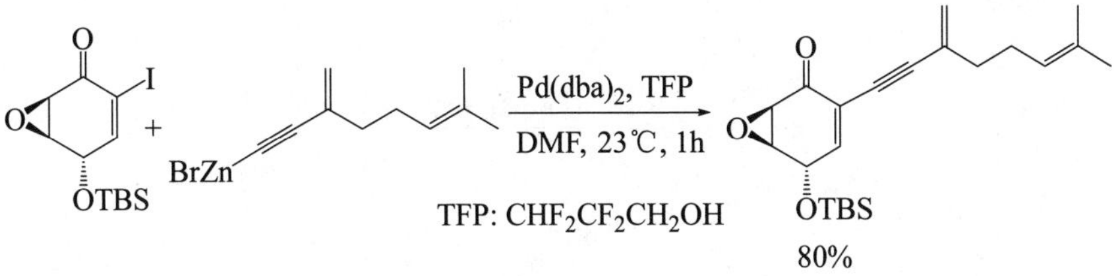

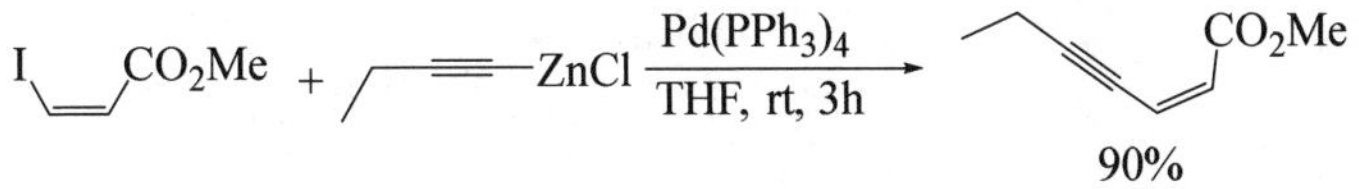

3.2.2.3 炔基-丙二烯基偶联

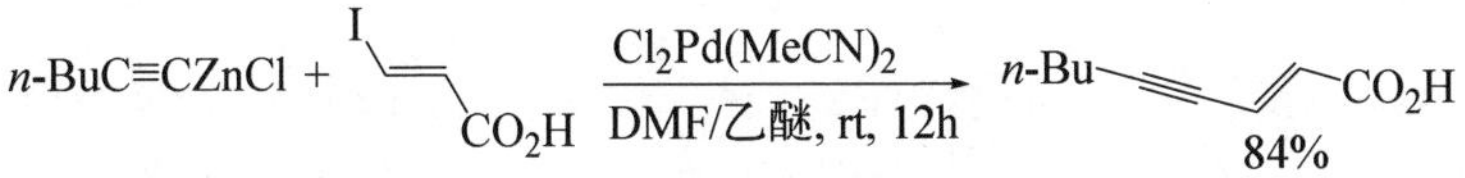

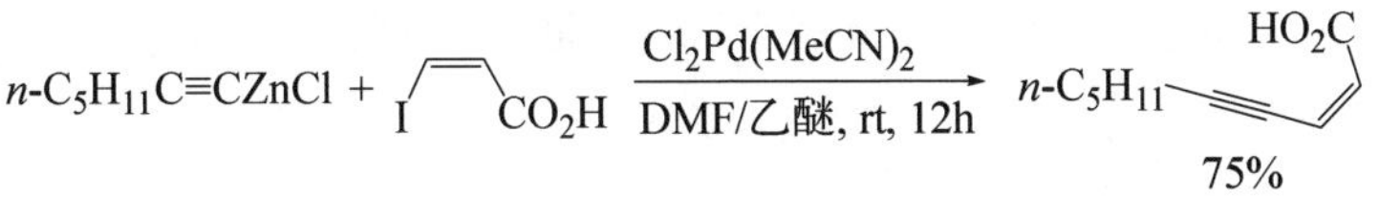

卤代丙二烯、卤代炔丙基与炔基氯化锌在钯催化下发生交叉偶联反应也可构建 C(sp)-C(sp²)键。理论上，当卤代丙二烯和卤代炔丙基转化为有机钯化合物时，有机钯化合物存在丙二烯-炔丙基间的动态平衡，因而在与炔基交叉偶联时既可生成丙二烯炔也可生成 1,4-二炔（图 3-3）。但在真正反应时全部或几乎全部生成丙二烯炔产物，因而此类交叉偶联反应提供了一条优异的合成丙二烯炔的途径[11]。

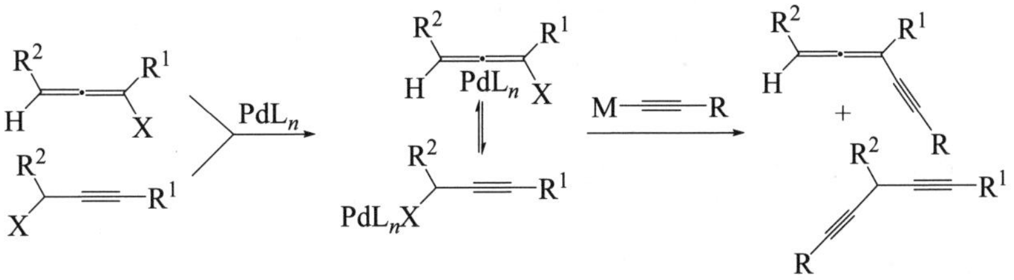

图 3-3 钯催化的丙二烯和炔丙基与金属炔基化合物的交叉偶联

在此类交叉偶联反应中，最适宜的金属炔化物是炔基卤化锌［反应式（23）～式（26）］；而炔丙基比丙二烯基更适合该偶联反应，除了炔丙基溴化物［反应式（25）］，甲磺酸酯［反应式（26）］等化合物也可作为反应底物进行反应[16]。

$$\text{CH}_2=\text{CBr} + \text{ClZn}-\!\!\!\equiv\!\!\!-\text{SiMe}_3 \xrightarrow[35℃, 0.5h]{Pd(PPh_3)_4, THF} \text{产物}\ 90\% \quad (23)$$

$$\text{Me}_2\text{C=CHBr} + \text{ClZn}-\!\!\!\equiv\!\!\!-\text{CH}_3 \xrightarrow[20℃, 1.5h]{Pd(PPh_3)_4, THF} \text{产物}\ 98\% \quad (24)$$

$$\text{BrCH}_2\text{C}\!\equiv\!\text{CH} + \text{ClZn}-\!\!\!\equiv\!\!\!-\text{Ph} \xrightarrow[20℃, 1.5h]{Pd(PPh_3)_4, THF} \text{产物}\ 95\% \quad (25)$$

$$n\text{-}C_5H_{11}\text{CH(OSO}_2\text{Me)C}\!\equiv\!\text{CH} + \text{ClZn}-\!\!\!\equiv\!\!\!-\text{H} \xrightarrow[20℃, 1.0h]{Pd(PPh_3)_4, THF} n\text{-}C_5H_{11}\text{产物}\ 95\% \quad (26)$$

3.2.2.4 炔基-酰基偶联

酰卤化合物也能与炔基卤化锌在钯催化下发生 Negishi 交叉偶联构建 C(sp)-C(sp³) 键，用于合成炔酮。由于羰基碳原子的高度亲电性以及作为离去基团的卤原子如氯容易离去，通常反应在温和条件下进行，而且各类酰基都能进行反应。以下反应均以良好到优异的收率得到相应的炔酮[17]。

$$n\text{-}C_5H_{11}\text{C}\!\equiv\!\text{CZnCl} + \text{CH}_3\text{-C(=O)-Cl} \xrightarrow[25℃, <6h]{Pd(PPh_3)_4, THF} n\text{-}C_5H_{11}\text{C}\!\equiv\!\text{C-C(=O)-CH}_3\ 97\% \quad (27)$$

$$n\text{-}C_5H_{11}\text{C}\!\equiv\!\text{CZnCl} + \text{CH}_3\text{CH=CHC(=O)Cl} \xrightarrow[25℃, <6h]{Pd(PPh_3)_4, THF} \text{产物}\ 89\% \quad (28)$$

$$\text{PhC}\!\equiv\!\text{CZnCl} + \text{CH}_2\!=\!\text{CHC(=O)Cl} \xrightarrow[10℃, 10min]{Pd(PPh_3)_4, THF} \text{产物}\ 56\% \quad (29)$$

$$n\text{-}C_5H_{11}\text{C}\!\equiv\!\text{CZnCl} + \text{ClC(=O)OCH}_3 \xrightarrow[10℃, <6h]{Pd(PPh_3)_4, THF} n\text{-}C_5H_{11}\text{C}\!\equiv\!\text{C-C(=O)-OCH}_3\ 72\% \quad (30)$$

3.2.3 C(sp³)-C(sp²)键的构建

3.2.3.1 芳基-烷基偶联

芳基卤代物与烷基锌有机化合物在钯或镍催化剂催化下进行交叉偶联反应生

成烷基芳香化合物。芳基卤代物包括取代的卤代苯和卤代杂芳基化合物，而烷基锌则可以是烷基卤化锌和二烷基锌。在此类反应中镍可以替代钯作为催化剂，有时甚至比钯具有更好的催化性能，而且对芳环上多种取代基有良好的耐受性[18]。

$$R \text{—} \underset{}{\underset{}{\text{Ar}}} \text{—I} + BrZn\text{—}CH_2CH_2\text{—}\underset{O}{\overset{O}{\underset{\diagdown}{\diagup}}} \xrightarrow[\text{THF, 23℃, ≤12h}]{\text{NiCl}_2(\text{Py})_4, \text{bpy}} R\text{—}\underset{}{\underset{}{\text{Ar}}}\text{—}CH_2CH_2\text{—}\underset{O}{\overset{O}{\underset{\diagdown}{\diagup}}} \quad (31)$$

R=4-MeO, 3.5h, 80%; 3-Me, 6h, 60%; 4-CF$_3$, 8h, 69%; 4-CO$_2$Me, 12h, 75%; 4-COMe, 3h, 66%; 4-CN, 3h, 83%; 3-Br, 2h, 75%; 2-COH, 8h, 52%; 2-CH$_2$CO$_2$H, 2h, 77%; 4-OTs, 9h, 65%

相同吸电子取代基时，反应的顺序是对位＞间位＞邻位；卤代吡啶不如卤代苯反应的结果好[19]。

$$NC\text{—}Ar\text{—}Br + \triangleright\text{—}ZnBr \xrightarrow[65℃, 18~24h]{\text{Pd(PPh}_3)_4, \text{THF}} NC\text{—}Ar\text{—}\triangleright \quad (32)$$

4-CN, 18h, 99%; 3-CN, 18h, 91%; 2-CN, 24h, 75%

$$\underset{N}{\text{Py}}\text{—}Br + \triangleright\text{—}ZnBr \xrightarrow[65℃, 24h]{\text{Pd(PPh}_3)_4, \text{THF}} \underset{N}{\text{Py}}\text{—}\triangleright \quad (33)$$

56%

在芳基卤代物与环丙基溴化锌交叉偶联反应中，环丙基可看作仲烷基，但环丙环上的三个碳原子是完全等同的，因而不涉及区域选择性问题。然而，当芳基卤代物与异丙基溴化锌反应时，由于反应过程中部分异丙基发生重排，因此除了生成芳基异丙烷外还生成部分芳基正丙烷，两个产物之间的比例与催化体系的配体有关[20]。

$$\underset{CN}{\text{Ar}}\text{—}Br + \underset{}{\diagdown}\text{ZnBr} \xrightarrow[\text{THF, rt, 0.5h}]{\text{Pd(OAc)}_2, \text{CPhos}} \underset{CN}{\text{Ar}}\text{—}iPr + \underset{CN}{\text{Ar}}\text{—}nPr \quad (34)$$

P1 　　　　P2
89%

CPhos: 2,2'-(Me$_2$N)$_2$-6,6'-PCy$_2$-biphenyl
XPhos: 2',4',6'-(i-Pr)$_3$-2-PCy$_2$-biphenyl

L	P1:P2
XPhos	25:75
CPhos	95:5

利用 Negishi 反应可以向多氮芳杂环如嘌呤骨架上引入各种烷基和苄基，表明该反应在药物合成上的潜在的巨大应用价值[21]。

$$\text{6-Cl-9-Bn-purine} + \text{R-ZnBr·LiCl} \xrightarrow[N_2, \text{rt}, 24h]{\text{Ni(acac)}_2, \text{THF}} \text{6-R-9-Bn-purine} \quad (35)$$

R=Bn, 98%; Me, 92%; n-C_5H_{11}, 80%;
环戊基, 71%; i-Pr, 40%

$$\text{6-Cl-9-(2',3',5'-tri-O-Ac-ribosyl)purine} + \text{PhCH}_2\text{ZnCl·LiCl} \xrightarrow[N_2, \text{rt}, 24h]{\text{Ni(acac)}_2, \text{THF}} \text{6-Bn-9-(2',3',5'-tri-O-Ac-ribosyl)purine} \quad (36)$$

3.2.3.2 烯基-烷基偶联

在芳基与烷基偶联相同或近似的反应条件下，烯基与烷基也能顺利交叉偶联，而且底物范围更广泛。无论是简单结构的卤代烯还是复杂结构的卤代烯都可以作为偶联底物进行反应；同样，无论是简单的烷基锌卤化物还是含官能团的烷基锌卤化物都可作为亲核试剂进行反应[22-24]。

$$\underset{\text{Me}}{\overset{n\text{-}C_5H_{11}}{>}}\!\!=\!\!\underset{\text{I}}{\overset{\text{H}}{<}} + \text{ClZnCH}_2\text{SiMe}_3 \xrightarrow{\text{Pd(PPh}_3)_4, \text{THF/Et}_2\text{O, rt}} \underset{\text{Me}}{\overset{n\text{-}C_5H_{11}}{>}}\!\!=\!\!\underset{\text{CH}_2\text{SiMe}_3}{\overset{\text{H}}{<}} \quad (37)$$
92%

$$\underset{\text{Me}}{\overset{n\text{-}C_5H_{11}}{>}}\!\!=\!\!\underset{\text{I}}{\overset{\text{H}}{<}} + \text{PhCH}_2\text{CH}_2\text{ZnBr} \xrightarrow{\text{Pd(PPh}_3)_4, \text{THF, 25℃, 3h}} \underset{\text{Me}}{\overset{n\text{-}C_5H_{11}}{>}}\!\!=\!\!\underset{\text{CH}_2\text{CH}_2\text{Ph}}{\overset{\text{H}}{<}} \quad (38)$$
83%

$$\text{farnesyl-ZnCl} + \text{3-Br-2(5H)-furanone} \xrightarrow{\text{PdCl}_2(\text{PPh}_3)_2, i\text{-Bu}_2\text{AlH}, \text{THF, 0~25℃, 3h}} \text{coupled product} \quad (39)$$
82%

共轭卤代烯烃也能与烷基氯化锌进行交叉偶联反应得到烷基化的共轭烯烃[25]。

$$C_5H_{11}\text{-C}\!\equiv\!\text{C-CH=CH-Cl} + n\text{-}C_8H_{17}\text{ZnCl} \xrightarrow{\text{PdCl}_2(\text{dppf})_2, \text{THF, 66℃, 7h}} C_5H_{11}\text{-C}\!\equiv\!\text{C-CH=CH-}C_8H_{17}\text{-}n \quad (40)$$

$$n\text{-}C_5H_{11}\underset{}{-\!\!\!=\!\!\!-}\diagup\!\!\!\diagdown Cl + \text{cyclohexyl-MgBr} \xrightarrow[\text{THF, 65°C, 3.5h}]{PdCl_2(dppf), ZnCl_2} n\text{-}C_5H_{11}\underset{}{-\!\!\!=\!\!\!-}\diagup\!\!\!\diagdown\text{-cyclohexyl} \quad 96\% \quad (41)$$

$$\text{4-}i\text{Pr-C}_6H_4\text{-CH=CH-CH=CH-CH=CH-Cl} + n\text{-}C_8H_{17}MgBr \xrightarrow[\text{THF, 65°C, 3.5h}]{PdCl_2(dppf), ZnCl_2} \text{产物} \quad 71\% \quad (42)$$

3.2.4 C(sp^3)–C(sp^3)键的构建

Negishi 催化体系催化的烷基与烷基间的交叉偶联，即 C(sp^3)–C(sp^3)键的构建，不像前述 C(sp^2)–C(sp^2)、C(sp)–C(sp^2)和 C(sp^3)–C(sp^2)键构建那么容易以及底物那么广泛。C(sp^3)–C(sp^3)键难以构建的原因在于：①催化循环中卤代烷比卤代芳基或烯基与金属氧化加成反应要难；②即使氧化加成真的发生，生成的烷基金属中间体发生分子内 β-氢消除比分子间金属转移反应要快得多。第一类 Negishi 烷基-烷基交叉偶联反应是低温下 Ni(acac)$_2$ 催化的含不饱和键的类碘代烷与二烷基锌间的交叉偶联，类碘代烷中的不饱和键与碘原子键含 3 或 4 个碳原子的连接链，不饱和键在反应过程中与镍配位，起稳定中间体作用[26]。以下是几个代表性的实例。

$$\text{I-(CH}_2)_n\text{-CH=CH-X} + R_2Zn \xrightarrow[-35°C, 0.5\sim18h]{Ni(acac)_2, THF/NMP} R\text{-(CH}_2)_n\text{-CH=CH-X} \quad (43)$$
$$n=3,4$$

底物	试剂	产率
I–CH$_2$CH$_2$–CH(Bu)–CH=CH$_2$	(n-C$_5$H$_{11}$)$_2$Zn	72%
I–CH$_2$CH$_2$–CH(Ph)–CH=CH$_2$	(t-BuCOO–(CH$_2$)$_3$)$_2$Zn	90%
I–(CH$_2$)$_3$–CH=CH–CO$_2$Et	(n-C$_5$H$_{11}$)$_2$Zn	83%
I–(CH$_2$)$_3$–CN	(n-C$_5$H$_{11}$)$_2$Zn	52% (48h)

对于类碘代烷结构中与碘间隔 3 或 4 个碳原子不含不饱和键的底物，可以加入间三氟甲基苯乙烯［反应式（44）］、苯乙酮［反应式（45）］或 4-氟苯乙烯［反应式

（46）～式（47）］作为促进剂以稳定反应中间体使反应得以顺利进行[26]。

$$\text{NC-C}_6\text{H}_4\text{-CO(CH}_2)_3\text{I} + (n\text{-C}_5\text{H}_{11})_2\text{Zn} \xrightarrow[3\text{-CF}_3\text{C}_6\text{H}_4\text{CH=CH}_2]{\text{Ni(acac)}_2, \text{THF/NMP}, -78\sim35℃, 2.5\text{h}} \text{NC-C}_6\text{H}_4\text{-CO(CH}_2)_3\text{-}n\text{-C}_5\text{H}_{11} \quad 81\% \quad (44)$$

$$\text{2-Thienyl-CO(CH}_2)_3\text{I} + (t\text{-BuCO}_2(\text{CH}_2)_3)_2\text{Zn} \xrightarrow[\text{PhCOCH}_3]{\text{Ni(acac)}_2, \text{THF/NMP}, -78\sim35℃, 2.5\text{h}} \text{2-Thienyl-CO(CH}_2)_3(\text{CH}_2)_3\text{O}_2\text{CBu-}t \quad 70\% \quad (45)$$

$$\text{BocHN-CH(CO}_2\text{Et)(CH}_2)_3\text{I} + (n\text{-C}_5\text{H}_{11})_2\text{Zn} \xrightarrow[4\text{-FC}_6\text{H}_4\text{CH=CH}_2]{\text{Ni(acac)}_2, \text{THF/NMP}, -30℃, 4\text{h}} \text{BocHN-CH(CO}_2\text{Et)(CH}_2)_3\text{-}n\text{-C}_5\text{H}_{11} \quad 56\% \quad (46)$$

$$\text{oxazolidinone-CH}_2\text{I} + (n\text{-C}_5\text{H}_{11})_2\text{Zn} \xrightarrow[4\text{-FC}_6\text{H}_4\text{CH=CH}_2]{\text{Ni(acac)}_2, \text{THF/NMP}, -30℃, 3\text{h}} \text{oxazolidinone-CH}_2\text{-}n\text{-C}_5\text{H}_{11} \quad 60\% \quad (47)$$

多齿氮配体与适宜的金属镍配合物结合，可使反应在室温范围内进行，并实现烷基锌与仲卤代烷间的交叉偶联，配体的引入抑制了中间体烷基镍的 β-氢消除[27]。

$$\text{R-X} + \text{R'ZnX'} \xrightarrow[\text{DMAC, rt, 20h}]{\text{Ni(cod)}_2, s\text{-Bu-Pybox}} \text{R-R'} \quad (48)$$

R-X	R'ZnX'	产率
Ts-N(piperidinyl)-Br	Me-CH(Me)-CH$_2$CH$_2$-ZnI	66%
Me$_2$CH-CH(I)-	EtO$_2$C-(CH$_2$)$_3$-ZnBr	62%
cyclopentyl-I	Et$_2$NC(O)-(CH$_2$)$_4$-ZnBr	78%
Phthalimido-(CH$_2$)$_n$-Br	Ph-(CH$_2$)$_3$-ZnBr	65%
Ph-CO-(CH$_2$)$_3$-I	dioxolanyl-CH$_2$CH$_2$-ZnBr	74%

s-Bu-Pybox: 2,6-bis[(4S)-4-butyl-4,5-dihydrooxazol-2-yl]pyridine

3.2.5 不对称 Negishi 反应

尽管 Negishi 交叉偶联反应底物范围广泛,但理论上只有能生成 C_2 对称轴的芳基与芳基交叉偶联,以及能构建手性碳中心的仲烷基与芳基或烯基、烷基的交叉偶联才有可能在手性诱导作用下进行不对称反应。上一节已经阐明多齿氮配体与过渡金属镍配合物可以构成有效的催化体系使得烷基锌能够与仲卤烷进行 Negishi 交叉偶联。人们自然意识到手性多齿氮配体与过渡金属的配位结合并协于其他助剂和反应条件的优化,能够实现不对称 Negishi 反应。这一预期首先在仲 α-溴代酰胺的烷基化反应得到实现。由手性配体 (R)-(i-Pr)-Pybox-Ni(cod)$_2$·glyme 构成的不对称催化体系在 1,3-二甲基-2-咪唑啉酮(DMI)/四氢呋喃混合溶剂中催化仲 α-溴代酰胺与有机锌卤化物的交叉偶联,高收率和高对映选择性地生成光学活性的 α-烷基酰胺,并具有广泛的官能团耐受性,对映选择性受有机锌卤化物的结构影响不大[28]。

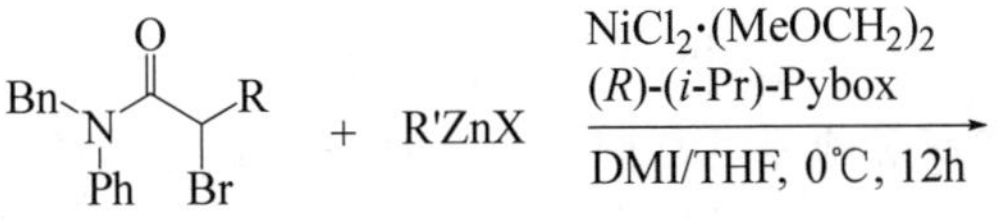

R	R'ZnX	收率/%	ee/%
Et	n-C$_6$H$_{13}$ZnBr	90	96
Et	MeZnI	90	91
Et	Ph(CH$_2$)$_3$ZnBr	84	96
n-Bu	Ph(CH$_2$)$_3$ZnBr	79	96
Et	(prenyl)ZnBr	78	95
Et	PhO(CH$_2$)$_4$ZnBr	77	96
Et	NC(CH$_2$)$_5$ZnBr	70	93

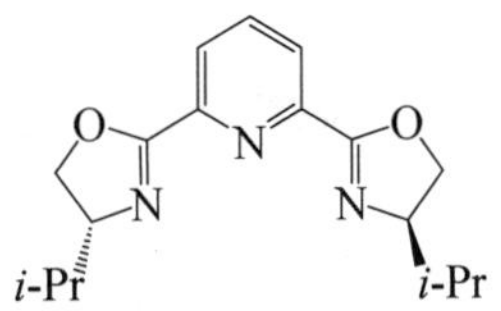

通过改变溶剂,在同样的催化体系下 α-溴代茚满类底物与烷基溴化锌交叉偶联高对映选择性地生成光学活性的 α-烷基茚满[29]。

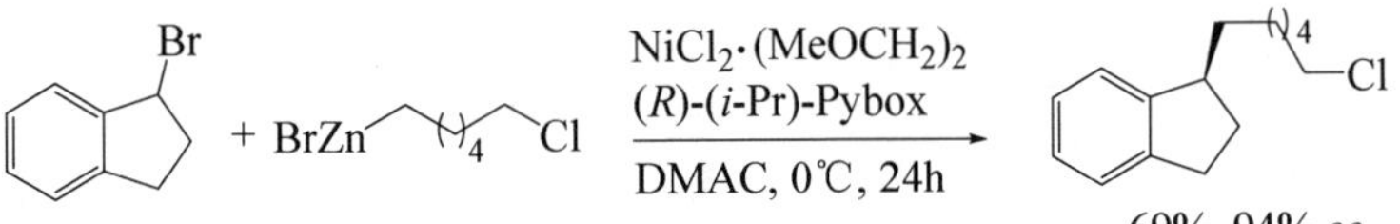

式 (50): 69%, 94% ee

式 (51): 82%, 91% ee

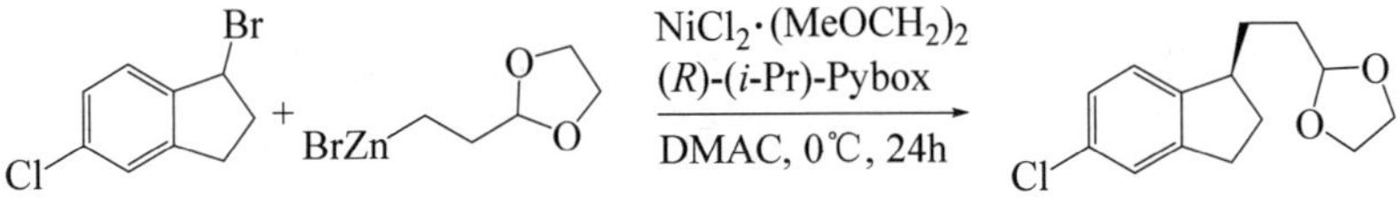

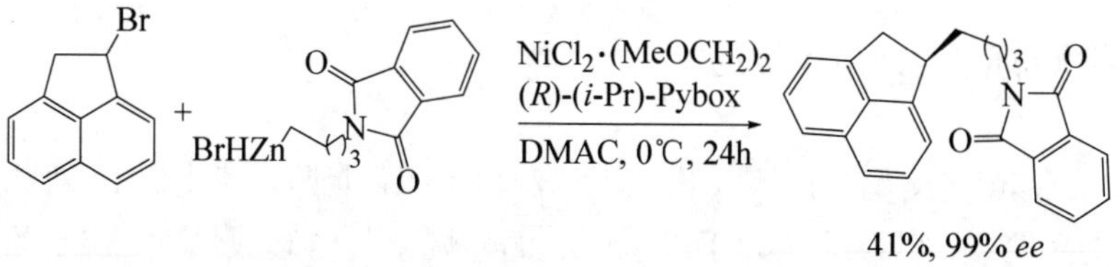

Br
BrHZn
O
N
O
3
NiCl2·(MeOCH2)2
(R)-(i-Pr)-Pybox
DMAC, 0℃, 24h
41%, 99% ee

(52)

在近似的催化体系下还可以实现烯丙基氯与烷基溴化锌间的不对称交叉偶联，生成光学活性的烷基取代烯烃衍生物[30]。

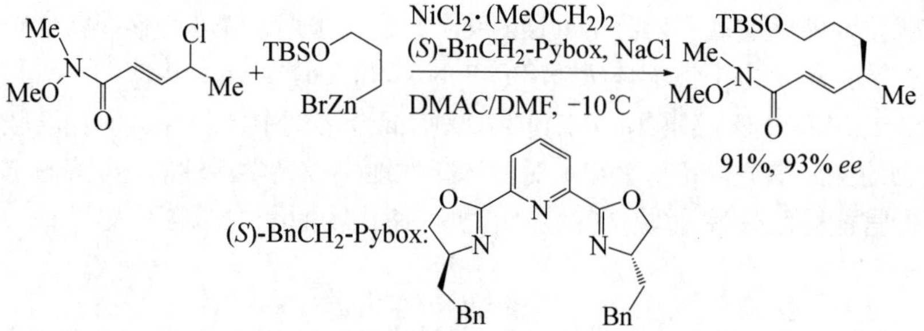

Me
MeO
Cl
Me
TBSO
BrZn
NiCl2·(MeOCH2)2
(S)-BnCH2-Pybox, NaCl
DMAC/DMF, −10℃
91%, 93% ee
(S)-BnCH2-Pybox:
Bn
Bn

(53)

一些不饱和键如羰基、烯基和氰基的α-卤代化合物与芳基或烯基锌化合物在与上述类似的不对称催化体系催化下反应得到相应的光学活性的交叉偶联产物[31]。

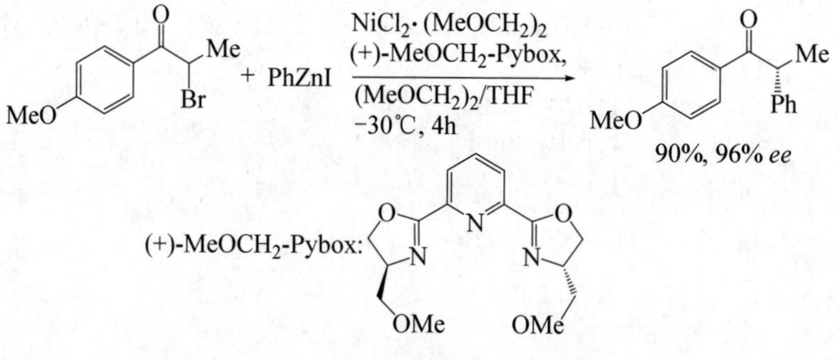

MeO
Me
Br
PhZnI
NiCl2·(MeOCH2)2
(+)-MeOCH2-Pybox,
(MeOCH2)2/THF
−30℃, 4h
Ph
90%, 96% ee
(+)-MeOCH2-Pybox:
OMe
OMe

(54)

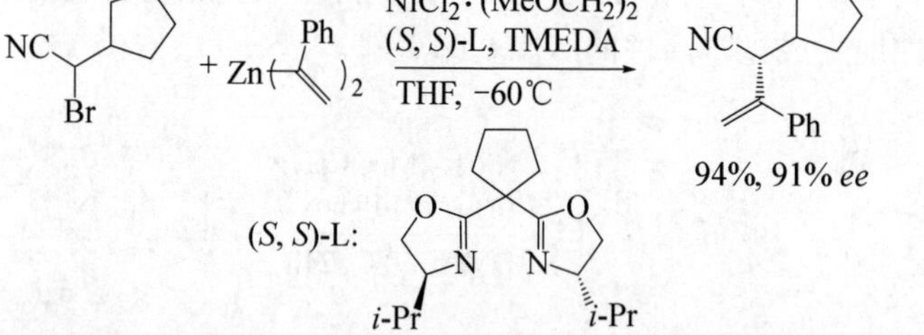

NC
Br
Zn
Ph
2
NiCl2·(MeOCH2)2
(S, S)-L, TMEDA
THF, −60℃
94%, 91% ee
(S, S)-L:
i-Pr
i-Pr

(55)

利用芳基与芳基的不对称 Negishi 交叉偶联构建 C(sp^2)-C(sp^2)键从而合成含 C_2 轴的光学活性化合物并不容易，但在特殊的不对称催化剂催化下并协于适宜的反应条件，也能够以较高的对映选择性得到光学活性的联二芳基化合物。其中基于二茂铁骨架的手性氮膦配体与钯结合可以催化 1-烷基-2-溴萘与二(2-烷基萘)锌交叉偶联,以较高的对映选择性生成光学活性的联二烷基萘,但底物范围相对较窄,烷基为甲基时结果最好[32]。

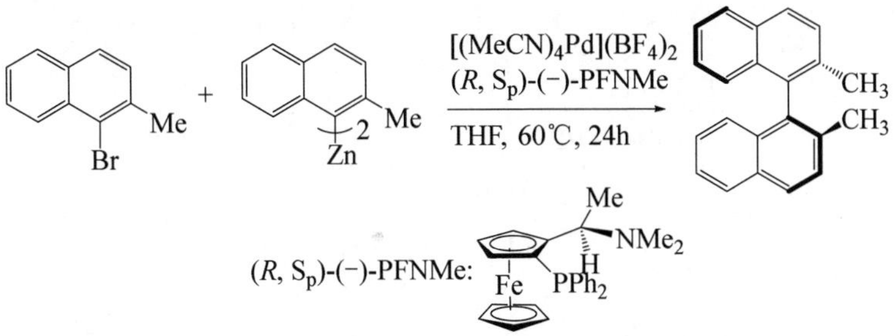

$$
\text{(56)}
$$

3.3 Negishi 反应在药物合成上的应用实例

Negishi 反应通常需要金属有机试剂制备有机锌化合物,反应条件苛刻,而且在有机锌化合物中 C-Zn 键是离子键,碳负离子的碱性和亲和性强,使得有机锌化合物与有机硼化合物相比官能团耐受性较差,因而相对于 Heck 和 Suzuki 反应,其在药物合成上应用较少。尽管如此,Negishi 反应在合成上有不可替代的作用。以下按碳-碳键的构建类型,介绍 Negishi 反应在一些药物和生物活性化合物合成上的应用。其中经由 Negishi 交叉偶联反应合成的产物编号为 NB-x,x 为整数,代表序号。

下面反应的最终目标产物是一个加强糖皮质激素受体的非甾体药物,利用 Negishi 芳基-芳基交叉偶联合成了关键中间体 NB-1。首先,间二甲氧基苯与正丁基锂作用生成芳基锂,然后与 $ZnCl_2$ 金属交换生成芳基氯化锌,生成的芳基氯化锌在 $PdCl_2(PPh_3)_2$ 催化下原位与芳基溴反应生成交叉偶联产物 NB-1;NB-1 经后续反应生成最终目标产物[33]。

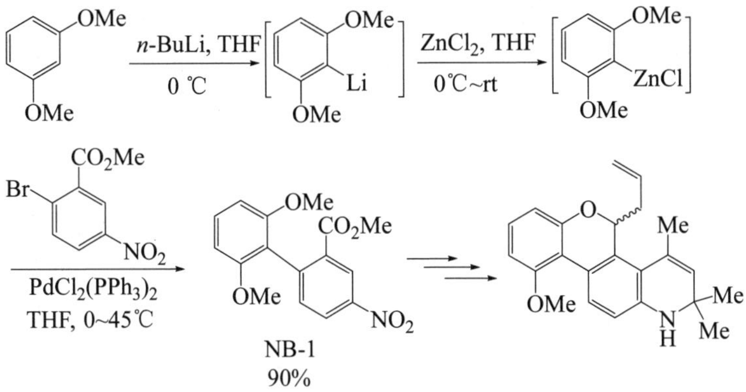

$$
\text{(57)}
$$

Negishi 芳基-杂芳基交叉偶联是合成一种 B-Raf 激酶抑制剂的关键步骤。首先,4-溴-1-氯-异喹啉用正丁基锂锂化,进而与 $ZnBr_2$ 金属交换生成 1-氯-4-异喹啉溴化锌;1-

氯-4-异喹啉溴化锌原位与 7-异喹啉三氟甲磺酸酯在标准 Negishi 催化反应体系下反应生成交叉偶联反应产物 NB-2，NB-2 经进一步转化生成 B-Raf 激酶抑制目标分子[34]。

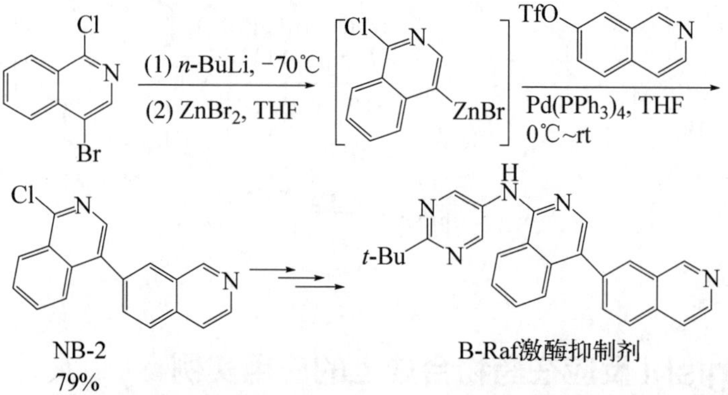

胸腺素 C(amythiamicin C)具有抗菌活性，其合成的关键步骤是两步 Negishi 杂芳基-杂芳基交叉偶联反应。第一步，碘代噻唑和 2,6-二溴吡啶基碘化锌在 $Pd_2(dba)_3$/tfp 催化下以良好的收率生成关键交叉偶联产物 3-取代-2,6-二溴吡啶（NB-3）；第二步，NB-3 与叔丁基噻唑-5-羧酸酯碘化锌在 $PdCl_2(PPh_3)_2$ 催化下交叉偶联以高区域选择性和中等的收率生成关键中间体 NB-4。所得 NB-4 再经后续转化得到胸腺素 C[35]。

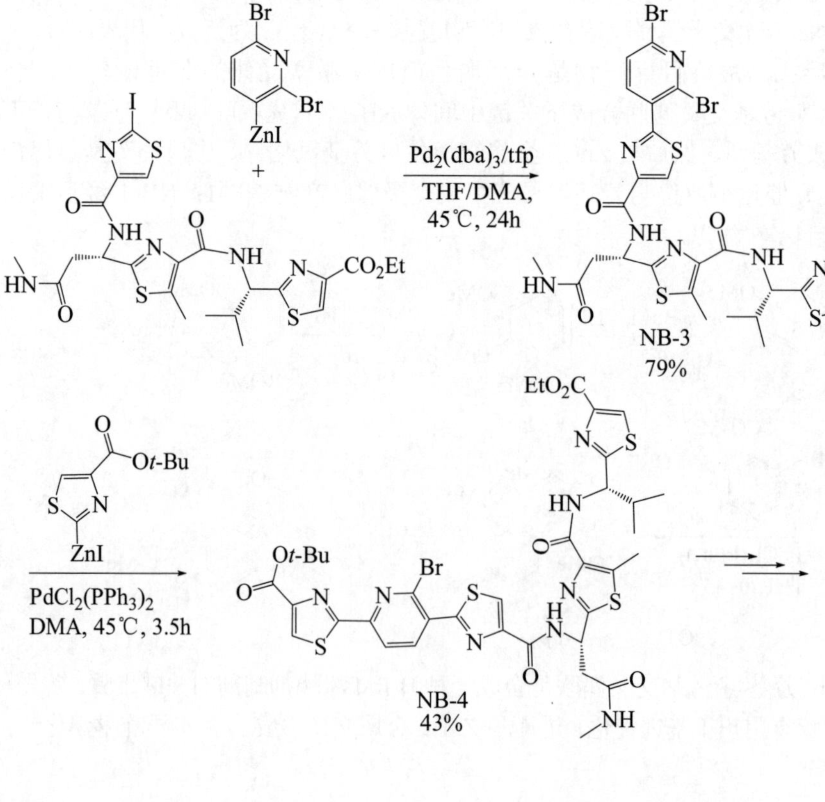

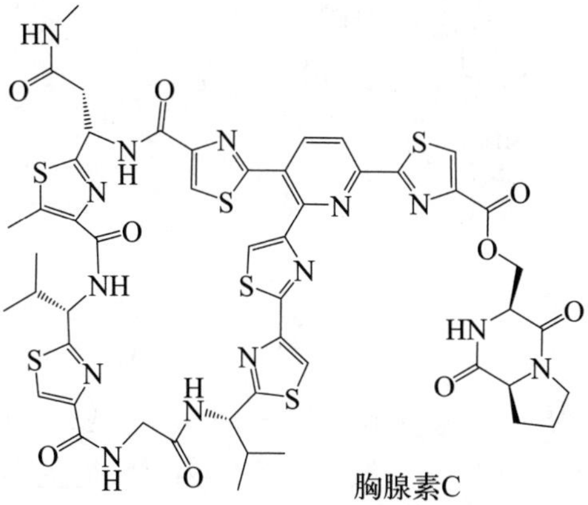

(59) 胸腺素C

1-(Z)-苍术素醇[1-(Z)-atractylodinol]是一种治疗风湿、消化紊乱、夜盲症和流感的药物，其合成的关键一步是Negishi杂芳基-烯基交叉偶联。采用烯基碲化物与α-呋喃氯化锌为偶联底物，在PdCl$_2$/CuI催化下以良好的收率生成关键中间体NB-5，进而转化为1-(Z)-苍术素醇[36]。

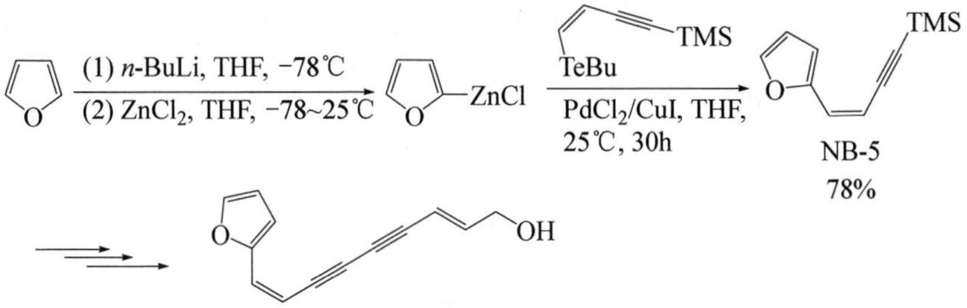

$$\text{furan} \xrightarrow[\text{(2) ZnCl}_2\text{, THF, }-78\sim25℃]{\text{(1) }n\text{-BuLi, THF, }-78℃} \text{2-furyl-ZnCl} \xrightarrow[\substack{\text{PdCl}_2\text{/CuI, THF,}\\25℃,\ 30\text{h}}]{\text{TeBu, TMS-enyne}} \text{NB-5 } 78\%$$

$$\xrightarrow{\quad\quad} \text{1-(Z)-苍术素醇} \tag{60}$$

brevisamide是一种具有生物活性的天然产物，利用Negishi烯基-烯基交叉偶联反应构建(E,E)-共轭二烯结构片段，实现其简捷合成。反应中，中间体烯基碘化锌与含全氢化吡喃环的碘代烯化合物在Pd(PPh$_3$)$_4$催化下顺利反应生成交叉偶联产物NB-6，NB-6不经纯化直接用樟脑磺酸（CSA）选择性脱除伯醇上的TBS保护基得到合成brevisamide的关键中间体NB-6'，两步收率58%，NB-6'经多步转化生成brevisamide [37]。

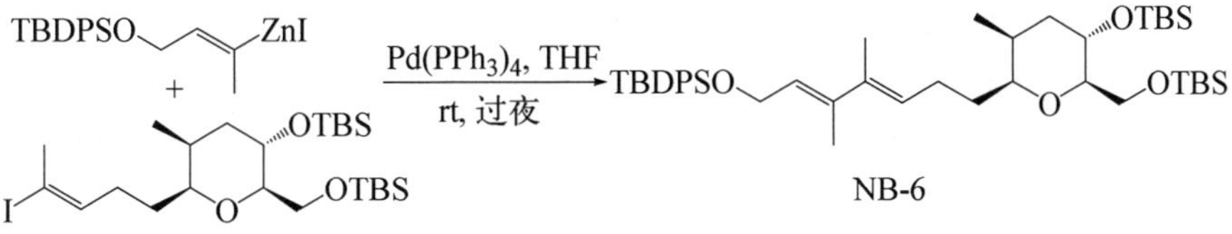

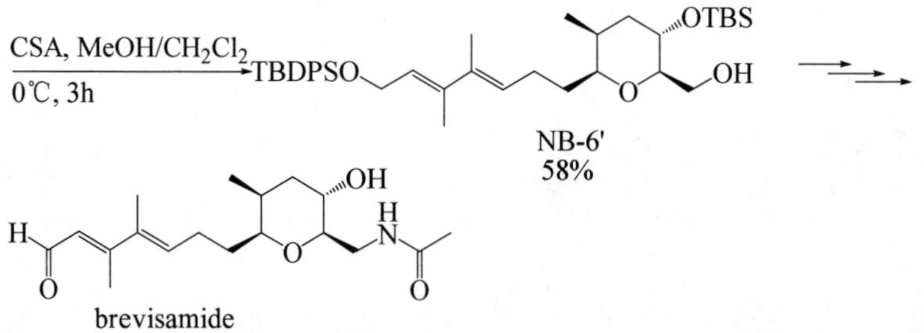

$$\xrightarrow[0\text{℃, 3h}]{\text{CSA, MeOH/CH}_2\text{Cl}_2}$$

NB-6'
58%

brevisamide

(61)

motuporin 是一种具有生物活性的海洋天然产物，其合成的关键步骤之一也是通过 Negishi 反应构建特定构型的共轭二烯结构片段。6-苯基-2-己炔衍生物先与 Cp_2ZrHCl 锆氢化，再与 $ZnCl_2$ 进行金属交换生成中间体烯基氯化锌，其与偶联伙伴碘代烯在 $Pd(PPh_3)_4$ 催化下交叉偶联以优异的收率生成共轭二烯片段 NB-7。NB-7 再经多步转化生成 motuporin[38]。

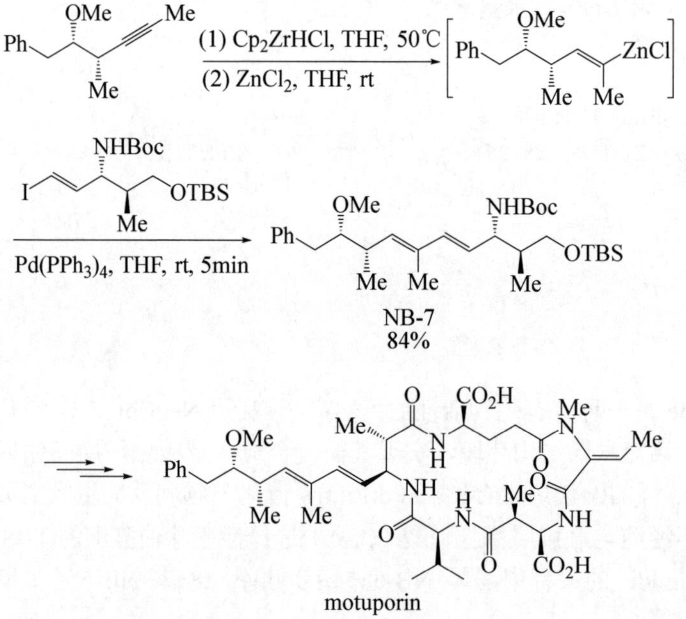

(1) Cp_2ZrHCl, THF, 50℃
(2) $ZnCl_2$, THF, rt

$Pd(PPh_3)_4$, THF, rt, 5min

NB-7
84%

motuporin

(62)

下面合成反应的目标分子是一种选择性结合 HIV-1 分离株的非核苷逆转录酶抑制剂。其合成的关键步骤之一是芳基与环丙基间的交叉偶联。首先，利用正丁基锂与氟代芳基甲醚选择性地交换反应生成芳基锂，进而与 $ZnBr_2$ 进行金属转化生成芳基溴化锌，芳基溴化锌与碘代环丙烷羧酸甲酯在钯催化剂催化下交叉偶联生成中间体 NB-8，反应过程中不会发生手性中心的消旋化[39]。

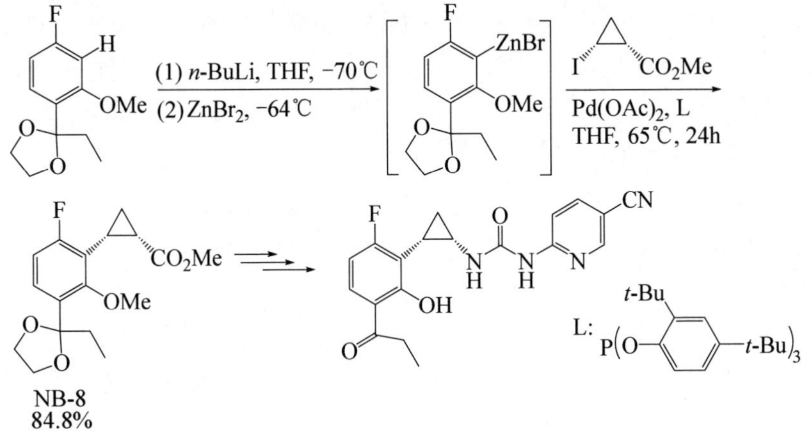

迪斯科莫利［(−)-discodermolide］是一种非常有应用前景的免疫抑制剂。其全合成的关键中间体 NB-9 由烷基碘化锌前体与碘代烯前体间的 Negishi 交叉偶联得到。交叉偶联反应在 Pd(PPh$_3$)$_4$ 催化下于乙醚溶剂中室温进行，其能否获得 NB-9 的关键因素之一是制备烷基氯化锌前体时，丁基锂与碘代烷的比例须严格控制在 3∶1，而且 ZnCl$_2$ 与碘代烷预先混合[40]。

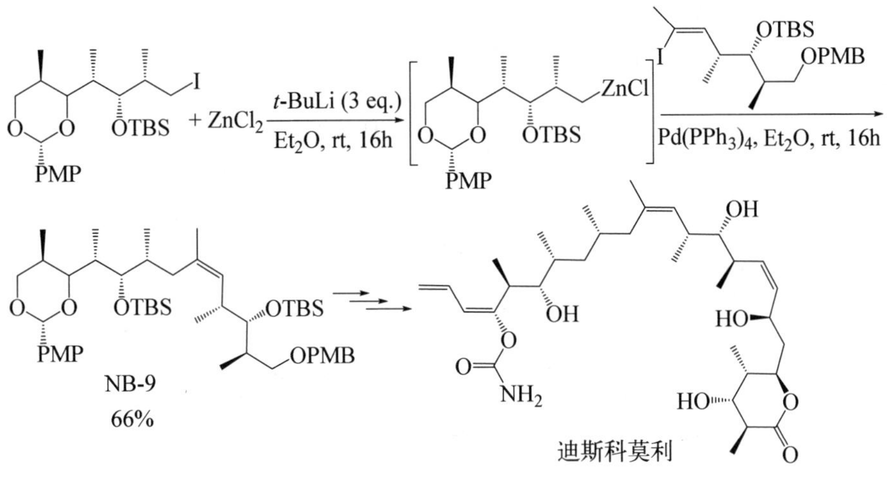

xerulin 是一种胆固醇生物合成抑制剂，很难从生物体分离得到纯的样品。然而，该化合物可以通过 5 步 Negishi 交叉偶联反应由容易得到的市售原料 (E)-1-溴丙烯、乙炔和丙炔酸合成，总收率 30%，立体选择性大于 96%。如反应式及 xerulin 的结构所示，所有七个连接乙炔衍生的 C═C 键和 C≡C 键的 C—C 单键都是由钯催化的包括 Negishi 和 Sonogashira 交叉偶联反应实现的。其中的重要步骤包括：①无须硅烷基保护-脱保护，由乙炔溴化锌直接合成端炔；②钯与锌双金属催化的由原位

第 3 章　Negishi 反应及其在药物合成上的应用　107

生成的烯基锆正离子衍生物进行烯基-烯基偶联；③钯催化的由 1-卤炔合成共轭二炔；④钯催化的 Sonogashira 烯-炔交叉偶联-羧金属化串联反应，即烯-炔偶联反应中间体与邻位羧基加成生成内酯[41]。这是一种少见的采用线型工艺路线，且主要步骤基于相同类型反应合成复杂分子的实例。

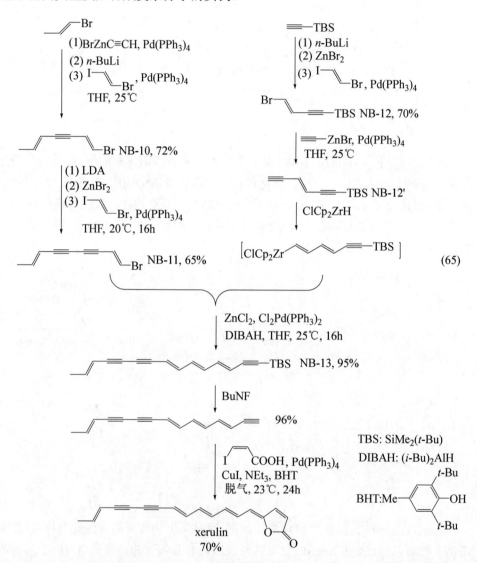

炔基与芳基的交叉偶联也常用来合成药物或天然产物。例如，我国科学家姚祝军教授课题组利用丙炔基氯化锌与芳基三氟甲酸酯间的 Negishi 交叉偶联反应来合成治疗周期性发热和疟疾药物 cassiarins A 和 B 合成的关键中间体 NB-14，反应条件温和，且收率较高[42]。

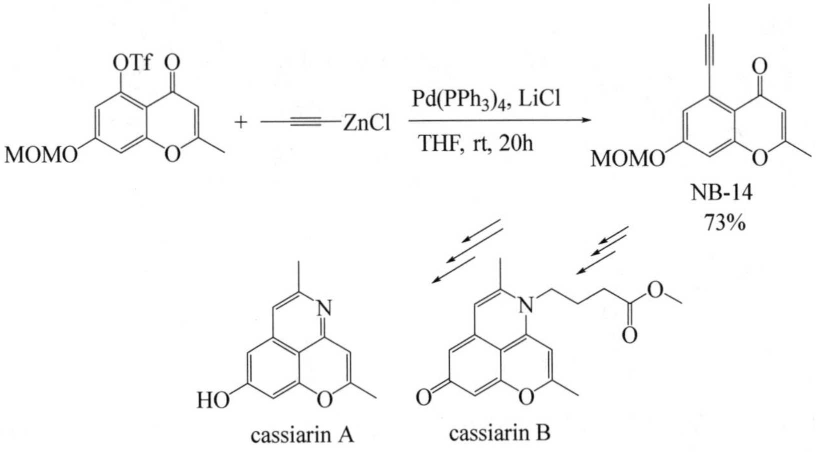

MaR1$_{n-3DPA}$ 是一种抗炎药物，其可通过 C(sp^3)-C(sp^3) 的 Negishi 交叉偶联反应合成。含多个不饱和键的溴化物与 4-溴化锌丁酸乙酯在催化剂 Pd-PEPPSI-IPr 催化下于室温下顺利反应，以良好的收率得到交叉偶联产物 NB-15，NB-15 作为前体经三步转化生成 MaR1$_{n-3DPA}$[43]。

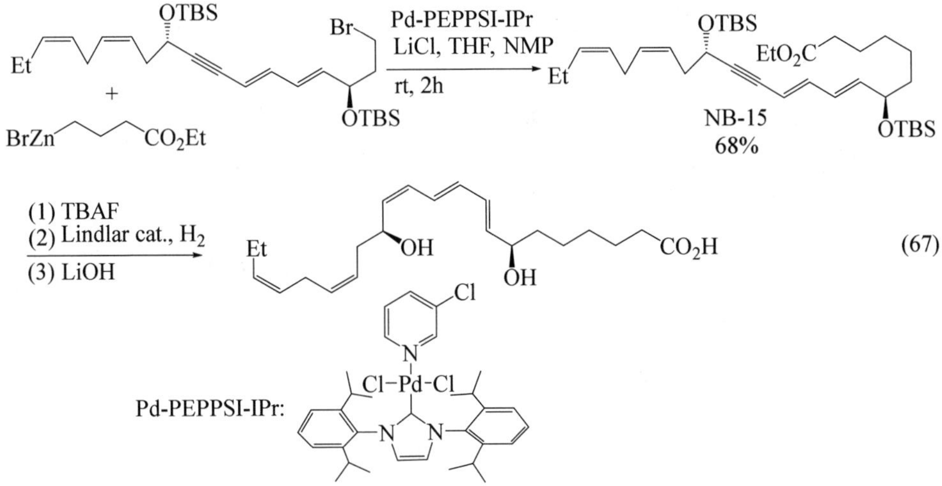

百部酰胺［(-)-stemoamide］是从百部科植物对叶百部根部分离得到的天然产物，有镇咳和驱虫的功效。其全合成所需的关键中间体 β，γ-不饱和酯（NB-16）的合成利用了钯催化的 Negishi 反应构建 sp^2-sp^3 碳-碳键。如反应式（68）所示，乙烯基碘与 Reformatsky 亲核试剂在标准的 Negishi 催化反应体系下反应以 78% 的收率顺利生成 NB-16，然后 NB-16 经多步反应生成百部酰胺[44]。

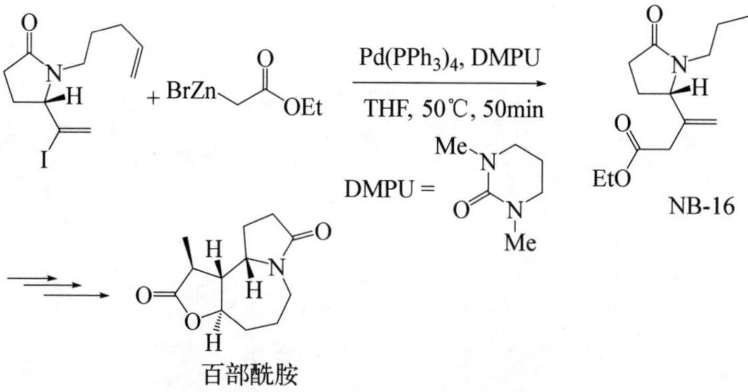

(68)

以上实例未涉及不对称 Negishi 反应。为进一步彰显 Negishi 反应在药物和生物活性化合物合成上的重要性,以下再介绍几个不对称 Negishi 反应的应用实例。

丙克拉莫［(S)-preclamol］是一种中枢多巴胺（DA）受体激动剂,可激活 DA 自身受体,同时作为突触后 DA 受体的拮抗剂,有望用于治疗精神分裂症和帕金森病等神经系统疾病。(S)-preclamol 有效的合成方法是基于手性双氮配体诱导钴催化的 α-溴代戊二酸二苄酯与间甲氧基苯基溴化锌的不对称 Negishi 交叉偶联反应形成 C(sp^2)-C(sp^3) 键构建手性中心,合成关键中间体 NB-17,NB-17 的收率高达 93%,对映体过量值为 87%。中间体 NB-17 经后续反应转化为丙克拉莫[45]。

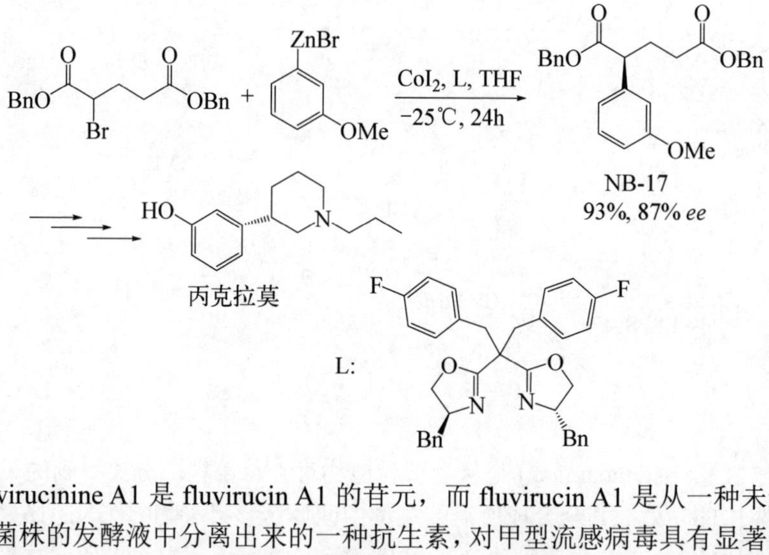

(69)

fluvirucinine A1 是 fluvirucin A1 的苷元,而 fluvirucin A1 是从一种未经鉴定的放线菌菌株的发酵液中分离出来的一种抗生素,对甲型流感病毒具有显著的抑制活性。fluvirucinine A1 合成的关键是两步镍催化的不对称 Negishi C(sp^3)-C(sp^3) 交叉偶联构建两个手性中心。第一步 Negishi 反应,共轭仲烯丙基氯与含缩醛基团的丙基氯化锌在手性配体 Pybox 诱导的镍催化下在 -10℃ 的标准条件下反应以优异的收率和对映选择性生成光学活性的交叉偶联产物 NB-18；第二步 Negishi 反应，NB-18 经多步反应衍生的溴化物与锌粉在单质碘存在下反应生成烷基溴化锌，烷基溴化锌原位与 4-氯-2-己烯酸乙酯在与第一步不对称 Negishi 反应相同的条件下反应以良好的收率、优异的对映和非对映选择性生成交叉偶联产物 NB-18'；NB-18'进一步转化生成 fluvirucinine A1[30]。

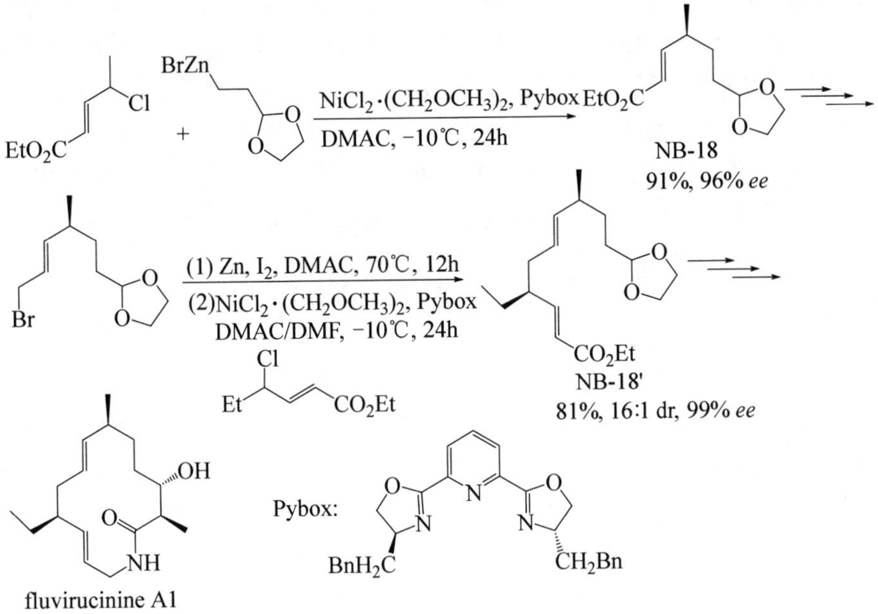

(70)

西那卡塞 [(*R*)-cinacalcet] 是用于治疗慢性肾病患者继发性甲状旁腺功能亢进症的药物，通常通过拆分方法得到其光学纯度的异构体，而利用钴催化的不对称 Negishi 反应可以实现西那卡塞更加高效的合成。如下反应路线，α-萘基溴化锌与 α-溴丙酸 4-甲氧基苄酯在手性二噁唑配体与 CoI$_2$ 构成的催化体系催化下以良好的收率和对映选择性生成交叉偶联产物 NB-19，NB-19 经多步转化生成关键中间体 (*R*)-1-(1-萘基) 乙胺。(*R*)-1-(1-萘基) 乙胺与 3-(3-三氟甲基) 丙酸缩合得到相应的酰胺，最终以四氢铝锂还原得到西那卡塞[46]。

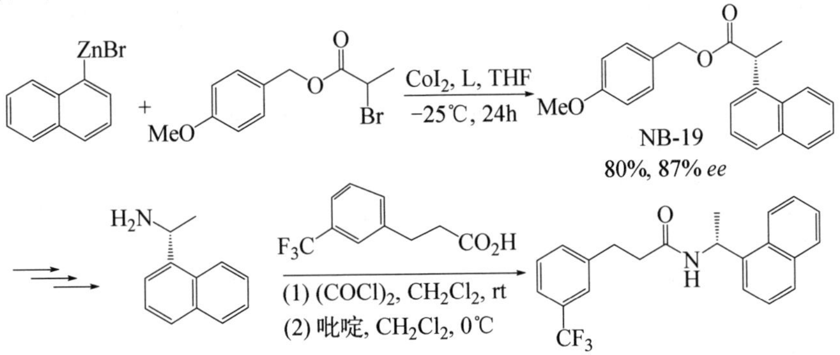

(−)-daphenylline 是从虎皮楠科植物中分离出的生物碱，其全合成是一挑战性课题，合成的关键是手性中心和三环体系的构建。Yamada 等利用不对称双噁唑-镍催化体系催化 6-甲氧基苯并-1-氯环戊烷与 4-溴化锌丁酸甲酯的不对称 Negishi 偶联，随后酯基水解，以 99%的对映体过量值生成 NB-20，成功构建了手性中心；NB-20 在三氟乙酸/三氟乙酸酐（TFA/TFAA）催化下进行分子内傅克酰基化反应生成三环中间体，该中间体再经多步转化最终生成 (−)-daphenylline[47]。

参考文献

[1] Baba S, Negishi E. A novel stereospecific alkenyl-alkenyl cross-coupling by a palladium-or nickel-catalyzed reaction of alkenylalanes with alkenyl halides[J]. Journal of the American Chemical Society, 1976, 98(21): 6729-6731.

[2] Negishi E, King A O, Okukado N. Selective carbon-carbon bond formation via transition metal catalysis. 3. A highly selective synthesis of unsymmetrical biaryls and diarylmethanes by the nickel-or palladium-catalyzed reaction of aryl-and benzylzinc derivatives with aryl halides[J]. The Journal of Organic Chemistry, 1977, 42(10): 1821-1823.

[3] Casares J A, Espinet P, Fuentes B, et al. Insights into the mechanism of the Negishi reaction: ZnRX versus ZnR_2 reagents[J]. Journal of the American Chemical Society, 2007, 129(12): 3508-3509.

[4] Charpentier B, Bernardon J M, Eustache J, et al. Synthesis, structure-affinity relationships, and biological activities of ligands binding to retinoic acid receptor subtypes[J]. Journal of Medicinal Chemistry, 1995, 38(26): 4993-5006.

[5] Gavryushin A, Kofink C, Manolikakes G, et al. An efficient Negishi cross-coupling reaction catalyzed by nickel (Ⅱ) and diethyl phosphite[J]. Tetrahedron, 2006, 62(32): 7521-7533.

[6] Negishi E I, Takahashi T, Baba S, et al. Nickel-or palladium-catalyzed cross coupling. 31. Palladium-or nickel-catalyzed reactions of alkenylmetals with unsaturated organic halides as a selective route to arylated alkenes and conjugated dienes: scope, limitations, and mechanism [J]. Journal of the American Chemical Society, 1987, 109(8): 2393-2401.

[7] Krasovskiy A, Lipshutz B H. Highly selective reactions of unbiased alkenyl halides and alkylzinc halides: Negishi-Plus couplings[J]. Organic Letters, 2011, 13(15): 3822-3825.

[8] Huang Z, Negishi E. Highly stereo-and regioselective synthesis of (Z)-trisubstituted alkenes via 1-bromo-1-alkyne hydroboration-migratory insertion-Zn-promoted iodinolysis and Pd-catalyzed organozinc cross-coupling [J]. Journal of the American Chemical Society, 2007, 129(47): 14788-14792.

[9] Wang G, Yin N, Negishi E. Highly stereoselective total synthesis of fully hydroxy-protected mycolactones A and B and their stereoisomerization upon deprotection [J]. Chemistry: A European Journal, 2011, 17(15): 4118-4130.

[10] King A O, Negishi E, Villani Jr F J, et al. A general synthesis of terminal and internal arylalkynes by the palladium-catalyzed reaction of alkynylzinc reagents with aryl halides[J]. The Journal of Organic Chemistry, 1978, 43(2): 358-360.

[11] Negishi E, Anastasia L. Palladium-catalyzed alkynylation[J]. Chemical Reviews, 2003, 103(5): 1979-2018.

[12] Negishi E, Okukado N, Lovich S F, et al. A method for the preparation of terminal and internal conjugated diynes via palladium-catalyzed cross-coupling[J]. Journal of Organic Chemistry, 1984, 49(14): 2629-2632.

[13] Negishi E, Tan Z, Liou S Y, et al. Strictly regiocontrolled α-monosubstitution of cyclic carbonyl compounds with alkynyl and alkyl groups via Pd-catalyzed coupling of cyclic α-iodoenones with organozincs[J]. Tetrahedron, 2000, 56(52): 10197-10207.

[14] Fürstner A, Dierkes T, Thiel O R, et al. Total synthesis of (−)-salicylihalamide[J]. Chemistry: A European Journal, 2001, 7(24): 5286-5298.

[15] Abarbri M, Parrain J L, Cintrat J C, et al. Efficient synthesis of conjugated (2E)-or (2Z)-en-4-ynoic acids and (2E, 4E)-or (2Z, 4E)-dienoic acids via palladium-catalysed cross coupling[J]. Synthesis, 1996, (1): 82-86.

[16] Ruitenberg K, Kleijn H, Westmijze H, et al. Organometal-mediated synthesis of conjugated allenynes, allenediynes, vinylallenes and diallenes[J]. Recueil des Travaux Chimiques des Pays-Bas, 1982, 101(11): 405-409.

[17] Negishi E, Bagheri V, Chatterjee S, et al. Palladium-catalyzed acylation of organozincs and other organometallics as a convenient route to ketones[J]. Tetrahedron Letters, 1983, 24(47): 5181-5184.

[18] Phapale V B, Guisán-Ceinos M, Buñuel E, et al. Nickel-catalyzed cross-coupling of alkyl zinc halides for the formation of $C(sp^2)$-$C(sp^3)$ bonds: scope and mechanism [J]. Chemistry: A European Journal, 2009, 15(46): 12681-12688.

[19] Coleridge B M, Bello C S, Leitner A. General and user-friendly protocol for the synthesis of functionalized aryl-and heteroaryl-cyclopropanes by Negishi cross-coupling reactions[J]. Tetrahedron Letters, 2009, 50(31): 4475-4477.

[20] Han C, Buchwald S L. Negishi coupling of secondary alkylzinc halides with aryl bromides and

chlorides[J]. Journal of the American Chemical Society, 2009, 131(22): 7532-7533.

[21] Wang D C, Niu H Y, Qu G R, et al. Nickel-catalyzed Negishi cross-couplings of 6-chloropurines with organozinc halides at room temperature[J]. Organic & Biomolecular Chemistry, 2011, 9(22): 7663-7666.

[22] Negishi E, Matsushita H, Kobayashi M, et al. A convenient synthesis of unsymmetrical bibenzyls homoallylarenes, and homopropargylarenes via palladium-catalyzed cross coupling[J]. Tetrahedron Letters, 1983, 24(36): 3823-3824.

[23] Negishi E, Luo F T, Rand C L. Stereo-and regioselective routes to allylic silanes[J]. Tetrahedron Letters, 1982, 23(1): 27-30.

[24] Kobayashi M, Negishi E. A palladium-catalyzed stereospecific substitution reaction of homoallylzincs with .beta.-bromo-substituted .alpha., .beta.-unsaturated carbonyl derivatives. A highly selective synthesis of Mokupalide [J]. The Journal of Organic Chemistry, 1980, 45(25): 5223-5225.

[25] Peyrat J F, Thomas E, L', Hermite N, et al. Versatile palladium (II)-catalyzed Negishi coupling reactions with functionalized conjugated alkenyl chlorides[J]. Tetrahedron Letters, 2003, 44(35): 6703-6707.

[26] Netherton M R, Fu G C. Nickel-catalyzed cross-couplings of unactivated alkyl halides and pseudohalides with organometallic compounds[J]. Advanced Synthesis & Catalysis, 2004, 346(13-15): 1525-1532.

[27] Zhou J, Fu G C. Cross-couplings of unactivated secondary alkyl halides: room-temperature nickel-catalyzed Negishi reactions of alkyl bromides and iodides[J]. Journal of the American Chemical Society, 2003, 125(48): 14726-14727.

[28] Fischer C, Fu G C. Asymmetric nickel-catalyzed Negishi cross-couplings of secondary α-bromo amides with organozinc reagents[J]. Journal of the American Chemical Society, 2005, 127(13): 4594-4595.

[29] Arp F O, Fu G C. Catalytic enantioselective Negishi reactions of racemic secondary benzylic halides[J]. Journal of the American Chemical Society, 2005, 127(30): 10482-10483.

[30] Son S, Fu G C. Nickel-catalyzed asymmetric Negishi cross-couplings of secondary allylic chlorides with alkylzincs[J]. Journal of the American Chemical Society, 2008, 130(9): 2756-2757.

[31] Lundin P M, Esquivias J, Fu G C. Catalytic asymmetric cross-couplings of racemic alpha-bromoketones with arylzinc reagents[J]. Angewandte Chemie International Edition, 2009, 48(1): 154-156.

[32] Genov M, Fuentes B, Espinet P, et al. Asymmetric Negishi reaction for sterically hindered couplings: synthesis of chiral binaphthalenes[J]. Tetrahedron: Asymmetry, 2006, 17(18): 2593-2595.

[33] Ku Y Y, Grieme T, Raje P, et al. A practical and scaleable synthesis of A-224817.0, a novel nonsteroidal ligand for the glucocorticoid receptor[J]. The Journal of Organic Chemistry, 2003, 68(8): 3238-3240.

[34] Denni-Dischert D, Marterer W, Bänziger M, et al. The synthesis of a novel inhibitor of B-Raf kinase[J]. Organic Process Research & Development, 2006, 10(1): 70-77.

[35] Ammer C, Bach T. Total syntheses of the thiopeptides amythiamicin C and D[J]. Chemistry: A European Journal, 2010, 16(47): 14083-14093.

[36] Oliveira J M, Zeni G, Malvestiti I, et al. Total synthesis of 1-(Z)-atractylodinol[J]. Tetrahedron Letters, 2006, 47(46): 8183-8185.

[37] Lee J, Panek J S. Total synthesis of brevisamide[J]. Organic Letters, 2009, 11(19): 4390-4393.

[38] Panek J S, Hu T. Asymmetric synthesis of (2S, 3S, 8S, 9S)-N-Boc ADDA: application of a palladium(0)-catalyzed cross-coupling reaction of trisubstituted olefins[J]. The Journal of Organic Chemistry, 1997, 62(15): 4914-4915.

[39] Cai S, Dimitroff M, McKennon T, et al. Process development on an efficient new convergent formal synthesis of MIV-150[J]. Organic Process Research & Development, 2004, 8(3): 353-359.

[40] Smith III A B, Qiu Y, Jones D R, et al. Total synthesis of (−)-discodermolide[J]. Journal of the American Chemical Society, 1995, 117(48): 12011-12012.

[41] Negishi E, Alimardanov A, Xu C. An efficient and stereoselective synthesis of xerulin via Pd-catalyzed cross coupling and lactonization featuring (E)-iodobromoethylene as a novel two-carbon synthon[J]. Organic Letters, 2000, 2(1): 65-67.

[42] Yao Y S, Yao Z J. Biomimetic total syntheses of cassiarins A and B[J]. The Journal of Organic Chemistry, 2008, 73(14): 5221-5225.

[43] Tungen J E, Aursnes M, Dalli J, et al. Total synthesis of the anti-inflammatory and pro-resolving lipid mediator MaR1$_{n-3\ DPA}$ utilizing an sp^3-sp^3 Negishi cross-coupling reaction[J]. Chemistry: A European Journal, 2014, 20(45): 14575-14578.

[44] Torssell S, Wanngren E, Somfai P. Total synthesis of (−)-stemoamide[J]. The Journal of Organic Chemistry, 2007, 72(11): 4246-4249.

[45] Zhou Y, Liu C, Wang L, et al. A concise enantioselective synthesis of (S)-preclamol via asymmetric catalytic Negishi cross-coupling reaction[J]. Synlett, 2019, 30(7): 860-862.

[46] Sun X, Wang X, Liu F, et al. Enantioselective synthesis of (R)-cinacalcet via cobalt-catalysed asymmetric Negishi cross-coupling[J]. Chirality, 2019, 31(9): 682-687.

[47] Yamada R, Adachi Y, Yokoshima S, et al. Total synthesis of (−)-daphenylline[J]. Angewandte Chemie International Edition, 2016, 55(20):6067-6070.

第 4 章
Kumada-Corriu 偶联反应及其在药物合成上的应用

4.1 Kumada-Corriu 偶联反应及其反应机理

4.1.1 Kumada-Corriu 偶联反应

Kumada-Corriu 偶联反应是指在过渡金属催化下格氏试剂与芳基、烯基卤代物或其类似物之间偶联构建碳-碳键的反应，过渡金属有镍、钯、铜和铁，通常为镍和钯。理论上，该反应是最廉价的交叉偶联反应，产生最少的废弃物。唯一缺点是由于格氏试剂的高反应活性使得该反应与许多官能团不相容。然而，该反应在工业上确实有着广泛的应用。Kumada-Corriu 偶联反应可由图 4-1 表示。

$$R-MgX + R'-X' \xrightarrow{M} R-R' + MgXX'$$

X=Cl, Br, I; X'=Cl, Br, I, OTf 等；
M=Ni, Pd, Cu, Fe 等；
R, R'=芳基、烯基、烷基

图 4-1 Kumada-Corriu 偶联反应

Kudama-Corriu 偶联反应分别由 Corriu 和 Kumada 于 1972 年独立报道[1,2]。该反应属于最先报道的几个催化交叉偶联反应之一。尽管其他几个交叉偶联反应，包括 Suzuki、Sonogashira、Stille、Hiyama、Negishi 的研究和应用远远超越了 Kumada 偶联反应，但其仍被持续应用于许多合成领域，例如在药物合成上用于阿利克仑（aliskiren）的合成，后者是一种高血压治疗药物。

实际上，关于格氏试剂与有机卤代物的催化偶联反应的研究可以追溯到 1941 年 Karasch 与 Fields 采用钴催化剂进行的相关偶联反应的研究[3]。1971 年，Tamura 和 Kochi 又改用银、铜和铁等催化剂进行了更详细的研究[4-6]，然而这些早期研究结果并不理想，其主要问题是发生自偶联反应导致交叉偶联反应产物收率过低。直到 1972 年 Corriu 和 Kumada 将镍类催化剂[1,2]，以及 1975 年 Murahashi[7]又将钯催化剂引入格氏试剂与有机卤代物的交叉偶联反应，反应的底物范围得以拓宽，使其在有机和药物合成上的应用更加广泛。

4.1.2 反应机理

不同金属催化的 Kudama-Corriu 偶联反应的机理细节上有所差异，但总的方向基本相同，以下以钯催化的 Kudama-Corriu 偶联反应为例进行介绍。钯催化的 Kudama-Corriu 反应机理与其他钯催化的交叉偶联反应的机理类似，催化循环如图 4-2 所示[8]：反应开始，有机卤代物（R'—X'）氧化加成到富电子的 L_2Pd^0 活性物种(**A**)生成配合物 $RL_2Pd^{II}X$(**B**)；**B** 与格氏试剂 R'MgX 进行金属交换生成杂金属有机配合物 $RL_2Pd^{II}R'$(**C**)；**C** 异构化为 **D**，使反式位的 R 和 R'转至邻位；最后中间体 **D** 发生还原消除生成交叉偶联产物 R—R'，并再生活性物种 **A**，完成催化循环。反应的决速步骤为 R'—X'对 L_2Pd^0 的氧化加成，钯催化的决速步骤的速率通常比镍催化的低。

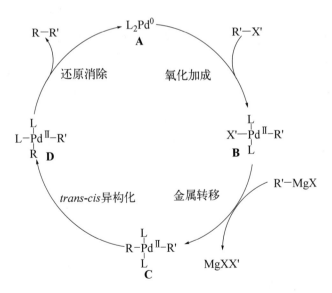

图 4-2　Kudama-Corriu 催化循环

尽管 Kudama-Corriu 反应条件下自偶联反应在很大程度上得到抑制，但反应底物自身的性质决定了此类副反应无法完全避免。这些自偶联反应有的来自金属有机物种的氧化二聚，氧化剂可以是空气中的氧分子或是反应中的亲电试剂；有些则来自有机卤代物经由 Wurtz-类型的还原二聚，还原剂可以是 Pd^0 物种或是制备金属有机试剂的金属单质残留。在配体-钯配位不牢固的情况下自偶联会更加明显，因此配体在 Kudama-Corriu 反应中起着举足轻重的作用。

4.2 基本的 Kudama-Corriu 偶联反应

4.2.1 格氏试剂与乙烯基的偶联

溴乙烯［反应式（1）～式（4）］、氯乙烯［反应式（5）、式（6）］衍生物均能在钯［反应式（1）、式（2）］、镍［反应式（3）～式（5）］、铁［反应式（6）］等过渡金属催化下与格氏试剂发生 Kudama-Corriu 反应，而格氏试剂可以是烷基［反应式（1）、式（6）］、烯基［反应式（2）］和芳基［反应式（3）～式（5）］格式试剂，反应结果与反应物的结构、催化剂的种类密切相关[1,7,9]。

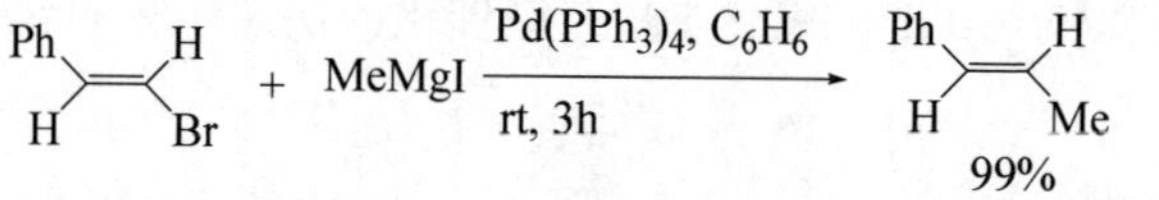

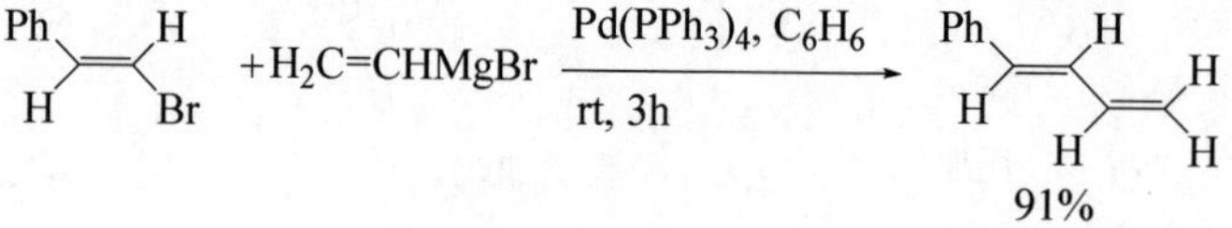

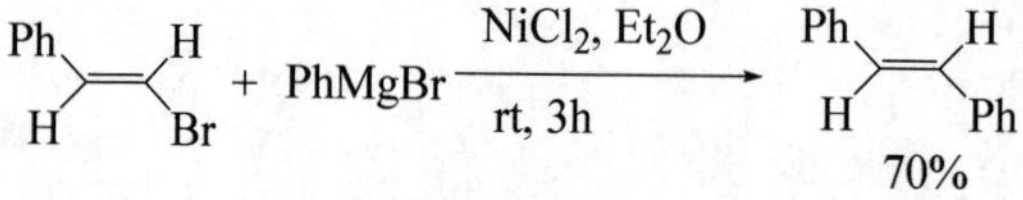

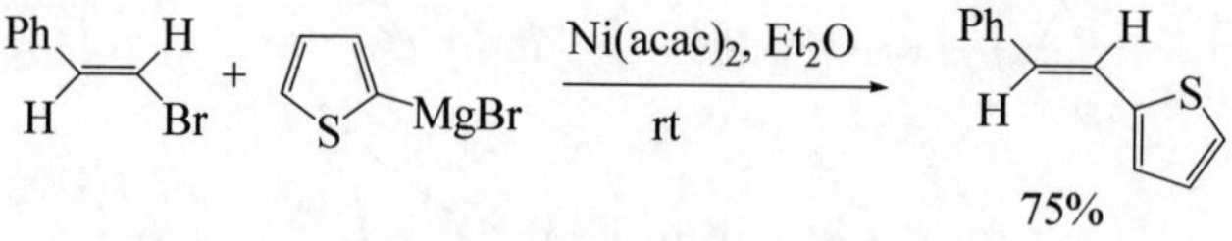

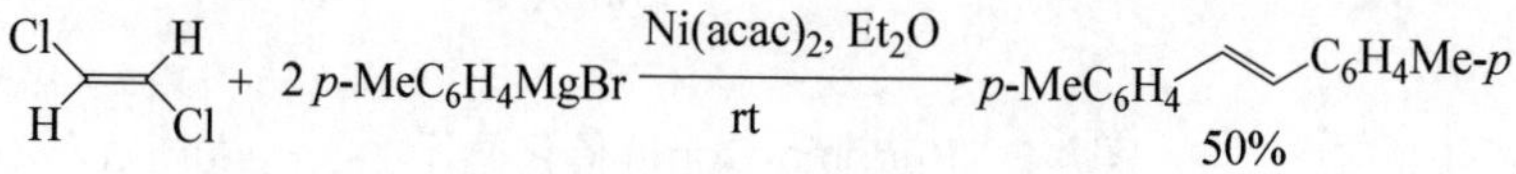

$$\text{Bu}_2\text{C=CHCl} + \text{BuMgCl} \xrightarrow[-5\sim0^\circ\text{C, 0.25h}]{\text{Fe(acac)}_3,\ \text{THF/NMP}} \text{Bu}_2\text{C=CHBu} \quad 85\% \tag{6}$$

上述反应产物烯烃的构型基本得到保持，但有时则发生反转。例如，下式中的碘代烯醇与乙烯基格氏试剂在催化剂 Pd(PPh$_3$)$_4$ 催化下反应，烯丙醇的构型得到保持，而在相同催化剂催化下与 2-丙烯基格氏试剂反应烯丙醇的构型则发生翻转[10]。

$$C_5H_{11}\text{-CH=CI-CH}_2\text{OH} + \text{BrMg-CH=CH}_2 \xrightarrow[\text{rt, 3h}]{\text{Pd(PPh}_3)_4,\ \text{THF}} \xrightarrow{\text{NH}_4\text{Cl(aq.)}} C_5H_{11}\text{-CH=C(CH=CH}_2)\text{-CH}_2\text{OH} \tag{7}$$

$$C_5H_{11}\text{-CH=CI-CH}_2\text{OH} + \text{BrMg-C(Me)=CH}_2 \xrightarrow[\text{rt, 24h}]{\text{Pd(PPh}_3)_4,\ \text{THF}} \xrightarrow{\text{NH}_4\text{Cl(aq.)}} \text{product} \tag{8}$$

手性的格氏试剂在 −78℃ 温度下构型能够保持，与溴乙烯在钯和镍催化剂催化下能得到构型保持的交叉偶联产物，而在铁和钴催化剂催化下则得到消旋的交叉偶联产物[11]。

$$\text{Ph-CH}_2\text{-*CH(MgCl)-CH}_2\text{CH}_3\ (90\%\ ee) + \text{CH}_2\text{=CHBr} \xrightarrow[-78^\circ\text{C, 5 d}]{(\text{dppf})\text{PdCl}_2,\ \text{THF}} \text{Ph-CH}_2\text{-*CH(CH=CH}_2)\text{-CH}_2\text{CH}_3\ (58\%,\ 89\%\ ee) \tag{9}$$

dppf: Fc(C$_5$H$_4$PPh$_2$)$_2$

偕二卤代烯烃可以发生双交叉偶联反应，其中 1,2-二氯乙烯与 2-(1,3-丁二烯基)溴化镁在镍催化剂催化下于室温下进行双交叉偶联反应生成树枝状烯烃[12]。

$$\text{CH}_2\text{=CCl}_2 + 2\ \text{CH}_2\text{=C(MgCl)-CH=CH}_2 \xrightarrow[\text{THF, rt, 24h}]{\text{Ni(dppp)Cl}_2,\ \text{PPh}_3} \text{产物} \quad 65\% \tag{10}$$

偕二氟烯烃与 1,4-或 1,5-双格氏试剂间的双交叉偶联反应可在镍催化剂催化下顺利进行，该反应可用来合成外环三取代的烯烃[13]。

$$\text{BnO-C}_6\text{H}_4\text{-CH=CF}_2 + \text{BrMg(CH}_2)_5\text{MgBr} \xrightarrow[\text{rt, 0.5~1h}]{\text{NiCl}_2(\text{dppp}),\ \text{THF}} \text{BnO-C}_6\text{H}_4\text{-CH=cyclopentylidene} \quad 92\% \tag{11}$$

$$\text{(thiophene-CH=CF}_2\text{)} + \text{BrMg-cycloheptyl} \xrightarrow[\text{rt, 0.5~1h}]{\text{NiCl}_2\text{(dppp), THF}} \text{product} \quad 66\% \tag{12}$$

在二茂铁基双齿膦配体存在下,钯可以催化对甲苯磺酸烯基酯与芳基或烷基格氏试剂的交叉偶联[14]。值得注意的是,对于环烯或芳基烯的对甲苯磺酸酯,反应容易进行,不发生重排 [反应式(13)~式(15)],而脂肪烯的对甲苯磺酸酯则发生重排,原因尚不清楚 [反应式(16)]。

$$\text{环己烯-OTs} + \text{ClMg-C}_6\text{H}_4\text{-CH}_3 \xrightarrow[\text{甲苯, 80℃, 4h}]{\text{Pd}_2\text{(dba)}_3, \text{L1}} \text{product} \quad 71\% \tag{13}$$

$$t\text{-Bu-环己烯-OTs} + \text{ClMg-}i\text{Bu} \xrightarrow[\text{甲苯, rt, 1h}]{\text{Pd}_2\text{(dba)}_3, \text{L1}} \text{product} \quad 82\% \tag{14}$$

$$\text{Ph-C(OTs)=CHCH}_3 + \text{ClMg-}i\text{Bu} \xrightarrow[\text{甲苯, 80℃, 16h}]{\text{Pd}_2\text{(dba)}_3, \text{L1}} \text{product} \quad 82\% \tag{15}$$

$$\text{(}i\text{Pr)C(OTs)=CHCH}_3 + \text{ClMg-C}_6\text{H}_4\text{-CH}_3 \xrightarrow[\text{甲苯, 80℃, 4h}]{\text{Pd}_2\text{(dba)}_3, \text{L2}} \text{product} \quad 43\% \tag{16}$$

Ferrocene-CH(CH₃)-P(t-Bu)₂ / PR₂ L1: R=Ph   L2: R=cHex

4.2.2 格氏试剂与芳基或杂芳基的偶联

各种芳基或杂芳基卤代物或磺酸酯均能与各类格氏试剂发生交叉偶联,无论从偶联伙伴哪一方来看底物范围都非常广泛,甚至含对格氏试剂敏感官能团的芳基卤代物等底物也能进行反应。而且,Kumada-Corriu 偶联反应特别适合于不对称双芳基或杂双芳基化合物的合成。与 Suzuki 偶联反应相比,格氏试剂作为交叉偶联伙伴的 Kumada-Corriu 反应具有步骤更简捷的优势,因为 Suzuki 偶联反应中的芳基硼酸多数情况下是由相应的格氏试剂作为前体合成的。钯、镍、铁等过渡金属,特别是在膦、氮配体存在下,均能催化此类反应的进行,其中 N-杂环卡宾(NHC)类配体与各种金属形成的催化体系催化性能更佳。

首先看一下芳基 [反应式（17）～式（19）] 或杂芳基卤代物 [反应式（20）] 与芳基格氏试剂交叉偶联反应的实例[8]。从这些实例可以看出，Kumada-Corriu 偶联反应既与底物的结构相关，也受催化体系、反应条件影响。

$$\text{2-MeO-C}_6\text{H}_4\text{-Cl} + \text{2-}i\text{Pr-C}_6\text{H}_4\text{-MgBr} \xrightarrow[\text{THF, 25℃, 15h}]{[(t\text{-Bu})_2\text{POH}]_2\text{PdCl}_2} \text{联芳基产物, 91\%} \quad (17)$$

$$\text{MeO}_2\text{C-Ar(Me)-I} + \text{2-Me-C}_6\text{H}_4\text{-MgBr} \xrightarrow[\text{THF, 0℃, 15h}]{[(t\text{-Bu})_2\text{POH}]_2\text{PdCl}_2} \text{联芳基产物, 86\%} \quad (18)$$

$$\text{X-C}_6\text{H}_4\text{-Cl} + \text{ArMgBr} \xrightarrow[\text{1,4-二氧六环/THF, 80℃, 1～5h}]{\text{Pd}_2(\text{dba})_3, \text{IPrHCl}} \text{X-C}_6\text{H}_4\text{-Ar} \quad (19)$$

IPrHCl: 1,3-双(2,6-二异丙基苯基)咪唑氯盐
dba: PhCH=CH-CO-CH=CHPh

X=Me, Ar=Ph, 99%
X=OH, Ar=Ph, 95%
X=MeO, Ar=Ph, 97%
X=MeO, Ar=2,4-二甲基苯基, 95%

$$\text{hetero-ArX} + \text{ArMgCl} \xrightarrow[\substack{\text{THF, }-40\sim+25℃\\5\sim18\text{h}}]{\text{Pd(dba)}_2, \text{dppf}} \text{hetero-Ar—Ar} \quad (20)$$

dppf: 1,1'-双(二苯基膦)二茂铁

4-Ph-吡啶 X=Br, 95%
6-Me-2-Ph-吡啶 X=Cl, 96%
7-Ph-喹啉 X=Cl, 80%
6-(PhO$_2$S)-2-Ph-吡啶 X=Br, 71%

2-Ph-嘧啶 X=Cl, 80%
2-Ph-吡嗪 X=Cl, 91%
4-(EtO$_2$C)-2-(4-(CO$_2$Me)苯基)吡啶 X=Br, 90%

与烯基对甲苯磺酸酯类似，在适宜的配体存在下，芳基或杂对甲苯磺酸酯也能与芳基格氏试剂进行交叉偶联生成联二芳基化合物，空间位阻大的 2,4,6-三甲苯基对甲苯磺酸酯反应困难[14]。

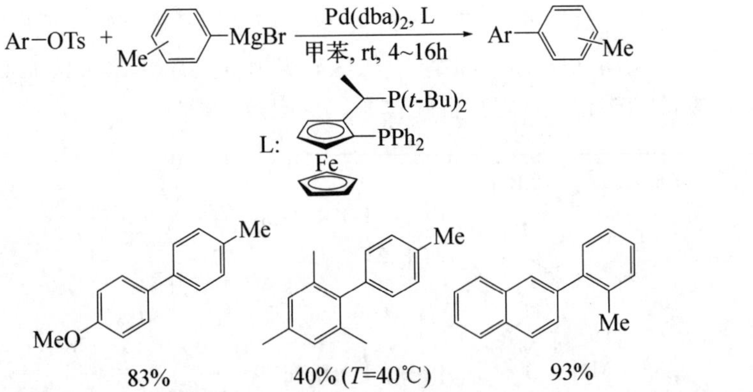

$$\text{Ar-OTs} + \text{Me-C}_6\text{H}_4\text{-MgBr} \xrightarrow[\text{甲苯, rt, 4~16h}]{\text{Pd(dba)}_2, \text{L}} \text{Ar-C}_6\text{H}_4\text{-Me} \qquad (21)$$

L: 二茂铁-CH(CH₃)-P(t-Bu)₂, PPh₂

- 83% (MeO-联苯-Me)
- 40% (T=40℃) (2,4,6-三甲基-联苯-Me)
- 93% (萘基-邻甲苯)

含吸电子基芳基和缺电子杂芳基对甲苯磺酸酯与芳基格氏试剂交叉偶联需要更高的温度和更长的反应时间。邻氯苯基对甲苯磺酸酯只是对甲苯磺酸基发生反应，表明对甲苯磺酸酯基比氯的反应活性高[15]。

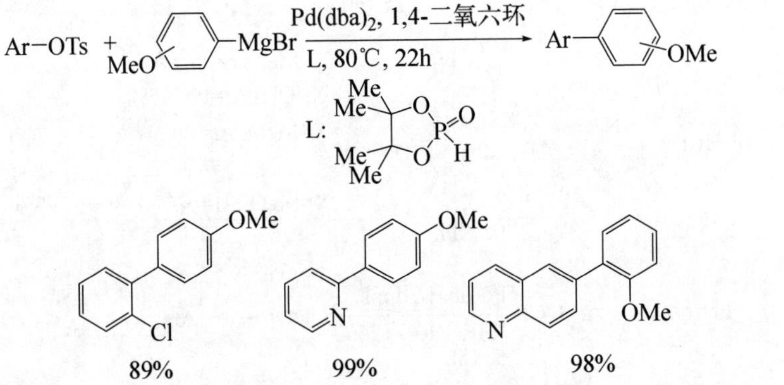

$$\text{Ar-OTs} + \text{MeO-C}_6\text{H}_4\text{-MgBr} \xrightarrow[\text{L, 80℃, 22h}]{\text{Pd(dba)}_2, \text{1,4-二氧六环}} \text{Ar-C}_6\text{H}_4\text{-OMe} \qquad (22)$$

L: 四甲基-二氧磷杂环戊烷-P(=O)H

- 89% (2-氯-4'-甲氧基联苯)
- 99% (2-(4-甲氧基苯基)吡啶)
- 98% (6-(2-甲氧基苯基)喹啉)

杂芳基格氏试剂与杂芳基卤代物间的交叉偶联生成联二杂芳基化合物，是药物合成中的重要反应。格氏试剂3-吡啶氯化镁与各类卤代杂环化合物在Ni(acac)₂/dppe催化下能够发生交叉偶联反应，并且溴代物反应结果通常更好[16,17]。

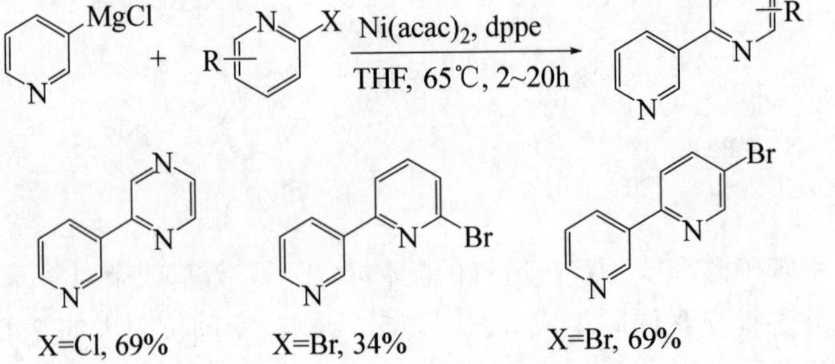

$$\text{3-吡啶-MgCl} + \text{R-杂环-X} \xrightarrow[\text{THF, 65℃, 2~20h}]{\text{Ni(acac)}_2, \text{dppe}} \text{3-吡啶-杂环-R} \qquad (23)$$

- X=Cl, 69% (3-吡啶-2-吡嗪)
- X=Br, 34% (3-吡啶-6-溴-2-吡啶)
- X=Br, 69% (3-吡啶-5-溴-2-吡啶)

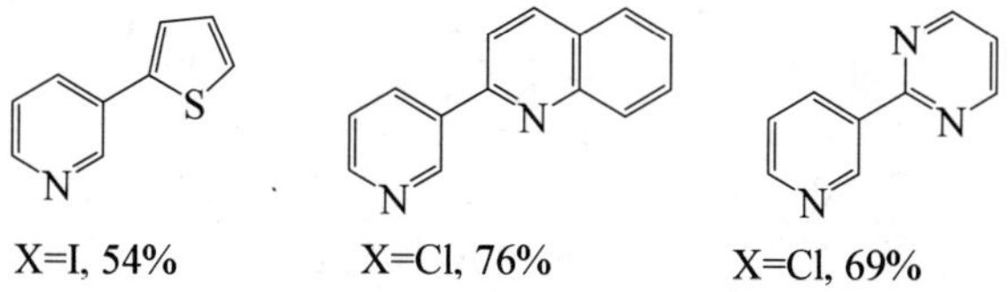

$$\text{X=I, 54\%} \qquad \text{X=Cl, 76\%} \qquad \text{X=Cl, 69\%}$$

与上述类似的催化剂 Ni(dppp)Cl$_2$ 可以催化富电子的噻吩间的偶联，生成 2,2'-联噻吩衍生物，需要注意的是烷基位置影响反应位点的空间位阻，进而影响不同 2-噻吩格氏试剂合成的难易程度，但对偶联反应影响不大[18]。

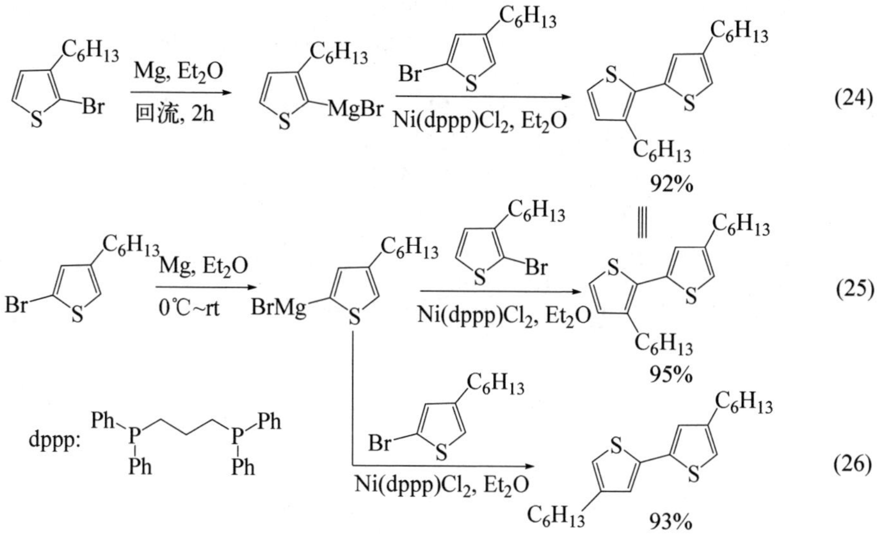

(24) (25) (26)

由于烯基格氏试剂难以制备，几乎没有芳基卤代物与烯基格氏试剂的反应。烷基格氏试剂容易制备，但格氏试剂通常为亲核试剂，能发生各种亲核加成和亲核取代反应。烷基镁物种与作为催化剂的过渡金属配合物在反应体系中会发生金属转移反应生成σ-烷基配合物，进而导致副反应的发生。更重要的是，在烷基 C2 位上带有质子的σ-烷基配合物会发生 β-氢消除，同时生成金属氢化物会引起还原反应的发生。另外，单电子转移或σ-烷基配合物中 M-C 键的均裂也会引起歧化反应的发生。令人欣慰的是，在多数情况下这些副反应可以通过加入适当的配体或者减慢格氏试剂的滴加速度得以抑制，因而（杂）芳基卤代物或其类似物与烷基格氏试剂的反应也很普遍[8]。以下是一些反应实例[19-22]。可以看到，就芳基一端而言，无论芳环上带有吸电子基还是给电子基，在适宜的配体和反应条件下，反应均能进行［反应式（28），式（31）］。就格氏试剂的结构而言，伯烷基、仲烷基和叔烷基都能作为偶联伙伴进行反应，但伯烷基反应结果最好［反应式（27）~式（29）］；叔烷基格氏试剂在 N-杂环卡宾（NHC）作配体存在时由于避免了叔丁基格氏试剂的消除反应，交叉偶联反应收率大大提高［反应式（31）］。

$$\text{3-CF}_3\text{-C}_6\text{H}_4\text{Br} + \text{EtMgCl} \xrightarrow[\text{THF, 20~25℃, 1h}]{\text{NiCl}_2,\ \text{Xantphos}} \text{3-CF}_3\text{-C}_6\text{H}_4\text{Et} \quad 75\% \qquad (27)$$

Xantphos: 4,5-bis(diphenylphosphino)-9,9-dimethylxanthene

$$\text{FC-C}_6\text{H}_4\text{-X} + n\text{-C}_6\text{H}_{13}\text{MgBr} \xrightarrow[\text{0~5℃, 5~10min}]{\text{Fe(acac)}_3,\ \text{THF, NMP}} \text{FC-C}_6\text{H}_4\text{-}n\text{-C}_6\text{H}_{13} \qquad (28)$$

FC=CO$_2$Me, X=Cl, 91%; FC=CO$_2$Me, X=OTf, 87%;
FC=CO$_2$Me, X=OTs, 83‰.; FC=CN, X=Cl, 91%;
FC=CN, X=OTf, 80%; FC=Me, X=OTf, 81%;
FC=SO$_3i$-Pr, X=Cl, 85%

$$\text{2-Cl-pyrimidine} + n\text{-C}_6\text{H}_{13}\text{MgBr} \xrightarrow[\text{0~5℃, 5~10min}]{\text{Fe(acac)}_3,\ \text{THF, NMP}} \text{2-}n\text{-C}_6\text{H}_{13}\text{-pyrimidine} \quad 93\% \qquad (29)$$

$$\text{Ph-C}_6\text{H}_4\text{-OR} + \text{R'MgCl} \xrightarrow[\text{THF, 65℃,}]{\text{FeF}_3\cdot 3\text{H}_2\text{O, IPr}} \text{Ph-C}_6\text{H}_4\text{-R'} \qquad (30)$$

IPr: 1,3-bis(2,6-diisopropylphenyl)imidazolium chloride

R = —SO$_2$NMe$_2$, Ph—C$_6$H$_4$—n-Bu 90%

R = —SO$_2$NMe$_2$, Ph—C$_6$H$_4$—Cy 77%

R = —SO$_2$NMe$_2$, Ph—C$_6$H$_4$—n-Bu 65% (i:n = 6.5:1)

R = —Ts, Ph—C$_6$H$_4$—i-Pr 50% (i:n = 1:10)

$$\text{R-C}_6\text{H}_4\text{-Br} + t\text{-BuMgCl} \xrightarrow[\text{THF, -10℃, 1.5h}]{\text{NiCl}_2\cdot 1.5\text{H}_2\text{O, L}} \text{R-C}_6\text{H}_4\text{-}t\text{-Bu} \qquad (31)$$

L: 1,3-dicyclohexylimidazolium tetrafluoroborate (Cy-N$^+$=N-Cy, BF$_4^-$)

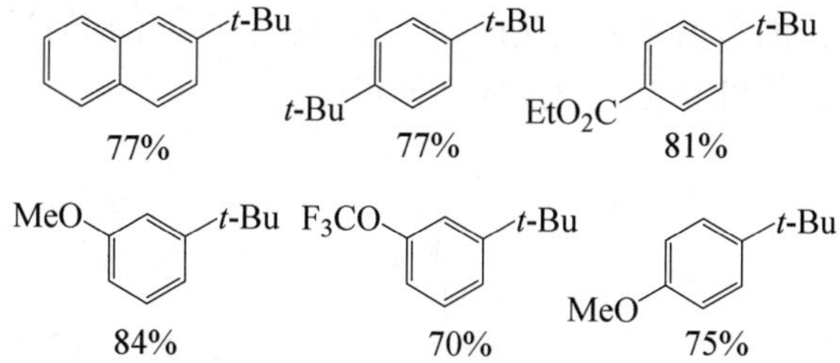

产物结构图：
- 2-叔丁基萘 77%
- 1,4-二叔丁基苯 77%
- 4-叔丁基苯甲酸乙酯 81%
- 3-甲氧基叔丁基苯 84%
- 3-三氟甲氧基叔丁基苯 70%
- 4-甲氧基叔丁基苯 75%

炔基格氏试剂在钯催化剂催化下能与碘代芳烃发生偶联反应[23]。当加入助剂 LiBr 时芳基三氟甲磺酸酯也能与炔基格氏试剂反应，但相同条件下芳基溴则几乎不反应，这与通常的 Kumada-Corriu 反应有所不同[24]。

$$\text{PhI} + \text{PhC}\equiv\text{CMgBr} \xrightarrow[\text{THF, 65℃, 0.5h}]{(\text{Ph}_3\text{P})_2\text{Pd(Ph)I}} \text{PhC}\equiv\text{CPh} \qquad (32)$$
84%

$$\text{4-BrC}_6\text{H}_4\text{OTf} + \text{RC}\equiv\text{CMgBr} \xrightarrow[\text{Et}_2\text{O, PhMe, 20~30℃, 1~20h}]{\text{PdCl}_2(\text{alaphos}), \text{LiBr}} \text{4-Br-C}_6\text{H}_4-\text{C}\equiv\text{CR} \qquad (33)$$

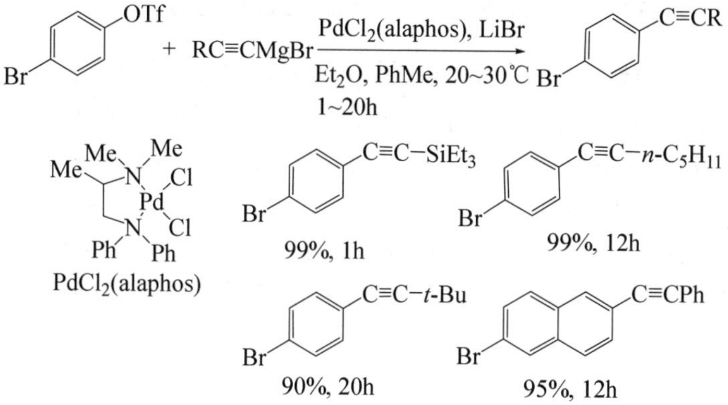

PdCl$_2$(alaphos) 结构（Me$_2$N-CH$_2$-CH$_2$-NPh$_2$ 与 PdCl$_2$ 配位）

产物：
- 4-Br-C$_6$H$_4$-C≡C-SiEt$_3$ 99%, 1h
- 4-Br-C$_6$H$_4$-C≡C-n-C$_5$H$_{11}$ 99%, 12h
- 4-Br-C$_6$H$_4$-C≡C-t-Bu   90%, 20h
- 6-溴-2-萘基-C≡CPh   95%, 12h

4.2.3 格氏试剂与烷基的偶联

相较于格氏试剂与 sp^2-杂化的烯基、芳基卤代物或其类似物间的众多的偶联反应，格氏试剂与 sp^3-杂化的烷基卤化物或其类似物间的偶联反应要少得多。其原因如图 4-3 所示[8]。首先，烷基卤化物在强碱性条件下容易消除卤化氢。再者，烷基卤化物与低价态的过渡金属不容易发生氧化加成反应，即使烷基卤化物与过渡金属配合物发生氧化加成，生成的烷基过渡金属配合物缺少 sp^2 碳原子中 p 轨道与过渡金属 d 轨道间的 p-d 稳定化作用，会发生竞争的 β-消除反应生成非目标产物烯烃，而生成的过渡金属氢化物又引起其他副反应的发生。其次，在生成交叉偶联产物的还原消除一步，烷基的还原消除相对较慢，也会引起副反应的发生。而且，通过向富电子过渡金属上的氧化加成使烷基卤化物得到活化在多数情况下属于单电子转移过程，最终导致自偶联产物的生成。还有，σ-烷基配合物中间体还可进行 M-C 键的均裂生成短寿命的自由基和歧化产物。亲电的烷基卤化物的反应活性顺序一般为伯烷基＞仲烷基，叔烷基卤化物反应活性很差。

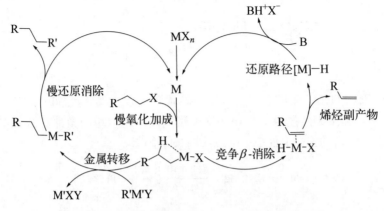

图 4-3 烷基卤化物与格氏试剂偶联及其竞争反应

尽管烷基卤化物与格氏试剂间的交叉偶联反应存在如上所述的诸多不利因素，但在适当的催化体系和反应条件下仍能进行反应，以可接受的收率得到交叉偶联反应产物。其中，芳基格氏试剂与伯烷基碘、伯烷基溴以及仲环烷基碘在一种 NN_2-钳形镍配合物与助剂四甲基乙二胺（TEMDA）构成的催化体系催化下在室温就能发生反应，以优良的收率得到交叉偶联产物，而且含官能团的烷基卤化物也能顺利反应[25]。TEMDA 可能通过与 Mg^{2+} 配位稳定格氏试剂并增强格氏试剂的亲核性能，从而抑制自偶联反应的发生。

$$R-X + ArMgX' \xrightarrow[THF, rt, 1h]{NN_2NiCl, TEMDA} R-Ar \quad (34)$$

X = I, X' = Cl, n-C_8H_{17}—〔Ph〕, 92%

X = I, X' = Br, n-C_8H_{17}—〔Ph〕—F, 98%

X = I, X' = Br, n-C_8H_{17}—〔Ph〕—OMe, 80%

X = I, X' = Br, n-C_8H_{17}—〔Py〕, 76%

X = Br, X' = I, 环己基-Ph, 65%

X = I, X' = Cl, (吲哚-CO$_2$Me, N-丙基Ph), 81%

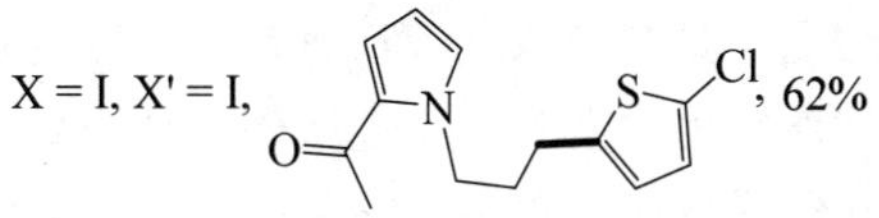

$X = I, X' = I$, [structure], 62%

在与上述相同的催化剂催化下，无须助剂，各种伯、仲烷基卤化物与烷基格氏试剂交叉偶联，构建 $C(sp^3)$-$C(sp^3)$ 键，而且含羰基底物中的羰基不与格氏试剂反应[26,27]。

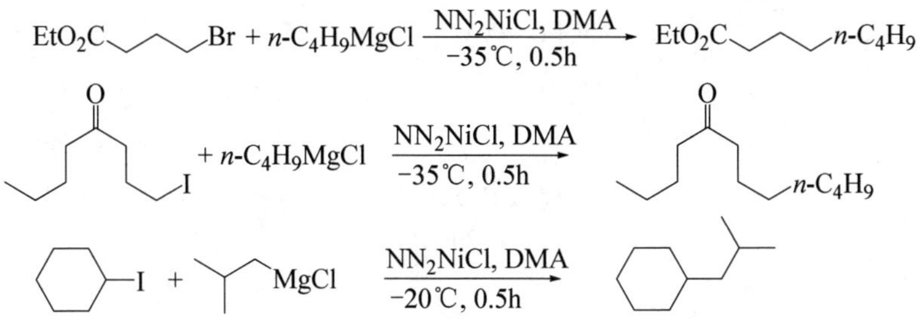

$$EtO_2C\text{—}\text{—}Br + n\text{-}C_4H_9MgCl \xrightarrow[-35℃, 0.5h]{NN_2NiCl, DMA} EtO_2C\text{—}\text{—}n\text{-}C_4H_9 \quad (35)$$

$$[\text{ketone-I}] + n\text{-}C_4H_9MgCl \xrightarrow[-35℃, 0.5h]{NN_2NiCl, DMA} [\text{ketone-}n\text{-}C_4H_9] \quad (36)$$

$$\text{Cy-I} + i\text{-BuMgCl} \xrightarrow[-20℃, 0.5h]{NN_2NiCl, DMA} \text{Cy-}i\text{Bu} \quad (37)$$

烷基对甲苯磺酸酯也能与烷基格氏试剂在铜催化剂催化下顺利地发生交叉偶联反应，如果仲烷基对甲苯磺酸酯是光学活性的，在特定的催化体系下反应按 S_N2 机理进行，反应后烷基的构型发生反转[28,29]。

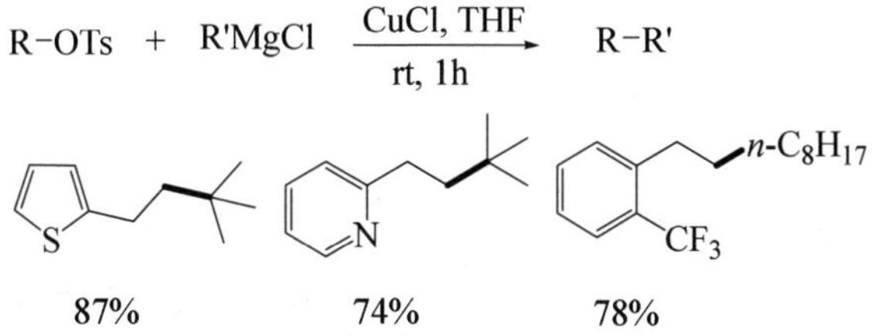

$$\text{R-OTs} + \text{R'MgCl} \xrightarrow[\text{rt, 1h}]{CuCl, THF} \text{R-R'} \quad (38)$$

87%　　74%　　78%

$$\underset{R^1}{\overset{OTs}{\text{C}}}\!R^2 + R^3\text{—MgBr} \xrightarrow[0℃, 24h]{CuI, TMEDA, LiOMe} \underset{R^1}{\overset{R^3}{\text{C}}}\!R^2 \quad (39)$$

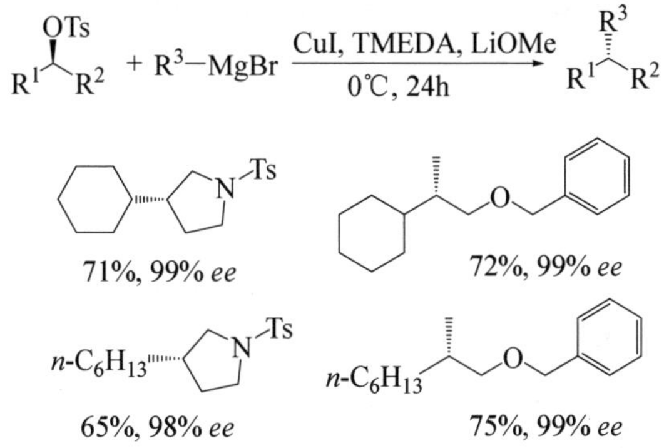

71%, 99% ee　　72%, 99% ee

65%, 98% ee　　75%, 99% ee

4.2.4　特殊的 Kumada-Corriu 偶联反应

表观上 Kumada-Corriu 偶联反应可看作亲核取代反应，亲核试剂是格氏试剂。

传统上醚、砜和氨甲酸酯不是亲核取代反应的底物，因为烃氧基、砜基和氨甲酰基不是好的离去基团。然而，在适宜的过渡金属催化体系催化下，上述各类化合物却能发生 Kumada-Corriu 偶联反应。

各种甲基醚与各类格氏试剂在适宜的过渡金属催化体系催化下能够进行反应得到相应的交叉偶联产物[30]。

$$\text{MeO-C}_6\text{H}_4\text{-CH}_2\text{CH}_2\text{NH}_2 + \text{PhMgBr} \xrightarrow[t\text{-AmOMe, 23℃, 15h}]{\text{NiCl}_2(\text{PhPCy}_2)_2,\ \text{Et}_2\text{O}} \text{Ph-C}_6\text{H}_4\text{-CH}_2\text{CH}_2\text{NH}_2 \quad 82\%} \tag{40}$$

$$(\text{4-MeO-C}_6\text{H}_4)\text{-(6-OMe-naphthyl)} + \text{MeMgBr} \xrightarrow[80℃, 20min]{\text{NiCl}_2(\text{PCy}_3)_2,\ \text{PhMe}} (\text{4-MeO-C}_6\text{H}_4)\text{-(6-Me-naphthyl)} \quad 90\% \tag{41}$$

$$\text{Naphthyl-CH}_2\text{OMe} + n\text{-C}_4\text{H}_9\text{MgBr} \xrightarrow[80℃, 12h]{\text{NiCl}_2(\text{dppf}),\ \text{PhMe}} \text{Naphthyl-CH}_2\text{-}n\text{-Bu} \quad 51\% \tag{42}$$

$$\text{Ar-CH(OR)-CH}_2\text{Me} + \text{MeMgI} \xrightarrow[\text{PhMe, rt, 24h}]{\text{Ni(cod)}_2,\ rac\text{-BINAP}} \text{Ar-CH(Me)-CH}_2\text{Me} \tag{43}$$

(6-MeO-naphthyl)-CH(Me)Et　70%, 85% ee 　　(benzofuran-2-yl)-CH(Me)Et　82%, 87% ee 　　(benzofuran-3-yl)-CH(Me)Et　87%, 95% ee

在镍催化剂催化下，芳基氨甲酸酯与各种格氏试剂进行交叉偶联反应得到多取代芳香化合物[31]。

$$\text{Ar-OCONEt}_2 + \text{RMgBr} \xrightarrow[\text{rt, 2~24h}]{\text{Ni(acac)}_2,\ \text{Et}_2\text{O}} \text{Ar-R} \tag{44}$$

BocHN-C₆H₄-Ph　73%　　MOMO-C₆H₄-CH₂TMS　82%　　vinyl-phenanthrene　55%

(4-TMS-quinolin-3-yl)-CH₂TMS　72%　　steroid-CH₂TMS　70%

我国科学家在本领域内做出过诸多贡献，例如李金恒教授发现 1-芳基-2-(4-甲基苯砜基) 乙酮与烷基、芳基格氏试剂在 $NiI_2(PPh_3)_2$/PCy_3 催化下以较高收率得到相应的交叉偶联反应产物[32]。

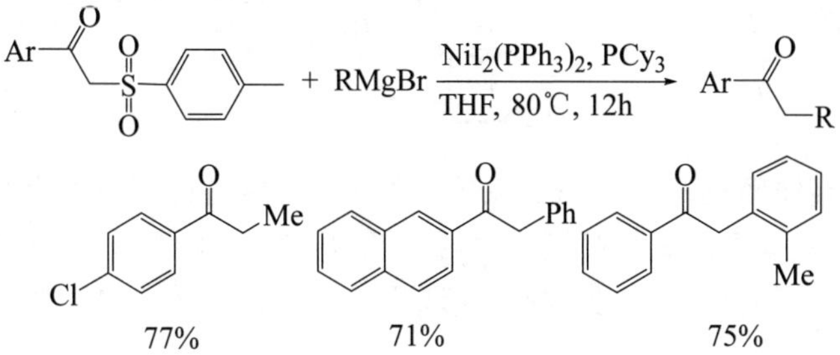

$$Ar\overset{O}{-}\overset{}{-}S(=O)_2-C_6H_4-CH_3 + RMgBr \xrightarrow{NiI_2(PPh_3)_2, PCy_3}{THF, 80℃, 12h} Ar\overset{O}{-}R \quad (45)$$

77% 71% 75%

施章杰教授也发现特戊酸烯醇酯在铁催化剂催化下在温和条件下也能够与烷基格氏试剂进行交叉偶联反应，反应耐受多种官能团[33]。

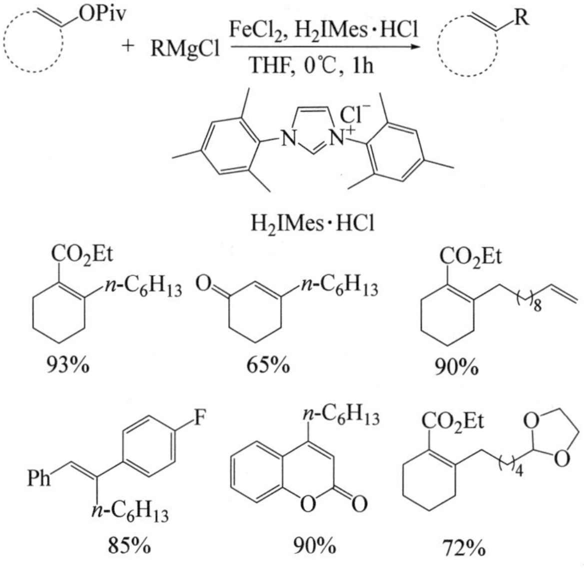

$$\text{烯醇}-OPiv + RMgCl \xrightarrow{FeCl_2, H_2IMes·HCl}{THF, 0℃, 1h} \text{烯烃}-R \quad (46)$$

$H_2IMes·HCl$

93% 65% 90%

85% 90% 72%

4.2.5 不对称 Kumada-Corriu 偶联反应

不对称 Kumada-Corriu 偶联反应具有挑战性，早期在溴乙烯与仲烷基格氏试剂交叉偶联反应中实现，反应在手性二茂铁基氨基膦配体与镍或钯配合物共同催化下进行，但对映选择性较低。分析认为，反应过程中配体的二甲氨基与格氏试剂中的镁配位形成非对映异构的配合物，立体选择性由催化循环中的金属转移一步决定[34]。

$$\text{(47)}$$

反应式(47)：PhCH(Me)MgCl + CH$_2$=CHBr $\xrightarrow{\text{PdCl}_2\text{L}^*,\text{ Et}_2\text{O}}{25℃,\ 60\sim70\text{h}}$ PhCH(Me)CH=CH$_2$，93%, 61% ee

L* = 二茂铁基 PPh$_2$/NMe$_2$/CH$_3$/PPh$_2$ 配体

后来，发现其他底物在适宜的手性配体存在下也能进行不对称交叉偶联反应。其中 α-溴代酮与芳基格氏试剂在氯化镍与手性双噁唑啉配体原位生成的手性配合物催化下反应以良好的收率和优良的对映选择性得到光学活性的交叉偶联产物[35]。

$$\text{ArC(O)CHBrR + Ar'MgBr} \xrightarrow[\text{DME, }-60℃,\ 16\sim32\text{h}]{\text{NiCl}_2\cdot\text{glyme, L}^*} \text{ArC(O)CH(R)(Ar')} \quad (48)$$

L*: 双噁唑啉配体 (4,4'-二苯基, gem-二甲基桥)

产物示例：

- 2-F-C$_6$H$_4$-C(O)-CH(Me)(Ph)：89%, 72% ee
- 3,4-methylenedioxyphenyl-C(O)-CH(Me)(Ph)：76%, 90% ee
- 2-thienyl-C(O)-CH(Me)(Ph)：91%, 87% ee
- Ph-C(O)-CH(Ph)CH$_2$CH$_2$N$_3$：72%, 80% ee
- Ph-C(O)-CH(Me)(2-CO$_2$Et-C$_6$H$_4$)：80%, 79% ee
- Ph-C(O)-CH(Me)(2-CN-C$_6$H$_4$)：91%, 95% ee
- Ph-C(O)-CH(Me)(2-OMe-C$_6$H$_4$)：76%, 92% ee
- Ph-C(O)-CH(Me)(5-(N-Boc-indolyl))：73%, 90% ee

在类似的手性催化体系催化下，对称的硫酸环酯与芳基格氏试剂的不对称偶联开环是对映选择性构建碳手性中心的重要手段，反应以优异的对映选择性（er）生成 R-构型的反应产物，具有广泛的官能团耐受性，多种官能团化的硫酸环酯都能进行反应[36]。

$$\text{环硫酸酯(R取代)} + \text{ArMgBr} \xrightarrow[\text{(2) H}_2\text{SO}_4\text{ (aq 20\%), rt, 24h}]{\text{(1) Ni(acac)}_2,\ (R,R)\text{-L, THF, }-25℃,\ 15\text{h}} \text{HOCH}_2\text{CH(R)Ar} \quad (49)$$

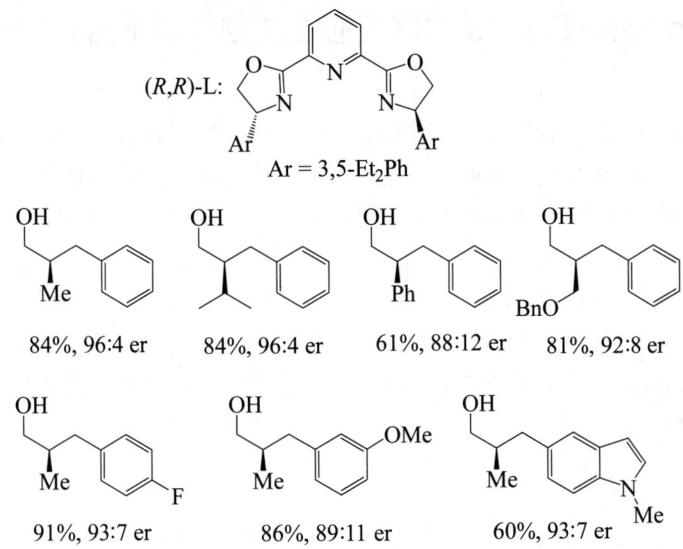

(R,R)-L: [pybox ligand structure]
Ar = 3,5-Et₂Ph

84%, 96:4 er   84%, 96:4 er   61%, 88:12 er   81%, 92:8 er

91%, 93:7 er   86%, 89:11 er   60%, 93:7 er

前面已经谈及光学活性的醚与格氏试剂的交叉偶联反应，反应被认为是按 S_N2 机理进行，因而产物的构型发生翻转［反应式（43）］，环醚也遵循这个原则。而对于含两个手性中心的芳基环醚如四氢呋喃和四氢吡喃，当与格氏试剂交叉偶联反应时，反应是非对映选择性的，反应位点的手性中心的构型仍然发生翻转，而另一手性中心的构型得到保持，而且反应的非对映选择性不受配体构型影响[37]。

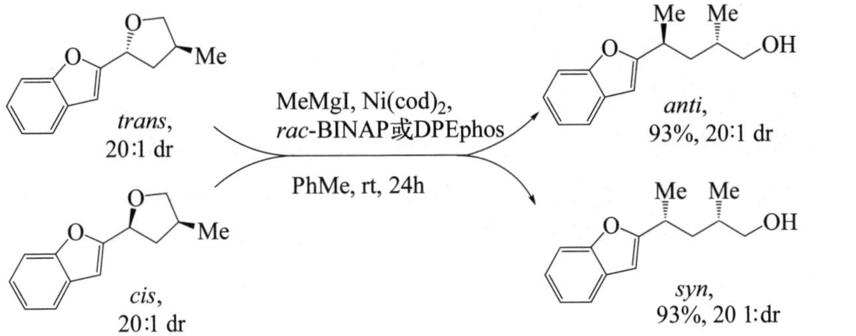

(50) trans, 20:1 dr → MeMgI, Ni(cod)₂, rac-BINAP 或 DPEphos, PhMe, rt, 24h → anti, 93%, 20:1 dr

(51) cis, 20:1 dr → syn, 93%, 20:1 dr

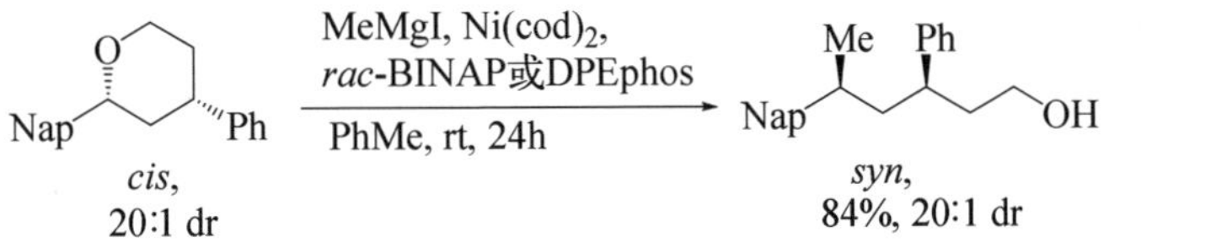

(52) cis, 20:1 dr + MeMgI, Ni(cod)₂, rac-BINAP 或 DPEphos, PhMe, rt, 24h → syn, 84%, 20:1 dr

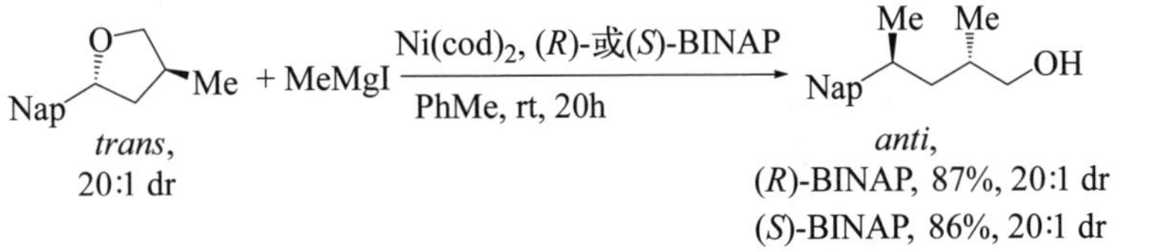

(53) trans, 20:1 dr + MeMgI, Ni(cod)₂, (R)- 或 (S)-BINAP, PhMe, rt, 20h → anti, (R)-BINAP, 87%, 20:1 dr; (S)-BINAP, 86%, 20:1 dr

4.3 Kumada-Corriu 偶联反应在药物合成上的应用实例

由基本反应可知，通过各种格氏试剂与多种卤化物或其类似物间的偶联反应可得到种类繁多的偶联产物，这些偶联产物往往是一些药物合成的关键中间体，因而为药物分子的合成提供了便捷的途径。以下将介绍 Kumada-Corriu 偶联反应在药物合成上的应用实例，为叙述方便，把 Kumada-Corriu 偶联产物编号为 KC-*x*，*x* 为整数，代表序号。

盐酸西那卡塞(cinacalcet hydrochloride)是一种拟钙剂，其合成借助了 Kumada-Corriu 偶联反应。首先，3-氯-*N*-[(1*R*)-1-(1-萘基)乙基]-2-丙烯-1-胺与 3-三氟甲基苯基溴化镁在 $Fe(acac)_3$ 催化下于 THF/NMP 混合溶剂中−5～0°C 反应以中等收率得到偶联产物 KC-1，KC-1 经催化加氢、成盐得盐酸西那卡塞[38]。

KC-1, 50%

盐酸西那卡塞

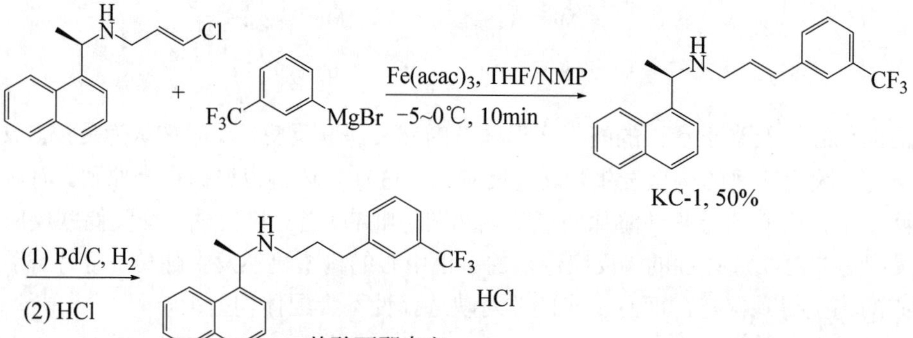

(54)

康普瑞汀（combretastatin A-4）是一种含顺二苯乙烯结构的天然产物，尽管结构简单，但它却显示出显著的生物活性，并与微管蛋白上的秋水仙碱位点强烈相互作用。康普瑞汀可借助多种交叉偶联反应合成，Kumada-Corriu 偶联反应是其中之一，优于其他合成路线。合成过程中，2-溴-1-(3,4,5-三甲氧基苯基) 乙烯与基团保护的芳基格氏试剂在铁催化剂催化下于温和条件下高立体选择性、高收率地生成关键中间体 KC-2，KC-2 经一步脱保护基 TBDMS 即得到康普瑞汀[39]。

KC-2, 61%

康普瑞汀

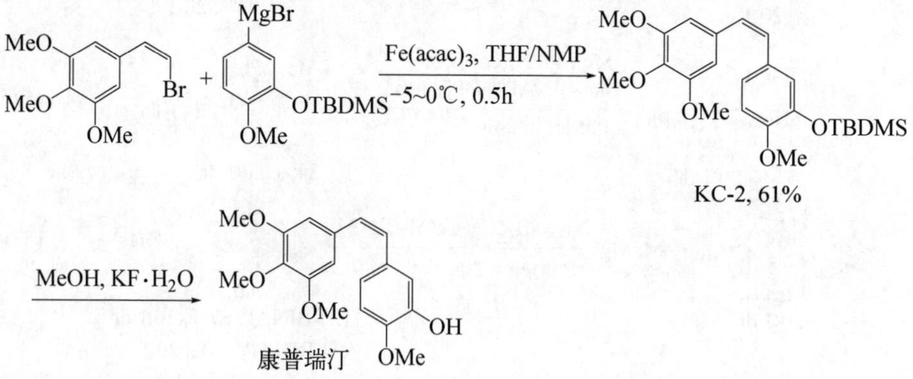

(55)

阿利吉伦（aliskiren）是新一代非肽类肾素阻滞药，其关键中间体 KC-3 是通过如下反应式所示的 Kumada-Corriu 偶联反应合成的。其中氯代烷格氏试剂需要在甲基氯化镁活化下才能生成，生成的卤代烷格氏试剂随后与偶联伙伴在铁催化剂催化下在温和条件下顺利发生交叉偶联反应生成关键中间体 KC-3，KC-3 经多步反应生成阿利吉伦[40]。

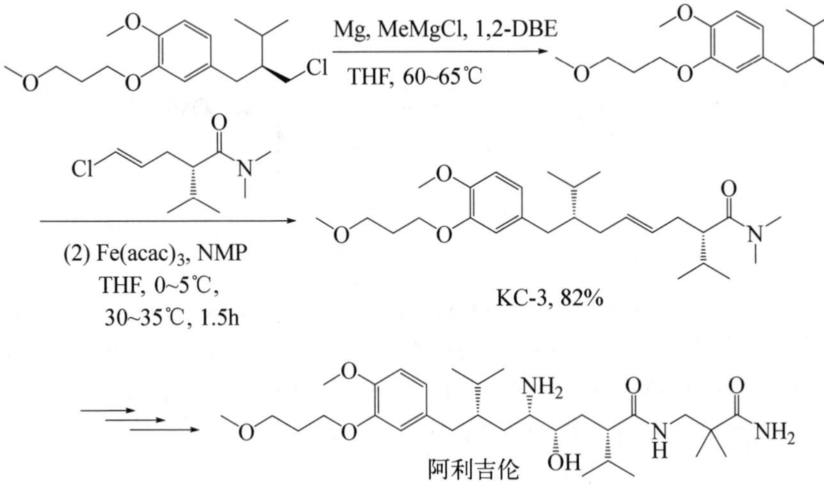

$$\text{(56)}$$

二氟尼柳（diflunisal），是一种非甾体抗炎药，能够缓解关节炎、类风湿性关节炎等引起的疼痛。其最简捷的合成路线是先由对甲氧基溴化镁与 2,4-二氟溴苯交叉偶联反应生成中间体 2,4-二氟-4'-甲氧基联苯（KC-4），KC-4 在酸性条件下脱甲基生成 4-(2,4-二氟苯基)苯酚，后者经 Kolbe-Schmitt 反应生成二氟尼柳[41]。

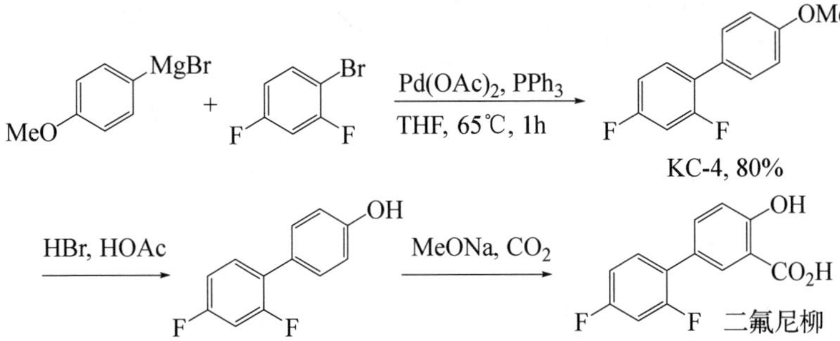

$$\text{(57)}$$

阿扎那韦（atazanavir）是一种 HIV-1 蛋白酶抑制剂，用于治疗艾滋病。阿扎那韦合成中所需的侧链中间体 4-(2-吡啶基)苯甲醛（KC-5）是通过 4-(二甲氧基)甲苯基溴化镁与 2-溴吡啶的 Kumada-Corriu 偶联反应合成的。一旦 KC-5 得到，经还原胺化等多步反应最终生成阿扎那韦[42]。

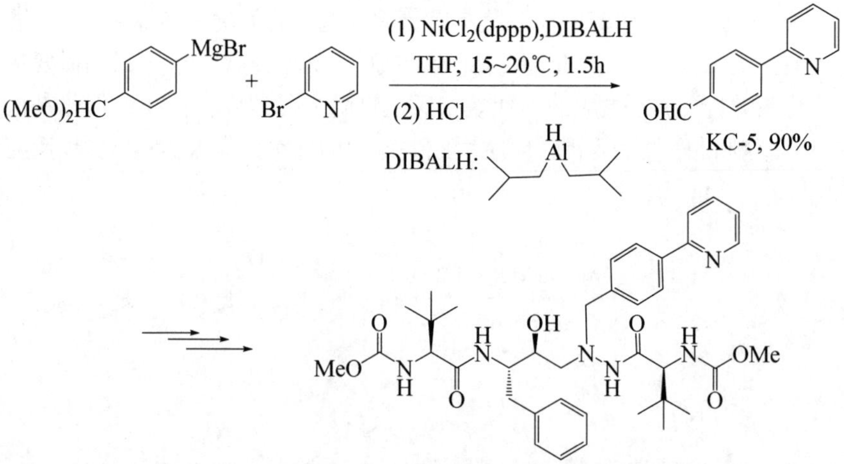

齐美定（zimelidine）是具有二环新结构的抗抑郁药，其前体（KC-6）可通过 Z-对甲苯磺酸芳基乙烯酯与 3-吡啶溴化镁的交叉偶联反应合成，催化体系为 $FeCl_3$/SIPr/TMEDA/Ti(OEt)$_4$/PhOMgX[43]。反应的特征是热力学不稳定的烯烃 Z-式构型得到保持。KC-6 经转化即可得到齐美定[44]。

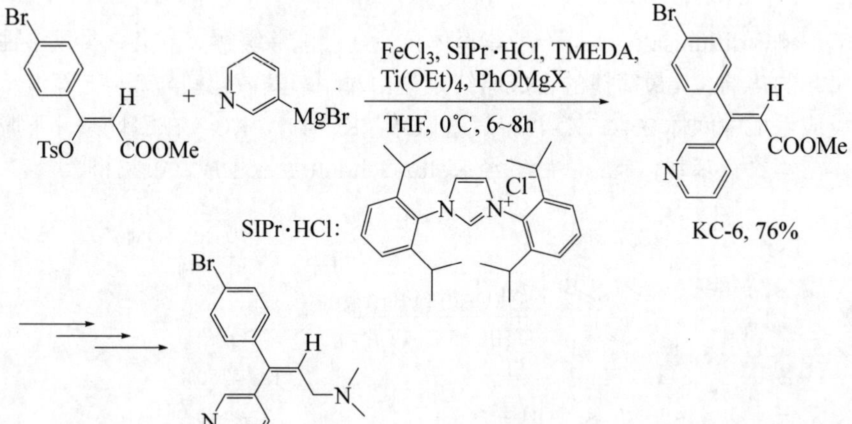

5,6-二取代苯并[cd]吲哚 AG341 是胸苷酸合成酶（TS）的有效抑制剂，因此具有潜在的抗癌治疗作用。其关键中间体 5-甲基苯并[cd]吲哚-2(1H)-酮（KC-7）可通过苯并[cd]吲哚-2(1H)-酮与甲基格氏试剂的交叉偶联反应合成。酰胺结构中的 N-H 质子不阻碍 Kumada-Corriu 偶联反应的进行[45]。KC-7 再经后续多步反应生成 AG341。

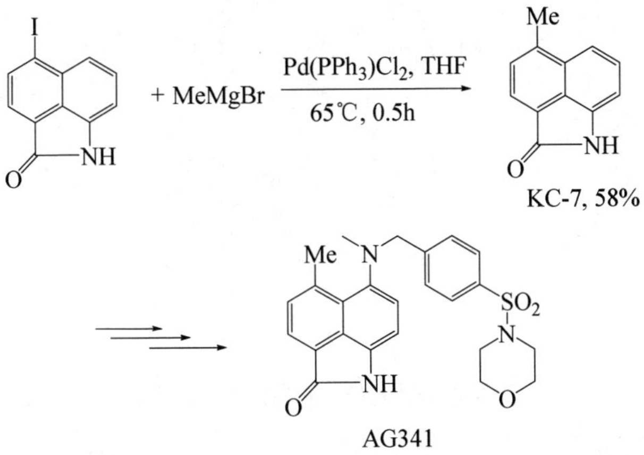

$$\text{(structure with I)} + \text{MeMgBr} \xrightarrow{\text{Pd(PPh}_3)_2\text{Cl}_2, \text{THF}}_{65℃, 0.5\text{h}} \text{KC-7, 58\%}$$

$$\xrightarrow{\Rightarrow\Rightarrow} \text{AG341} \tag{60}$$

ST1535 是一种高选择性腺苷 A2A 受体配体拮抗剂，被认为是治疗帕金森病的良好临床候选药物，其结构属于 2-烷基-6-氨基-9-甲基-8-三唑嘌呤。尽管其可由 Stille 反应和 Suzuki-Miyaura 反应合成，但均存在条件苛刻、成本高、试剂毒性大等诸多缺点。采用 Kumada-Corriu 偶联反应合成 ST1535 避免了以上缺点。实际合成过程中，前体 2-氯嘌呤与正丁基格氏试剂在 Fe(acac)$_3$ 催化下在 THF/NMP 混合溶剂中于室温下反应 1h，以 91%的收率得到交叉偶联产物中间体 KC-8，中间体 KC-8 再经转化生成 ST1535[46]。

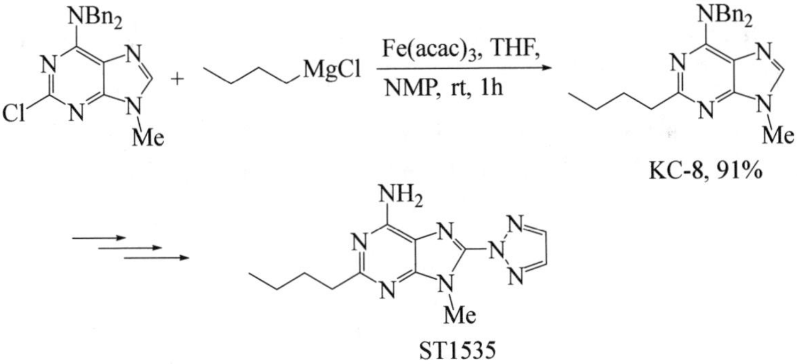

$$\text{2-Cl-purine} + \text{butyl-MgCl} \xrightarrow[\text{NMP, rt, 1h}]{\text{Fe(acac)}_3, \text{THF},} \text{KC-8, 91\%}$$

$$\xrightarrow{\Rightarrow\Rightarrow} \text{ST1535} \tag{61}$$

BILN 2061 是一种 HCV NS3 蛋白酶抑制剂，其合成中的最基本中间体 7-氯-1-辛烯（KC-9）则是通过 1-溴-4-氯丁烷与烯丙基溴化镁在四氯铜酸二锂（Li$_2$CuCl$_4$）催化下交叉偶联合成的，反应选择性地发生在溴的一端[47]。7-氯-1-辛烯得到后，再经多步反应得到 BILN 2061。

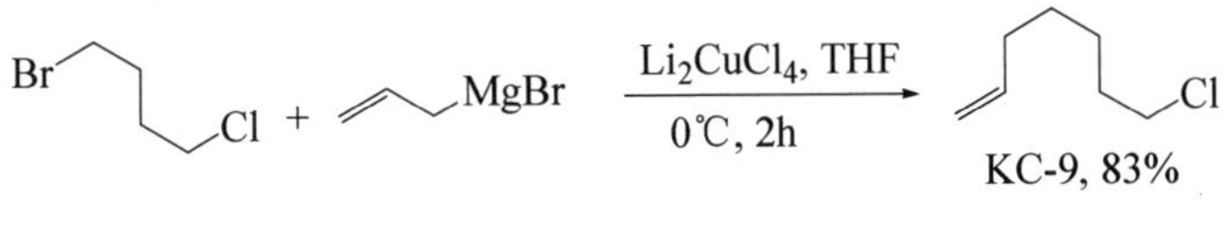

$$\text{Br-(CH}_2)_4\text{-Cl} + \text{allyl-MgBr} \xrightarrow[0℃, 2\text{h}]{\text{Li}_2\text{CuCl}_4, \text{THF}} \text{KC-9, 83\%}$$

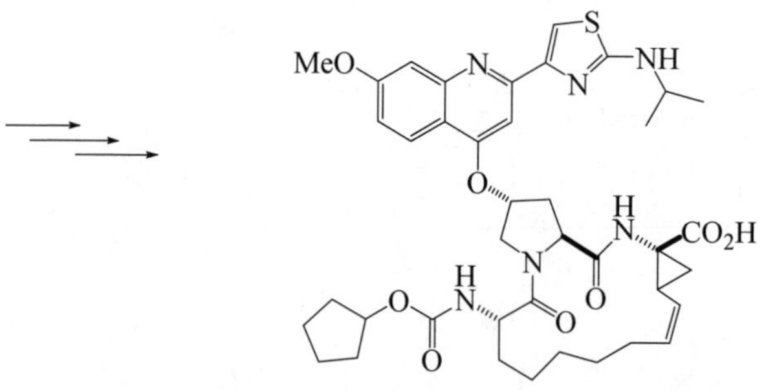

MeO
N
S
N
NH
O
H
N
CO2H
N
O
H
N
O
O
O
O
BILN 2061

(62)

（7*Z*，11*Z*，13*E*）-十六碳三烯醛是柑橘叶潜蝇的性信息素主要成分，其简短合成路线如反应式（63）所示。在该合成路线中，乙基格氏试剂与溴代十四碳四烯间的Kumada-Corriu偶联是最关键的一步，反应在镍催化剂催化下于温和条件下进行，高选择性、高收率地生成*E*-构型的乙基化产物（KC-10）[48]。

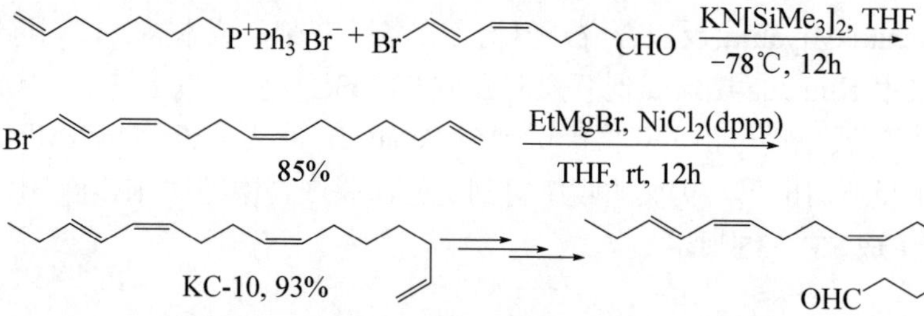

P+Ph3 Br− + Br
CHO
KN[SiMe3]2, THF
−78℃, 12h
Br
85%
EtMgBr, NiCl2(dppp)
THF, rt, 12h
KC-10, 93%
OHC

(63)

他莫昔芬（tamoxifen）是治疗乳腺癌最常用的处方药之一，其结构存在空间拥挤，但利用钯催化的氨基甲酸烯醇酯与格氏试剂4-[2-(二甲氨基)-乙氧基]苯基溴化镁的Kumada-Corriu偶联反应，两种构型的氨基甲酸烯醇酯均能顺利反应，空间位阻小的*E*-异构体反应更容易，而且反应后烯烃的构型得到完全保持[49]。

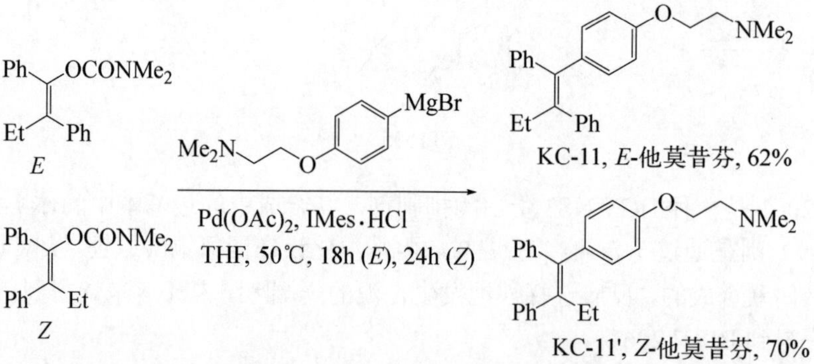

Ph
OCONMe2
Et
Ph
E
MgBr
Me2N
O
Pd(OAc)2, IMes·HCl
THF, 50℃, 18h (E), 24h (Z)
Ph
OCONMe2
Ph
Et
Z
O
NMe2
Ph
Et
Ph
KC-11, E-他莫昔芬, 62%
O
NMe2
Ph
Ph
Et
KC-11', Z-他莫昔芬, 70%

(64)

利用烯醇三氟甲磺酸酯与甲基格氏试剂的交叉偶联反应可以合成天然的倍半萜(–)-*α*-荜澄茄烯[(–)-*α*-cubebene]（KC-11），反应在$Fe(acac)_3$催化下顺利进行，得到的偶联产物KC-12进一步转化可得到差向异构的4-*epi*-荜澄茄醇[50]。

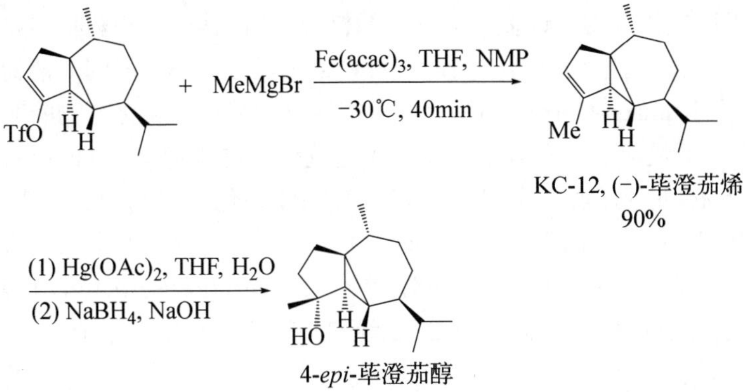

別红藻氨酸［(+)-allokainic acid］具有显著的生物活性，影响哺乳动物中枢神经系统中的神经兴奋性活动，其合成受到广泛关注，其中 2-溴丙烯格氏试剂与仲溴代烷衍生物的交叉偶联生成中间体 KC-13 是其中关键一步，反应以廉价易得无毒的氯化铁为催化剂，彰显该路线的优势，但由于无法避免吡咯烷环开环的开环反应，收率偏低[51]。中间体 KC-12 经多步转化生成别红藻氨酸。

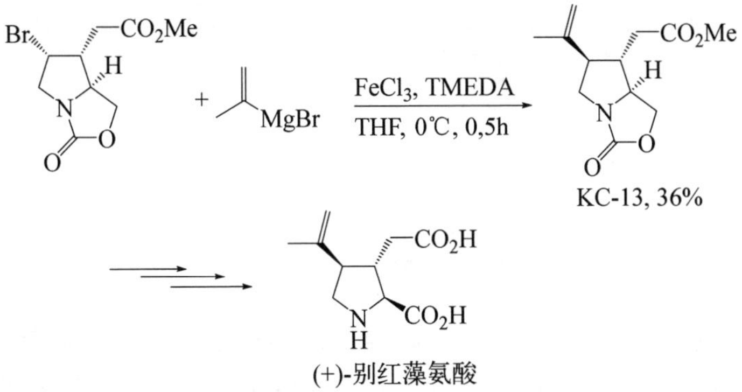

胆甾烯是生物活性化合物，对其母核进行修饰有可能获得更多的生物活性化合物或新药。碘代胆甾烯与炔基格氏试剂在溴化亚铁催化下于 THF/NMP 混合溶剂中反应以中等的收率得到炔基官能团化的胆甾烯，并且倾向于构型保持，非对映选择性 dr 为 7∶1[52]。

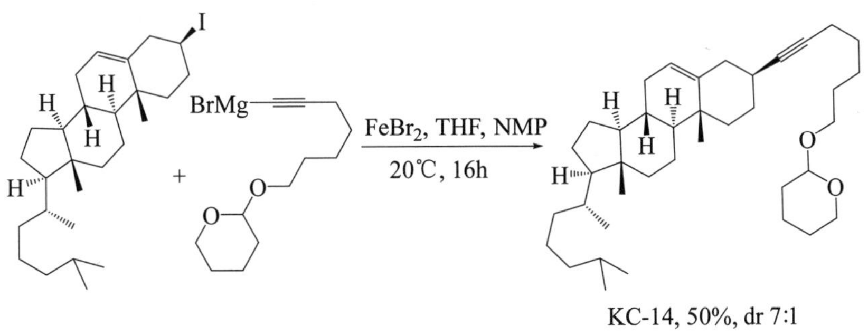

第 4 章　Kumada-Corriu 偶联反应及其在药物合成上的应用

三芳基甲烷类化合物是许多药物分子的核心结构,其中一些药物用以治疗癌症、细菌感染和糖尿病,通常通过傅-克(Friedel-Crafts)反应合成,得到外消旋混合物。利用烷基-芳基 Kumada-Corriu 偶联反应可实现三芳基甲烷类化合物的对映选择性合成。例如,以下反应的目标分子(TM-1)是抗癌药物他莫昔芬(tamoxifen)的类似物,对乳腺癌细胞有杀伤作用。首先,对映体强化的醚与对甲氧基溴化镁在 Ni(cod)$_2$/dpph[1,6-双(二苯膦基)己烷]催化下进行对映选择性的交叉偶联反应得到构型翻转的三芳基甲烷 KC-15,反应的对映专一性(enantiospecificity, es)高达 90%,KC-15 的对映体过量值达 77%[53]。KC-15 经两步转化得到目标分子。

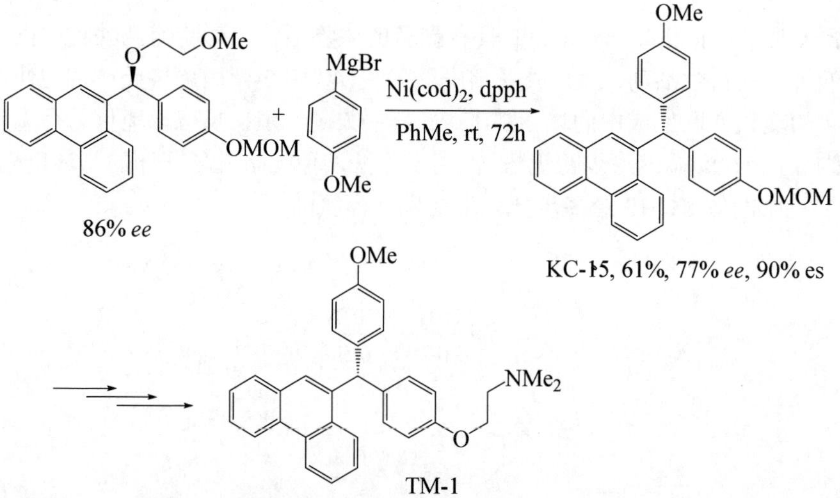

与上一合成类似,烷基-烷基 Kumada-Corriu 偶联反应也是立体专一性的,可用来合成苯唑噻吩抗失眠药中间体。如下反应所示,对映体强化的二芳甲基甲基醚与甲基格氏试剂交叉偶联构建手性季碳中心得到反式构型强化的二芳基乙烷 KC-16,KC-16 再经 3 步转化最终得到抗失眠药物 TM-2。

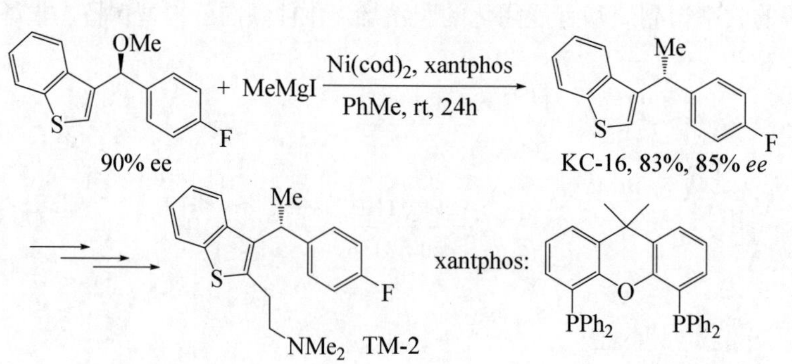

利用手性双噁唑啉配体诱导钴催化 α-溴代羧酸酯与烯基格氏试剂的不对称偶联，成功实现了加州红鳞片信息素（California red scale pheromone）的合成。合成反应起始于外消旋的 2-溴-5-己烯酸苄酯与 2-丙烯溴化镁的不对称交叉偶联，反应在手性双噁唑林配体-$CoCl_2$ 催化体系催化下以中等收率和优秀的对映选择性生成 (S)-2-(1-甲基乙烯基)-5-己烯酸苄酯（KC-17），KC-17 再经选择性还原、氧化等多步反应最终得到加州红鳞片信息素[54]。

(70)

参考文献

[1] Corriu R J P, Masse J P. Activation of Grignard reagents by transition-metal complexes. A new and simple synthesis of trans-stilbenes and polyphenyls[J]. Journal of the Chemical Society, Chemical Communications, 1972, (3): 144a.

[2] Tamao K, Sumitani K, Kumada M. Selective carbon-carbon bond formation by cross-coupling of Grignard reagents with organic halides. Catalysis by nickel-phosphine complexes[J]. Journal of the American Chemical Society, 1972, 94(12): 4374-4376.

[3] Kharasch M S, Fields E K. Factors determining the course and mechanisms of grignard reactions. Ⅳ. The effect of metallic halides on the reaction of aryl grignard reagents and organic halides1[J]. Journal of the American Chemical Society, 1941, 63(9): 2316-2320.

[4] Kochi J K, Tamura M. Mechanism of the silver-catalyzed reaction of Grignard reagents with alkyl halides[J]. Journal of the American Chemical Society, 1971, 93(6): 1483-1485.

[5] Kochi J K, Tamura M. Alkylcopper(Ⅰ) in the coupling of Grignard reagents with alkyl halides[J]. Journal of the American Chemical Society, 1971, 93(6): 1485-1487.

[6] Tamura M, Kochi J K. Vinylation of Grignard reagents. Catalysis by iron[J]. Journal of the American Chemical Society, 1971, 93(6): 1487-1489.

[7] Yamamura M, Moritani I, Murahashi S I. The reaction of σ-vinylpalladium complexes with alkyllithiums. Stereospecific syntheses of olefins from vinyl halides and alkyllithiums[J]. Journal of Organometallic Chemistry, 1975, 91(2): C39-C42.

[8] Knappke C E I, von Wangelin A J. 35 years of palladium-catalyzed cross-coupling with Grignard reagents: how far have we come ?[J]. Chemical Society Reviews, 2011, 40(10): 4948-4962.

[9] Cahiez G, Avedissian H. Highly stereo-and chemoselective iron-catalyzed alkenylation of organomagnesium compounds[J]. Synthesis, 1998, 1998(8): 1199-1205.

[10] Gamez P, Ariente C, Goré J, et al. Synthesis of cross-conjugated dienic alcohols[J]. Tetrahedron, 1998, 54(49): 14825-14834.

[11] Hölzer B, Hoffmann R W. Kumada-Corriu coupling of Grignard reagents, probed with a chiral Grignard reagent[J]. Chemical Communications, 2003, (6): 732-733.

[12] Bojase G, Payne A D, Willis A C, et al. One-step synthesis and exploratory chemistry of [5]dendralene[J]. Angewandte Chemie, 2008, 120(5): 924-926.

[13] Dai W, Zhang X, Zhang J, et al. Synthesis of exocyclic trisubstituted alkenes via nickel-catalyzed Kumada-type cross-coupling reaction of gem-difluoroalkenes with di-Grignard reagents[J]. Advanced Synthesis & Catalysis, 2016, 358(2): 183-187.

[14] Limmert M E, Roy A H, Hartwig J F. Kumada coupling of aryl and vinyl tosylates under mild conditions[J]. The Journal of Organic Chemistry, 2005, 70(23): 9364-9370.

[15] Ackermann L, Althammer A. Air-stable PinP(O)H as preligand for palladium-catalyzed Kumada couplings of unactivated tosylates[J]. Organic Letters, 2006, 8(16): 3457-3460.

[16] Tamao K, Kodama S, Nakajima I, et al. Nickel-phosphine complex-catalyzed Grignard coupling——II : grignard coupling of heterocyclic compounds[J]. Tetrahedron, 1982, 38(22): 3347-3354.

[17] Heravi M M, Hajiabbasi P. Recent advances in Kumada-Tamao-Corriu cross-coupling reaction catalyzed by different ligands[J]. Monatshefte für Chemie-Chemical Monthly, 2012, 143: 1575-1592.

[18] El-Shehawy A A, Abdo N I, Ahmed A, et al. A selective and direct synthesis of 2-bromo-4-alkylthiophenes: convenient and straightforward approaches for the synthesis of head-to-tail(HT) and tail-to-tail(TT) dihexyl-2, 2'-bithiophenes[J]. Tetrahedron Letters, 2010, 51(34): 4526-4529.

[19] Roques N, Saint-Jalmes L. Efficient access to 3-alkyl-trifluoromethylbenzenes using Kumada's coupling reaction[J]. Tetrahedron Letters, 2006, 47(20): 3375-3378.

[20] Fürstner A, Leitner A. Iron-catalyzed cross-coupling reactions of alkyl-grignard reagents with aryl chlorides, tosylates, and triflates[J]. Angewandte Chemie International Edition, 2002, 41(4): 609-612.

[21] Joshi-Pangu A, Wang C Y, Biscoe M R. Nickel-catalyzed Kumada cross-coupling reactions of tertiary alkylmagnesium halides and aryl bromides/triflates[J]. Journal of the American Chemical Society, 2011, 133(22): 8478-8481.

[22] Agrawal T, Cook S P. Iron-catalyzed cross-coupling reactions of alkyl Grignards with aryl sulfamates and tosylates[J]. Organic Letters, 2013, 15(1): 96-99.

[23] Sekiya A, Ishikawa N. The cross-coupling of aryl halides with Grignard reagents catalyzed by iodo(phenyl)bis(triphenylphosphine) palladium(II)[J]. Journal of Organometallic Chemistry, 1976, 118(3): 349-354.

[24] Kamikawa T, Hayashi T. Dichloro [(2-dimethylamino) propyldiphenylphosphine] palladium(II) ($PdCl_2$(alaphos)): an efficient catalyst for cross-coupling of aryl triflates with alkynyl Grignard reagents1[J]. The Journal of Organic Chemistry, 1998, 63(24): 8922-8925.

[25] Vechorkin O, Proust V, Hu X. functional group tolerant Kumada-Corriu-Tamao coupling of nonactivated alkyl halides with aryl and heteroaryl nucleophiles: catalysis by a nickel pincer complex permits the coupling of functionalized Grignard reagents[J]. Journal of the American Chemical Society,

2009, 131(28): 9756-9766.

[26] Vechorkin O, Hu X. Nickel-catalyzed cross-coupling of non-activated and functionalized alkyl halides with alkyl Grignard reagents[J]. Angewandte Chemie International Edition, 2009, 48(16): 2937-2940.

[27] Vechorkin O, Csok Z, Scopelliti R, et al. Nickel complexes of a pincer amidobis (amine) ligand: synthesis, structure, and activity in stoichiometric and catalytic C-C bond-forming reactions of alkyl halides[J]. Chemistry: A European Journal, 2009, 15(15): 3889-3899.

[28] Ren P, Stern L A, Hu X. Copper-catalyzed cross-coupling of functionalized alkyl halides and tosylates with secondary and tertiary alkyl grignard reagents[J]. Angewandte Chemie International Edition, 2012, 51(36): 9110-9113.

[29] Yang C T, Zhang Z Q, Liang J, et al. Copper-catalyzed cross-coupling of nonactivated secondary alkyl halides and tosylates with secondary alkyl Grignard reagents[J]. Journal of the American Chemical Society, 2012, 134(27): 11124-11127.

[30] Li W N, Wang Z L. Kumada-Tamao-Corriu cross-coupling reaction of O-based electrophiles with Grignard reagents via C-O bond activation[J]. RSC Advances, 2013, 3(48): 25565-25575.

[31] Sengupta S, Leite M, Raslan D S, et al. Nickel(0)-catalyzed cross coupling of aryl O-carbamates and aryl triflates with Grignard reagents. Directed ortho metalation-aligned synthetic methods for polysubstituted aromatics via a 1, 2-dipole equivalent[J]. The Journal of Organic Chemistry, 1992, 57(15): 4066-4068.

[32] Wu J C, Gong L B, Xia Y, et al. Nickel-catalyzed kumada reaction of tosylalkanes with Grignard reagents to produce alkenes and modified arylketones[J]. Angewandte Chemie International Edition, 2012, 51(39): 9909-9913.

[33] Li B J, Xu L, Wu Z H, et al. Cross-coupling of alkenyl/aryl carboxylates with Grignard reagent via Fe-catalyzed C-O bond activation[J]. Journal of the American Chemical Society, 2009, 131(41): 14656-14657.

[34] Hayashi T, Konishi M, Fukushima M, et al. Asymmetric synthesis catalyzed by chiral ferrocenylphosphine-transition metal complexes. 2. Nickel-and palladium-catalyzed asymmetric Grignard cross-coupling[J]. Journal of the American Chemical Society, 1982, 104(1): 180-186.

[35] Lou S, Fu G C. Nickel/bis(oxazoline)-catalyzed asymmetric Kumada reactions of alkyl electrophiles: cross-couplings of racemic α-bromoketones[J]. Journal of the American Chemical Society, 2010, 132(4): 1264-1266.

[36] Eno M S, Lu A, Morken J P. Nickel-catalyzed asymmetric Kumada cross-coupling of symmetric cyclic sulfates[J]. Journal of the American Chemical Society, 2016, 138(25): 7824-7827.

[37] Tollefson E J, Hanna L E, Jarvo E R. Stereospecific nickel-catalyzed cross-coupling reactions of benzylic ethers and esters[J]. Accounts of Chemical Research, 2015, 48(8): 2344-2353.

[38] Mukhtar S, Nair D S, Medhane R R, et al. Processes for the preparation of cinacalcet: U.S. Patent 8,759,586[P]. 2014-06-24.

[39] Camacho-Davila A A. Kumada-Corriu cross coupling route to the anti-cancer agent combretastatin A-4[J]. Synthetic Communications, 2008, 38(21): 3823-3833.

[40] Gangula S, Neelam U K, Baddam S R, et al. Investigation of a Kumada cross coupling reaction for large-scale production of (2S, 7R, E)-2-isopropyl-7-(4-methoxy-3-(3-methoxypropoxy) benzyl)-N, N, 8-trimethylnon-4-enamide[J]. Organic Process Research & Development, 2015, 19(3): 470-475.

[41] Giordano G, Coppi L, Minisci F. Process for the preparation of 5-(2,4-difluorophenyl)-salicylic acid: U.

S. Patent 5,312,975[P]. 1994-05-17.

[42] Fan X, Song Y L, Long Y Q. An efficient and practical synthesis of the HIV protease inhibitor atazanavir via a highly diastereoselective reduction approach[J]. Organic Process Research & Development, 2008, 12(1): 69-75.

[43] Wei Y M, Ma X D, Wang L, et al. Iron-catalyzed stereospecific arylation of enol tosylates using Grignard reagents[J]. Chemical Communications, 2020, 56(7): 1101-1104.

[44] Savage S, McClory A, Zhang H, et al. Synthesis of selective estrogen receptor degrader GDC-0810 via stereocontrolled assembly of a tetrasubstituted all-carbon olefin[J]. The Journal of Organic Chemistry, 2018, 83(19): 11571-11576.

[45] Marzoni G, Varney M D. An Improved large-scale synthesis of benz [cd] indol-2 (1H)-one and 5-methylbenz [cd] indol-2 (1H)-one[J]. Organic Process Research & Development, 1997, 1(1): 81-84.

[46] Bartoccini F, Piersanti G, Armaroli S, et al. Development of a practical and sustainable strategy for the synthesis of ST1535 by an iron-catalyzed Kumada cross-coupling reaction[J]. Tetrahedron Letters, 2014, 55(7): 1376-1378.

[47] Wang X, Zhang L, Smith-Keenan L L, et al. Efficient synthesis of (S)-2-(cyclopentyloxycarbonyl)-amino-8-nonenoic acid: key building block for BILN 2061, an HCV NS3 protease inhibitor[J]. Organic Process Research & Development, 2007, 11(1): 60-63.

[48] Ma Z, Yang X, Zhang Y, et al. Concise syntheses of (7Z, 11Z, 13E)-hexadecatrienal and (8E, 18Z)-tetradecadienal[J]. Synlett, 2012, 2012(4): 581-584.

[49] Chen Z, So C M. Pd-catalyzed cross-coupling of highly sterically congested enol carbamates with Grignard reagents via C-O bond activation[J]. Organic Letters, 2020, 22(10): 3879-3883.

[50] Fürstner A, Hannen P. Platinum-and gold-catalyzed rearrangement reactions of propargyl acetates: total syntheses of (−)-α-cubebene,(−)-cubebol, sesquicarene and related terpenes[J]. Chemistry: A European Journal, 2006, 12(11): 3006-3019.

[51] Yamada K, Sato T, Hosoi M, et al. Stereoselective formal synthesis of (+)-allokainic acid via thiol-mediated acyl radical cyclization[J]. Chemical and Pharmaceutical Bulletin, 2010, 58(11): 1511-1516.

[52] Cheung C W, Ren P, Hu X. Mild and phosphine-free iron-catalyzed cross-coupling of nonactivated secondary alkyl halides with alkynyl grignard reagents[J]. Organic Letters, 2014, 16(9): 2566-2569.

[53] Taylor B L H, Harris M R, Jarvo E R. Synthesis of enantioenriched triarylmethanes by stereospecific cross-coupling reactions[J]. Angewandte Chemie International Edition, 2012, 51(31): 7790-7793.

[54] Zhou Y, Wang L, Yuan G, et al. Cobalt-bisoxazoline-catalyzed enantioselective cross-coupling of α-bromo esters with alkenyl Grignard reagents[J]. Organic Letters, 2020, 22(11): 4532-4536.

第 5 章
Sonogashira 反应及其在药物合成上的应用

5.1 Sonogashira 反应及机理

5.1.1 Sonogashira 反应

经典的 Sonogashira 反应是指端炔与芳基、烯基卤化物或其类似物在碱存在下由催化剂钯和助催化剂铜(Ⅰ)共同化发生交叉偶联形成 $C(sp^2)$-$C(sp)$ 键的反应，反应在室温下进行。反应如图 5-1 所示。

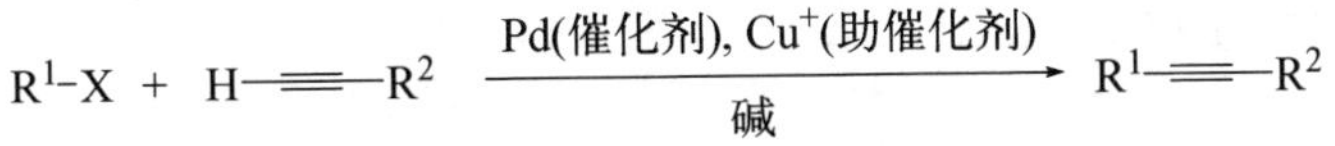

$$R^1-X + H\!\!=\!\!\!=\!\!R^2 \xrightarrow[\text{碱}]{\text{Pd(催化剂), Cu}^+\text{(助催化剂)}} R^1\!\!=\!\!\!=\!\!R^2$$

R^1=芳基、杂芳基、乙烯基
R^2=芳基、杂芳基、烯基、烷基、SiR_3
X=I、Br、Cl、OTf

图 5-1 经典的 Sonogashira 反应

该反应源自 Sonogashira 和 Hagihara 等于 1975 年的一个经典实验，即端炔与芳基、烯基卤化物在有机胺溶剂中在 $PdCl_2(PPh_3)_2$ 和 CuI 共同催化下于室温下发生交叉偶联生成芳基炔和烯-炔共轭化合物[1]。实际上，在此发现半年之前，Cassar、Heck 等[2,3]分别发现在无 CuI 存在下，单独的钯催化剂可催化上述反应在高温下进行。显然，铜的引入明显地降低了反应温度，使反应的底物适应范围、官能团的耐受范围大大拓宽，极大地提高反应的实用性。与此同时，铜的引入也导致反应体系复杂，对环境不友好，反应物料难以回收。而且，从反应的角度看，由端炔原位生成的炔酮在空气气氛下经常会发生自身偶联污染目标产物。为克服这些问题，人们后来相继开发了多种无铜存在的反应方法，这些方法通常称为无铜 Sonogashira 偶联,而不叫 Cassar 或 Heck 偶联,似乎有些不公。同时,经典的 Sonogashira 反应需要过量的有机胺,对环境也有所破坏,一些新的催化反应体系也应运而生,以避免有机胺的使用。近年来,以非贵金属替代贵重金属钯构建 $C(sp^2)$-$C(sp)$ 键的反应已取得显著进展,尽管这些催化剂已非钯-铜催化剂组合,但此类反应仍称作 Sonogashira 反应。

在 Sonogashira 反应中,sp^2 活性物种的活性顺序为:乙烯基碘≥乙烯基三氟甲磺酸酯>乙烯基溴>乙烯基氯>芳基碘>芳基三氟甲磺酸酯≥芳基溴≫芳基氯。因此,贵重的乙烯基或芳基碘反应进行顺利;溴代物则需要被活化即缺电子的条件下才能进行,没被活化的溴代物反应困难;最廉价的氯代物如果不被强烈活化反应更加困难,也就是说氯代物为底物的交叉偶联反应是一个挑战性问题[4]。

5.1.2 反应机理

详细的 Sonogashira 反应机理仍不清楚,但学界普遍认可钯-铜催化的 Sonogashira 反应由两个催化循环构成:钯催化循环和铜催化循环(图 5-2)。在钯催化循环中,钯催化剂前体在反应条件下被活化形成活性物种[Pd^0L_2](**A**)。需要注意的是,**A** 的具体结构依赖于反应条件,通常在简单膦配体如三苯基膦(PPh_3)存在下 **A** 的配位数 n 是 2,而空间位阻大的膦配体如三邻甲苯基膦配体 $P(o\text{-tol})_3$ 存在下 n 等于 1[5],并且一些实验结果指出在一些阴离子或卤负离子存在下钯负离子物种[L_2Pd^0Cl]$^-$可能是真正的催化剂活性物种[6]。一旦 **A** 生成,芳基或乙烯基卤化物(R^1—X)与[Pd^0L_2]活性物种发生氧化加成生成中间体[$Pd^{II}R^1L_2X$](**B**),此步反应被认为是整个反应的速率决定步骤。紧接着,中间体 **B** 与铜催化循环中的炔铜中间体(**F**)进行金属转移反应生成中间体[$Pd^{II}L_2R^1(C≡CR^2)$](**C**)。**C** 经过顺-反异构化生成 **C'**,**C'** 进行还原消除生成交叉偶联炔烃产物并再生催化活性物种 **A**。

此钯催化循环与其他交叉偶联反应的催化循环很接近,但铜催化循环则很少见。一般认为,碱协助 CuX(**D**)与炔形成炔-铜 π-配合物(**E**),使得端炔质子的酸性增强,从而使 **E** 脱质子转化为炔铜中间体(**F**)。**F** 与钯催化循环中的中间体进行金属交换再生 **D**。

关于无铜存在下钯催化的 Sonogashira 反应的机理一直存在争论,直到 2018 年 Košmrlj 等才证明反应按两个互相联系的 Pd^0/Pd^{II} 催化循环进行[7](图 5-3)。Pd^0 催化循环与钯-铜催化循环中的钯循环类似,催化循环起始于芳基或乙烯基卤代物(或三氟甲磺酸酯)向 Pd^0 活性物种 **A** 的氧化加成形成中间体 **B**,**B** 与 Pd^{II} 循环中的双炔基钯(**F**)进行金属转移反应生成烯基(或芳基)和炔基共同配位的钯配合物中间体 **C**,**C** 通过还原消除生成偶联产物炔并再生 Pd^0 活性物种 **A**。

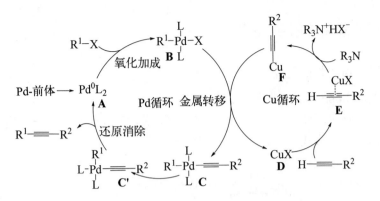

图 5-2 钯-铜催化 Sonogashira 反应的催化循环

在 Pd^{II} 催化循环中，端炔烃被体系中的碱活化，在温和条件下既可以与钯物种形成单炔配合物 D，也可进一步与单炔配合物 D 结合形成双炔配合物 F；Pd^0 催化循环中的中间体 B 与 Pd^{II} 催化循环中的 F 进行金属转移反应形成中间体 C，并再生单炔配合物 D，完成 Pd^{II} 催化循环。需要注意的是，在形成催化循环中的各中间体时有机胺可以作为配体与膦配体竞争配位，结果依赖于胺和膦配体与金属配位的竞争程度，因而选用不同的有机胺会对催化循环产生动态且复杂的作用。

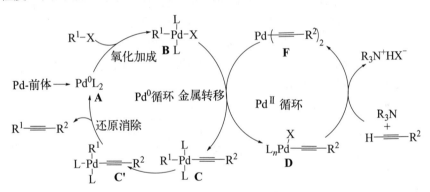

图 5-3 钯催化 Sonogashira 反应的催化循环

5.2 基本的 Sonogashira 反应

如前所述，经典的 Sonogashira 反应是在钯-铜共同催化下在有机胺存在下各种芳基、乙烯基卤代物或其类似物与端炔之间的交叉偶联反应。随着研究的不断深入，反应底物、催化剂的种类以及反应体系不断拓展，反应众多，无法完全、准确阐述，以下将按催化剂的不同介绍一些基本的 Sonogashira 反应。

5.2.1 经典催化体系下的反应

(1) 芳基卤代物的反应

在经典的 Sonogashira 催化体系下，芳基碘亲电底物可与含不同取代基的端炔进行反应［反应式（2）、式（3）］；从端炔方面看，乙炔的两端均能与碘苯反应生成二苯基乙炔［反应式（1）］[1]。

$$2PhI + HC\equiv CH \xrightarrow[Et_2NH, rt, 6h]{PdCl_2(PPh_3)_2, CuI} \underset{85\%}{Ph-C\equiv C-Ph} \quad (1)$$

$$PhI + HC\equiv CCH_2OH \xrightarrow[Et_2NH, rt, 3h]{PdCl_2(PPh_3)_2, CuI} \underset{80\%}{PhC\equiv CCH_2OH} \quad (2)$$

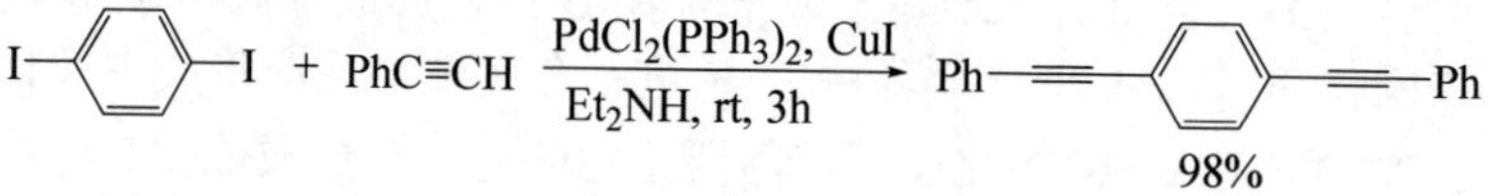

$$I-\!\!\!\bigcirc\!\!\!-I + PhC\equiv CH \xrightarrow[Et_2NH, rt, 3h]{PdCl_2(PPh_3)_2, CuI} \underset{98\%}{Ph-\!\!\equiv\!\!-\!\!\!\bigcirc\!\!\!-\!\!\equiv\!\!-Ph} \quad (3)$$

含各种取代基的芳基碘也能与端炔在经典的 Sonogashira 催化体系下反应得到官能团化的芳基炔类化合物，预示 Sonogashira 反应在药物及天然产物合成上具有广泛应用。例如，苯环上同时含吸电子基（甲酰基）和给电子基（甲氧基）的碘代苯与取代的苯乙炔在经典的钯-铜催化剂催化下顺利发生交叉偶联生成多取代双芳基乙炔，由于所生成的芳基炔炔基的邻位存在甲氧基，因而炔基和甲氧基在碘作用下进行环合反应生成苯并呋喃结构[8]。

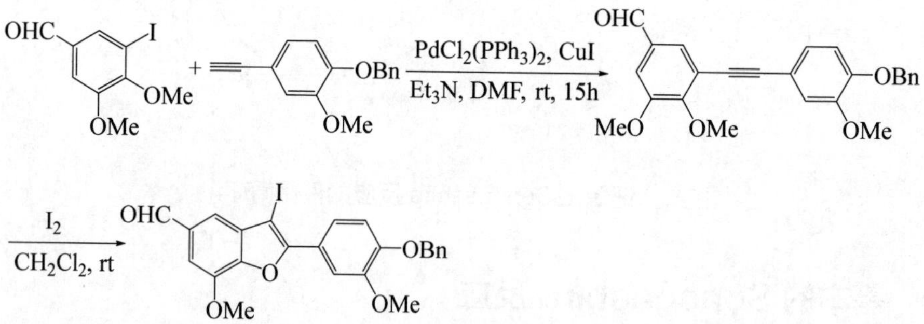

（4）

无论是富电子的芳杂环碘代物还是缺电子的芳杂环碘代物都能在经典的 Sonogashira 催化体系下进行反应，而且耐受各种官能团。例如，含有各种取代基的苯并呋喃碘代物与苯乙炔在温和条件下进行偶联，目标产物的收率随取代基的不同有所变化，但多数情况变化不大[9]。

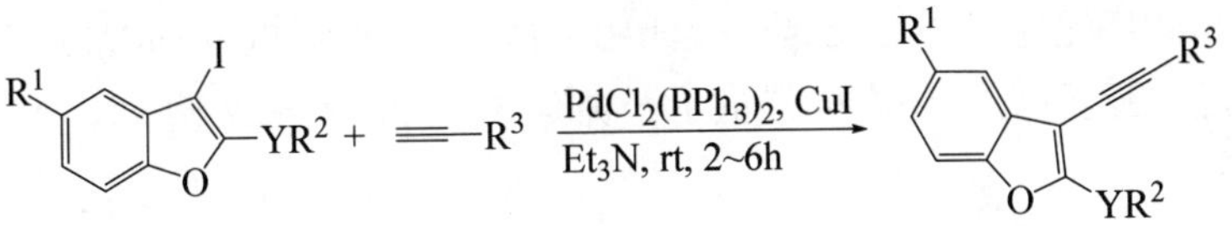

R^1	YR^2	R^3	收率/%
H	SMe	Ph	99
H	SMe	CMe_2OH	93
Me	SMe	Ph	75
F	SMe	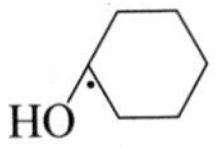 HO-环己基	91
H	SePh	CH_2CH_2OH	55 (60℃, 16h)

3-碘吡啶与十一碳-10-炔醇在 $PdCl_2(PPh_3)_2$ 和 CuBr 共同催化下以优异的收率生成交叉偶联产物[10]。2,4-二氨基-6-碘嘧啶与端炔也能在近似的反应体系下反应生成炔基嘧啶衍生物,需要注意的是相同条件下氯代物不反应,需要通过卤素交换将氯化物转化为碘代物[11]。

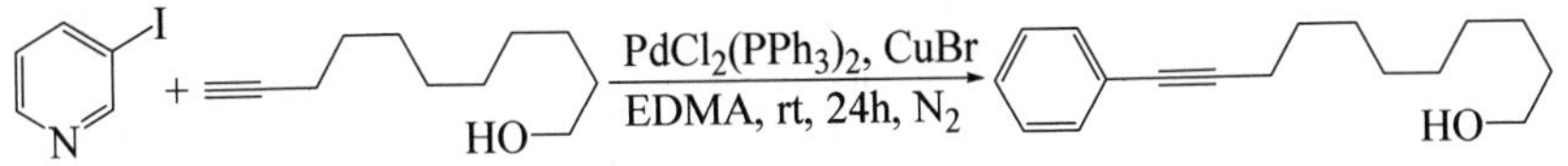

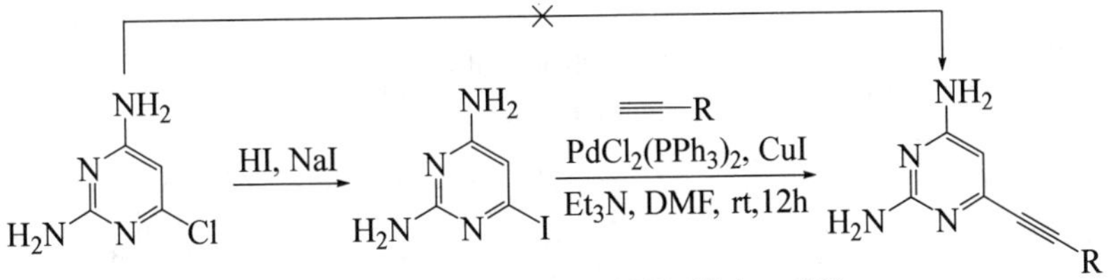

与芳基碘活性相近的芳基三氟甲磺酸酯在适当的催化反应体系下可以替代碘代物顺利进行 Sonogashira 反应。例如,3-(2-丁酰氨基吡啶)三氟甲磺酸酯在 $Pd(PPh_3)_4$/CuI 催化下顺利与各种端炔反应生成炔基吡啶,而在相同条件下 $PdCl_2(PPh_3)_2$/CuI 的催化效果不佳;n-Bu_4NI 作为助剂显著提高目标产物的收率[12]。

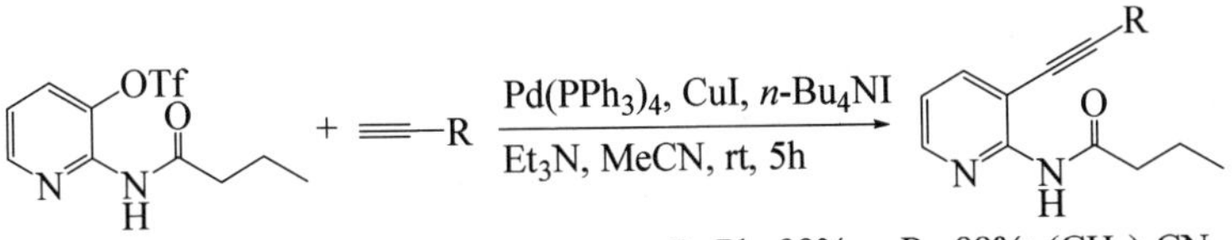

与吡啶三氟甲磺酸酯类似，哒嗪三氟甲磺酸酯也能进行 Sonogashira 反应，为以哒嗪环为母核的药物分子的合成提供了一种有效途径，反应实例如下[13]。

$$\text{(底物)} + \text{HO}\!-\!\!\equiv\!\! \xrightarrow[\text{55℃, 4h}]{\text{Pd(PPh}_3)_4,\text{ CuI}\atop i\text{-Pr}_2\text{NEt, DMF}} \text{(产物)} \quad 94\% \tag{9}$$

多数情况下芳基溴代物可以替代价格昂贵的芳基碘代物进行 Sonogashira 反应，这在有机和药物合成上非常重要，但有时需要适当提高反应温度或是引入适当的配体等。如下反应所示，各种取代的溴苯与苯乙炔在加热条件下发生交叉偶联生成1,2-二芳基乙炔［反应式（10）］。由于加热会导致中间体炔的自偶联反应，在氢气气氛下可抑制该副反应的发生。在相近的催化反应体系下 4-溴吡啶与苯乙炔进行交叉偶联以优良的收率得到交叉偶联目标产物［反应式（11）］[14]；溴代咪唑衍生物也能与苯乙炔交叉偶联生成相应的目标产物［反应式（12）］。

$$\text{Ph}\!-\!\!\equiv\!\!\text{H} + \text{Br}\!-\!\text{Ar-R} \xrightarrow[\text{3h, N}_2+\text{H}_2]{\text{PdCl}_2(\text{PPh}_3)_2,\text{ CuI}\atop \text{Et}_3\text{N, DMF, 100℃}} \text{Ph}\!-\!\!\equiv\!\!\text{Ar-R} \tag{10}$$

R=4-CN, 98%; 4-CHO, 98%; 4-COCH$_3$, 96%

$$\text{R-Ar}\!-\!\!\equiv\!\!\text{H} + \text{Br}\!-\!\text{Py} \xrightarrow[\text{8h, N}_2+\text{H}_2]{\text{PdCl}_2(\text{PPh}_3)_2,\text{ CuI}\atop \text{Et}_3\text{N, MeCN, 80℃}} \text{R-Ar}\!-\!\!\equiv\!\!\text{Py} \tag{11}$$

R=4-H, 88%; 2,4-di-Me, 89%; 4-OMe, 91%; 4-NMe$_2$, 94%; 4-NEt$_2$, 95%; 4-Me, 85%

$$\text{(溴咪唑)} + \text{Ph}\!-\!\!\equiv\!\!\text{H} \xrightarrow[\text{80℃, 24h}]{\text{PdCl}_2(\text{PPh}_3)_2,\text{ CuI,}\atop \text{DMF, 哌啶}} \text{(产物)} \quad 74\% \tag{12}$$

大空间位阻、给电子的膦配体如 P(t-Bu)$_3$ 可以促进芳基溴与端炔的 Sonogashira 反应，使反应在室温下进行，提高目标产物的选择性[15]。

$$\text{R-Ar-Br} + \equiv\!\!\text{R}^1 \xrightarrow[\text{CuI, HN}(i\text{-Pr})_2,\text{ 1,4-二氧六环,}\atop \text{rt, 0.5~15h}]{\text{Pd(PhCN)}_2\text{Cl}_2,\text{ P}(t\text{-Bu})_3} \text{R-Ar}\!-\!\!\equiv\!\!\text{R}^1 \tag{13}$$

R, R^1=H, Ph, 94%; H, CMe$_2$OH, 91%; 4-OMe, Ph, 94%
4-OMe, CMe$_2$OH, 95%; 4-NMe$_2$, Ph, 94%;
4-COMe, CMe$_2$OH, 95%; 2-Me, Ph, 94%

利用大空间位阻、给电子的膦配体结合钯催化剂、碱和溶剂的调整，并提高反应温度，芳基氯甚至都可以进行 Sonogashira 反应得到炔基化芳基化合物；碱在该反应起重要作用，常用的有机胺二异丙胺导致反应失败，而选用无机碱碳酸钠反应结果最佳[16]。

$$\text{R}\underset{}{\overset{}{\bigcirc}}\text{—Cl} + \equiv\text{—R}^1 \xrightarrow[\text{CuI, Na}_2\text{CO}_3/\text{DMSO, 二甲苯,}]{\text{Na}_2\text{PdCl}_4, (1\text{-Ad})_2\text{PBn·HBr}} \text{R}\underset{}{\overset{}{\bigcirc}}\text{—}\equiv\text{—R}^1 \quad (14)$$

$(1\text{-Ad})_2\text{PBn}:$ (Ad)$_2$P-CH$_2$Ph

R, R^1= 4-COMe, Ph, 90%; 4-COMe, n-C$_6$H$_{13}$, 92%;
4-NO$_2$, Ph, 96%; 4-NO$_2$, i-Pr$_3$Si, 89%;
4-CF$_3$, i-Pr$_3$Si, 89%; H, i-Pr$_3$Si, 76%; 4-MeO,
Ph, 75%; 4-Me, Ph, 70%; 4-Me, i-Pr$_3$Si,71%

严重缺电子的芳杂环氯代物无须膦配体的调控在加热条件下与端炔进行 Sonogashira 反应。例如，6-氯-9-四氢吡喃基嘌呤在经典 Sonogashira 催化反应体系和加热条件下与端炔交叉偶联得到炔基化的嘌呤衍生物[17]。

$$\text{(6-Cl purine-THP)} + \equiv\text{—CH(Me)CH}_2\text{OTBDS} \xrightarrow[\text{i-Pr}_2\text{NEt, DMF, 80°C, 23h}]{\text{Pd(PPh}_3)_2\text{Cl}_2, \text{CuI}} \text{(purine-THP)—}\equiv\text{—CH(Me)CH}_2\text{OTBDS} \quad 55\% \quad (15)$$

（2）乙烯基卤代物的反应

在 Sonogashira 反应中，乙烯基卤代物比芳基卤代物要活泼，乙烯基溴在与芳基碘的 Sonogashira 反应相同反应条件下就能与各种端炔反应，产物的收率很高，而且烯烃的构型在反应中得到保持，实例如下[1]。

$$\text{CH}_2=\text{CHBr} + \equiv\text{—Ph} \xrightarrow[\text{Et}_2\text{NH, rt, 3h}]{\text{PdCl}_2(\text{PPh}_3)_2, \text{CuI}} \text{CH}_2=\text{CH—}\equiv\text{—Ph} \quad 91\% \quad (16)$$

$$\text{PhCH=CHBr} + \equiv\text{—Ph} \xrightarrow[\text{Et}_2\text{NH, rt, 3h}]{\text{PdCl}_2(\text{PPh}_3)_2, \text{CuI}} \text{PhCH=CH—}\equiv\text{—Ph} \quad 90\% \quad (17)$$

$$\text{PhCH=CHBr} + \equiv\text{—CH}_2\text{OH} \xrightarrow[\text{Et}_2\text{NH, rt, 3h}]{\text{PdCl}_2(\text{PPh}_3)_2, \text{CuI}} \text{PhCH=CH—}\equiv\text{—CH}_2\text{OH} \quad 70\% \quad (18)$$

$$\text{(cyclohexenyl-Br)} + \equiv\text{—Ph} \xrightarrow[\text{Et}_2\text{NH, rt, 3h}]{\text{PdCl}_2(\text{PPh}_3)_2, \text{CuI}} \text{(cyclohexenyl)—}\equiv\text{—Ph} \quad 95\% \quad (19)$$

$$\text{PhCH=CHBr} + \equiv \xrightarrow[\text{Et}_2\text{NH, rt, 6h}]{\text{PdCl}_2(\text{PPh}_3)_2, \text{CuI}} \text{PhCH=CH—}\equiv\equiv\text{—CH=CHPh} \quad 95\% \quad (20)$$

甚至氯代烯烃、共轭氯代烯烃都能在室温下与端炔在经典的催化反应体系下顺利发生 Sonogashira 反应[18]。

$$C_5H_{11}\diagup\diagdown Cl + \equiv-C_5H_{11} \xrightarrow[\text{哌啶, rt, 0.5h}]{PdCl_2(PhCN)_2, CuI} C_5H_{11}\diagup\diagdown-\equiv-C_5H_{11} \qquad (21)$$

$$C_5H_{11}\underset{OH}{\diagup}(\ )_n Cl + \equiv-R^1 \xrightarrow[\text{哌啶, rt, 4h}]{PdCl_2(PhCN)_2, CuI} C_5H_{11}\underset{OH}{\diagup}(\ )_n \equiv-R_1 \qquad (22)$$

R, n=C$_5$H$_{11}$, 1, 95%; SiMe$_3$, 1, 97%; CH$_2$OH,
1, 84%; C$_5$H$_{11}$, 2, 75%; SiMe$_3$, 2, 74%

含各种官能团的乙烯基卤代物在经典的 Sonogashira 催化反应下与端炔交叉偶联生成官能团化的烯-炔衍生物，彰显 Sonogashira 反应在药物和天然产物合成上应用前景。例如，含砜基[19]和芳硒基[20]的乙烯基碘代物在 Pd(PPh$_3$)$_4$/CuI 催化下于哌啶中室温下顺利反应，收率受取代基的影响很小，而且烯烃的构型得到保持。

$$\underset{H}{\overset{R}{\diagdown}}C=C\underset{I}{\overset{SO_2Ar}{\diagup}} + \equiv-R^1 \xrightarrow[\text{哌啶, rt, 40min}]{Pd(PPh_3)_4, CuI} \underset{H}{\overset{R}{\diagdown}}C=C\underset{\equiv-R^1}{\overset{SO_2Ar}{\diagup}} \qquad (23)$$

R, Ar, R^1=n-C$_4$H$_9$, Ph, n-C$_4$H$_9$, 88%; Ph, Ph, Ph, 90%;
CH$_3$OCH$_2$, 4-CH$_3$C$_6$H$_4$, Ph, 87%; n-C$_4$H$_9$, Ph,
HOCH$_2$, 83%; n-C$_4$H$_9$, Ph, SiMe$_3$, 86%

$$\underset{H}{\overset{ArSe}{\diagdown}}C=C\underset{I}{\overset{H}{\diagup}} + \equiv-R^1 \xrightarrow[\text{哌啶, rt, 3h}]{Pd(PPh_3)_4, CuI} \underset{H}{\overset{ArSe}{\diagdown}}C=C\underset{\equiv-R}{\overset{H}{\diagup}} \qquad (24)$$

Ar, R=Ph, n-C$_4$H$_9$, 88%; 4-ClC$_6$H$_4$, Me$_3$Si, 87%;
4-MeC$_6$H$_4$, Me$_3$Si, 88%; Ph, SiMe$_3$, 85%;
Ph, Ph, 89%; Ph, CH$_3$OCH$_2$, 90%

3-碘-2-丁烯酸乙酯与甲基炔丙基醚顺利地反应生成相应的交叉偶联产物[4]。

$$\underset{CO_2Et}{\overset{Me}{\diagdown}}C=CHI + \equiv-OMe \xrightarrow[\text{CuI, Et}_3\text{N, rt}]{Pd(PPh_3)_2Cl_2} MeO-\equiv-\underset{CO_2MEt}{\overset{Me}{\diagdown}}C=CH \qquad (25)$$

68%

呋喃糖和吡喃糖衍生的卤代-exo-烯糖在 Pd(PPh$_3$)$_4$/CuI 催化下进行 Sonogashira 反应生成碳水化合物衍生的烯-炔化合物，底物中所有手性中心的构型不发生变化，长碳链端炔反应速率较慢[21]。

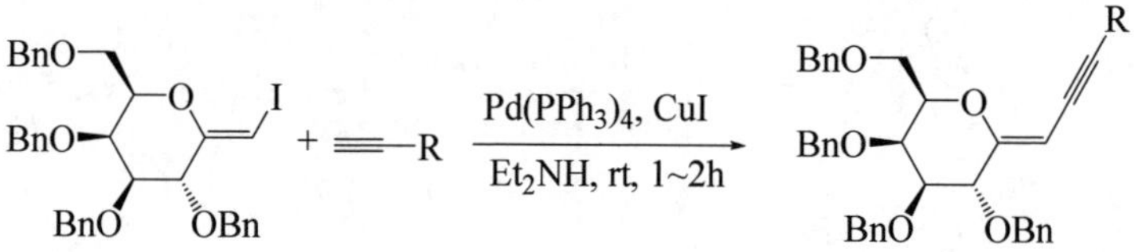

$$\text{(26)}$$

R=Ph, 92%; Me$_3$Si, 84%;
(CH$_2$)$_9$CH$_3$, 86%

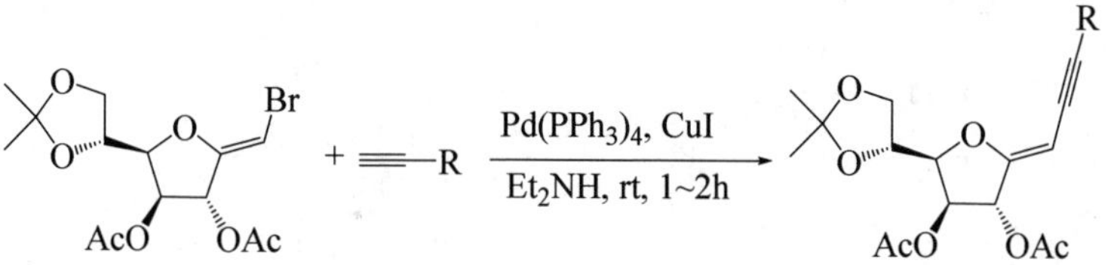

$$\text{(27)}$$

R=Ph, 88%; Me$_3$Si, 87%;
(CH$_2$)$_9$CH$_3$, 93% (24h)

在类似的催化体系下溴代烯丙基醚类化合物也能进行 Sonogashira 反应，多数情况下反应能顺利进行，但含吸电子基的芳基溴代烯丙基醚与空间位阻大的端炔反应不易进行，原因不详[22]。

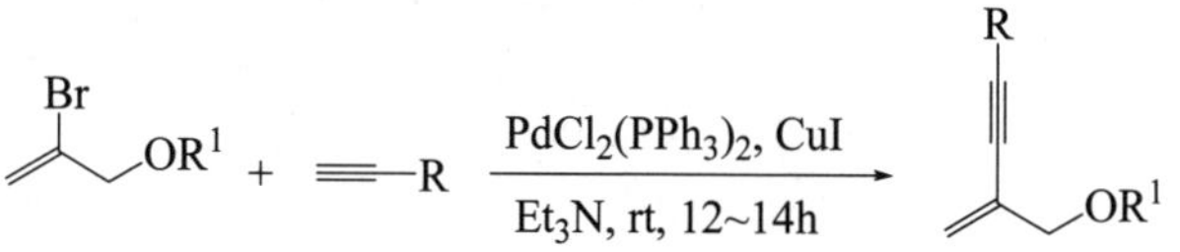

$$\text{(28)}$$

R^1, R=4-NO$_2$C$_6$H$_4$, t-Bu, 29%; 4-MeOC$_6$H$_4$, n-Pen,
84%; 4-MeOC$_6$H$_4$, t-Bu, 91%; Ph, t-Bu, 94%;
4-MeOC$_6$H$_4$CH$_2$, n-Pen, 70%

（3）酰氯的反应

在经典 Sonogashira 催化体系下，酰氯也能与端炔发生交叉偶联生成 1-炔基酮，1-炔基酮是有机合成，特别是杂环合成的非常有用的中间体。通过对有机碱或反应温度的调变，各种酰氯都能作为亲电底物进行反应[23]。

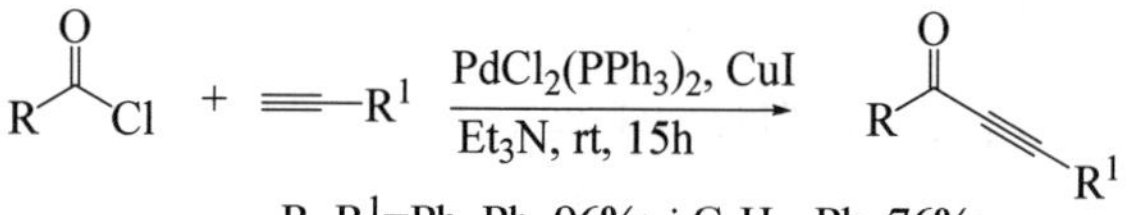

$$\text{(29)}$$

R, R^1=Ph, Ph, 96%; i-C$_3$H$_7$, Ph, 76%;
Ph, n-C$_4$H$_9$, 81%

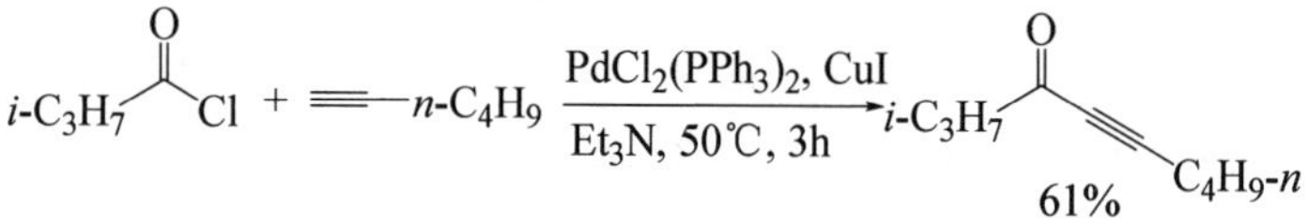

$$\text{(30)}$$

$$t\text{-}C_4H_9\text{-}CO\text{-}Cl + \text{HC≡C-Ph} \xrightarrow[\text{Et}_3\text{N, 吡啶, rt, 15h}]{\text{PdCl}_2(\text{PPh}_3)_2, \text{CuI}} t\text{-}C_4H_9\text{-}CO\text{-C≡C-Ph} \quad 79\% \tag{31}$$

$$(\text{H}_3\text{C})_2\text{N-CO-Cl} + \text{HC≡C-Ph} \xrightarrow[\text{Et}_3\text{N, 90℃, 6h}]{\text{PdCl}_2(\text{PPh}_3)_2, \text{CuI}} (\text{H}_3\text{C})_2\text{N-CO-C≡C-Ph} \tag{32}$$

$$R\text{-CO-Cl} + \text{HC≡C-Ph} \xrightarrow[\text{Et}_3\text{N, C}_6\text{H}_6, \text{rt, 15h}]{\text{PdCl}_2(\text{PPh}_3)_2, \text{CuI}} R\text{-CO-C≡C-Ph} \tag{33}$$

R=PhCH=CH, 75%; 2-呋喃基, 80%

酰氯为亲电底物的 Sonogashira 反应还能在十二烷基硫酸钠（SLS）作为相转移催化剂的无机碱性水溶液中进行，反应更加绿色环保[24]。

$$\text{4-MeC}_6\text{H}_4\text{-CO-Cl} + \text{HC≡C-(1-萘基)} \xrightarrow[\substack{\text{K}_2\text{CO}_3, \text{H}_2\text{O, SLS}\\ 65℃, 4\text{h}}]{\text{PdCl}_2(\text{Ph}_3)_2, \text{CuI}} \text{4-MeC}_6\text{H}_4\text{-CO-C≡C-(1-萘基)} \quad 97\% \tag{34}$$

5.2.2 单纯钯催化的 Sonogashira 反应

经典的 Sonogashira 催化反应体系需要卤化亚铜作为助催化剂，不仅对环境造成污染，而且反应必须在无氧条件下进行以避免端炔的自偶联副反应的发生，造成成本提高。克服这些弊端的首选措施是开发无须铜的单纯钯催化体系。经过多年的研究，人们发展了许多无铜钯催化体系，使得大部分在经典 Sonogashira 催化体系下芳基或杂芳基的炔基化反应也可在无铜存在下进行，这些无铜催化体系通常由传统的钯催化剂与适当的助剂或配体构成。由于与经典 Sonogashira 催化体系下的反应基本相同，本节不对单纯钯催化的 Sonogashira 反应过多介绍，仅通过几个反应实例，了解单纯钯催化的 Sonogashira 反应。

有些无铜存在的钯催化体系很简单，无须特殊的助剂或配体。例如，各种取代的芳基碘代物在以 Pd(OAc)$_2$ 为催化剂，NaOH 作碱的乙腈-水混合溶剂中与苯乙炔等顺利反应生成相应的交叉偶联产物［反应式（35）］，但相应的溴代物则需三苯基膦作配体，哌啶作碱反应才能进行［反应式（36）］[25]。

$$R^1\text{-}C_6H_4\text{-}I + \equiv\text{-}R^2 \xrightarrow[\text{H}_2\text{O/Me}_2\text{CO, 60℃, 1h}]{\text{Pd(OAc)}_2, \text{NaOH}} R^1\text{-}C_6H_4\text{-}\equiv\text{-}R^2 \quad (35)$$

$R^1, R^2 = NH_2, Ph, 88\%$; Br, Ph, 68%;
OMe, $n\text{-}C_4H_9$, 92%;
NO_2, HO-cyclohexyl, 94%

$$R^1\text{-}C_6H_4\text{-}Br + \equiv\text{-}R^2 \xrightarrow[\text{H}_2\text{O/Me}_2\text{CO, 60℃, 12~24h}]{\text{PdCl}_2, \text{PPh}_3, \text{哌啶}} R^1\text{-}C_6H_4\text{-}\equiv\text{-}R^2 \quad (36)$$

$R^1, R^2 = NO_2, Ph, 85\%$; OMe, Ph, 67%;
NO_2, $n\text{-}C_4H_9$, 70%;
NO_2, HO-cyclohexyl, 72%

大多数无铜存在的钯催化体系则需要有机配体协助，配体种类多种多样，按配位原子分为硫配体、氮配体和膦配体等。这些配体通常原位与钯配位形成配合物，配体的电子和空间性能影响所形成的钯物种的催化活性。下面反应式中的双齿硫配体，其结构中的两个噻吩环在反应体系与钯反式弱配位形成活性物种，能够使芳基或杂芳基溴或碘代物与端炔偶联[26]。

$$\text{Ar-X} + \equiv\text{-R} \xrightarrow[\text{MeCN, 80℃, 17h}]{\text{PdCl}_2(\text{PPh}_3)_2, \text{L, Et}_3\text{N}} \text{Ar}\text{-}\equiv\text{-R} \quad (37)$$

L = 1,2-双(噻吩基乙炔基)苯

Ar-X, R = PhI, 环己烯基, 95%;
$p\text{-}C_6H_4Br$, Ph, 82%;
2-噻吩基Br, Ph, 99%

以下是一个氮配体协助钯催化的 Sonogashira 反应实例。在 $Pd(OAc)_2$ 为催化剂，2-氨基嘧啶-4,6-二醇为配体的催化体系下，各种芳基溴或碘代物能够在温和条件下顺利地与端炔交叉偶联生成相应的交叉偶联产物[27]。

$$R\text{-}C_6H_4\text{-}X + \equiv\text{-}R^1 \xrightarrow[\text{MeCN, rt, 5~24h}]{\text{Pd(OAc)}_2, \text{L, Cs}_2\text{CO}_3} R\text{-}C_6H_4\text{-}\equiv\text{-}R^1 \quad (38)$$

L = 2-氨基-4,6-二羟基嘧啶

R, X, R^1 = 4-OMe, I, $n\text{-}C_8H_{17}$, 87% (7h);
2-NO_2, I, Ph, 95% (6h);
4-NO_2, Br, Ph, 95% (20h);
4-OMe, Br, Ph, 81% (24h)

相较于硫、氮配体，膦配体是更普遍用于构建 Sonogashira 反应的纯钯催化体系，膦配体既可以是简单的三苯基膦，也可以是电子和空间性能独特的其他膦配体。例如，由 Pd(OAc)$_2$、PPh$_3$ 和 NaOAc 构成的催化体系可以催化磺酰氨基乙炔或酰氨基乙炔与取代的碘代苯交叉偶联，NaOAc 作为助剂对反应的顺利进行起重要作用[28]。

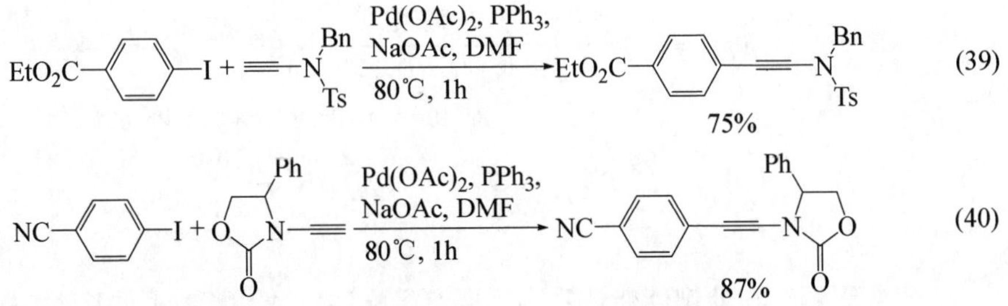

$$\text{EtO}_2\text{C-C}_6\text{H}_4\text{-I} + \text{HC≡C-N(Bn)(Ts)} \xrightarrow[\text{80℃, 1h}]{\text{Pd(OAc)}_2\text{, PPh}_3\text{, NaOAc, DMF}} \text{EtO}_2\text{C-C}_6\text{H}_4\text{-C≡C-N(Bn)(Ts)} \quad 75\% \tag{39}$$

$$\text{NC-C}_6\text{H}_4\text{-I} + \text{oxazolidinone-C≡CH} \xrightarrow[\text{80℃, 1h}]{\text{Pd(OAc)}_2\text{, PPh}_3\text{, NaOAc, DMF}} \text{NC-C}_6\text{H}_4\text{-C≡C-oxazolidinone} \quad 87\% \tag{40}$$

配体 Cy*Phine 空间位阻大，并且磷原子上两个环己基具有比苯基更强的给电子能力，使得这个配体与钯配位形成对空气和水稳定的配合物（图 5-4），该配合物作为催化剂能够催化芳基氯与端炔交叉偶联[29]。

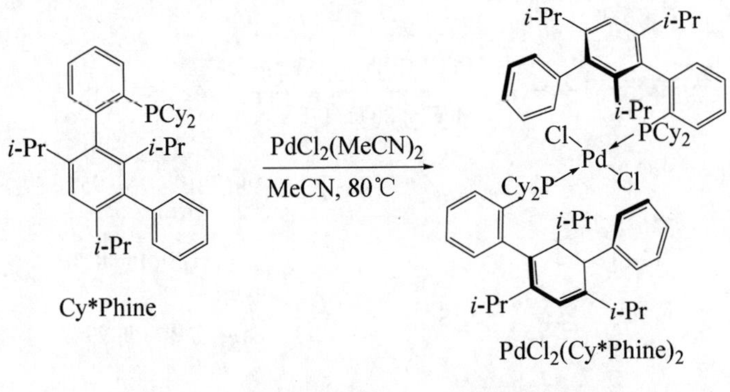

图 5-4 膦配体 Cy*Phine 及其钯配合物

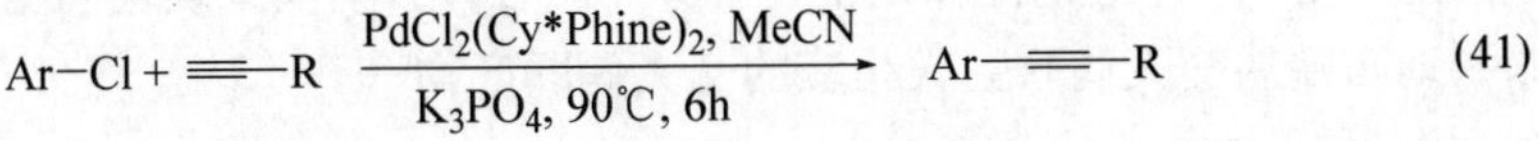

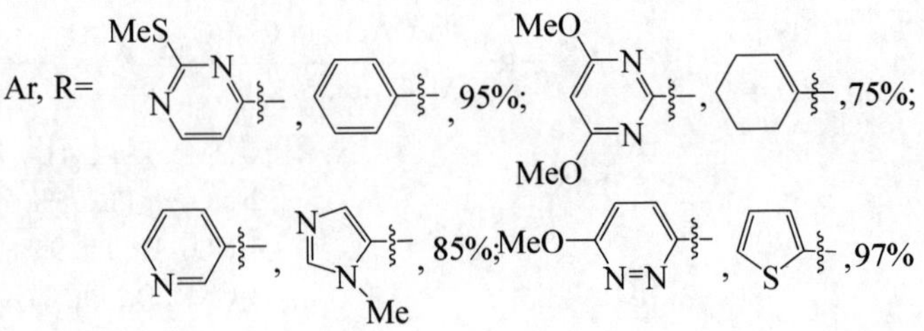

$$\text{Ar-Cl} + \text{HC≡C-R} \xrightarrow[\text{K}_3\text{PO}_4\text{, 90℃, 6h}]{\text{PdCl}_2(\text{Cy*Phine})_2\text{, MeCN}} \text{Ar-C≡C-R} \tag{41}$$

Ar, R= 2-(MeS)嘧啶-4-基, 苯基, 95%; 4,6-(MeO)$_2$嘧啶-2-基, 环己基, 75%; 吡啶-3-基, 1-甲基咪唑-5-基, 85%; 6-MeO-哒嗪-3-基, 噻吩-2-基, 97%

5.2.3 非贵金属催化的 Sonogashira 反应

基于钯的催化剂或催化体系在 Sonogashira 反应中具有催化活性和选择性高、底物适用性广的优势，但其昂贵的价格使得钯催化的 Sonogashira 反应在大规模生产上受制约。采用非贵重金属替代钯是消除前述制约的途径之一，因此，人们已发展了包括铜、镍、锌、铁和钴等非贵重金属催化剂用于 Sonogashira 反应[30-33]。其中，铁、钴属高丰度金属，而且无毒或低毒，因而基于铁、钴的 Sonogashira 反应的催化剂更具潜力，因此以下只对铁、钴催化的 Sonogashira 反应进行简单的介绍。但从以下实例可以看出，非贵重金属催化剂普遍存在活性较低、反应底物适用范围较窄的问题。

$FeCl_3$ 被证明是一种良好的铁基催化剂前体，它与氮配体结合可形成催化体系，催化各种底物顺利进行 Sonogashira 反应。例如，$FeCl_3$、双齿氮配体邻菲咯啉和弱碱 K_3PO_4 构成的催化体系在四丁基溴化铵（TBAB）存在下，可以实现不同取代的碘苯与苯乙炔的交叉偶联，但需要较高的反应温度和较长的反应时间，显然比钯基催化剂的催化活性差[34]。

$$R\text{-}C_6H_4\text{-}I + \equiv\text{-}Ph \xrightarrow[130℃, 48h]{FeCl_3,\ 邻菲咯啉\ K_3PO_4,\ TBAB,\ H_2O} R\text{-}C_6H_4\text{-}\equiv\text{-}Ph \quad (42)$$

R=4-CH_3CO, 87%; 4-NO_2, 80%;
3-CF_3, 83%; 4-NH_2, 71%;
4-CH_3O, 76%; 2-CH_3, 74%

$FeCl_3$ 通过与二氧化硅负载的三乙烯二胺的静电作用得到可循环使用的固相化催化剂，该催化剂在 Cs_2CO_3 作碱的 DMF/H_2O 混合溶剂中显示较高的催化活性，无论是给电子基还是吸电子基取代的碘苯都能与苯乙炔反应生成相应的交叉偶联产物[35]。

$$R\text{-}C_6H_4\text{-}I + \equiv\text{-}Ph \xrightarrow[DMF/H_2O, 110℃, 24h]{cat.,\ Cs_2CO_3} R\text{-}C_6H_4\text{-}\equiv\text{-}Ph \quad (43)$$

cat.: SiO_2-Si(OMe)-CH_2CH_2CH_2-N(+)(N) $FeCl_4^-$

R=H, 81%; CH_3, 83%;
CH_3O, 78%; NO_2, 80%

钴基催化剂多为各种载体负载的氯化亚钴（$CoCl_2$）或钴纳米离子。例如，以亚胺为连接键和官能团的多孔有机聚合物（POP）负载的 $CoCl_2$（Co@imine-POP）在 KOH 的聚乙二醇（PEG）溶液中对取代的碘代苯与苯乙炔的交叉偶联反应有较好的催化性能[36]。

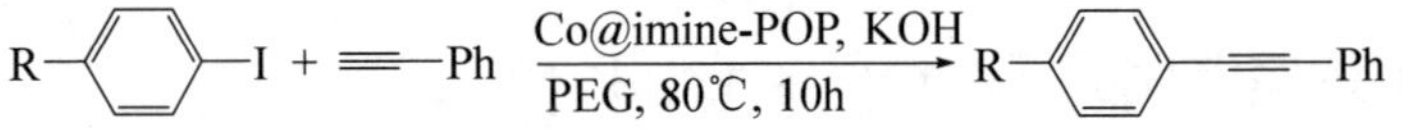

$$R-\text{C}_6\text{H}_4-I + \text{HC≡C}-Ph \xrightarrow[\text{PEG, 80°C, 10h}]{\text{Co@imine-POP, KOH}} R-\text{C}_6\text{H}_4-\text{C≡C}-Ph \quad (44)$$

R = H, 72%; CH₃O, 69%
NO₂, 80%; CH₃CO, 78%

5.3 Sonogashira 反应在药物合成上的应用实例

由上一节基本反应可以看到，Sonogashira 反应可以在温和条件下高效生成共轭芳-炔（arenynes）、共轭烯-炔（enynes）和共轭炔-酮（ynones）。这些化合物都是重要的有机中间体，表明 Sonogashira 反应在药物、天然产物或生物活性化合物合成上有着无可争辩的地位。芳基卤化物与端炔构建共轭芳-炔的反应被广泛应用于各种药物分子的合成，而卤代烯烃与端炔构建共轭烯-炔的反应常用于天然产物等生物活性物质的合成。以下大致按卤代物结构不同介绍一些合成实例，为了叙述方便，其中经由 Sonogashira 反应合成的产物编号为 SD-x，x 为整数，代表序号。

2-氯-4-三氟甲基苯甲腈与叔丁基乙炔的 Sonogashira 反应被成功用于 TRPV1 受体拮抗剂的合成[37]。通常芳基氯在 Sonogashira 反应中活性低，但苯环上的吸电子的三氟甲基和氰基对底物起强烈的活化作用，使得反应在经典催化剂组合和大空间位阻膦配体（Davephos）存在下顺利进行。

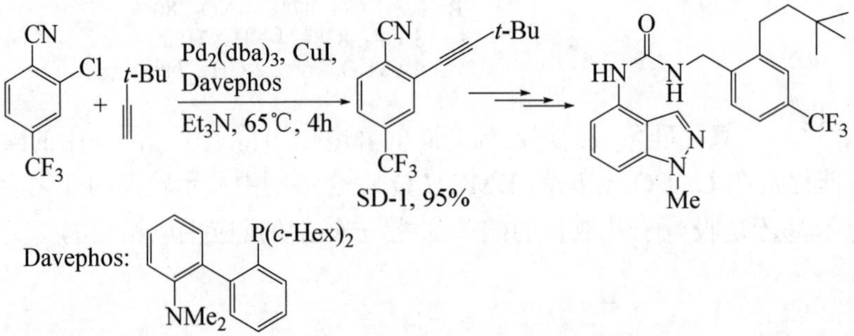

(45)

反叶酸（antifolate）是一种抗癌制剂，其大规模合成的起始步骤是经典钯-铜催化体系催化下的 3-丁炔-1-醇与 4-溴苯甲酸酯的 Sonogashira 交叉偶联。反应液经过滤去除铵盐，滤液再经螯合树脂除钯，重结晶得固体偶联产物 SD-2，SD-2 经后续多步转化生成反叶酸[38]。

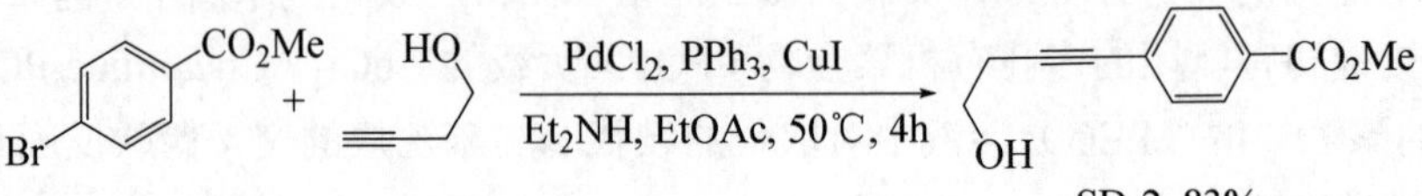

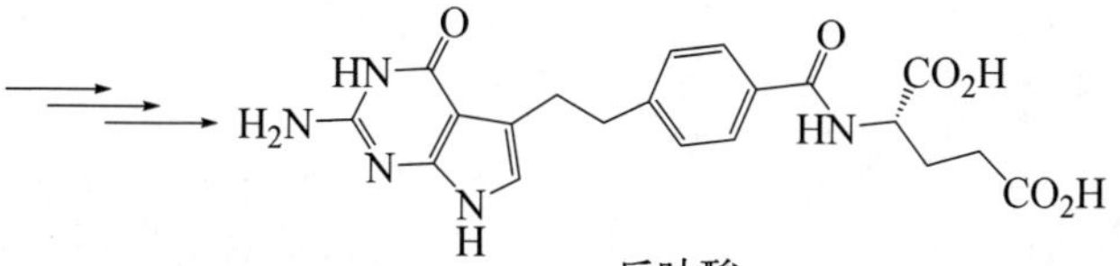

(46) 反叶酸

下面反应的目标分子也是一种抗癌药物,其可抑制人类表皮细胞增殖。其简捷合成包括"一锅"两步 Sonogashira 反应合成关键中间体。首先,2-溴对苯二甲醚与 2-甲基-3-丁炔-2-醇在 PdCl$_2$、PPh$_3$ 和 CuI 构成的催化体系催化下发生第一步 Sonogashira 反应,反应完毕,在减压下浓缩以除去过量的 2-甲基-3-丁炔-2-醇,得 SD-3 浓缩液;将 SD-3 浓缩液加入 6-溴-4-乙基苯并嘧啶、四丁基溴化铵的甲苯溶液,再加入饱和 NaOH 水溶液,加热回流进行第二步 Sonogashira 反应生成 SD-3',无须补加催化剂。在第二步 Sonogashira 反应中,SD-3 在反应条件下原位生成 2-乙炔基对苯二甲醚,其再与 6-溴-4-乙基苯并嘧啶进行交叉偶联反应。最后,SD-3'经选择性加氢转化为目标分子[39]。

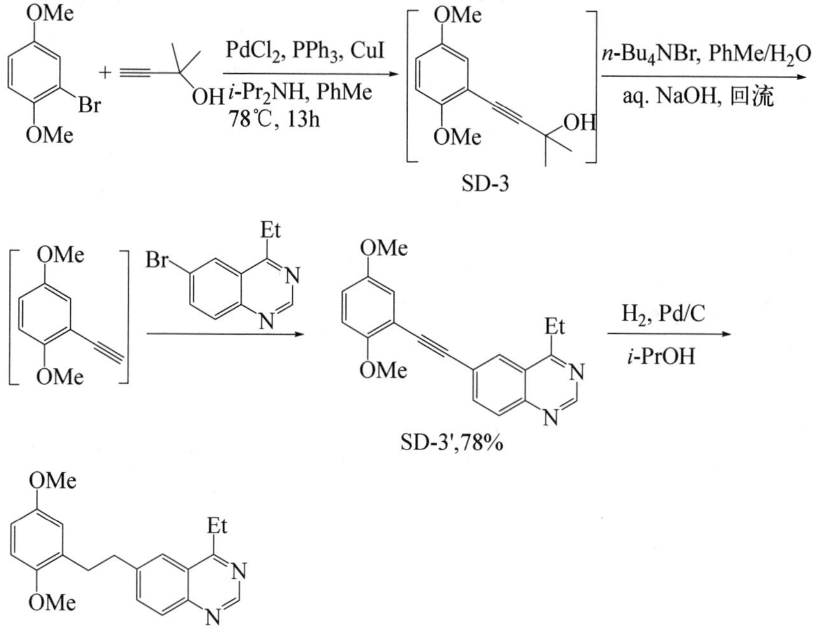

(47)

多官能团、复杂结构的芳基碘与复杂结构端炔的 Sonogashira 反应被用于合成一种组织蛋白酶 S 的半胱氨酸蛋白酶抑制剂,该药物分子还对各种免疫疾病有潜在的疗效。如反应式(48)所示,多官能团的芳基碘与 4-(4'-氯苄氨甲基)苯乙炔在经典钯-铜催化剂催化下顺利进行 Sonogashira 反应,以良好的收率得到目标分子 SD-4[40]。

SD-4, 85%

(48)

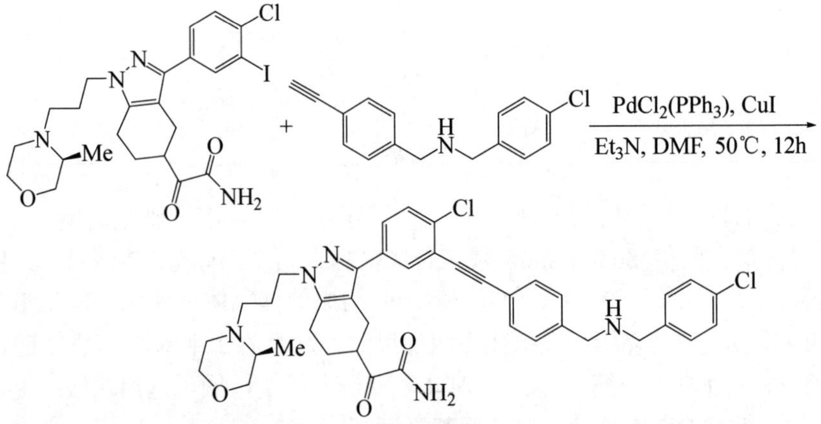

3-芳基-3-苯并咪唑丙烯酰胺类化合物具有杀菌、抗病毒活性。利用 2,5-二氟苯基三氟甲磺酸酯与 *N*-甲基丙炔酰胺的 Sonogashira 反应高效生成交叉偶联产物 SD-5，SD-5 作为关键中间体与碘代苯并咪唑衍生物进行选择性 Heck-Mizorogi 反应生成一种 3-芳基-3-苯并咪唑丙烯酰胺类抗人鼻病毒药物（antirhinoviral agent）[41]。

antirhinoviral agent

(49)

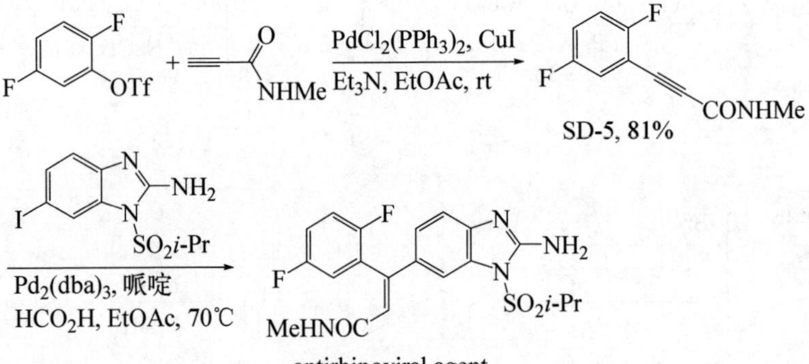

STA-5312 是一种新型的微管抑制剂，具有抗肿瘤活性。在 STA-5312 合成路线中，2-溴吡啶与 3-(4'-氰基苯基)-1-丙炔间的 Sonogashira 偶联是关键一步。反应在经典的钯-铜催化体系催化下于 THF 溶剂中 60°C 顺利进行，交叉偶联产物 SD-5 的收率为 80%。关键中间体 SD-6 进一步转化生成 STA-5312[42]。

SD-6, 80%

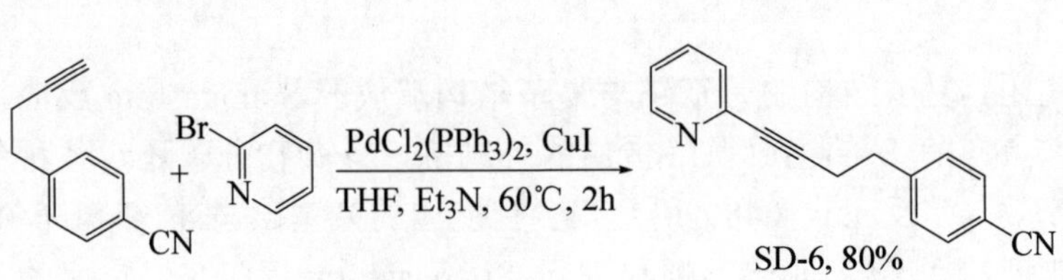

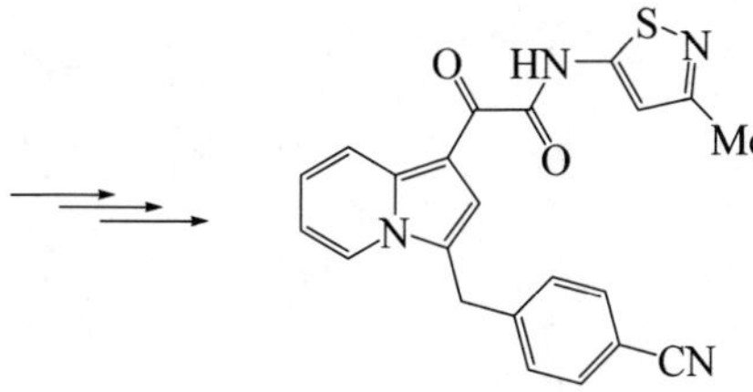

恩尿嘧啶（eniluracil）是一种有效的能使尿嘧啶还原酶失活的酶抑制剂，它可保护5-氟尿嘧啶（5-Fu）不被迅速代谢和破坏，称作抗癌药 5-Fu 的增效剂。其最简捷高效的合成路线是 5-碘尿嘧啶与三甲基硅基乙炔进行 Sonogashira 偶联生成 5-(三甲基硅基乙炔基) 尿嘧啶 SD-7，再经氢氧化钠水溶液脱三甲基硅基得到恩尿嘧啶。交叉偶联反应在 $PdCl_2(PPh_3)_2$-CuI 催化下于室温顺利进行，以优异的收率得到恩尿嘧啶[43]。

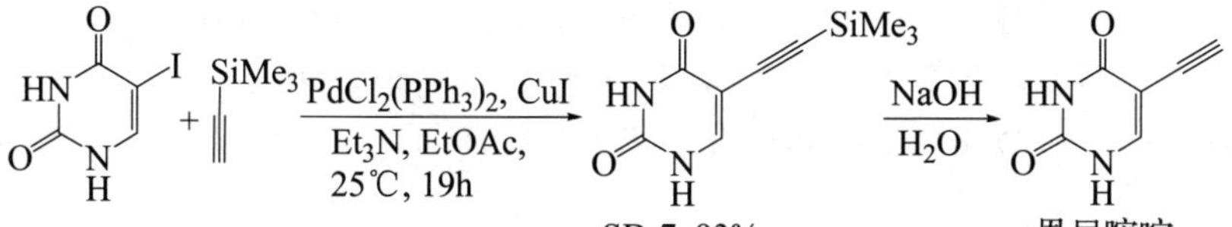

碘代烯烃与端炔的 Sonogashira 偶联多用于天然产物关键中间体的合成。其代表性实例之一是 (−)-disorazole C1 合成中两次烯-炔结构片段的构建，两次 Sonogashira 反应在相同的催化剂体系和反应条件下进行，而且均以几乎定量的收率得到交叉偶联产物 SD-8 和 SD-8'[44]。

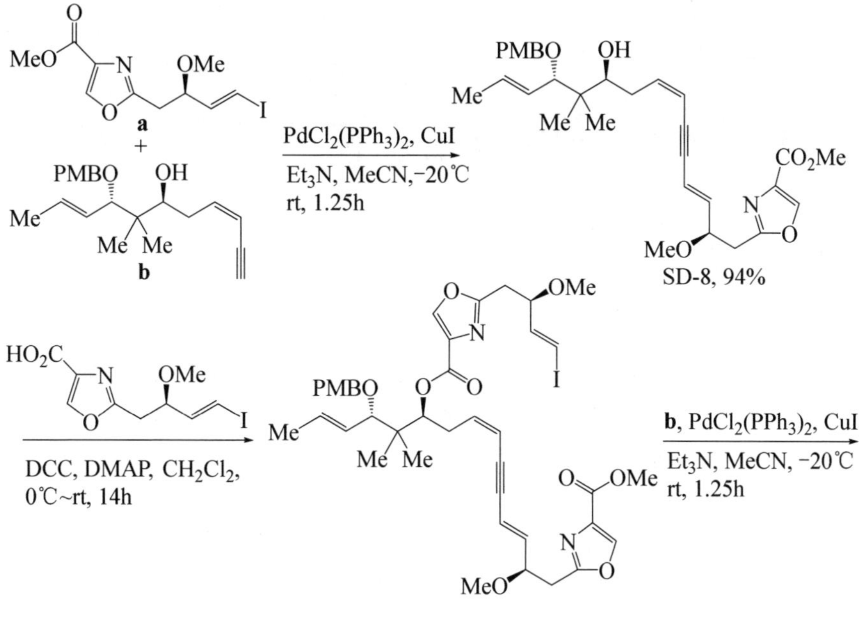

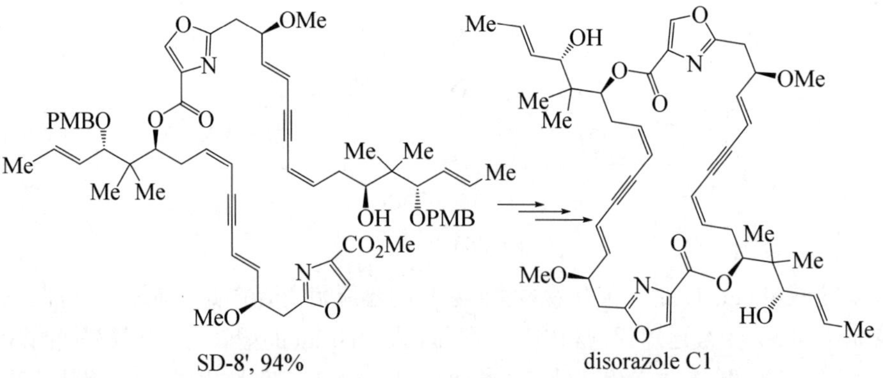

卡奇霉素（calicheamicin γ_1）是一种特殊结构的天然产物，属于烯-二炔类抗癌抗生素，其结构含有顺式炔-烯-炔结构片段，其合成具有挑战性，采用 Sonogashira 反应成功解决了这一难题。钯-铜催化体系催化的端炔与(Z)-(4-氯-3-丁烯-1-炔基)三甲基硅烷的交叉偶联得到含顺式炔-烯-炔结构片段的中间体 SD-9。由于反应在较低温度下进行，使得反应底物上的各种官能团以及烯烃的 Z-构型得到保持[45,46]。

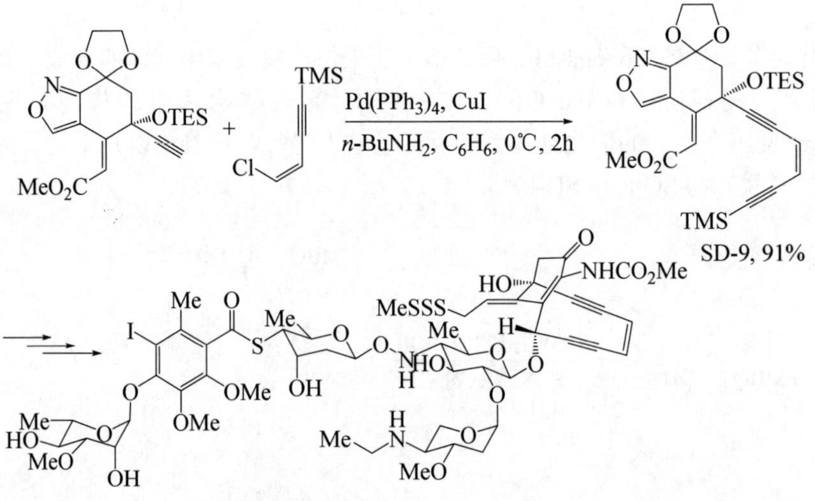

paracentrone 属多烯天然产物，碘代共轭烯烃衍生物与含官能团的端炔的 Sonogashira 反应为其合成提供了一种便捷的路线。合成路线起始于 3-甲基-2,4,6-辛三烯酸乙酯与环己基乙炔衍生物的交叉偶联，反应在简单的钯-铜催化反应体系下进行，以良好的收率生成交叉偶联产物 SD-10，由于反应条件温和使得乙炔基底物上的环氧环并不与二异丙胺发生加成开环反应。关键中间体 SD-10 经后续多步反应最终生成 paracentrone[47]。

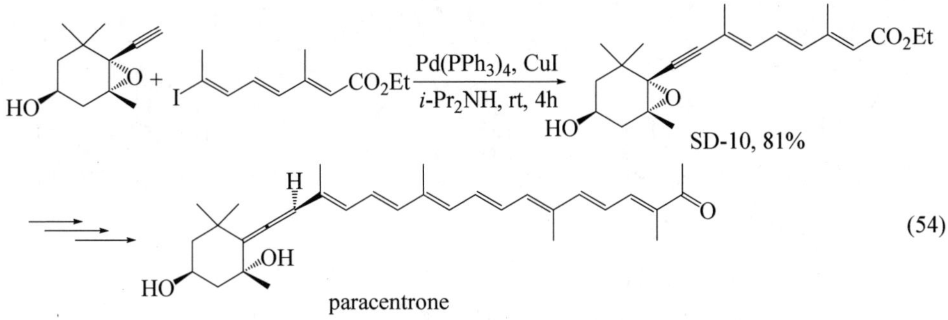

maduropeptin chromophore 是含有一个双环[7.3.0]烯二炔和一个旋转异构的 15 元柄型大环内酰胺结构单元的结构高度复杂的天然产物，显示强效的抗肿瘤和抗菌活性。在其全合成过程中，Sonogashira 偶联被成功用于双环[7.3.0]烯二炔结构单元的构建。多官能团化环戊烯基三氟甲磺酸酯与多官能团化的端炔在经典钯-铜催化剂催化下以定量的收率生成交叉偶联产物 SD-11，SD-11 经多步反应转化为目标产物 maduropeptin chromophore[48]。

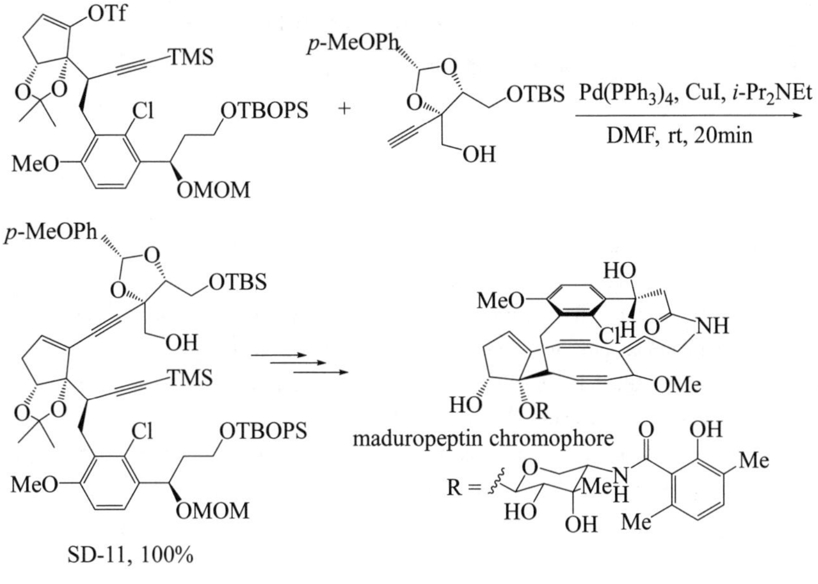

在基本反应部分，没有谈及分子内 Sonogashira 反应，因为通过 Sonogashira 反应形成的烯-炔结构在后续成环步骤存在张力，只有在构建大于 15 元环的大环时反应才能进行，但通常反应收率也较低。只有为数不多的利用 Sonogashira 反应合成大环天然产物的实例，其中包括海洋天然产物 penarolide sulfate A1。首先碘代烯和端炔构筑单元在 EDC 和 HOBt 缩合剂作用下偶联生成分子内同时含有端炔和碘代烯基团的分子内 Sonogashira 反应的前体，该前体在经典钯-铜催化体系催化下以 35%的收率得到 30 元大环化合物 SD-12，SD-12 进一步转化生成 penarolide sulfate A1[49]。

penarolide sulfate A1

参考文献

[1] Sonogashira K, Tohda Y, Hagihara N. A convenient synthesis of acetylenes: catalytic substitutions of acetylenic hydrogen with bromoalkenes, iodoarenes and bromopyridines[J]. Tetrahedron Letters, 1975, 16(50): 4467-4470.

[2] Cassar L. Synthesis of aryl-and vinyl-substituted acetylene derivatives by the use of nickel and palladium complexes[J]. Journal of Organometallic Chemistry, 1975, 93(2): 253-257.

[3] Dieck H A, Heck F R. Palladium catalyzed synthesis of aryl, heterocyclic and vinylic acetylene derivatives[J]. Journal of Organometallic Chemistry, 1975, 93(2): 259-263.

[4] Chinchilla R, Nájera C. The Sonogashira reaction: a booming methodology in synthetic organic chemistry[J]. Chemical Reviews, 2007, 107(3): 874-922.

[5] Stambuli J P, Bühl M, Hartwig J F. Synthesis, characterization, and reactivity of monomeric, arylpalladium halide complexes with a hindered phosphine as the only dative ligand[J]. Journal of the American Chemical Society, 2002, 124(32): 9346-9347.

[6] Amatore C, Jutand A. Anionic Pd (0) and Pd (Ⅱ) intermediates in palladium-catalyzed Heck and cross-coupling reactions[J]. Accounts of Chemical Research, 2000, 33(5): 314-321.

[7] Gazvoda M, Virant M, Pinter B, et al. Mechanism of copper-free Sonogashira reaction operates through palladium-palladium transmetallation[J]. Nature Communications, 2018, 9(1): 4814.

[8] Bang H B, Han S Y, Choi D H, et al. Facile total synthesis of benzo[b]furan natural product XH-14[J]. Synthetic Communications, 2009, 39(3): 506-515

[9] Manarin F, Roehrs J A, Brandao R, et al. Synthesis of 3-alkynyl-2-(methylsulfanyl) benzo[b]furans via Sonogashira cross-coupling of 3-iodo-2-(methylsulfanyl) benzo[b]furans with terminal alkynes[J].

Synthesis, 2009(23): 4001-4009.
[10] Krauss J, Bracher F. New total synthesis of niphatesine C and norniphatesine C based on a Sonogashira reaction[J]. Archiv der Pharmazie: An International Journal Pharmaceutical and Medicinal Chemistry, 2004, 337(7): 371-375.
[11] Hocková D, Holý A, Masojídková M, et al. An efficient synthesis of cytostatic mono and bis-alkynylpyrimidine derivatives by the Sonogashira cross-coupling reactions of 2, 4-diamino-6-iodopyrimidine and 2-amino-4, 6-dichloropyrimidine[J]. Tetrahedron, 2004, 60(23): 4983-4987.
[12] Sun L P, Huang X H, Wang J X. A facile and efficient synthesis of 7-azaindoles[J]. Journal of the Chinese Chemical Society, 2007, 54(3): 569-573.
[13] González-Gómez J C, Uriarte E. Efficient preparation of 2-substituted pyridazino [4, 3-*h*] psoralen derivatives[J]. Synlett, 2003(14): 2225-2227.
[14] Elangovan A, Wang Y H, Ho T I. Sonogashira coupling reaction with diminished homocoupling[J]. Organic Letters, 2003, 5(11): 1841-1844.
[15] Hundertmark T, Littke A F, Buchwald S L, et al. Pd (PhCN)$_2$Cl$_2$/P(*t*-Bu)$_3$: a versatile catalyst for Sonogashira reactions of aryl bromides at room temperature[J]. Organic Letters, 2000, 2(12): 1729-1731.
[16] Köllhofer A, Pullmann T, Plenio H. A versatile catalyst for the Sonogashira coupling of aryl chlorides[J]. Angewandte Chemie International Edition, 2003, 42(9): 1056-1058.
[17] Berg T C, Gundersen L L, Eriksen A B, et al. Synthesis of optically active 6-alkynyl-and 6-alkylpurines as cytokinin analogs and inhibitors of 15-lipoxygenase; studies of intramolecular cyclization of 6-(hydroxyalkyn-1-yl) purines[J]. European Journal of Organic Chemistry, 2005(23): 4988-4994.
[18] Doucet H, Hierso J C. Palladium-based catalytic systems for the synthesis of conjugated enynes by Sonogashira reactions and related alkynylations[J]. Angewandte Chemie International Edition, 2007, 46(6): 834-871.
[19] Hao W Y, Jiang J W, Cai M Z. A facile stereospecific synthesis of (*Z*)-2-sulfonyl-substituted 1, 3-enynes via Sonogashira coupling of (*E*)-α-iodovinyl sulfones with 1-alkynes[J]. Chinese Chemical Letters, 2007, 18(7): 773-776.
[20] Fang X, Jiang M, Hu R, et al. Facile stereoselective synthesis of (*E*)-1-arylseleno-substituted 1, 3-enynes and their applications in synthesis of (*E*)-enediynes[J]. Synthetic Communications, 2008, 38(23): 4170-4181.
[21] Gómez A M, Barrio A, Pedregosa A, et al. Sonogashira couplings of halo-and epoxy-halo-exo-glycals: concise entry to carbohydrate-derived enynes[J]. European Journal of Organic Chemistry, 2010(15): 2910-2920.
[22] Kutsumura N, Niwa K, Saito T. Novel one-pot method for chemoselective bromination and sequential Sonogashira coupling[J]. Organic Letters, 2010, 12(15): 3316-3319.
[23] Tohda Y, Sonogashira K, Hagihara N. A convenient synthesis of 1-alkynyl ketones and 2-alkynamides[J]. Synthesis, 1977(11): 777-778.
[24] Chen L, Li C J. A remarkably efficient coupling of acid chlorides with alkynes in water[J]. Organic Letters, 2004, 6(18): 3151-3153.
[25] Shi S, Zhang Y. Palladium-catalyzed copper-free Sonogashira coupling reaction in water and acetone[J]. Synlett, 2007(12): 1843-1850.
[26] Atobe S, Sonoda M, Suzuki Y, et al. Copper-free Sonogashira coupling reaction using a trans-spanning 1, 2-bis (2-thienylethynyl) benzene ligand[J]. Chemistry Letters, 2011, 40(9): 925-927.

[27] Li J H, Zhang X D, Xie Y X. Efficient and copper-free Sonogashira cross-coupling reaction catalyzed by Pd(OAc)$_2$/pyrimidines catalytic system[J]. European Journal of Organic Chemistry, 2005(20):4256-4259.

[28] Wakamatsu H, Takeshita M. Copper-free Sonogashira cross-coupling of ynamides: easy access to various substituted ynamides from nonsubstituted ynamides[J]. Synlett, 2010(15): 2322-2344.

[29] Yang Y, Lim J F Y, Chew X, et al. Palladium precatalysts containing meta-terarylphosphine ligands for expedient copper-free Sonogashira cross-coupling reactions[J]. Catalysis Science & Technology, 2015, 5(7): 3501-3506.

[30] Amrutha S, Radhika S, Anilkumar G. Recent developments and trends in the iron-and cobalt-catalyzed Sonogashira reactions[J]. Beilstein Journal of Organic Chemistry, 2022, 18(1): 262-285.

[31] Beletskaya I P, Latyshev G V, Tsvetkov A V, et al. The nickel-catalyzed Sonogashira-Hagihara reaction[J]. Tetrahedron Letters, 2003, 44(27): 5011-5013.

[32] Asha S, Thomas A M, Ujwaldev S M, et al. A novel protocol for the Cu-catalyzed Sonogashira coupling reaction between aryl halides and terminal aalkynes using trans-1, 2-diaminocyclohexane ligand[J]. ChemistrySelect, 2016, 1(13): 3938-3941.

[33] Thankachan A P, Sindhu K S, Krishnan K K, et al. An efficient zinc-catalyzed cross-coupling reaction of aryl iodides with terminal aromatic alkynes[J]. Tetrahedron Letters, 2015, 56(41): 5525-5528.

[34] Sindhu K S, Thankachan A P, Thomas A M, et al. Iron-catalyzed Sonogashira type cross-coupling reaction of aryl iodides with terminal alkynes in water under aerobic conditions[J]. ChemistrySelect, 2016, 1(3): 556-559.

[35] Hajipour A R, Abolfathi P, Tavangar-Rizi Z. Iron-catalyzed cross-coupling reaction: heterogeneous palladium and copper-free Heck and Sonogashira cross-coupling reactions catalyzed by a reusable Fe (Ⅲ) complex[J]. Applied Organometallic Chemistry, 2018, 32(6): e4353.

[36] Hajipour A R, Khorsandi Z. Pd/Cu-free Heck and Sonogashira coupling reactions applying cobalt nanoparticles supported on multifunctional porous organic hybrid[J]. Applied Organometallic Chemistry, 2020, 34(4): e5398.

[37] Yu S, Haight A, Kotecki B, et al. Synthesis of a TRPV1 receptor antagonist[J]. The Journal of Organic Chemistry, 2009, 74(24): 9539-9542.

[38] Barnett C J, Wilson T M, Kobierski M E. A practical synthesis of multitargeted antifolate LY231514[J]. Organic Process Research & Development, 1999, 3(3): 184-188.

[39] Königsberger K, Chen G P, Wu R R, et al. A practical synthesis of 6-[2-(2, 5-dimethoxyphenyl) ethyl]-4-ethylquinazoline and the art of removing palladium from the products of Pd-catalyzed reactions[J]. Organic Process Research & Development, 2003, 7(5): 733-742.

[40] Deng X, Liang J T, Peterson M, et al. Practical synthesis of a cathepsin S inhibitor: route identification, purification strategies, and serendipitous discovery of a crystalline salt form[J]. The Journal of Organic Chemistry, 2010, 75(6): 1940-1947.

[41] Hay L A, Koenig T M, Ginah F O, et al. Palladium-catalyzed hydroarylation of propiolamides. A regio- and stereo-controlled method for preparing 3, 3-diarylacrylamides[J]. The Journal of Organic Chemistry, 1998, 63(15): 5050-5058.

[42] Li H, Xia Z, Chen S, et al. Development of a practical synthesis of STA-5312, a novel indolizine oxalylamide microtubule inhibitor[J]. Organic Process Research & Development, 2007, 11(2): 246-250.

[43] Cooke J W B, Bright R, Coleman M J, et al. Process research and development of a dihydropyrimidine

dehydrogenase inactivator: large-scale preparation of eniluracil using a Sonogashira coupling[J]. Organic Process Research & Development, 2001, 5(4): 383-386.

[44] Wipf P, Graham T H. Total synthesis of (−)-disorazole C1[J]. Journal of the American Chemical Society, 2004, 126(47): 15346-15347.

[45] Nicolaou K C, Hummel C W, Pitsinos E N, et al. Total synthesis of calicheamicin. gamma. 1I[J]. Journal of the American Chemical Society, 1992, 114(25): 10082-10084.

[46] Smith A L, Hwang C K, Pitsinos E, et al. Enantioselective total synthesis of (-)-calicheamicinone[J]. Journal of the American Chemical Society, 1992, 114(8): 3134-3136.

[47] Murakami Y, Nakano M, Shimofusa T, et al. Total synthesis of paracentrone, C31-allenic apo-carotenoid[J]. Organic & Biomolecular Chemistry, 2005, 3(8): 1372-1374.

[48] Komano K, Shimamura S, Inoue M, et al. Total synthesis of the maduropeptin chromophore aglycon[J]. Journal of the American Chemical Society, 2007, 129(46): 14184-14186.

[49] Mohapatra D K, Bhattasali D, Gurjar M K, et al. First asymmetric total synthesis of penarolide sulfate A1[J]. European Journal of Organic Chemistry, 2008(36):6213-6224.

第6章 Stille 反应及其在药物合成上的应用

6.1 Stille 反应及其机理

6.1.1 Stille 反应

Stille 反应是指在过渡金属特别是钯催化剂催化下有机锡化合物与碳亲电试剂的交叉偶联反应。尽管 Stille 不是第一个研究此类反应的有机化学家，但由于其对该合成反应持续、广泛的研究，并揭示了反应机理，因此化学家将该反应称作 Stille 反应。Stille 反应可用图 6-1 表示。其中，R^1X 可以是溴代物、碘代物和磺酸酯如三氟甲磺酸酯；R^2 一般为不饱和的有机基团，如(乙)烯基、芳基、杂芳基、炔基和烯丙基，有时也可以是烷基；R^3 是不发生反应的配体，通常是丁基或甲基。偶联反应的亲电试剂是有机卤代物，如溴代物或碘代物，以及类卤代物磺酸酯如三氟甲磺酸酯，特殊情况下也可以是其他离去基团。

$$R^1X + R^2Sn(R^3)_3 \xrightarrow{ML_n} R^1-R^2 + (R^3)_3SnX$$

M = Pd, Mn, Ni, Cu等；
R^1X=溴代物、碘代物和磺酸酯，如三氟甲磺酸酯等；
R^2=(乙)烯基、芳基、杂芳基、炔基和烯丙基，或烷基；
R^3=丁基、甲基等。

图 6-1 Stille 反应式及底物范围

在 Stille 反应中，有机锡试剂作为交叉偶联伙伴之一提供亲核基团具有如下优势：有机锡试剂在空气和湿气下稳定，容易纯化和储存；反应过程对空气和湿气不敏感，有时氧气和水甚至促进反应进行；有机锡试剂不与常见的官能团反应，因此在利用 Stille 反应合成时无需官能团保护。最初的 Stille 反应通常在较高的反应温度下进行,通过向反应体系引入配体和一价铜盐使反应条件更加温和。Stille 反应的缺点是有机锡化合物通常具有较高的毒性,但三丁基锡试剂的毒性远远低于三甲基锡和三乙基锡衍生物的毒性。毒理学家规定药品中的锡含量上限是 $20×10^{-6}$。在 Stille 反应体系中引入 CsF 或 Bu_4NF 可以有效清除有机锡残留[1]。鉴于 Stille 反应条件温和、底物适用性广泛,有机锡残留得到有效控制,其在药物合成和生物活性的天然产物合成中具有更广泛的应用,特别适合于含有多官能团的复杂分子的合成。事实上,Stille 反应已成功应用于许多含有敏感官能团的环体系的构建[2]。

6.1.2 反应机理

与其他钯催化的交叉偶联反应类似,Stille 反应机理也可以简单概括为如图 6-2 所示的三步催化循环,详细的反应机理参见相关文献[3]。

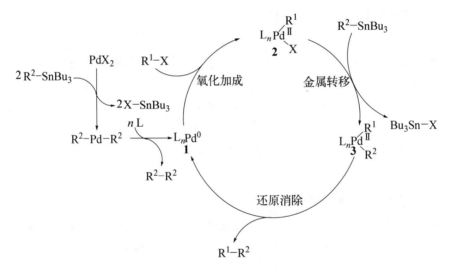

图 6-2 钯催化 Stille 反应的催化循环

当钯催化剂以二价钯盐的形式加入时,有机锡烷快速将钯盐还原为 Pd^0 物种,其立刻进入催化循环。当然,钯催化剂也可以直接以 Pd^0 形式加入,但以何种价态的催化剂前体加入似乎对反应速率影响不大。通常催化剂的活性物种是由金属钯配合物原位生成的 14 电子配合物 L_nPd^0(**1**)。催化循环的第一步是氧化加成,即反应底物 R^1—X 加成到活性物种 **1** 生成 16 电子的中间体 **2**,此时钯的价态由 0 价变为 +2 价;紧接着中间体 **2** 与 R^2—$SnBu_3$ 发生金属转移反应生成另一关键中间体 R^1—Pd—R^2,副产 Bu_3SnX;最后,R^1—Pd—R^2 发生还原消除反应生成交叉偶联产物 R^1—R^2,并再生催化剂活性物种 Pd^0。

在上面催化循环中，中间体 **2** 中 R^1 与 Pd 通过 σ 键结合，即 **2** 为 σ 配合物，因此当 R^1 为甲基以外的烷基时 **2** 会发生 β-氢消除，Stille 反应不易控制，但通过适宜膦配体的调控，当 R^1 为烷基时 Stille 反应也能进行[4]。在金属转移基元反应中，有机锡试剂中的有机基团 R^2 转移到 Pd 生成中间体 **3**，而锡烷基与中间体 **2** 中的卤素离子或磺酰氧基结合生成 Bu_3SnX。从有机合成的角度看，有机锡试剂中不同结构特征的有机基团 R^2 向金属 Pd 转移的速率不同，迁移速率为：炔基>烯基>芳基>烯丙基≈苄基>>烷基。正是因为烷基由锡向钯转移的速率远远低于其他基团，所以其他结构复杂基团（烯基或芳基）与三甲基或三丁基的混合锡烷可用作 Stille 反应的亲核偶联伙伴。中间体 **2** 与有机锡试剂的金属转移是催化循环中决定反应速率的步骤[5]。

6.2 基本的 Stille 反应

6.2.1 基本的分子间 Stille 反应

Stille 反应相较于其他过渡金属（尤其是钯）催化构建碳-碳键的反应具有更广泛的底物范围，反应众多。这是因为在交叉偶联伙伴中，亲电试剂 R^1—X 几乎包括了所有结构特征的卤化物或类卤化物，而亲核的锡试剂的有机基团则是除了烷基以外的所有结构特征的有机基团，并且反应条件温和，官能团耐受性广泛。为了较容易地了解 Stille 反应，以下按交叉偶联伙伴或反应底物的结构分类介绍一些基本的 Stille 反应。

6.2.1.1 卤代烯烃或烯基磺酸酯为亲电底物

卤代烯烃作为亲电底物的 Stille 反应较少，反应的活性顺序为 RCl<RBr<RI。碘代烯烃通常在室温下反应，溴代烯烃在加热下反应，而氯代烯烃反应往往需要其他促进剂如 CuI 等。由于碘代烯烃的交叉偶联反应条件温和，通常是高立体选择的；溴代烯烃需要在加热下反应，有时会发生明显的 Z/E 异构化。在相同反应温度下，碘代烯烃反应时间更短，收率更高[6]。

$$Ph\diagup\!\!\!\diagdown^I + n\text{-}Bu_3Sn\diagdown\!\!\!\diagup \xrightarrow[\text{DMF, 25℃, 0.1h}]{(MeCN)_2PdCl_2} Ph\diagup\!\!\!\diagdown\!\!\!\diagup \quad (1)$$
$$85\%$$

$$Ph\diagup\!\!\!\diagdown^{Br} + n\text{-}Bu_3Sn\diagdown\!\!\!\diagup^{SiMe_3} \xrightarrow[\text{DMF, 25℃, 23h}]{(MeCN)_2PdCl_2} Ph\diagup\!\!\!\diagdown\!\!\!\diagup^{SiMe_3} \quad (2)$$
$$56\%$$

下面的反应进一步说明碘代烯烃比溴代烯烃反应活性高，反应后底物上的溴被保留，而碘则完全反应[7]。

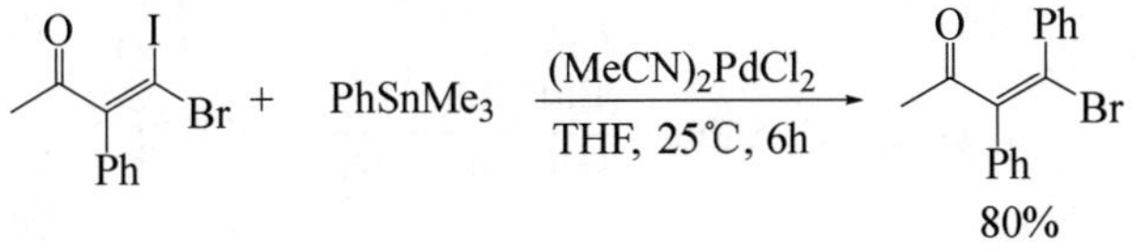

$$\underset{\underset{Ph}{|}}{\overset{O}{\underset{\|}{C}}}\overset{I}{\underset{Br}{C}} + PhSnMe_3 \xrightarrow[THF, 25℃, 6h]{(MeCN)_2PdCl_2} \underset{Ph}{\overset{O}{\underset{\|}{C}}}\overset{Ph}{\underset{Br}{C}} \quad 80\%$$ (3)

相同反应条件下，引入适宜的配体，不但可提高反应的速率，而且还可提高反应的立体选择性[8]。

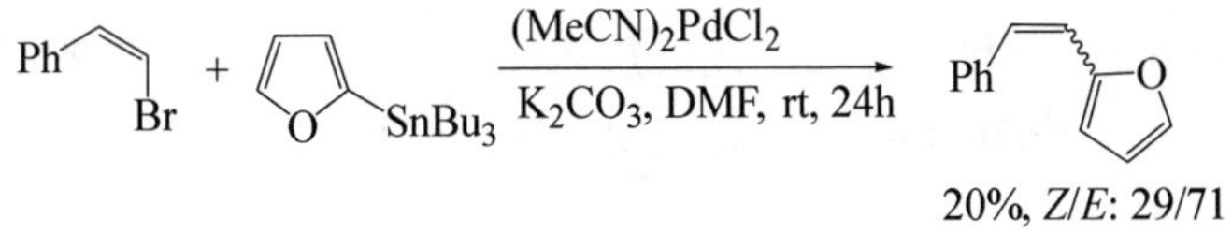

$$Ph\text{-CH=CH-Br} + \text{furan-SnBu}_3 \xrightarrow[K_2CO_3, DMF, rt, 24h]{(MeCN)_2PdCl_2} Ph\text{-CH=CH-furan} \quad 20\%, Z/E: 29/71$$ (4)

$$Ph\text{-CH=CH-Br} + \text{furan-SnBu}_3 \xrightarrow[K_2CO_3, DMF, 45℃, 2h]{Pd_2(dba)_3, P(o\text{-tol})_3} Ph\text{-CH=CH-furan} \quad 96\%, Z/E: 96/4$$ 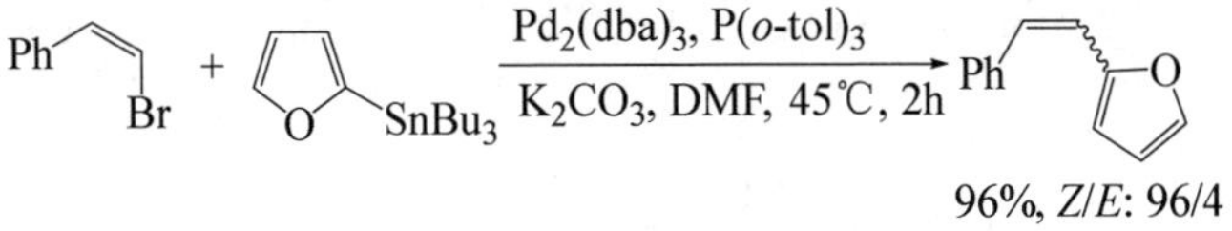 (5)

烯基三氟甲磺酸酯可替代碘代或溴代烯烃进行反应，使得此类 Stille 反应在有机合成上更有应用价值，因为相较于卤代烯烃，高构型异构体纯度的三氟甲磺酸酯更容易获得。应该注意的是，烯基三氟甲磺酸酯为底物的 Stille 反应通常需要加入过量的 LiCl，以促进反应的进行[9,10]。

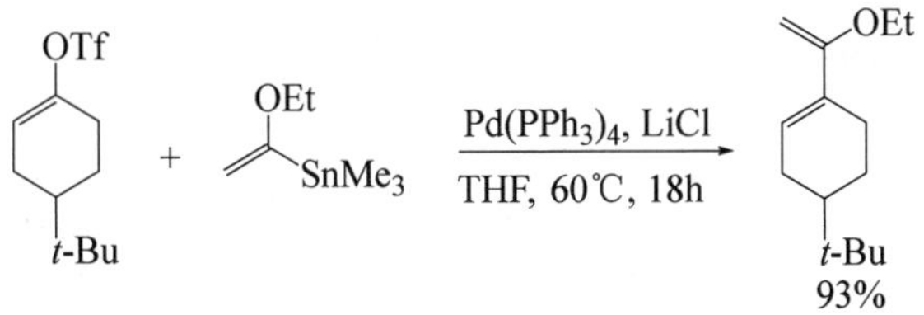

(6) $t\text{-Bu}$-cyclohexenyl-OTf + CH$_2$=C(OEt)SnMe$_3$ $\xrightarrow[THF, 60℃, 18h]{Pd(PPh_3)_4, LiCl}$ product 93%

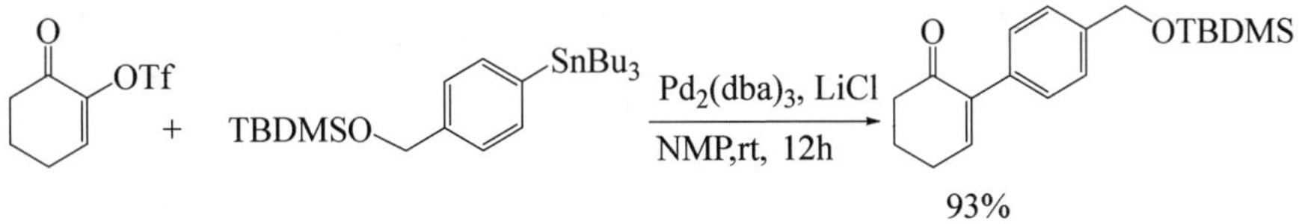

(7) cyclohexenone-OTf + TBDMSO-C$_6$H$_4$-CH$_2$-SnBu$_3$ $\xrightarrow[NMP, rt, 12h]{Pd_2(dba)_3, LiCl}$ product 93%

氯代环丁烯二酮衍生物与芳基或炔基三丁基锡在钯催化剂催化下进行交叉偶联反应，环丁烯二酮骨架结构不受影响[11]。

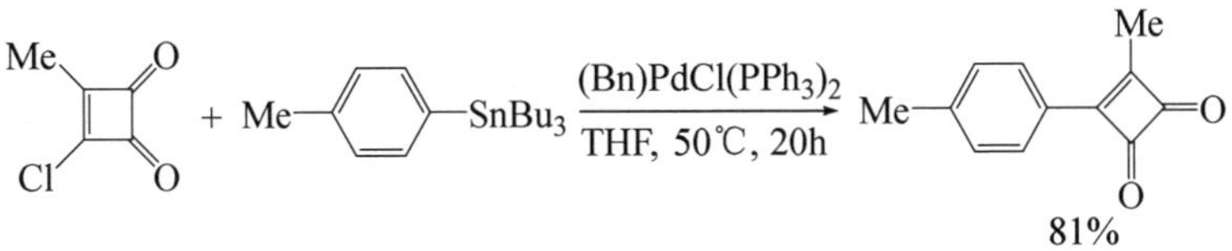

(8) Me-cyclobutenedione-Cl + Me-C$_6$H$_4$-SnBu$_3$ $\xrightarrow[THF, 50℃, 20h]{(Bn)PdCl(PPh_3)_2}$ product 81%

$$\text{Me-环丁烯二酮-Cl} + n\text{-BuC≡C-SnBu}_3 \xrightarrow[\text{THF, 50℃, 20h}]{\text{(Bn)PdCl(PPh}_3)_2} n\text{-BuC≡C-环丁烯二酮-Me} \quad 50\% \quad (9)$$

6.2.1.2 芳基或杂芳基卤化物为亲电底物

芳基溴和芳基碘类化合物与许多有机锡烷反应高收率地生成相应的交叉偶联产物。相较于碘代物，溴代物需要较高的反应温度，而氯代物需要碱性助剂的存在。此外，卤原子对位的吸电子基可以促进反应进行[12,13]。

$$\text{Me-C}_6\text{H}_4\text{-Br} + \text{Me}_4\text{Sn} \xrightarrow[\text{65℃, 1~20h}]{\text{Pd(PPh}_3)_4, \text{HMPA}} \text{Me-C}_6\text{H}_4\text{-Me} \quad (10)$$

$$\text{Ph-Br} + \text{Ph}_4\text{Sn} \xrightarrow[\text{65℃, 1~20h}]{\text{Pd(PPh}_3)_4, \text{HMPA}} \text{Ph-Ph} \quad (11)$$

$$\text{MeO-C}_6\text{H}_4\text{-X} + \text{PhSn}(n\text{-Bu})_3 \xrightarrow[\text{KF, 1,4-二氧六环, 80℃}]{\text{Pd(OAc)}_2 (10^{-4}\%\text{摩尔分数}), L} \text{MeO-C}_6\text{H}_4\text{-Ph} \quad (12)$$

L: DABCO

X=Br, 27% (24h)
X=I, 85% (16h)

$$\text{O}_2\text{N-C}_6\text{H}_4\text{-X} + \text{PhSn}(n\text{-Bu})_3 \xrightarrow[\text{Bu}_4\text{NF, 1,4-二氧六环, 80℃, 16h}]{\text{Pd(OAc)}_2 (10^{-4}\%\text{摩尔分数}), L} \text{O}_2\text{N-C}_6\text{H}_4\text{-Ph} \quad (13)$$

L: DABCO

X=Br, 95%
X=Cl, 62%

与前述烯基三氟甲磺酸酯一样，当芳基三氟甲磺酸酯作为亲电底物时，反应需要 LiCl 作为助剂，其作用是氧化加成反应后 Cl⁻取代三氟甲磺酰氧负离子，使反应顺利进行。例如，对乙酰苯基三氟甲磺酸酯与苯基三甲基锡在四(三苯基膦)合钯催化下顺利交叉偶联生成 4-苯基苯乙酮[14]。除了乙酰基外，含其他羰基、硝基、酰胺和醚等官能团的三氟甲磺酸酯同样顺利地进行 Stille 反应，说明反应的官能团耐受性广泛。

$$\text{MeC(O)-C}_6\text{H}_4\text{-OTf} + \text{PhSnMe}_3 \xrightarrow[\text{1,4-二氧六环, 98℃, 4h}]{\text{Pd(PPh}_3)_4, \text{LiCl}} \text{MeC(O)-C}_6\text{H}_4\text{-Ph} \quad 95\% \quad (14)$$

其他芳基和杂芳基的卤代物或三氟甲磺酸酯作为亲电底物也能进行 Stille 反应,尤其是杂芳基参与的反应,拓宽了 Stille 反应在药物合成中的应用范围。一些反应实例如下[15-17]。

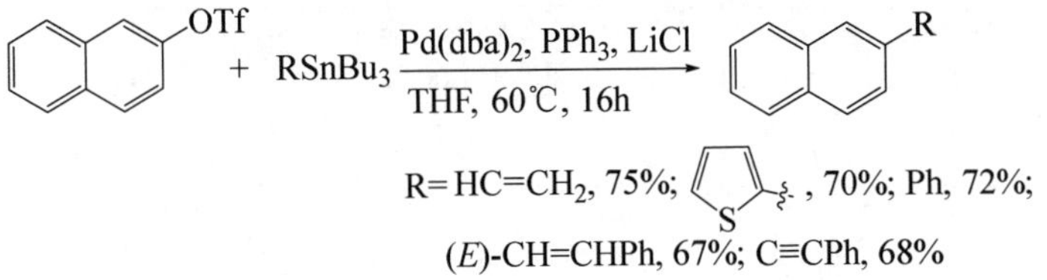

$$\text{Naphthyl-OTf} + RSnBu_3 \xrightarrow[\text{THF, 60℃, 16h}]{Pd(dba)_2, PPh_3, LiCl} \text{Naphthyl-R} \quad (15)$$

R= HC=CH$_2$, 75%; 噻吩基, 70%; Ph, 72%;
(E)-CH=CHPh, 67%; C≡CPh, 68%

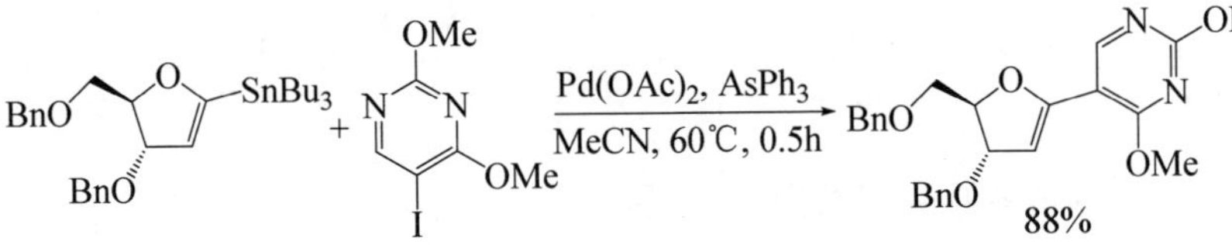

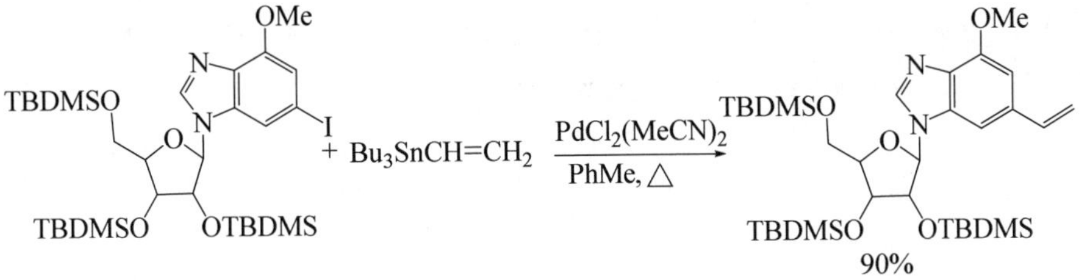

$$(16)\ \text{糖基-SnBu}_3 + \text{碘代二甲氧基嘧啶} \xrightarrow[\text{MeCN, 60℃, 0.5h}]{Pd(OAc)_2, AsPh_3} \text{偶联产物, 88\%}$$

$$(17)\ \text{碘代苯并咪唑核苷} + Bu_3SnCH=CH_2 \xrightarrow[\text{PhMe, △}]{PdCl_2(MeCN)_2} \text{乙烯基产物, 90\%}$$

杂环与杂环之间也可通过 Stille 反应交叉偶联,为双杂环这种重要的有机单元的合成提供了有效的途径[18]。在下面的反应实例中,二溴咪唑衍生物与三甲基锡-4-吡啶区域选择性地交叉偶联使咪唑环与 4-吡啶基偶联,剩下的溴代咪唑基进一步与三丁基锡-2-呋喃进行第二次 Stille 反应,形成多杂环化合物。

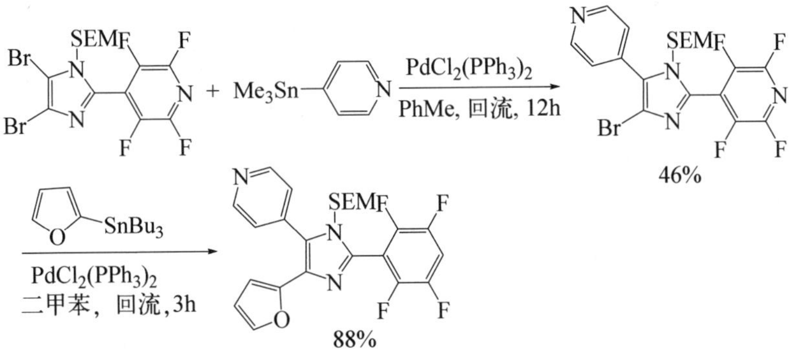

$$\text{二溴咪唑衍生物} + Me_3Sn\text{-吡啶} \xrightarrow[\text{PhMe, 回流, 12h}]{PdCl_2(PPh_3)_2} \text{单偶联中间体 46\%}$$

$$\xrightarrow[\text{二甲苯, 回流, 3h}]{\text{呋喃-SnBu}_3,\ PdCl_2(PPh_3)_2} \text{双偶联产物 88\%} \quad (18)$$

通常缺电子的杂芳基锡烷与多杂原子芳环很难交叉偶联,但在催化剂 Pd-

PEPPSI-IPent 催化下多种缺电子的杂芳基（噻唑、噁唑、哌嗪等）锡烷与 2-氯吡啶或 2-氯哌嗪高效偶联生成双杂环化合物。

$$\underset{Y=C,\ N}{\text{2-Cl-pyridine}} + \text{HetAr-SnBu}_3 \xrightarrow[\text{1,4-二氧六环, 60℃, 32h}]{\text{Pd-PEPPSI-IPent}} \text{HetAr-pyridine} \quad (19)$$

产物收率：75%、98%、90%

催化剂：Pd-PEPPSI-IPent

6.2.1.3 苄基、烯丙基或其类似物为亲电底物

在 Stille 反应中，溴化苄通常是很好的亲电底物，能与各种有机锡试剂在钯催化剂催化下反应，以良好的收率得到相应的交叉偶联产物；在 KF 等助剂存在下氯化苄也能顺利进行反应。以下是一些反应实例[12,19,20]。

$$\text{PhCH}_2\text{Br} + \text{Bu}_3\text{SnCH=CH}_2 \xrightarrow[\text{65℃, 1~20h}]{\text{Pd(PPh}_3)_4,\ \text{HMPA}} \text{PhCH}_2\text{CH=CH}_2 \quad (20)$$
100%

$$\text{PhCH}_2\text{Br} + \text{Bu}_3\text{Sn-indole(N-CH}_2\text{OCH}_2\text{CH}_2\text{SiMe}_3\text{)} \xrightarrow[\text{THF, 60℃, 3h}]{\text{Pd}_2(\text{dba})_3,\ \text{P(2-furyl)}_3} \text{PhH}_2\text{C-indole} \quad (21)$$
95%

$$\text{3-MeC}_6\text{H}_4\text{CH}_2\text{Cl} + (\text{Bu})_3\text{Sn-C}_6\text{H}_4\text{-4-Me} \xrightarrow[\text{1,4-二氧六环, 80℃, 3h}]{\text{Pd(OAc)}_2,\ \text{dppf, KF}} \text{产物} \quad (22)$$
99%

含 β-氢的苄溴在反-丁烯二腈存在下与四甲基锡顺利发生交叉偶联，避免 β-氢消除，但换成其他烷基锡试剂反应不能进行[21]。

$$\text{PhCH(Br)Me} + \text{Me}_4\text{Sn} \xrightarrow[\text{HMPA, 60℃, 113h}]{\text{Pd(bpy), NC-CH=CH-CN}} \text{PhCH(Me)Me} \quad (23)$$
77%

与溴化苄结构类似的杂芳甲基溴也能与锡试剂发生反应生成相应的交叉偶联产物，显示出 Stille 在药物及天然产物合成上的应用价值[22]。

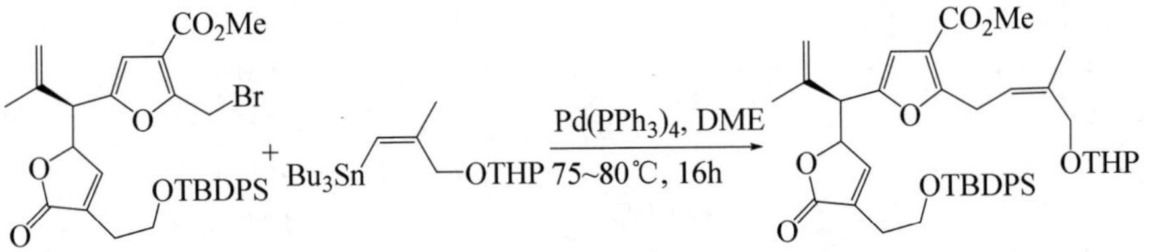

$$\text{(24)}$$

与 α-三氟甲基苄基氯化物类似，α-三氟甲基-3-噻吩甲基氯在钯催化体系催化下能够与烯丙基类锡试剂反应生成同时含三氟甲基和烯丙基的噻吩衍生物，为含氟生物活性化合物的合成提供一条有效的途径[23]。

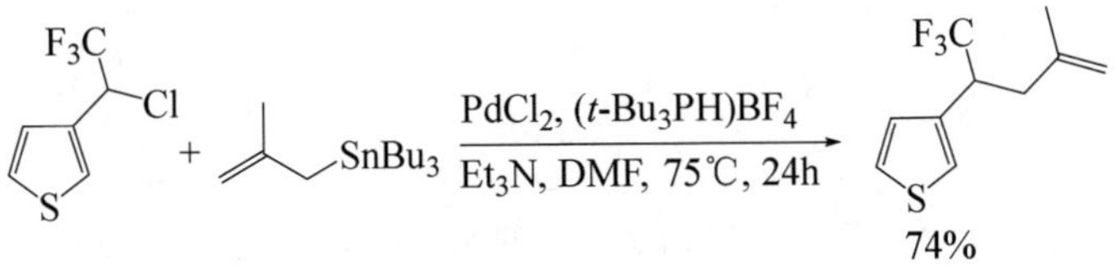

$$\text{(25)}$$

通常不管是烯丙基溴还是烯丙基氯都能在钯催化剂催化下与烯基或芳基锡试剂反应以高收率生成 1,4-二烯。不同于过渡金属催化的许多其他交叉偶联反应，此类反应条件温和，能够耐受各种有机官能团，包括酯基[反应式（26）]、氰基[反应式（30）]、羟基[反应式（27）]，甚至醛羰基[反应式（29）]。反应过程中烯基锡试剂一侧碳碳双键的构型保持不变[反应式（26）、式（27）]。而烯丙基一侧，由于反应经 η^3-烯丙基钯中间体进行，偶联反应既可发生在 α-位[反应式（26）～式（28），式（30）、式（31）]也可发生在 γ-位[反应式（29）]，但通常发生在烯丙基伯碳上而非仲碳上。也就是说，此类反应具有立体专一性和区域选择性，能够获得具有多种官能团的交叉偶联产物[24,25]。

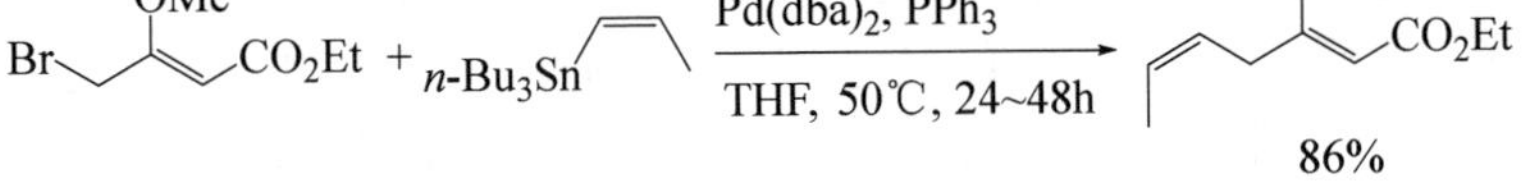

$$\text{(26)}$$

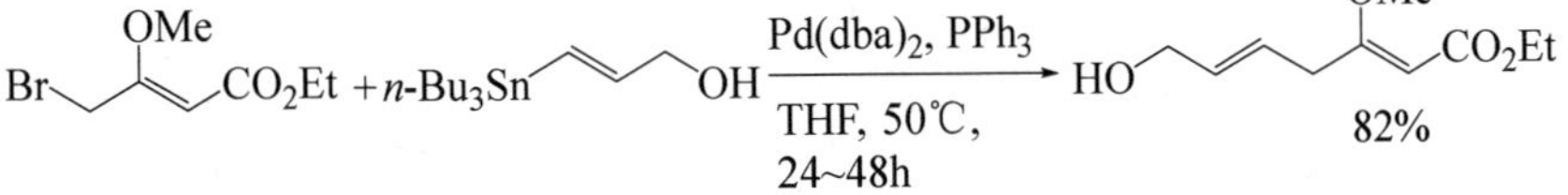

$$\text{(27)}$$

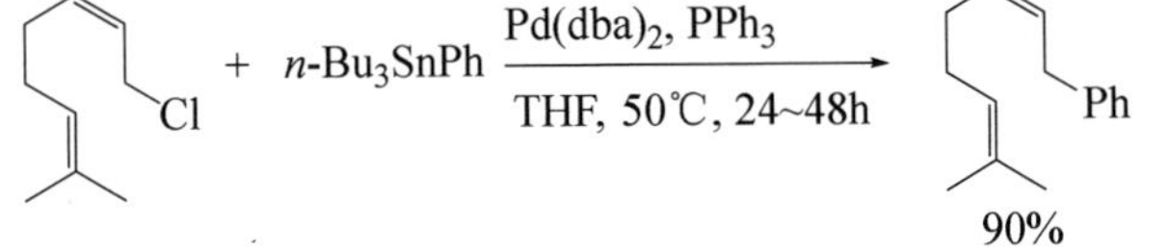

$$\text{(28)}$$

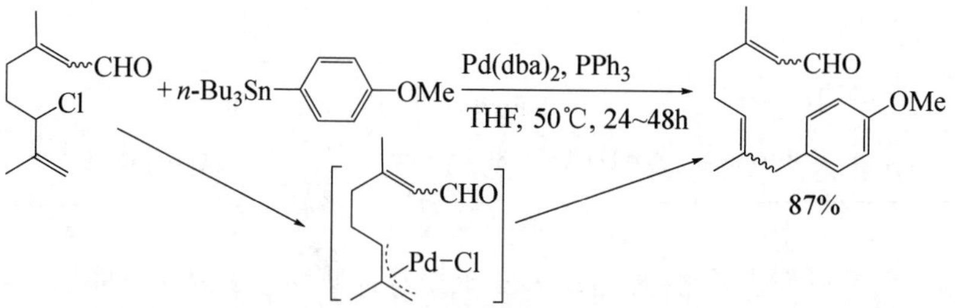

烯丙基乙酸酯可以替代烯丙基卤代物进行 Stille 反应，具有广泛的底物适用性，通常无须膦配体，但需要 LiCl 的存在，反应发生在烯丙基较少取代基的碳上[26]。

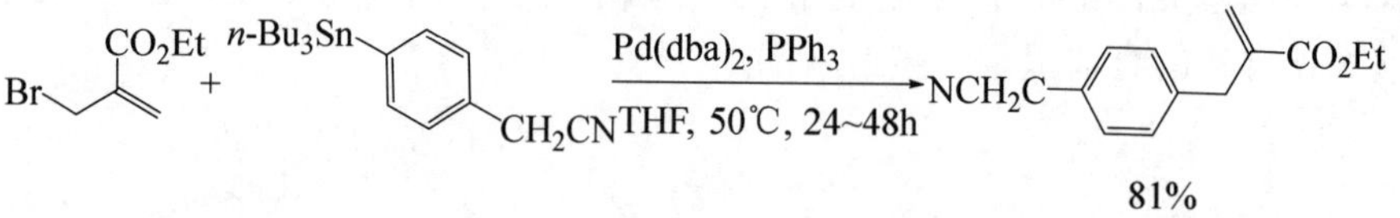

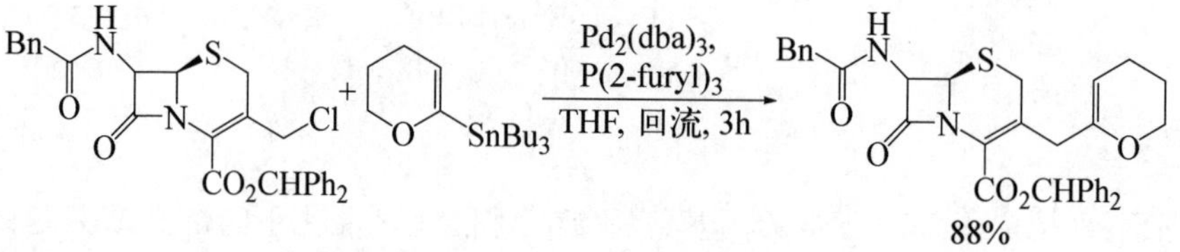

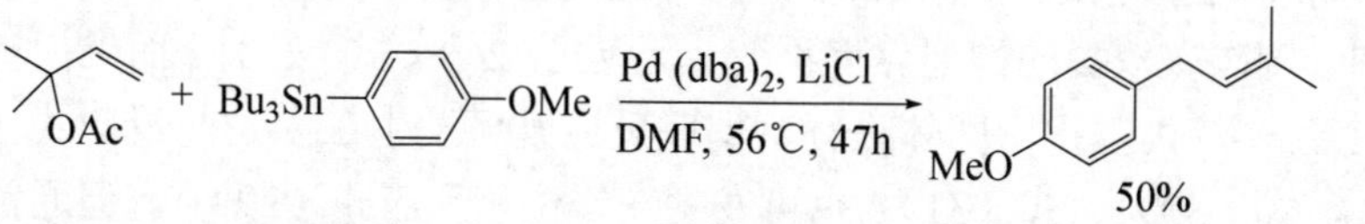

6.2.1.4 酰氯作为亲电底物

酰氯作为亲电底物与有机锡试剂可以进行 Stille 反应用于合成酮 [反应式 (34)～式 (38)] 或醛 [反应式 (33)]，反应几乎耐受所有官能团，如硝基 [反应式 (34)]、氰基 [反应式 (37)]、酯基 [反应式 (38)]，甚至甲酰基 [反应式 (38)] 等。反应根据催化剂不同选用不同的溶剂，但通常在极性溶剂如 HMPA、DMF 和 $CHCl_3$ 中比在芳烃溶剂中反应效果更好。此类反应有时会发生脱羰基，在 CO 气氛下反应会抑制脱羰基的发生；通过优化催化体系，降低反应温度也可减弱或抑制脱羰基的发生。有机锡试剂中其他有机基团脱离三甲基或三丁基锡与酰基结合的活性顺序为：炔基＞烯基＞芳基＞烯丙基、苄基＞烷基。以下是一些反应实例[27-30]。

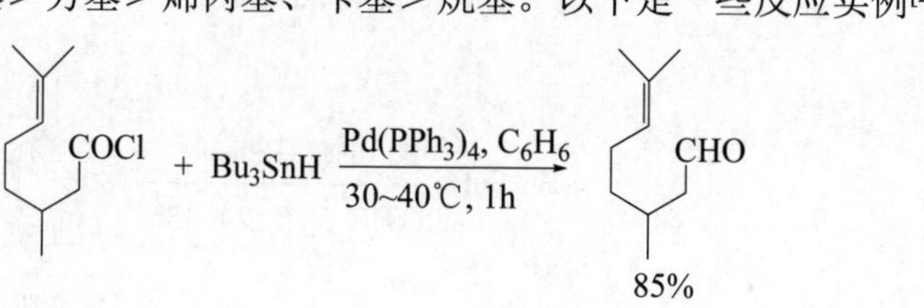

$$\text{O}_2\text{N-C}_6\text{H}_4\text{-COCl} + \text{Me}_4\text{Sn} \xrightarrow[\text{CHCl}_3, 65℃, 24h]{\text{BnPd(PPh}_3)_2\text{Cl}} \text{O}_2\text{N-C}_6\text{H}_4\text{-COCH}_3 \quad (34)$$

$$\text{PhCOCl} + n\text{-Bu}_3\text{Sn-CH=CH-CH}_3 \xrightarrow[\text{CHCl}_3, 65℃, 4.5h]{\text{BnPd(PPh}_3)_2\text{Cl}} \text{PhCO-CH=CH-CH}_3 \quad 74\% \quad (35)$$

$$\text{PhCOCl} + \text{Me}_3\text{SnC}\equiv\text{CPr-}n \xrightarrow[\text{CHCl}_3, 65℃, 23h]{\text{BnPd(PPh}_3)_2\text{Cl}} \text{PhCO-C}\equiv\text{C-Pr-}n \quad 70\% \quad (36)$$

$$\text{Bu}_3\text{Sn-(indole)} + \text{NC-C}_6\text{H}_4\text{-COCl} \xrightarrow[110℃, 16h]{\text{Pd(PPh}_3)_4, \text{PhMe}} \text{NC-C}_6\text{H}_4\text{-CO-(indole)} \quad 75\% \quad (37)$$

$$i\text{-BuCO-Cl} + \text{Bu}_3\text{Sn-(furan)-CHO} \xrightarrow[\text{PhMe/HMPA}, 100℃, 3h]{\text{PdCl}_2(\text{PPh}_3)_2, \text{对苯二酚}} i\text{-BuCO-O-(furan)-CHO} \quad 70\% \quad (38)$$

6.2.1.5 其他亲电底物

尽管卤代烷作为亲电底物不像以上亲电底物那样活泼，反应不很广泛，但仍有一些反应实例，反应的催化体系须有强给电子基配体的存在，这从一个侧面说明配体在过渡金属催化的交叉偶联反应中的重要性[31]。此外，镍催化剂在强碱协助下可以催化仲溴代烷与芳基锡试剂的交叉偶联[32]。

$$\text{morpholine-CO-(CH}_2)_4\text{-Br} + \text{Bu}_3\text{Sn-CH=CH-CH}_2\text{CH}_2\text{-Ph} \xrightarrow[\text{Me}_4\text{NF, MS, THF, rt, 24h}]{[(\eta\text{-C}_3\text{H}_5)\text{PdCl}]_2, \text{PCy(pyrr)}_2} \text{morpholine-CO-(CH}_2)_4\text{-CH=CH-CH}_2\text{CH}_2\text{-Ph} \quad 79\% \quad (39)$$

PCy(pyrr)$_2$: 环己基-P(吡咯烷基)$_2$

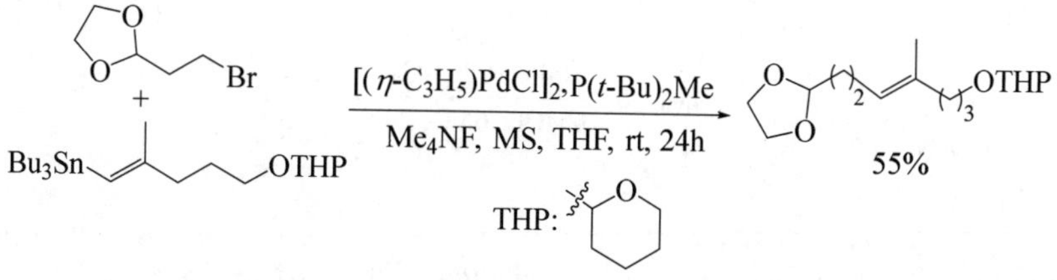

$$\text{(40)}$$

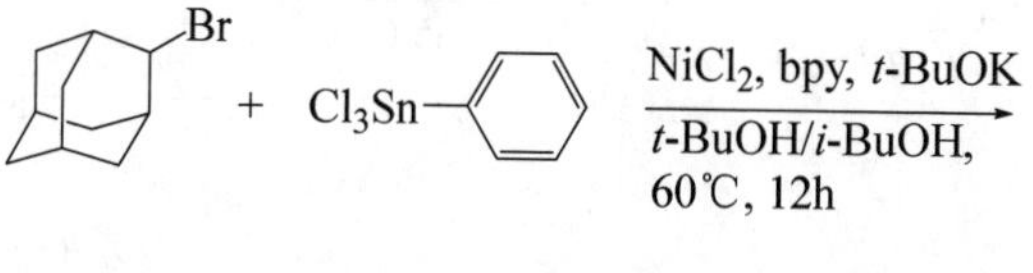

$$\text{(41)}$$

α-卤代羰基化合物的亲电性较强,反应特殊。当以钯-膦配合物为催化剂时,其可以与烯丙基或酰甲基锡反应,但并非生成 Stille 交叉偶联产物,而是锡试剂作为亲核试剂进攻亲电底物上的羰基,然后环合生成环氧化物或氧杂环化合物[33];当无膦配体存在时则按正常的 Stille 反应生成相应的交叉偶联产物[34]。

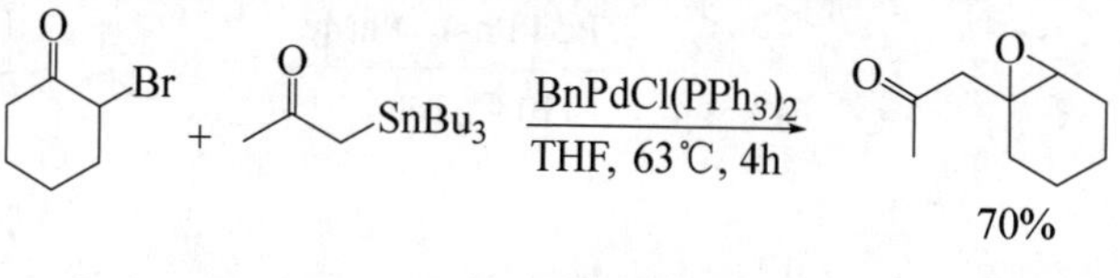

$$\text{(42)}$$

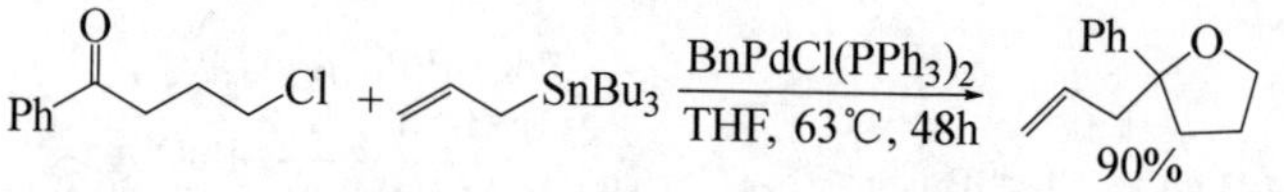

$$\text{(43)}$$

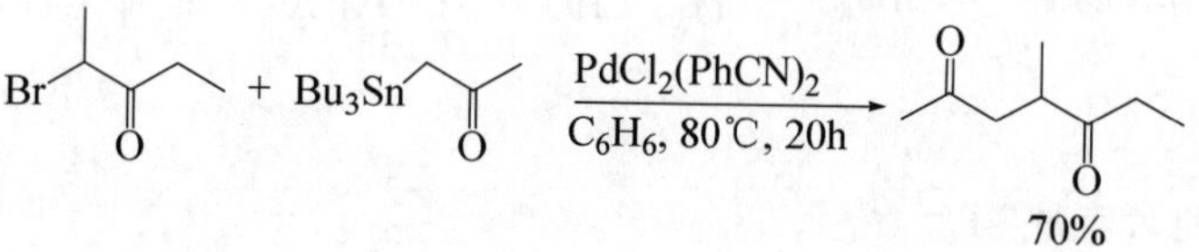

$$\text{(44)}$$

尽管没有系统地对卤代炔烃作为亲电底物的 Stille 反应进行研究,但其已在合成上得到应用。如下反应所示,含双炔基碘的亲电底物与 1,2-二(三甲锡基)乙烯进行两次交叉偶联高收率生成结构复杂的多环化合物,表明碘代炔与锡试剂交叉偶联反应容易进行的程度[35]。

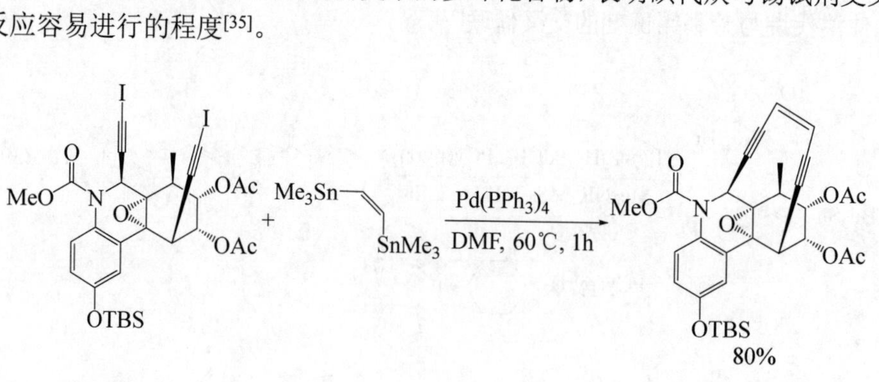

$$\text{(45)}$$

6.2.1.6 Stille 羰基化反应

Stille 羰基化反应是指 Stille 交叉偶联反应伙伴在一氧化碳存在下发生一氧化碳插入生成酮类化合物的反应，其代表式如下：

$$R^1-X + CO + R^2SnR^3_3 \xrightarrow{Pd(0)} R^1\text{-CO-}R^2 + R^3_3SnX \tag{46}$$

亲电基团 R^1 和亲核基团 R^2 包括烯基、芳基、杂芳基和烯丙基。Stille 羰基化反应为酮的合成提供了一条有效的途径，因为有机卤代物来源非常广泛，而反应条件也比较温和，而且所有与常规、非羰基化 Stille 反应相容的官能团也与该反应相容。

碘代烯和烯基锡烷在中性、温和的条件（40～50℃，0.1～0.3MPa CO）下顺利进行。主要副反应是烯基之间没有插入 CO 而直接偶联，提高 CO 的压力可抑制该副反应的发生[36]。反应的另一问题是 Z-式构型烯烃的构型不能保持，异构化可发生于亲电烯烃一侧，也可发生于锡试剂一侧的烯基[37]。

$$n\text{-Bu-CH=CH-I} + Bu_3Sn\text{-CH=CH}_2 \xrightarrow[\text{THF, 40℃, 12h}]{\text{CO (0.3MPa), PdCl}_2(\text{PPh}_3)_2} n\text{-Bu-CH=CH-CO-CH=CH}_2 \quad 70\% \tag{47}$$

$$n\text{-Bu-CH=CH-I} + Bu_3SnH \xrightarrow[\text{THF, 50℃, 3h}]{\text{CO (0.1 MPa), Pd(PPh}_3)_4} n\text{-Bu-CH=CH-CHO} \quad 88\% \tag{48}$$

氯化苄以及各种三氟甲磺酸酯也可作为亲电底物进行反应，进一步拓展了底物范围，使该反应更具应用价值[38,39]。

$$\text{(芳基-CH}_2\text{Cl)} + \text{CH}_2\text{=CH-SnBu}_3 \xrightarrow[\text{P(2-furyl)}_3, \text{HMPA}, 80℃, 1.5h]{\text{CO(0.1MPa), Pd(PPh}_3)_4} \text{产物} \quad 74\% \tag{49}$$

$$R^1\text{-OTf} + \underset{\text{TMS}}{\overset{\text{SnR}^4_3}{\text{C=C}}}R^2 \xrightarrow[65℃, 24\sim72h]{\substack{\text{Pd(PPh}_3)_4, \text{CuI, LiCl} \\ \text{CO(0.1MPa), THF}}} R^1\text{-CO-C(}R^2\text{)=CH-TMS} \tag{50}$$

产物：98%，81%，72%

6.2.2 分子内 Stille 反应

作为 Stille 交叉偶联反应亲电试剂的各种卤代物、三氟甲磺酸酯等比较稳定,均能耐受合成有机锡试剂的反应条件,或是有机锡试剂耐受生成亲电试剂的转化条件,这就为在同一有机化骨架中同时引入 Stille 反应所需的两个交叉偶联伙伴的结构片段提供了保证,使得分子内 Stille 反应成为可能。下面这些反应实例说明几乎所有分子间交叉偶联伙伴同样能作为分子内交叉偶联反应伙伴,从而利用分子内 Stille 反应可构建大小不等的各种环状化合物。

含溴乙烯基的 β-三甲锡基-β,γ-烯羧酸酯在 Pd(PPh$_3$)$_4$ 催化下进行分子内 Stille 反应以很高的收率得到含张力的四元环类化合物,而且反应中所有烯基的构型得到保持[40]。

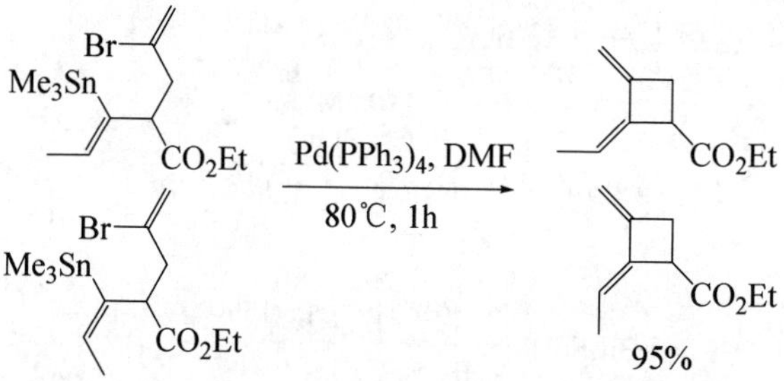

$$\text{(51)}$$

在如下结构的底物中偶联伙伴基团间隔五个碳,发生分子内 Stille 反应生成五元碳环,并与已有的碳环结合形成稠环化合物,而且已有碳环的大小决定所生成稠环化合物的张力,小环稠合张力大影响反应进行,但反应仍能进行[41]。此类反应为一些天然产物的合成提供了有效途径。

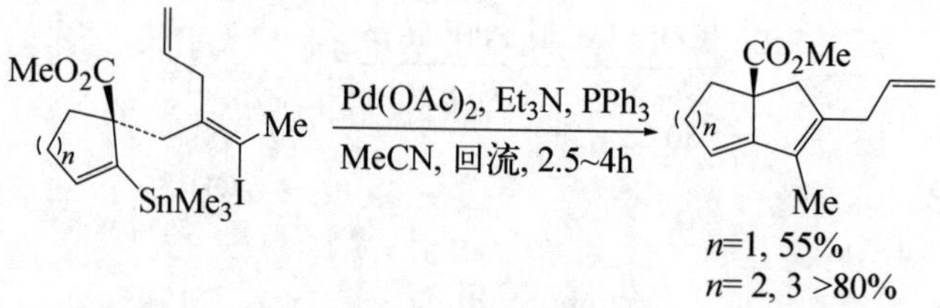

$$\text{(52)}$$

$n=1, 55\%$
$n=2, 3 >80\%$

含三丁基锡基的烷氧基甲酰氯在钯催化下于室温长时间反应以中等收率生成六元环的戊内酯衍生物[42]。

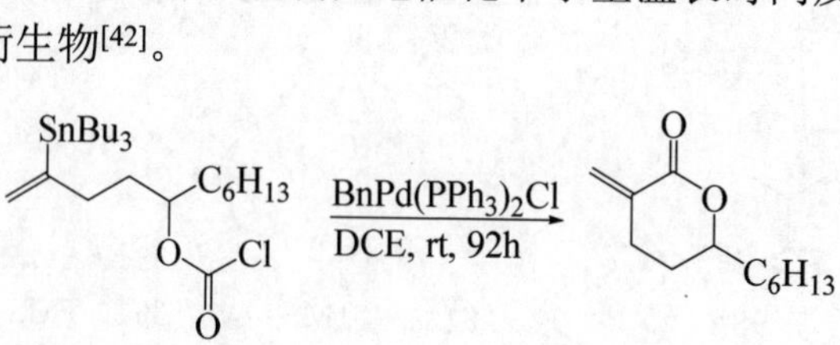

$$\text{(53)}$$

通过醚键连接的芳基碘或芳基溴与三丁基烯结构片段在钯催化剂催化下通过

分子内 Stille 反应得到环醚骨架的七元环化合物[43]。

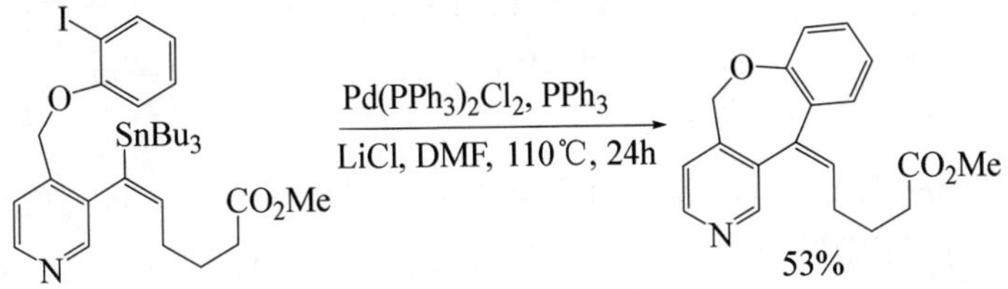

$$\text{(54)}$$

Pd(PPh$_3$)$_2$Cl$_2$, PPh$_3$ / LiCl, DMF, 110℃, 24h ； 53%

苄基溴与三丁基锡基烯作为偶联伙伴也能顺利进行分子内交叉偶联。如下结构的喹啉衍生物底物通过分子内 Stille 反应以优异的收率得到氮杂环多环化合物[44]。

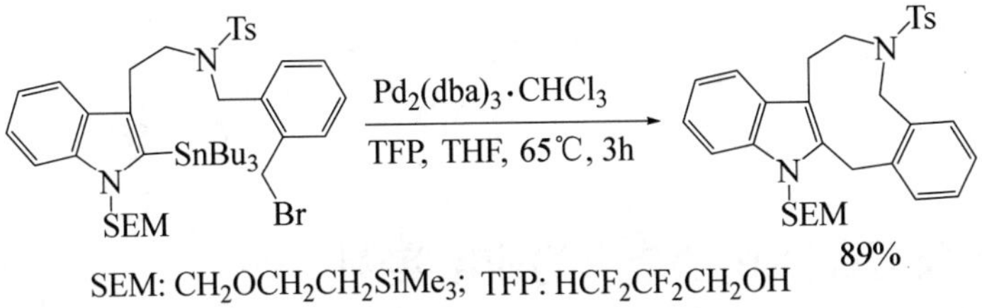

$$\text{(55)}$$

Pd$_2$(dba)$_3$·CHCl$_3$ / TFP, THF, 65℃, 3h ； 89%

SEM: CH$_2$OCH$_2$CH$_2$SiMe$_3$； TFP: HCF$_2$CF$_2$CH$_2$OH

利用同时含酰氯与三丁基锡基烯的酯类化合物的分子内 Stille 反应可以合成中环或大环内酯化合物，需要注意的是三丁基锡基烯的构型是否保持决定于生成环状产物构型的稳定性而非反应底物中三丁基锡基烯构型的稳定性。因此，十二元环产物中碳碳双键的构型为 Z-式，尽管反应底物中碳碳双键的构型为热力学稳定的 E-式；十六元环产物中碳碳双键的构型为 E-式，而反应底物中碳碳双键的构型为 Z-式[45]。反应结果也说明三丁基锡基烯的在反应过程中能够发生异构化。

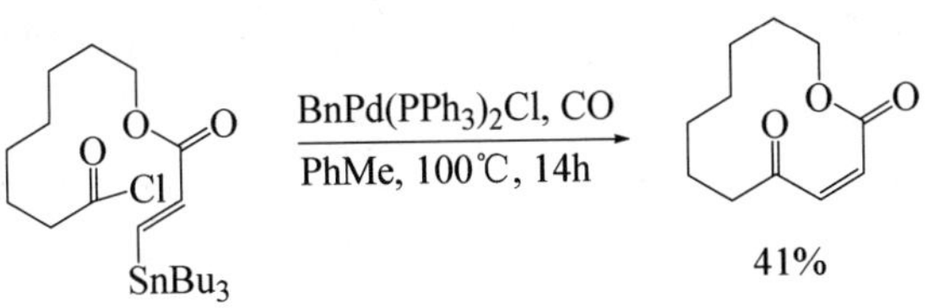

$$\text{(56)}$$

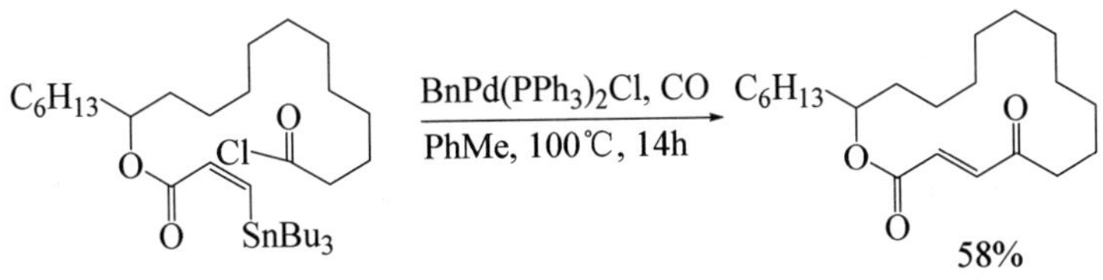

BnPd(PPh$_3$)$_2$Cl, CO / PhMe, 100℃, 14h ； 41%

$$\text{(57)}$$

BnPd(PPh$_3$)$_2$Cl, CO / PhMe, 100℃, 14h ； 58%

三丁锡基炔作为亲核基团与溴乙烯基团的分子内交叉偶联提供了一条合成烯-炔共轭大环化合物的有效途径[46]。

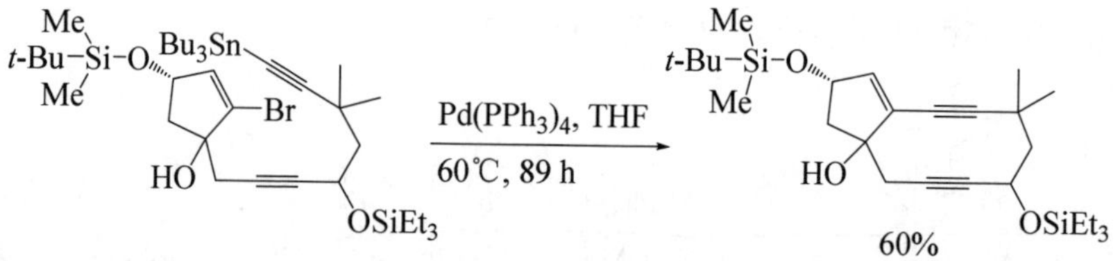

$$\text{(58)}$$

含有三氟甲磺酸乙烯酯基和三丁基锡基烯的长链酯在一个压力的 CO 存在下经钯催化发生分子内羰基化交叉偶联反应生成大环酮内酯，收率与目标产物环的大小有一定的关系[47]。

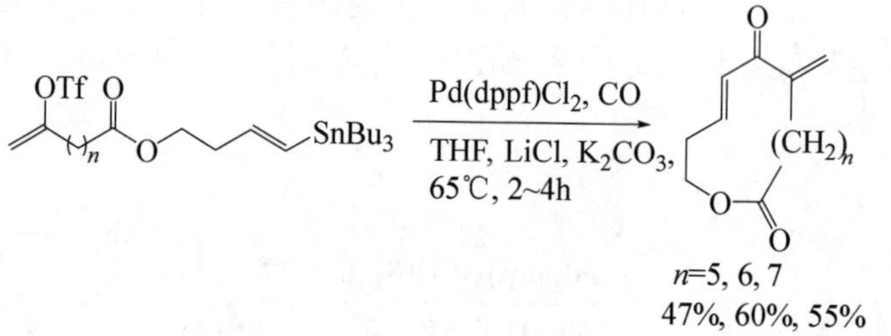

$$\text{(59)}$$

n=5, 6, 7
47%, 60%, 55%

6.3　Stille 反应在药物合成上的应用实例

Stille 反应中的交叉偶联伙伴有机卤代物或三氟甲磺酸酯和有机锡试剂均来源广泛、对空气和水稳定，且耐受各种官能团，理应在药物合成上起重要作用，但有机锡试剂的毒性限制了其在药物合成上的更大应用。尽管如此，Stille 反应在一些药物合成上无法替代，特别是容易在同一有机骨架引入交叉偶联伙伴基团，使其在环状天然产物和生物活性物质的合成上有着无可比拟的实用性。下面将按交叉偶联伙伴类别或交叉偶联产物的结构特征，介绍 Stille 反应在非环状或环状药物和生物活性化合物合成上的应用实例。为了叙述方便，其中经由 Stille 交叉偶联反应合成的产物编号为 SC-x，x 为整数，代表序号。

萘吡霉素（napyradiomycins）是一类抗生素，可抑制革兰氏阳性菌，而 (±)-A80915G 是其中一员，其合成路线为：基于卤代芳基的反应活性不同，2-溴-6-氯-3-碘-1,4-二甲氧基苯选择性与不同烯丙基锡试剂分别进行两次 Stille 反应得到关键中间体 SC-1 和 SC-1'，SC-1' 再经后续转化得到 A80915G[48]。

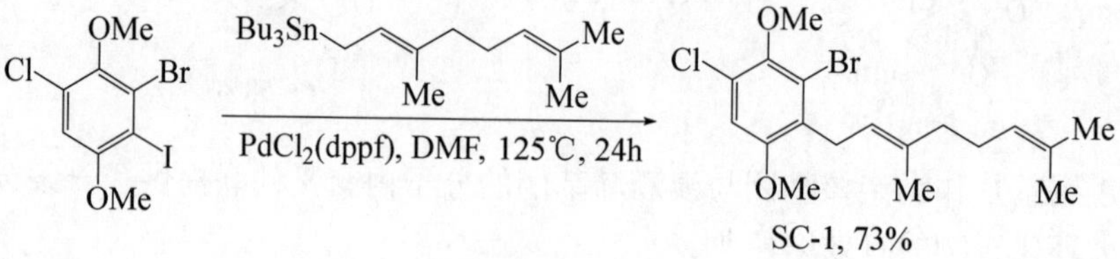

SC-1, 73%

SC-1', 67%

(±)-A80915G

(60)

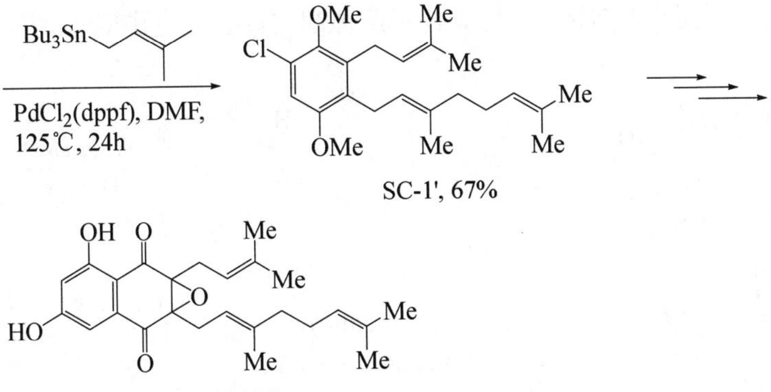

奥泽沙星（ozenoxacin）是一种非氟喹诺酮抗生素菌剂，用于治疗脓疱疮。其合成路线之一是以 2,6-二溴甲苯为起始原料经与环丙胺缩合、缩合产物与乙氧基亚甲基丙二酸二乙酯缩合环合得到喹诺酮溴代物；喹诺酮溴代物与三丁锡基吡啶衍生物在钯催化剂催化下交叉偶联以良好的收率生成关键中间体 SC-2，SC-2 再转化为奥泽沙星[49]。

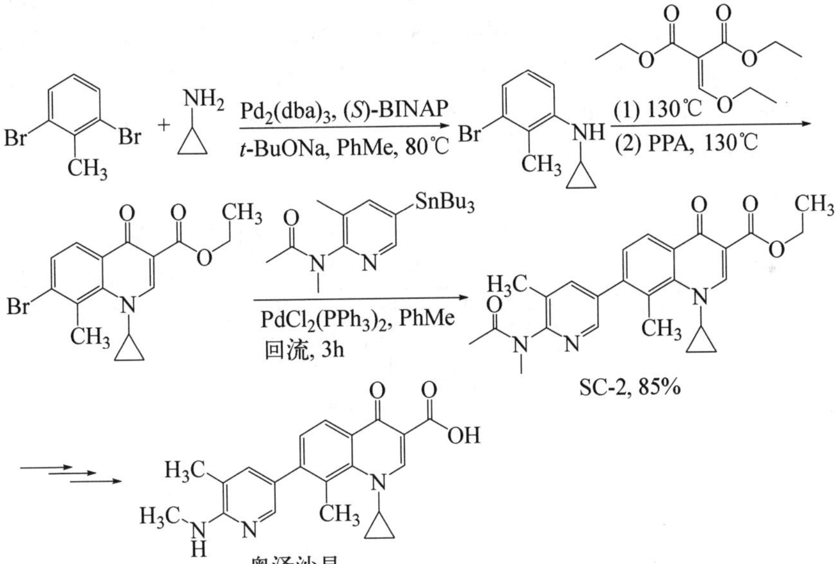

SC-2, 85%

奥泽沙星

(61)

下面反应的目标产物是一种 β-内酰胺衍生物，具有抗革兰氏阳性菌病原体活性。其合成路线的关键步骤之一是 β-内酰胺母核的羟甲基化。β-内酰胺单环重氮衍生物由两步反应生成中间体 β-内酰胺母核的三氟甲磺酸酯，该中间体与羟甲基三丁基锡在 $Pd(dba)_2$、$P(2\text{-furyl})_3$ 和 $ZnCl_2$ 构成的催化体系下顺利进行 Stille 反应生成交叉偶联产物 SC-3，SC-3 再经多步转化生成目标产物。在此 Stille 反应中，无膦配体

时反应不进行，而且 P(2-furyl)$_3$ 比其他膦配体起更好的促进作用[50]。

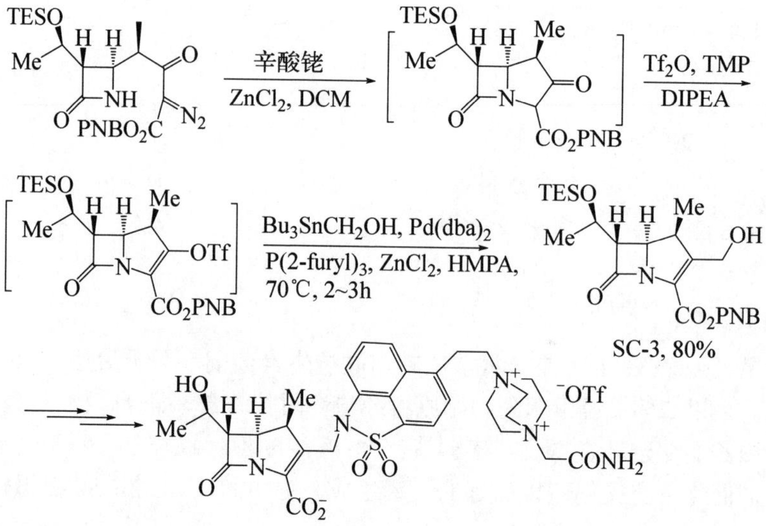

$$\text{(62)}$$

下面是一种血管内皮生长因子受体（VEGFR）激酶抑制剂的合成路线，该化合物具有明显的抗肿瘤活性。文献调研发现 Stille 反应是咪唑环与噻吩吡啶耦合的最可靠的方法。首先，等摩尔的碘代噻吩吡啶衍生物与 2-三丁锡基-1-甲基咪唑在 Pd(PPh$_3$)$_4$ 催化下于 DMF 中顺利偶联得到交叉偶联产物 SC-4，SC-4 进一步转化生成 VEGFR 激酶抑制剂目标分子[51]。

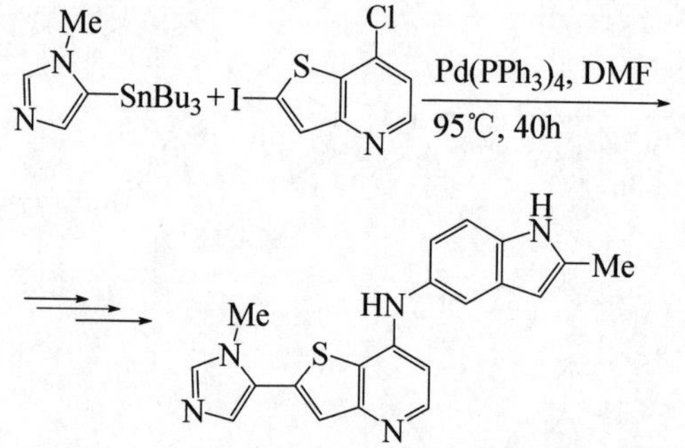

$$\text{(63)}$$

顺-头孢丙烯(cis-cefprozil)是一种口服抗生素,其首条合成路线采用 Wittg 反应，但存在严重的 E/Z 选择性问题。利用烯基有机锡试剂参与的 Stille 反应顺式构型得到保持的优势，解决了上述难题，并开启了交叉偶联反应在 β-内酰胺抗生素合成应用的先河[52,53]。如反应式所示，7-ACA 衍生物的三氟甲磺酸酯与丙烯基三丁基锡在 Pd(OAc)$_2$ 催化下于 NMP 中非常顺利地反应生成交叉偶联产物 SC-5，再经转化、脱保护得到顺-头孢丙烯。

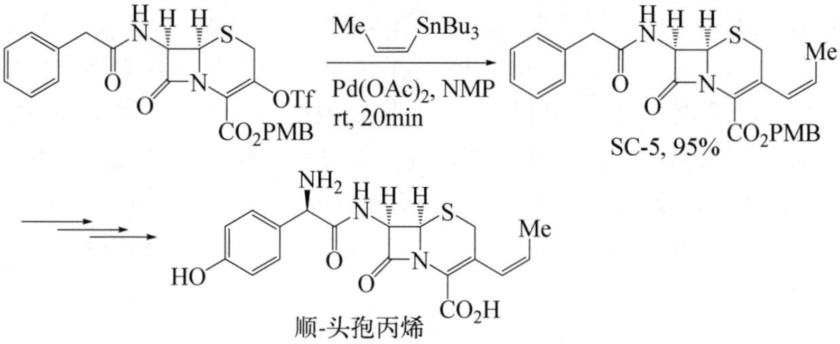

利用炔基锡试剂与 7-ACA 衍生物甲磺酸酯的交叉偶联，可以合成头孢克肟的乙炔类似物，该类似物具有一定的抗菌活性[54]。

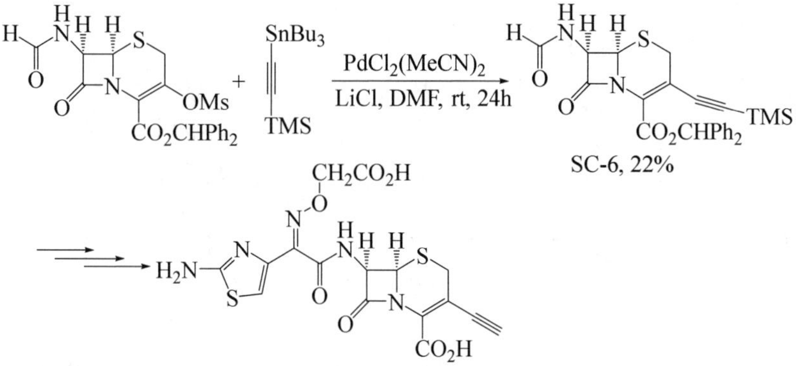

SN29751 是一种生物还原性低氧选择性抗癌化合物，其骨架结构为 1,2,4-苯并三嗪-1,4-二氧化物，在三嗪环上引入乙基是其合成的关键步骤。如反应式所示，3-氯-6-氟苯并三嗪-1-氧化物与四乙基锡进行 Stille 反应生成 3-乙基-6-氟苯并三嗪-1-氧化物（SC-7），反应需要在微波下进行。SC-7 再进行烷氧基化、氧化最终生成 SN29751[55]。

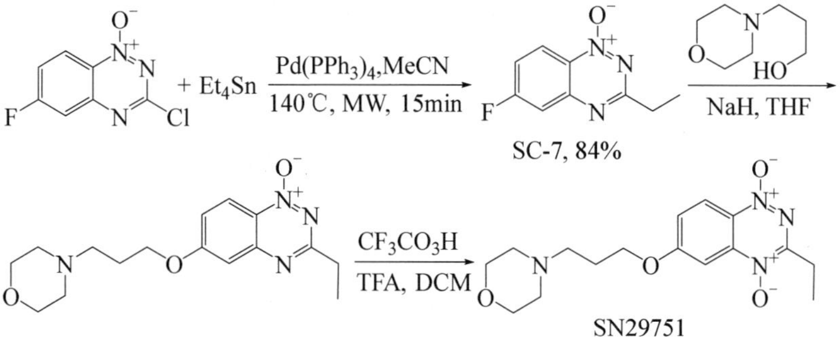

Stille 反应更广泛用于天然产物的全合成,例如其在 isohericenone J 合成中起关键作用。isohericenone J 是一种苯并呋喃酮衍生物,最早从一种蘑菇中分离得到。如反应式所示在 CsF 协助下 Pd(PPh$_3$)$_2$Cl$_2$ 催化溴代苯并呋喃酮衍生物与偶联伙伴 3,7-二甲基-2,6-辛二烯基三丁基锡反应以优异的收率生成交叉偶联产物 SC-8,CsF 起稳定反应作用。该反应是一个富电子、双邻位取代芳基溴化物高效进行 Stille 反应的成功范例。SC-8 在樟脑磺酸(CSA)催化脱保护基甲氧基甲基(MOM)得 isohericenone J[56]。

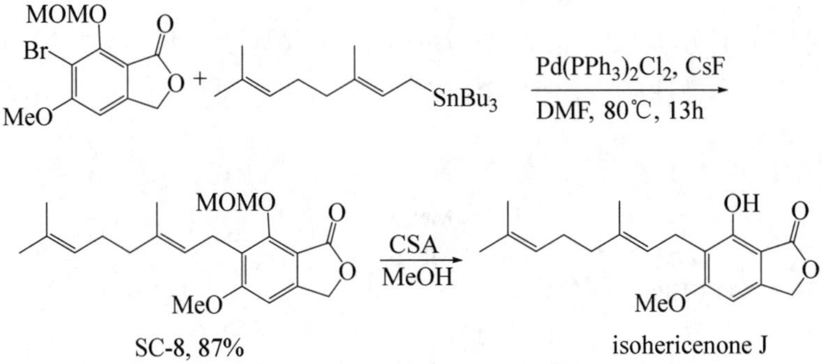

$$\text{(67)}$$

磷烯菌素(fostriecin)是从链霉菌发酵液中分离出的一种天然磷酸酯类化合物,具有抗肿瘤活性。中国科学院广州生物医药与健康研究院张健存研究员课题组实现了该分子的全合成,关键中间体为 SC-9,该中间体可由如下反应式中所示的碘代烯与丁二烯基三丁基锡在简单钯催化剂催化下交叉偶联得到,通过此 Stille 反应构建了 Z,Z,E-三烯醇结构单元。该中间体一旦合成,经后续反应转化生成目标产物磷烯菌素[57]。

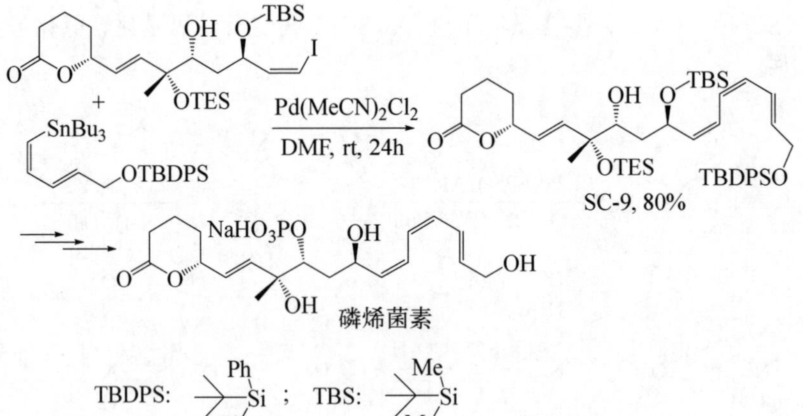

$$\text{(68)}$$

具有生物活性的海洋类胡萝卜素 pyrrhoxanthin 的合成路线包括三步 Stille 反应。第一步，Stille 反应涉及同时含烯基溴与 α,β-不饱和-α-溴内酯亲电试剂与烯基锡试剂之间的交叉偶联，反应在 $Pd(PPh_3)_4$ 与 CuI 共同催化下选择性地发生在 α,β-不饱和-α-溴内酯反应生成 SC-10；第二步，在 $Pd_2(dba)_3 \cdot CHCl_3$ 与 $P(2\text{-furyl})_3$ 催化体系下剩下的烯基溴反应位再与二（三丁锡基）三烯空间位阻小的端基反应位选择性反应生成 SC-10'；SC-10'不经分离直接利用剩下的内烯基锡反应位与不多见的炔基溴进行交叉偶联生成 pyrrhoxanthin，两步总收率 38%。反应路线所涉及的三步 Stille 反应因涉及不同交叉偶联伙伴间的选择性反应，因而需针对不同反应基团的反应特性选择不同催化剂和反应条件[58]。

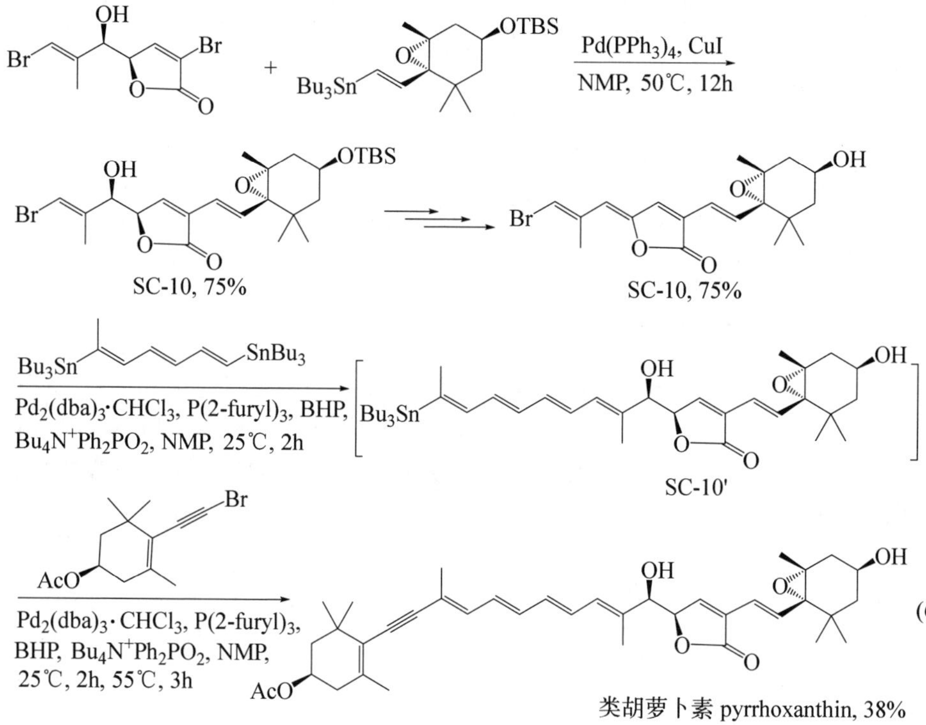

除了通过分子间 Stille 反应合成众多药物和生物活性分子，分子内 Stille 反应在构筑中到大的各类碳环和杂环天然产物中成为最可靠的手段。chivosazole F 是从纤维素堆囊菌中分离出的天然产物，是一种有效的癌细胞抑制剂，结构上属大环内酯类化合物。该化合物环合一步利用了前体中烯基碘与烯基锡之间的分子内 Stille 反应，反应在简单的钯催化剂 $PdCl_2(PhCN)_2$ 催化下生成 SC-11，随后采用 HF-吡啶脱保护得 chivosazole F，两步总收率 18%，反应后交叉偶联前体烯烃的构型得到保持[59]。

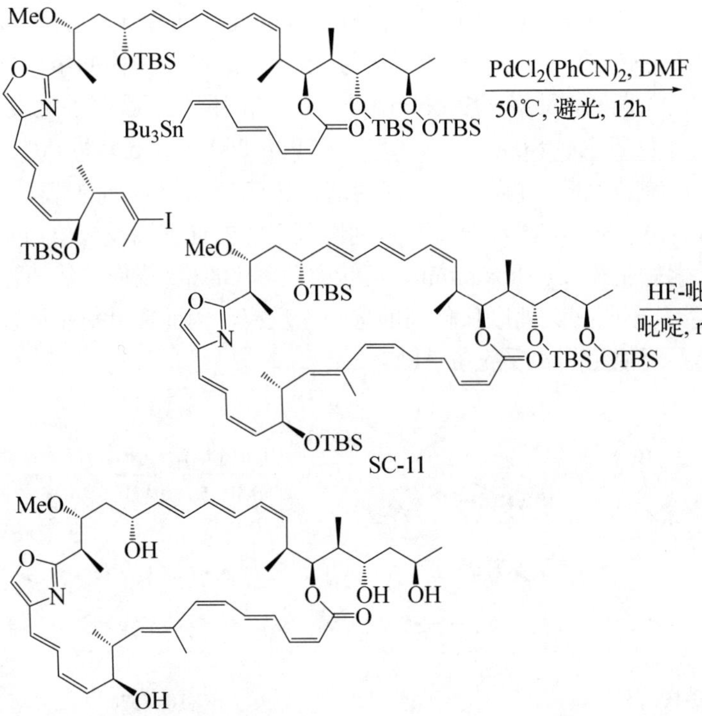

vicenistatin 是一种大环内酰胺糖苷，有抗癌活性。含有乙烯基碘和烯基三丁锡的五烯化合物作为前体，在催化反应体系[$Pd_2(dba)_3 \cdot CHCl_3$, $AsPh_3$, i-Pr_2NEt, DMF]下环合构建大环内酰胺结构 SC-12，作为关键中间体的 SC-12 再经后续反应生成 vicenistatin[60]。

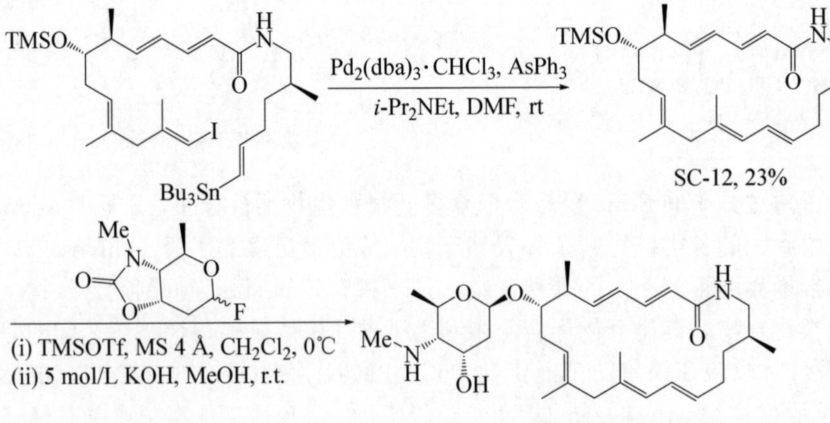

分子间与分子内 Stille 反应的结合进一步显示了其在大环天然产物合成上的强力作用，最具说服力的实例是雷帕霉素（rapamycin）的合成。雷帕霉素是一种免疫抑制剂并具有抗真菌活性。其合成借助两次 Stille 选择性交叉偶联，实现大环的精准构建，即双乙烯基碘前体与反-1,2-二（烯基三丁基锡）前体先各利用一个偶联基团选择性进行一次分子间交叉偶联反应生成中间体 SC-13，反应在 PdCl$_2$(MeCN)$_2$ 催化下于室温、DMF/THF 混合溶剂中进行，i-Pr$_2$NEt 作助剂。SC-13 结构中同时含有一个乙烯基碘和一个烯基三丁基锡偶联基团，因此进行第二次交叉偶联，即分子内 Stille 反应实现环合，并生成全反式三烯结构，整个反应无基团保护和脱保护[61]。之所以处于远端的两个交叉偶联基团能够反应，显然是因为钯催化活性中心的模板效应（template effect）[62]。

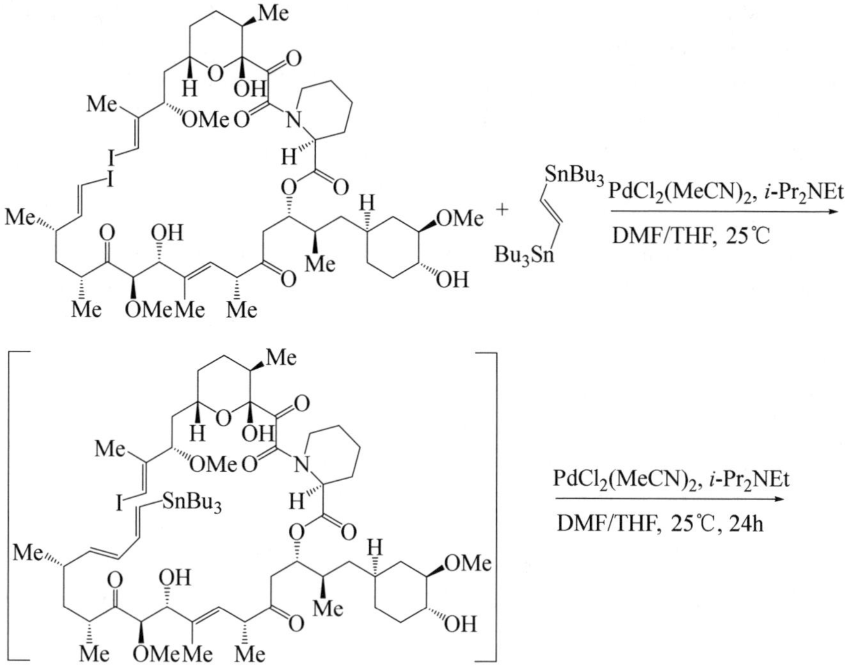

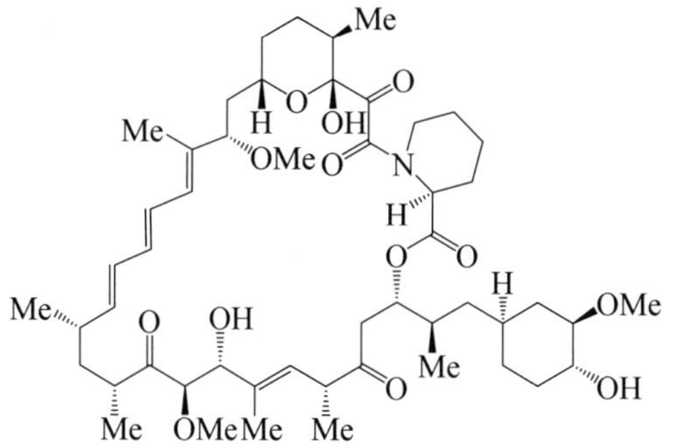

参考文献

[1] Le Grognec E, Chretien J M, Zammattio F, et al. Methodologies limiting or avoiding contamination by organotin residues in organic synthesis[J]. Chemical Reviews, 2015, 115(18): 10207-10260.

[2] Heravi M M, Mohammadkhani L. Recent applications of Stille reaction in total synthesis of natural products: an update[J]. Journal of Organometallic Chemistry, 2018, 869: 106-200.

[3] Espinet P, Echavarren A M. The mechanisms of the Stille reaction[J]. Angewandte Chemie International Edition, 2004, 43(36): 4704-4734.

[4] Tang H, Menzel K, Fu G C. Ligands for palladium-catalyzed cross-couplings of alkyl halides: use of an alkyldiaminophosphane expands the scope of the Stille reaction[J]. Angewandte Chemie International Edition, 2003, 42(41): 5079-5082.

[5] Duncton M A J, Pattenden G. The intramolecular Stille reaction[J]. Journal of the Chemical Society, Perkin Transactions 1, 1999, (10): 1235-1246.

[6] Stille J K, Groh B L. Stereospecific cross-coupling of vinyl halides with vinyl tin reagents catalyzed by palladium[J]. Journal of the American Chemical Society, 1987, 109(3): 813-817.

[7] Angara G J, Bovonsombat P, McNeils E. Formation of β, β-dihaloenones from halogenated tertiary alkynols[J]. Tetrahedron letters, 1992, 33(17): 2285-2288.

[8] Lu G, Voigtritter K R, Cai C, et al. Ligand effects on the stereochemistry of Stille couplings, as manifested in reactions of Z-alkenyl halides[J]. Chemical Communications, 2012, 48(69): 8661-8663.

[9] Kwon H B, McKee B H, Stille J K. Palladium-catalyzed coupling reactions of (. alpha.-ethoxyvinyl) trimethylstannane with vinyl and aryl triflates[J]. The Journal of Organic Chemistry, 1990, 55(10): 3114-3118.

[10] Farina V, Roth G P. Catalyst tailoring for palladium-mediated cross coupling of arylstannanes with vinyl triflates[J]. Tetrahedron Letters, 1991, 32(34): 4243-4246.

[11] Liebeskind L S, Wang J. Synthesis of substituted cyclobutenediones by the palladium catalyzed cross-coupling of halocyclobutenediones with organostannanes[J]. Tetrahedron Letters, 1990, 31(30): 4293-4296.

[12] Milstein D, Stille J K. Palladium-catalyzed coupling of tetraorganotin compounds with aryl and benzyl halides. Synthetic utility and mechanism[J]. Journal of the American Chemical Society, 1979, 101(17): 4992-4998.

[13] Li J H, Liang Y, Wang D P, et al. Efficient Stille cross-coupling reaction catalyzed by the Pd(OAc)$_2$/DABCO catalytic system[J]. The Journal of Organic Chemistry, 2005, 70(7): 2832-2834.

[14] Echavarren A M, Stille J K. Palladium-catalyzed coupling of aryl triflates with organostannanes[J]. Journal of the American Chemical Society, 1987, 109(18): 5478-5486.

[15] Crisp G T, Papadopoulos S. Palladium-catalyzed coupling of naphthyl triflates with organostannanes[J]. Australian Journal of Chemistry, 1988, 41(11): 1711-1715.

[16] Zhang H C, Brakta M, Daves Jr G D. Preparation of 1-(tri-n-butylstannyl) furanoid glycals and their use in palladium-mediated coupling reactions[J]. Tetrahedron Letters, 1993, 34(10): 1571-1574.

[17] Nair V, Turner G A, Chamberlain S D. Novel approaches to functionalized nucleosides via palladium-catalyzed cross coupling with organostannanes[J]. Journal of the American Chemical Society, 1987, 109(23): 7223-7224.

[18] Zhao D, You J, Hu C. Recent progress in coupling of two heteroarenes[J]. Chemistry: A European Journal, 2011, 17(20): 5466-5492.

[19] Palmisano G, Santagostino M. 2-(Tributylstannyl)-1-{[2-(trimethylsilyl) ethoxy] methyl}-1*H*-indole: synthesis and use as a 1*H*-indol-2-yl-anion equivalent[J]. Helvetica Chimica Acta, 1993, 76(6): 2356-2366.

[20] Nichele T Z, Monteiro A L. Synthesis of diarylmethane derivatives from Stille cross-coupling reactions of benzylic halides[J]. Tetrahedron Letters, 2007, 48(42): 7472-7475.

[21] Sustmann R, Lau J, Zipp M. Alkylation of aralkyl bromides with tetra alkyl tin compounds in presence of (2, 2′-bipyridine) fumaronitrile palladium (0)[J]. Tetrahedron letters, 1986, 27(43): 5207-5210.

[22] Rayner C M, Astles P C, Paquette L A. Total synthesis of furanocembranolides. 2. Macrocyclization studies culminating in the synthesis of a dihydropseudopterolide and gorgiacerone. Related furanocembranolide interconversions[J]. Journal of the American Chemical Society, 1992, 114(10): 3926-3936.

[23] Punna N, Harada K, Shibata N. Stille cross-coupling of secondary and tertiary α-(trifluoromethyl)-benzyl chlorides with allylstannanes[J]. Chemical Communications, 2018, 54(52): 7171-7174.

[24] Stille J K. Palladium catalyzed coupling of organotin reagents with organic electrophiles[J]. Pure and Applied Chemistry, 1985, 57(12): 1771-1780.

[25] Bateson J H, Burton G, Elsmere S A, et al. Synthesis of some cephem spiroacetal-lactones[J]. Synlett, 1994, (2): 152-154.

[26] Del Valle L, Stille J K, Hegedus L S. Palladium-catalyzed coupling of allylic acetates with aryl-and vinylstannanes[J]. The Journal of Organic Chemistry, 1990, 55(10): 3019-3023.

[27] Four P, Guibe F. Palladium-catalyzed reaction of tributyltin hydride with acyl chlorides. A mild, selective, and general route to aldehydes[J]. The Journal of Organic Chemistry, 1981, 46(22): 4439-4445.

[28] Labadie J W, Tueting D, Stille J K. Synthetic utility of the palladium-catalyzed coupling reaction of acid chlorides with organotins[J]. The Journal of Organic Chemistry, 1983, 48(24): 4634-4642.

[29] Cherry K, Lebegue N, Leclerc V, et al. Efficient synthesis of 5-and 6-tributylstannylindoles and their reactivity with acid chlorides in the Stille coupling reaction[J]. Tetrahedron Letters, 2007, 48(33): 5751-5753.

[30] Jousseaume B, Kwon H, Verlhac J, et al. 3(or 5)-Formyl-2-furan(or pyrrole)carboxylates and 3(4 or 5)-formyl-2-furan(or pyrrole)-carboxamides via alkoxycarbonylation or carbamoylation of stannylated formyl heterocycles[J]. Synlett, 1993, (2): 117-118.

[31] Frisch A C, Beller M. Catalysts for cross-coupling reactions with non-activated alkyl halides[J]. Angewandte Chemie International Edition, 2005, 44(5): 674-688.

[32] Powell D A, Maki T, Fu G C. Stille cross-couplings of unactivated secondary alkyl halides using monoorganotin reagents[J]. Journal of the American Chemical Society, 2005, 127(2): 510-511.

[33] Pri-Bar I, Pearlman P S, Stille J K. Synthesis of substituted cyclic ethers from halo ketones and halo aldehydes by palladium-catalyzed coupling with organotin reagents[J]. The Journal of Organic Chemistry, 1983, 48(24): 4629-4634.

[34] Kosugi M, Takano I, Sakurai M, et al. Palladium or ruthenium catalyzed reaction of α-halo ketones with tributyltin enolates: preparation of unsymmetrical 1,4-diketones [J]. Chemistry Letters, 1984, 13(7): 1221-1224.

[35] Shair M D, Yoon T, Danishefsky S J. A remarkable cross coupling reaction to construct the enediyne linkage relevant to dynemicin A: synthesis of the deprotected ABC system[J]. The Journal of Organic Chemistry, 1994, 59(14): 3755-3757.

[36] Goure W F, Wright M E, Davis P D, et al. Palladium-catalyzed cross-coupling of vinyl iodides with organostannanes: synthesis of unsymmetrical divinyl ketones[J]. Journal of the American Chemical Society, 1984, 106(21): 6417-6422.

[37] Baillargeon V P, Stille J K. Palladium-catalyzed formylation of organic halides with carbon monoxide and tin hydride[J]. Journal of the American Chemical Society, 1986, 108(3): 452-461.

[38] Couladouros E A, Mihou A P, Bouzas E A. First total synthesis of trans-and cis-resorcylide: remarkable hydrogen-bond-controlled, stereospecific ring-closing metathesis[J]. Organic Letters, 2004, 6(6): 977-980.

[39] Wu X F, Neumann H, Beller M. Palladium-catalyzed carbonylative coupling reactions between Ar-X and carbon nucleophiles[J]. Chemical Society Reviews, 2011, 40(10): 4986-5009.

[40] Piers E, Lu Y F. Deconjugation-alkylation of ethyl 3-(trimethylstannyl)-2-alkenoates. Stereocontrolled synthesis of ethyl 2-alkylidene-3-methylenecyclobutanecarboxylates[J]. The Journal of Organic Chemistry, 1988, 53(4): 926-927.

[41] Piers E, Skerlj R T. Regioselctive germyl–stannylation of α, β-acetylenic esters[J]. Journal of the Chemical Society, Chemical Communications, 1987, (13): 1025-1026.

[42] Adlington R M, Baldwin J E, Gansäuer A, et al. A study of the intramolecular Stille cross coupling reaction of vinylstannyl chloroformates: application to the synthesis of α-methylene lactones[J]. Journal of the Chemical Society, Perkin Transactions 1, 1994, (13): 1697-1701.

[43] Finch H, Pegg N A, Evans B. The synthesis of a conformationally restrained, combined thromboxane antagonist/synthase inhibitor using an intramolecular 'Stille'- or 'Grigg'-palladium-catalysed cyclisation strategy[J]. Tetrahedron Letters, 1993, 34(51): 8353-8356.

[44] Palmisano G, Santagostino M. 2-Modified tryptamines by Sn-Pd transmetallation-coupling process[J]. Synlett, 1993(10): 771-773.

[45] Baldwin J E, Adlington R M, Ramcharitar S H. Intramolecular palladium-catalysed cross coupling; a direct route to γ-oxo-α, β-unsaturated macrocycles[J]. Tetrahedron, 1992, 48(14): 2957-2976.

[46] Hirama M, Gomibuchi T, Fujiwara K, et al. Synthesis and DNA cleaving abilities of functional neocarzinostatin chromophore analogs. Base discrimination by a simple alcohol[J]. Journal of the American Chemical Society, 1991, 113(26): 9851-9853.

[47] Stille J K, Su H, Hill D H, et al. Synthesis of large-ring keto lactones by the homogeneous and polymer-supported palladium-catalyzed carbonylative coupling of esters having vinyl triflate and vinylstannane termini[J]. Organometallics, 1991, 10(6): 1993-2000.

[48] Takemura S, Hirayama A, Tokunaga J, et al. A concise total synthesis of (±)-A80915G, a member of the napyradiomycin family of antibiotics[J]. Tetrahedron Letters, 1999, 40(42): 7501-7505.

[49] Hayashi K, Kito T, Mitsuyama J, et al. Quinolonecarboxylic acid derivatives or salts thereof: U.S. Patent 6335447[P]. 2002-01-01.

[50] Yasuda N, Yang C, Wells K M, et al. Preparation of crystallinep-nitrobenzyl 2-hydroxymethyl carbapenem as a key intermediate for the anti-MRS carbapenem L-786,392[J]. Tetrahedron Letters, 1999, 40(3): 427-430.

[51] Ragan J A, Raggon J W, Hill P D, et al. Cross-coupling methods for the large-scale preparation of an imidazole-thienopyridine: synthesis of [2-(3-methyl-3H-imidazol-4-yl)-thieno [3, 2-b] pyridin-7-yl]-(2-methyl-1H-indol-5-yl)-amine[J]. Organic Process Research & Development, 2003, 7(5): 676-683.

[52] Baker S R, Roth G P, Sapino C. Palladium in cephalosporin chemistry: mild triflate couplings in the

absence of phosphines and halide donors[J]. Synthetic Communications, 1990, 20(14): 2185-2189.
[53] Roth G P, Sapino C. Palladium in cephalosporin chemistry: an inexpensive triflate replacement for palladium acetate mediated coupling reactions[J]. Tetrahedron Lett, 1991, 32(33): 4073-4076.
[54] Barrett D, Terasawa T, Okuda S, et al. Studies on β-lactam antibiotics synthesis and antibacterial activity of novel C-3 alkyne-substituted cephalosporins[J]. The Journal of Antibiotics, 1997, 50(1): 100-102.
[54] Pchalek K, Hay M P. Stille coupling reactions in the synthesis of hypoxia-selective 3-alkyl-1, 2, 4-benzotriazine 1, 4-dioxide anticancer agents[J]. The Journal of Organic Chemistry, 2006, 71(17): 6530-6535.
[56] Kobayashi S, Ando A, Kuroda H, et al. Rapid access to 6-bromo-5, 7-dihydroxyphthalide 5-methyl ether by a $CuBr_2$-mediated multi-step reaction: concise total syntheses of hericenone J and 5′-deoxohericenone C (Hericene A)[J]. Tetrahedron, 2011, 67(47): 9087-9092.
[57] Li D, Zhao Y, Ye L, et al. A formal total synthesis of fostriecin by a convergent approach[J]. Synthesis, 2010, (19): 3325-3331.
[58] Burghart J, Brückner R. Total synthesis of naturally configured pyrrhoxanthin, a carotenoid butenolide from plankton[J]. Angewandte Chemie International Edition, 2008, 47(40): 7664-7668.
[59] Brodmann T, Janssen D, Kalesse M. Total synthesis of chivosazole F[J]. Journal of the American Chemical Society, 2010, 132(39): 13610-13611.
[60] Fukuda H, Nakamura S, Eguchi T, et al. Concise total synthesis of vicenistatin[J]. Synlett, 2010, (17): 2589-2592.
[61] Nicolaou K C, Chakraborty T K, Piscopio A D, et al. Total synthesis of rapamycin[J]. Journal of the American Chemical Society, 1993, 115(10): 4419-4420.
[62] Nicolaou K C, Bulger P G, Sarlah D. Palladium-catalyzed cross-coupling reactions in total synthesis/metathesis reactions in total synthesis[J]. Angewandte Chemie International Edition, 2005, 44(29): 4442-4489.

第 7 章
Hiyama 偶联反应及其在药物合成上的应用

7.1 Hiyama 偶联反应及其反应机理

7.1.1 Hiyama 偶联反应

Hiyama 偶联反应是过渡金属（主要是钯）催化的有机硅烷与有机卤化物或其类似物之间的交叉偶联反应，在有机合成中用于构建碳-碳键，具有很高的化学和区域选择性。Hiyama 偶联反应可由图 7-1 表示。

$$R-Si(R^1R^2R^3) + R'-X \xrightarrow{M, F^-} R-R'$$

R=芳基、烯基、炔基
R'=芳基、烯基、炔基、烷基
R^1, R^2, R^3= Cl, F, OH, OM, 烷基, 烷氧基
X=Cl, Br, I, OTf 等
M=Pd, Ni, Rh, Cu 等

图 7-1 Hiyama 偶联反应

Hiyama 偶联反应发展的初衷是为了解决有机金属试剂在偶联反应中使用不便、反应的选择性不高等问题。首个 Hiyama 偶联反应并非由 Hiyama 报道的，而是由 Kumada 报道的。Kumada[1]发现，有机氟硅酸盐在氯化钯催化下能发生自偶联反应，例如苯乙烯基五氟硅酸钾在氯化钯催化下以中等收率生成 1,4-二苯基-1,3-丁二烯。后来，Hiyama 等[2]发现有机硅烷等许多有机硅试剂在氟离子活化下能够与有机卤化物在钯盐催化下顺利反应形成碳-碳键得到交叉偶联产物，成为有机合成上构建碳-碳键的又一有效手段。

$$K_2[Ph\text{—}CH=CH\text{—}SiF_5] \xrightarrow[\text{rt, 3h}]{PdCl_2, THF} Ph\text{—}CH=CH\text{—}CH=CH\text{—}Ph \quad (1)$$
$$54\%$$

由于有机硅化学的高度发展，各种有机硅试剂的制备已相当成熟，而且由于硅资源非常丰富，各种硅试剂廉价易得。有机硅试剂中碳-硅键的极性较小，与其他金属有机试剂相比非常稳定，并显示弱的亲核性能，因而，基于有机硅试剂的交叉偶联反应具有更高的化学选择性。也正是由于有机硅试剂中碳-硅键较小的极性，反应需要活化剂活化，活化剂通常为氟离子，有时为碱。活化剂与硅试剂形成五价硅中心，使碳-硅键极性增强以便在金属转移一步发生断裂。

7.1.2 反应机理

钯催化的 Hiyama 偶联反应的机理如图 7-2 所示：首先，有机卤化物 R'—X 氧化加成到催化剂分子 PdL_2(**A**) 生成中间体 R'PdL_2X(**B**)，与此同时硅试剂 R—SiR^{1}R^{2}R^3 被负离子活化生成五配位硅中间体[R—SiR^{1}R^{2}R^{3}F]$^-$NBu$_4^+$(**D**)；中间体 **B** 与 **D** 发生金属转移反应生成中间体 RR'PdL_2(**C**)；中间体 **C** 进行还原消除生成偶联产物 R—R'，同时再生催化剂 **A**，完成催化循环。

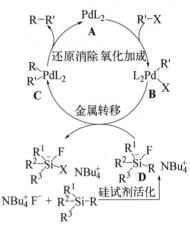

图 7-2　Hiyama 偶联反应的机理

7.2 基本的 Hiyama 偶联反应

有机硅试剂在过渡金属催化下可以与各种有机卤化物或其类似物进行交叉偶联反应生成各种杂化形态的碳原子相连的碳-碳键，包括 C(sp)-C(sp)、C(sp)-C(sp^2)、C(sp^2)-C(sp^2)、C(sp^2)-C(sp^3) 和 C(sp^3)-C(sp^3)，因而能够构建各种各样的碳骨架结构，如二炔、烯炔、芳烯、联芳基、烯丙基芳基和烷基芳烃化合物。下面将按碳原子杂化类型分类介绍基本的 Hiyama 偶联反应。

7.2.1 分子间 Hiyama 偶联反应

7.2.1.1　C(sp)-C(sp)键的构建

在中性条件下，一价铜能催化芳甲基三甲基硅烷与芳基乙炔氯化物发生交叉偶联生成非对称二炔化合物，硅试剂无须活化剂活化[3]。从以下实例可以推断，改变硅试剂或芳基乙炔氯化物芳环上的取代基，可以得到多种二芳基二炔化合物。

$$4\text{-MeO-}C_6H_4\!\!=\!\!=\!\!\text{SiMe}_3 + \text{Ph}\!\!=\!\!=\!\!\text{Cl} \xrightarrow[80℃, 48h]{\text{CuI, DMF}} 4\text{-MeO-}C_6H_4\!\!=\!\!=\!\!=\!\!=\!\!\text{Ph} \quad (2)$$
$$90\%$$

$$\begin{array}{c}4\text{-MeO-}C_6H_4\!\!=\!\!=\!\!\text{SiMe}_3 \\ + \\ 4\text{-Cl-}C_6H_4\!\!=\!\!=\!\!\text{Cl}\end{array} \xrightarrow[80℃, 48h]{\text{CuI, DMF}} 4\text{-MeO-}C_6H_4\!\!=\!\!=\!\!=\!\!=\!\!C_6H_4\text{-Cl-4} \quad (3)$$
$$95\%$$

$$\begin{array}{c}4\text{-CN-}C_6H_4\!\!=\!\!=\!\!\text{SiMe}_3 \\ + \\ 4\text{-MeCO-}C_6H_4\!\!=\!\!=\!\!\text{Cl}\end{array} \xrightarrow[80℃, 48h]{\text{CuI, DMF}} 4\text{CN-}C_6H_4\!\!=\!\!=\!\!=\!\!=\!\!C_6H_4\text{-COMe-4} \quad (4)$$
$$93\%$$

7.2.1.2　C(sp)-C(sp^2)键的构建

在活化剂三(二甲氨基)锍二氟三甲基硅酸（TASF）存在下，η^3-烯丙基氯化钯二聚体可催化炔基三甲基硅烷与卤代烯烃在温和条件下顺利发生偶联反应生成烯炔，反应后烯烃的构型得到保持[4]。

$$R\!\!=\!\!=\!\!\text{SiMe}_3 + \text{Br}\diagdown\!\!\diagup\text{Ph} \xrightarrow[\text{TASF, THF, rt}]{[\text{PdCl}(\eta_3\text{-}C_3H_5)]_2} R\!\!=\!\!=\!\!\diagdown\!\!\diagup\text{Ph} \quad (5)$$

TASF: [S(NMe$_2$)$_3$]$^+$[SiMe$_3$F$_2$]$^-$　R = Ph, 83%; n-C$_5$H$_{11}$, 86%; HOCH$_2$, 84%

如前所述，相较于其他金属有机试剂如锡试剂，硅试剂的反应活性低，多数情况下需要活化试剂活化才能与卤代烯烃进行交叉偶联，因而利用该活性差可实现 Stille 反应与 Hiyama 反应串联得到烯炔烯化合物。如反应式（6）所示，Stille 反应完成后，无须分离，直接加入另一个卤代烯烃和活化剂 TASF 在相同条件下进一步进行 Hiyama 反应，反应后两个烯烃的构型都得到保持[5]。

$$\text{Bu}_3\text{Sn}\!\!=\!\!=\!\!\text{SiMe}_3 \xrightarrow[\text{THF, 50℃}]{R^1\text{-I, Pd(PPh}_3)_4} [R^1\!\!=\!\!=\!\!\text{SiMe}_3] \xrightarrow[\text{TASF, THF, 50℃}]{R^2\text{-I, Pd(PPh}_3)_4} R^1\!\!=\!\!=\!\!R^2 \quad (6)$$

产物示例：

- n-C$_4$H$_9$ 取代烯基-炔-烯-n-C$_6$H$_{13}$，85%
- 环己烯基-炔-烯-n-C$_6$H$_{13}$，72%
- n-C$_8$H$_{17}$/n-C$_4$H$_9$ 取代烯基-炔-烯-n-C$_6$H$_{13}$，85%
- Ph-烯-炔-烯-n-C$_6$H$_{13}$，47%
- n-C$_6$H$_{13}$-烯-炔-烯-n-C$_6$H$_{13}$，68%
- n-C$_6$H$_{13}$-烯-炔-烯-n-C$_6$H$_{13}$，70%

芳基卤化物与炔基硅试剂在铜或钯催化下可发生交叉偶联反应用于合成芳炔，反应通常在碱性介质中进行[6,7]。在反应式（8）中，邻碘苯酚先在钯催化剂催化下发生 Hiyama 交叉偶联，紧接着进行分子内环合得到生物活性的苯并呋喃骨架结构。

$$Ph-\!\!\!\equiv\!\!\!-SiMe_3 + I-\!\!\!\!\!\!\bigcirc\!\!\!\!\!\!-OMe \xrightarrow[120℃, 12h]{CuCl, PPh_3, PhCO_2K, DMI} Ph-\!\!\!\equiv\!\!\!-\!\!\!\!\!\!\bigcirc\!\!\!\!\!\!-OMe \quad 89\% \quad (7)$$

$$Ph-\!\!\!\equiv\!\!\!-Si(OEt)_3 + \underset{I}{\underset{OH}{\bigcirc}} \xrightarrow[80℃, 4h]{cat., NaOH, 1,4-二氧六环, H_2O} \text{苯并呋喃-2-苯基} \quad 57\% \quad (8)$$

cat.: (mesityl)N=C(Cy)-Pd(Cl)_2-C≡N(mesityl)

芳基磺酸酯替代芳基卤化物亦可与硅试剂进行交叉偶联反应，但需要钯和铜共同催化反应才能顺利进行[8]。

$$R^1-\!\!\!\equiv\!\!\!-SiMe_3 + \underset{R^2}{\bigcirc}-OTf \xrightarrow[DMF, 80℃, 3\sim24h]{Pd(PPh_3)_4, CuCl} R^1-\!\!\!\equiv\!\!\!-\underset{R^2}{\bigcirc} \quad (9)$$

$R^1=C_6H_5$, R^2=4-MeCO, 97%; R^1=4-MeO–C_6H_4, R^2=4-MeCO, 99%;
R^1=4-MeO–C_6H_4, R^2=H, 64%; R^1=4-MeO–C_6H_4, R^2=4-CN, 99%;
R^1=4-MeCO–C_6H_4, R^2=4-CN, 99%; R^1=4-MeO–C_6H_4, R^2=2,4-F_2, 82%;
R^1= 噻吩基, R^2=4-MeCO, 60%; R^1= 噻吩基, R^2 = 4-CN, 79%

7.2.1.3　C(sp^2)-C(sp^2)键的构建

7.2.1.3.1　烯芳基

由活化剂 TASF 活化并在钯催化剂催化下，乙烯基三甲基硅烷与芳基卤化物交叉偶联生成芳基乙烯化合物[4]。该交叉偶联反应可耐受芳基上的各类官能团，如羰基、硝基、氨基等。

$$\text{CH}_2\!\!=\!\!\text{CH}-SiMe_3 + Ar-I \xrightarrow[HMPA, 50℃,]{(\eta^3-C_3H_5PdCl)_2, TASF} \text{CH}_2\!\!=\!\!\text{CH}-Ar \quad (10)$$

1-乙烯基萘 98%；4-甲基苯乙烯 89%；4-硝基苯乙烯 83%

H_2N-〈Ar〉-CH=CH₂ O_2N-〈Ar〉-CH=CH₂ CH₂=CH-〈Ar〉-CH=CH₂
85%   86%   84%

当乙烯基上带有脂肪族取代基时，此类乙烯基三甲基硅烷不与芳基卤化物发生上述交叉偶联反应，可能是由于取代基的供电子性能导致不能有效形成五配位的硅物种所致。当乙烯基三甲基硅烷上的一个或两个甲基被氟离子取代时交叉偶联反应则顺利进行，而三个甲基都被氟离子取代时，反应又不进行。一个或两个氟离子取代的烯基硅试剂能够形成五配位的硅物种被认为是反应得以进行的原因；而三氟硅试剂中更多的氟可能会强化 C—Si 键从而阻止烯基硅烷与钯配合物的金属转移或是倾向于形成六配位的硅物种，该硅物种在温和条件下不进行反应[9]。需要注意的是，根据底物的不同需要选用不同的催化剂或活化剂。

$$n\text{-}C_6H_{13}\text{-}CH=CH\text{-}SiMe_{3-n}F_n + \text{1-iodonaphthalene} \xrightarrow{(\eta^3\text{-}C_3H_5PdCl)_2, \text{TASF, THF, 50°C}} \text{naphthyl-}CH=CH\text{-}n\text{-}C_6H_{13} \quad (11)$$

n=0, 24h, 0%; n=1, 10h, 81%;
n=2, 48h, 74%; n=3, 24h, 0%

$$n\text{-}C_6H_{13}\text{-}CH=CH\text{-}SiMe_2F + \text{3-iodobenzyl acetate} \xrightarrow{(\eta^3\text{-}C_3H_5PdCl)_2, \text{TASF, THF, 50°C, 24h}} \text{产物} \quad (12)$$
85%

$$n\text{-}C_8H_{17}\text{-}CH=CH\text{-}SiMe_2F + \text{4-iodoacetophenone} \xrightarrow{Pd(Ph_3)_4, \text{TASF, DMF, 50°C, 12h}} \text{产物} \quad (13)$$
79%

$$n\text{-}C_4H_9\text{-}C(=CH_2)\text{-}SiMe_2F + \text{3-iodobenzyl acetate} \xrightarrow{(\eta^3\text{-}C_3H_5PdCl)_2, \text{TBAF, THF, 50°C, 22h}} \text{产物} \quad (14)$$
83%

环烯硅氧烷在活化剂四丁基氟化铵（TBAF）存在和钯催化剂催化下与芳基碘交叉偶联，同时开环生成羟基得到羟基化的芳基烯[10]。

$$\text{cyclic silyl ether} + \text{aryl iodide (OMe, Me, OMe)} \xrightarrow{Pd_2(dba)_3, \text{TBAF, THF, 50°C, 18h}} \text{产物} \quad (15)$$
92%

烯基与芳基卤化物的交叉偶联还可以通过分子内羟基活化实现。如反应式(16)所示，Hiyama 设计了一种烯基和芳基(2-羟甲基苯基)二甲基硅烷化合物，这种稳定的四取代的有机硅试剂含有一个作为活化基的羟基，该羟基立于硅试剂中硅原子所在苯环邻位亚甲基上，通过在硅原子上引入不同的烯基再与卤代物交叉偶联得到芳基烯类化合物。反应除了钯催化剂，还需要 CuI 和 K_2CO_3 存在，副产物苯并二甲基环戊硅氧烷很容易与偶联产物分离，并可循环使用[11,12]。

$$\underset{R^2}{\overset{R_1}{>}}{=}\underset{SiMe_2}{\overset{R_3}{<}}\text{—Ar（OH)}+\text{I—Ar}\xrightarrow[\text{DMSO, 35~80℃, 12h}]{\underset{\text{THF, 0℃~rt}}{\overset{R_1}{R^2}{>}{=}\overset{R_3}{<}\text{MgBr}}\atop\text{PdCl}_2\text{, L, CuI, K}_2\text{CO}_3}\underset{R^2}{\overset{R_1}{>}}{=}\underset{Ar}{\overset{R_3}{<}}+\text{(副产物)}\quad(16)$$

L: $(\underset{O}{\bigcirc})_3P$, 邻-(CH=NCy)C₆H₄PPh₂

$n\text{-}C_6H_{13}$—CH=CH—(2-噻吩基) 99%
$n\text{-}C_6H_{13}$—CH=CH—(3-吡啶基) 80%
$n\text{-}C_6H_{13}$—CH=CH—C₆H₄—CN 93%
NC—(CH₂)₃—CH=CH—C₆H₄—CN 95%
CH₂=C(Ph)—C₆H₄—OMe 80%

硅原子上含有羟基[反应式（17）~式（18）]、氯[反应式（19）~式（23）]的烯基硅试剂也能与芳基碘交叉偶联生成相应的芳基烯，但氯原子时需要用氢氧化钠活化[反应式（19）~式（23）]；此时，芳基卤化物也可以是芳基氯[反应式（22）]和芳基溴[反应式（19）~式（21），式（23）]，但相同条件下三甲基硅试剂则不反应[13,14]。

$$n\text{-}C_5H_{11}\text{—CH=CH—Si(Me)}_2\text{OH} + \text{Ph—I} \xrightarrow[\text{THF, rt, 10min}]{\text{Pd(dba)}_3\text{, TBAF}} \text{Ph—CH=CH—}n\text{-}C_5H_{11}\quad 91\% \quad(17)$$

$$n\text{-}C_5H_{11}\text{—CH=CH—Si(Me)}_2\text{OH}\;(Z) + \text{Ph—I} \xrightarrow[\text{THF, rt, 10min}]{\text{Pd(dba)}_3\text{, TBAF}} \text{Ph—CH=CH—}n\text{-}C_5H_{11}\;(Z) \quad 90\% \quad(18)$$

$$n\text{-}C_4H_9\text{—CH=CH—SiMeCl}_2 + \text{4-Br—C}_6H_4\text{—CN} \xrightarrow[\text{THF, 60℃, 12h}]{\text{Pd(OAc)}_2\text{, NaOH}} n\text{-}C_4H_9\text{—CH=CH—C}_6H_4\text{—CN} \quad 80\% \quad(19)$$

$$n\text{-}C_6H_{13}\diagdown\text{SiEtCl}_2 + \underset{\text{COCH}_3}{\overset{\text{Br}}{\bigcirc}} \xrightarrow[\text{THF, 60°C, 12h}]{\text{Pd(OAc)}_2, \text{NaOH}} n\text{-}C_6H_{13}\diagdown\underset{\text{COCH}_3}{\bigcirc} \quad (20)$$

70%

$$n\text{-}C_4H_9\diagdown\text{SiMeCl}_2 + \underset{\text{CF}_3}{\overset{\text{Br}}{\bigcirc}} \xrightarrow[\text{THF, 60°C, 12h}]{\text{Pd(OAc)}_2, \text{NaOH}} n\text{-}C_4H_9\diagdown\underset{\text{CF}_3}{\bigcirc} \quad (21)$$

80%

$$n\text{-}C_4H_9\diagdown\text{SiMeCl}_2 + \underset{\text{CF}_3}{\overset{\text{Cl}}{\bigcirc}} \xrightarrow[\text{THF, 80°C, 12h}]{\text{Pd(OAc)}_2, \text{NaOH}} n\text{-}C_4H_9\diagdown\underset{\text{CF}_3}{\bigcirc} \quad (22)$$

95%

$$\text{Me}_3\text{Si}\diagdown\text{SiMeCl}_2 + \text{Br}\underset{\text{CF}_3}{\overset{\text{CF}_3}{\bigcirc}} \xrightarrow[\text{THF, 60°C, 36h}]{\text{Pd(OAc)}_2, \text{NaOH}} \text{Me}_3\text{Si}\diagdown\underset{\text{CF}_3}{\overset{\text{CF}_3}{\bigcirc}} \quad (23)$$

74%

如果硅试剂不含氯原子，其与芳基溴或芳基氯进行 Hiyama 交叉偶联反应需要在较高的温度下进行，有时还需要微波协助。例如，对氯苯乙酮与乙烯基三甲氧基硅烷之间的交叉偶联反应需要在 TBAF 活化和(η^3-C_3H_5PdCl)$_2$ 与膦配体 L 原位生成的配合物催化下，于 110°C 并于微波辐射才能以优异的收率得到对乙烯基苯乙酮[15]。

$$\underset{\text{COCH}_3}{\overset{\text{Cl}}{\bigcirc}} + \diagdown\text{Si(OMe)}_3 \xrightarrow[\text{MW, 110°C, 18min}]{(\eta^3\text{-}C_3H_5\text{PdCl})_2, \text{L, TBAF}} \underset{\text{COCH}_3}{\bigcirc}\diagdown \quad (24)$$

L: Cy$_2$P-N⟨⟩N-Me   95%

对溴苯甲酸、3-溴吡啶与乙烯基三甲氧基硅烷则以 NaOH 为活化剂、PEG-2000 为相转移催化剂，在 Pd(OAc)$_2$ 催化下于 140°C 进行交叉偶联以优异的收率分别生成对乙烯基苯甲酸和 3-乙烯基吡啶[16]。

$$\underset{\text{COOH}}{\overset{\text{Br}}{\bigcirc}} + \diagdown\text{Si(OMe)}_3 \xrightarrow[\text{PEG-2000, 140°C, 3h}]{\text{Pd(OAc)}_2, \text{NaOH, H}_2\text{O}} \underset{\text{COOH}}{\bigcirc}\diagdown \quad (25)$$

97%

$$\underset{N}{\overset{\text{Br}}{\bigcirc}} + \diagdown\text{Si(OMe)}_3 \xrightarrow[\text{PEG-2000, 140°C, 3h}]{\text{Pd(OAc)}_2, \text{NaOH, H}_2\text{O}} \underset{N}{\bigcirc}\diagdown \quad (26)$$

96%

(E)-或（Z)-烯基（甲基）环丁硅烷被证明是好的亲核试剂，在钯催化剂催化下与芳基碘在室温下就能发生交叉偶联反应[17]。此类硅试剂可以由相应的烯基铝与氯（甲基）环丁硅烷进行取代反应或是炔基甲基环丁硅烷加氢制备。烯基甲基环丁硅烷在反应条件下先水解生的成烯基（甲基）（丙基）硅醇才是真正的反应活性物种。

$$\begin{array}{c}R^1\text{—Si(Me)(环丁基)} + Ar\text{—I} \xrightarrow{Pd(dba)_2, Bu_4N^+F^-}{THF, rt, 0.17\sim3h} R^1\text{—Ar} \end{array} \quad (27)$$

经 H_2O 水解 → $R^1\text{—Si(OH)(Me)(n-Pr)}$

$n\text{-}C_5H_{11}$—CH=CH—C_6H_5 (E), 10min, 91%
$n\text{-}C_5H_{11}$—CH=CH—C_6H_5 (Z), 10min, 90%
$n\text{-}C_5H_{11}$—CH=CH—(2-噻吩基), 3h, 89%
$n\text{-}C_5H_{11}$—CH=CH—(2-噻吩基) (Z), 3h, 85%
$n\text{-}C_5H_{11}$—CH=CH—(1-萘基), 0.5h, 91%

7.2.1.3.2　1,3-二烯

当与硅试剂进行交叉偶联反应时，烯基卤化物与芳基卤化物的性质类似，因而多数情况下用烯基卤化物替代芳基卤化物反应都能顺利进行生成 1,3-二烯，而且交叉偶联底物的构型得到保持[4,9,13]。

$$CH_2=CH\text{—}SiMe_3 + I\text{—}CH=CH\text{—}C_6H_{13}\ (Z) \xrightarrow{(\eta^3\text{-}C_3H_5PdCl)_2,\ TASF}{P(OEt)_3,\ THF,\ 50℃} CH_2=CH\text{—}CH=CH\text{—}C_6H_{13}\quad 76\% \quad (28)$$

$$CH_2=CH\text{—}SiMe_3 + I\text{—环己烯基} \xrightarrow{(\eta^3\text{-}C_3H_5PdCl)_2,\ TASF}{P(OEt)_3,\ THF,\ 50℃} CH_2=CH\text{—环己烯基}\quad 100\% \quad (29)$$

$$CH_2=CH\text{—}SiMe_3 + I\text{—}CH=CH(CH_2)COCH_3 \xrightarrow{(\eta^3\text{-}C_3H_5PdCl)_2,\ TASF}{P(OEt)_3,\ THF,\ 50℃} CH_2=CH\text{—}CH=CH(CH_2)COCH_3\quad 67\% \quad (30)$$

$$CH_3CH=CH\text{—}SiMe_3 + I\text{—}CH=CH\text{—}C_6H_{13} \xrightarrow{(\eta^3\text{-}C_3H_5PdCl)_2,\ TASF}{P(OEt)_3,\ THF,\ 50℃} CH_3CH=CH\text{—}CH=CH\text{—}C_6H_{13}\quad 78\% \quad (31)$$

$$n\text{-}C_8H_{17}\text{—}CH=CH\text{—}SiMe_2F + I\text{—}CH=CH\text{—}n\text{-}C_6H_{13} \xrightarrow{Pd(PPh_3)_4,\ TASF}{DMF,\ 50℃,\ 14h} n\text{-}C_8H_{17}\text{—}CH=CH\text{—}CH=CH\text{—}n\text{-}C_6H_{13}\quad 84\% \quad (32)$$

$$n\text{-}C_8H_{17}\text{—}CH=CH\text{—}SiMe_2F + I\text{—}CH=CH\text{—}n\text{-}C_6H_{13}\ (Z) \xrightarrow{Pd(PPh_3)_4,\ TASF}{DMF,\ 50℃,\ 19h} n\text{-}C_8H_{17}\text{—}CH=CH\text{—}CH=CH\text{—}n\text{-}C_6H_{13}\quad 74\% \quad (33)$$

$$n\text{-}C_5H_{11}\overset{Me\ Me}{\underset{}{\text{Si}}}\text{OH} + I\diagup(CH_2)_4OH \xrightarrow[\text{THF, rt, 1.5h}]{Pd(dba)_2,\ TBAF} n\text{-}C_5H_{11}\diagup\diagup(CH_2)_4OH \quad 91\% \quad (34)$$

$$n\text{-}C_5H_{11}\overset{Me\ Me}{\underset{}{\text{Si}}}\text{OH} + I\diagup(CH_2)_4OH \xrightarrow[\text{THF, rt, 1.5h}]{Pd(dba)_2,\ TBAF} \underset{72\%}{n\text{-}C_5H_{11}\diagup\diagup(CH_2)_4OH} \quad (35)$$

7.2.1.3.3 联二芳基

芳基硅试剂与芳基卤化物在钯催化剂催化下交叉偶联生成联二芳基化合物。最早发现芳基硅试剂硅原子上必须带有两个氟原子才能反应，而且由于甲基有可能参加偶联反应，硅原子上的烷基最好为乙基或丙基，活化剂可以用廉价的氟化钾替代常用的 TBAF；随后发现改变催化体系并提高反应温度，硅原子含氯的硅试剂与芳基溴也能进行交叉偶联，反应过程中 KF 先与硅试剂中的氯发生交换[18]。

$$Ar^1\text{-SiRF}_2 + I\text{-}Ar^2 \xrightarrow[\text{DMF, 70~100°C}]{(\eta^3\text{-}C_3H_5PdCl)_2,\ KF} Ar^1\text{-}Ar^2 \quad (36)$$

R=Et, 70°C, 10h, 81% R=Et, 100°C, 49h, 89%

R=Et, 100°C, 21h, 67% R=n-Pr, 100°C, 28h, 52%

R=n-Pr, 100°C, 43h, 47% R=n-Pr, 100°C, 15h, 83%

$$Ar^1\text{-SiEtCl}_2 + Br\text{-}Ar^2 \xrightarrow[\text{KF, DMF, 120°C, 18h}]{Pd(OAc)_2,\ P(o\text{-tol})_3} Ar^1\text{-}Ar^2 \quad (37)$$

$$\downarrow \text{KF, DMF} \quad [Ar^1\text{-SiEtF}_2]$$
60°C, 3h

92%   83%   88%

86%   69%

相对于上述卤代硅试剂，芳基三甲氧基硅烷更容易得到，在 TBAF 和 Pd(dba)$_2$ 存在下其可以与芳基碘或芳基溴交叉偶联生成联二芳基化合物；当反应体系中引入膦配体或 N-杂卡宾配体时芳基氯也能反应［反应式（41）］[19,20]。

$$\text{Ph-Si(OMe)}_3 + \text{Me-C}_6\text{H}_4\text{-I} \xrightarrow[\text{DMF, 95℃, 2h}]{\text{Pd(dba)}_2,\ \text{TBAF}} \text{Ph-C}_6\text{H}_4\text{-Me} \quad 90\% \qquad (38)$$

$$\text{Ph-Si(OMe)}_3 + \text{MeCO-C}_6\text{H}_4\text{-Br} \xrightarrow[\text{DMF, 95℃, 2h}]{\text{Pd(dba)}_2,\ \text{TBAF}} \text{Ph-C}_6\text{H}_4\text{-COMe} \quad 78\% \qquad (39)$$

$$\text{Ph-Si(OMe)}_3 + \text{MeO-C}_6\text{H}_3(\text{Cl}) \xrightarrow[\text{DMF, 87℃, 27h}]{\text{Pd(dba)}_2,\ \text{TBAF, L}} \text{Ph-C}_6\text{H}_4\text{-OMe} \quad 71\% \qquad (40)$$

L = 2-(二环己基膦基)联苯 (PCy$_2$)

$$\text{Ph-Si(OMe)}_3 + \text{O}_2\text{N-C}_6\text{H}_4\text{-Cl} \xrightarrow[\text{PhMe, 120℃, 5h}]{\text{NHC-Pd, TBAF}} \text{Ph-C}_6\text{H}_4\text{-NO}_2 \quad 92\% \qquad (41)$$

NHC-Pd: [双(1,3-二(2,6-二异丙基苯基)咪唑-2-亚基)二氯化钯]-哌嗪桥联二聚体结构

2-噻吩基三乙氧基硅烷与 4-三氟甲基碘苯在 CuI 催化下就能进行交叉偶联反应，以优异的收率生成 2-(4-三氟甲基苯基)噻吩。反应在 CsF 存在下进行，CsF 起稳定中间体[CuAr]的作用并作为氟源[21]。

$$\text{2-Th-Si(OEt)}_3 + \text{F}_3\text{C-C}_6\text{H}_4\text{-I} \xrightarrow[\text{120℃, 24h}]{\text{CuI, CsF, DMF}} \text{F}_3\text{C-C}_6\text{H}_4\text{-2-Th} \quad 94\% \qquad (42)$$

芳基硅醇、芳基硅二醇和芳基硅三醇可以由芳基锂试剂与氯硅烷反应生成相应的芳基氯硅烷水解制备。三种类型的芳基硅醇均可在 Ag$_2$O 活化和钯催化剂催化下与芳基碘发生交叉偶联生成联二芳基化合物，反应结果与硅醇的种类、芳基碘的结构和反应条件如反应时间相关[22]。

$$\text{Ar}^1\text{-SiR}_{(3-n)}(\text{OH})_n + \text{Ar}^2\text{-I} \xrightarrow[\text{THF, 60℃, 5~36h}]{\text{Pd(PPh}_3)_4,\ \text{Ag}_2\text{O}} \text{Ar}^1\text{-Ar}^2 \qquad (43)$$

MeO—⟨C6H4⟩—⟨C6H5⟩
R=Me, *n*=1, 36h, 80%

MeO—⟨C6H4⟩—⟨C6H4⟩—Me
R=Me, *n*=1, 36h, 75%

MeO—⟨C6H4⟩—⟨C6H4⟩—CF3
R=Me, *n*=1, 36h, 54%

⟨C6H5⟩—⟨C6H4⟩—OMe
R=Me, *n*=1, 36h, 35%

⟨C6H5⟩—⟨C6H4⟩—OMe
R=Me, *n*=2, 12h, 93%

⟨C6H5⟩—⟨C6H4⟩—OMe
R=Et, *n*=2, 12h, 80%

⟨C6H5⟩—⟨C6H4⟩—OMe
R=0, *n*=3, 12h, 83%

⟨C6H5⟩—⟨C6H4⟩—COMe
R=0, *n*=3, 12h, 97%

芳基三甲基硅烷不容易与芳基卤化物交叉偶联，加入 CuI 作为助催化剂则可以使反应顺利进行。例如，以 TBAF 为活化剂并辅以 PPh$_3$ 配体，1-氟-4-碘苯与 4-氯吡啶三甲基硅烷在 PdCl$_2$(PPh$_3$)$_2$ 和 CuI 共同催化下于 DMF 中室温反应顺利生成交叉偶联产物[23]。

$$\text{4-Cl-2-(SiMe}_3\text{)-pyridine} + \text{I—C}_6\text{H}_4\text{—F} \xrightarrow[\text{TBAF, DMF, rt, 12h}]{\text{PdCl}_2(\text{PPh}_3)_2, \text{CuI, PPh}_3} \text{4-Cl-2-(4-F-C}_6\text{H}_4\text{)-pyridine} \quad 95\%$$ (44)

与前述用于烯基化的（2-羟甲基苯基）二甲基硅烷（HOMSi）硅试剂类似，芳基（2-羟甲基苯基）二甲基硅烷是很好的芳基化试剂，与芳基卤化物或类似物在碱性介质、钯催化下很容易进行交叉偶联，其中芳基卤化物类似物 1-咪唑基硫酸 2-萘酯与 4-甲氧苯基 HOMSi 硅试剂反应几乎定量地生成 2-(4-甲氧基苯基) 萘[24]。

$$\text{2-naphthyl-OSO}_2\text{-imidazolyl} + \text{2-(HOCH}_2\text{)C}_6\text{H}_4\text{-Si(Me)}_2\text{-C}_6\text{H}_4\text{-OMe} \xrightarrow[\text{K}_2\text{CO}_3, 65^\circ\text{C, 18h}]{\text{Pd(dppf)Cl}_2, \text{H}_2\text{O, DMSO}} \text{2-(4-MeO-C}_6\text{H}_4\text{)-naphthalene} + \text{benzosiloxole}$$ (45)

与烯基环丁硅烷硅试剂类似，芳基环丁硅烷试剂与芳基卤化物在钯催化下也能进行交叉偶联反应生成联二芳基化合物，引入给电子、空间位阻大的三叔丁基膦配体以减弱芳基卤化物的自二聚反应[25]。

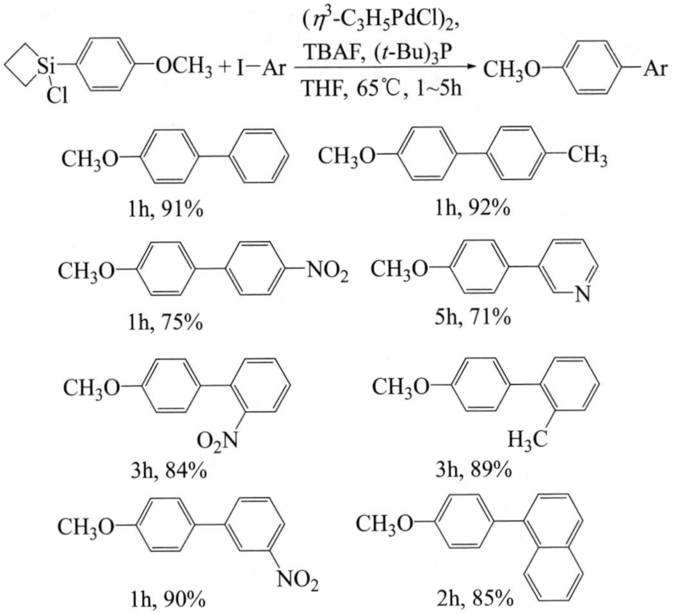

$$\text{Si-}\underset{\text{Cl}}{\bigcirc}\text{-}\underset{}{\bigcirc}\text{-OCH}_3 + \text{I-Ar} \xrightarrow[\text{THF, 65℃, 1~5h}]{(\eta^3\text{-C}_3\text{H}_5\text{PdCl})_2, \text{TBAF, }(t\text{-Bu})_3\text{P}} \text{CH}_3\text{O-}\underset{}{\bigcirc}\text{-Ar} \quad (46)$$

CH₃O—⟨⟩—⟨⟩
1h, 91%

CH₃O—⟨⟩—⟨⟩—CH₃
1h, 92%

CH₃O—⟨⟩—⟨⟩—NO₂
1h, 75%

CH₃O—⟨⟩—⟨N⟩
5h, 71%

CH₃O—⟨⟩—⟨⟩
        O₂N
3h, 84%

CH₃O—⟨⟩—⟨⟩
           H₃C
3h, 89%

CH₃O—⟨⟩—⟨⟩
          NO₂
1h, 90%

CH₃O—⟨⟩—⟨naphthyl⟩
2h, 85%

7.2.1.3.4　C(sp²)-C(sp³)键的构建
（1）烷基化芳香化合物

前述作为硅试剂活化剂的 TASF 自身可作为甲基化硅试剂与各种芳基卤化物在钯催化剂催化下进行交叉偶联反应生成芳甲基化合物[26]。

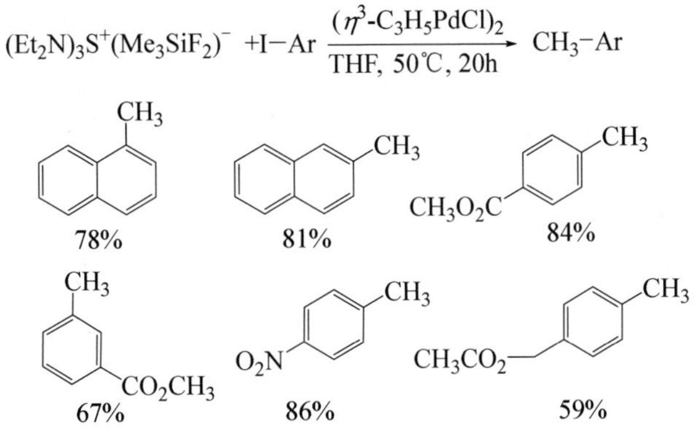

$$(\text{Et}_2\text{N})_3\text{S}^+(\text{Me}_3\text{SiF}_2)^- + \text{I-Ar} \xrightarrow[\text{THF, 50℃, 20h}]{(\eta^3\text{-C}_3\text{H}_5\text{PdCl})_2} \text{CH}_3\text{-Ar} \quad (47)$$

1-甲基萘 78%

2-甲基萘 81%

CH₃O₂C—⟨⟩—CH₃ 84%

⟨⟩—CH₃
 CO₂CH₃
67%

O₂N—⟨⟩—CH₃
86%

CH₃CO₂CH₂—⟨⟩—CH₃
59%

其他烷基硅试剂在 TBAF 活化和钯催化下也能与芳基卤化物进行交叉偶联生成烷基化芳香化合物，而且烷基上还可含其他官能团[27]。

$$\text{Ar-X} + \text{Y-(CH}_2)_n\text{-SiF}_3 \xrightarrow[\text{THF, 100℃, 8~48h}]{\text{Pd(PPh}_3)_4\text{, TBAF}} \text{Y-(CH}_2)_n\text{-Ar} \quad (48)$$

CH₃O—⟨ ⟩—n-C₆H₁₃   CH₃O—⟨ ⟩—n-C₆H₁₃
X=I, 24h, 61%   X=Br, 37h, 63%

CH₃O—⟨ ⟩—CH₂CH₂Ph   CH₃O—⟨ ⟩—CH₂CH₂CO₂CH₃
X=Br, 34h, 71%   X=I, 8h, 77%

naphthyl-n-C₆H₁₃   quinolinyl-CH₂CH₂CO₂CH₃
X=Br, 24h, 62%   X=Br, 8h, 65%

CH₃O₂C-furyl-CH₂CH₂CO₂CH₃   CH₃O—⟨ ⟩—(CH₂)₃CN
X=Br, 48h, 36%   X=I, 24h, 86%

在 CuI 和邻菲咯啉配体存在下，三氟甲基三乙基硅烷与芳基碘高效反应生成三氟甲基化的芳香化合物[28]。例如，三氟甲基三乙基硅烷与 4-碘苯甲酸乙酯在氟化钾、CuI 和邻菲咯啉配体存在下在温和条件下以优异的收率生成 4-三氟甲基苯甲酸乙酯。在相近的催化反应体系下，二氟甲基三甲基硅烷与芳基碘反应得到二氟甲基化的芳香化合物[29]。

$$\text{CF}_3\text{-SiEt}_3 + \text{I-}\langle\rangle\text{-CO}_2\text{Et} \xrightarrow[\text{NMP, DMI, 60℃, 24h}]{\text{CuI, phen, KF}} \text{CF}_3\text{-}\langle\rangle\text{-CO}_2\text{Et} \quad (49)$$
89%

$$\text{CHF}_2\text{-SiMe}_3 + \text{I-}\langle\rangle\text{-}n\text{-Bu} \xrightarrow[\text{120℃, 24h}]{\text{CuI, CsF, NMP}} \text{CHF}_2\text{-}\langle\rangle\text{-}n\text{-Bu} \quad (50)$$
90%

去甲麻黄碱（norephedrine，NPD）与镍原位生成的配合物特别适合于各种结构的仲烷基与芳基三氟化硅交叉偶联生成结构繁多的仲烷基取代的芳基化合物[30]。

$$\begin{matrix}R^1\\R^2\end{matrix}\!\!\!\!>\!\!\text{X} + \text{F}_3\text{Si-Ar} \xrightarrow[\text{DMA, 60℃, 16h}]{\text{NiCl}_2\cdot(\text{MeOCH}_2)_2,\ \text{NPD}\atop\text{LiHMDS, H}_2\text{O, CsF}} \begin{matrix}R^1\\R^2\end{matrix}\!\!\!\!>\!\!\text{Ar} \quad (51)$$

NPD: Ph-CH(OH)-CH(NH₂)-Me   LiHMDS: (Me₃Si)₂NLi

cyclohexyl-Ph   cycloheptyl-⟨ ⟩-F   Ph-CH(Me)-Ph
X=I, 94%   X=Br, 59%   X=Br, 65%

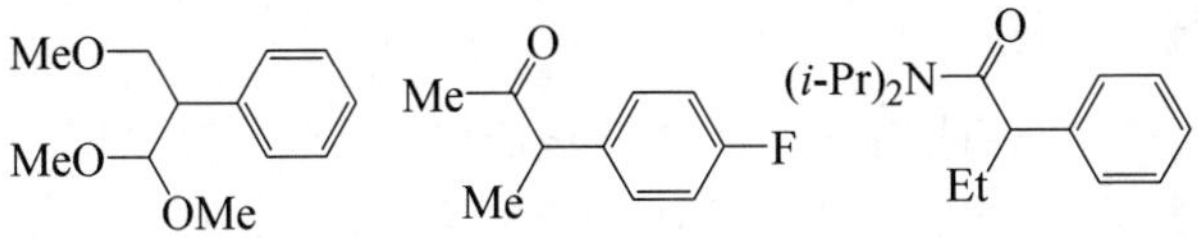

X=Br, 86%   X=Br, 78%   X=Br, 83%

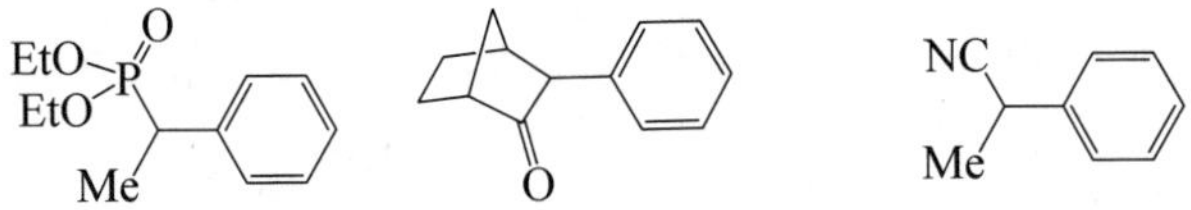

X=Br, 92%   X=Cl, 80%   X=Cl, 88%

（2）芳基烯丙基化合物

烯丙基三氟化硅与芳基卤化物或三氟甲磺酸酯进行交叉偶联发生在 γ-碳，生成芳基烯丙基化合物[31]。该反应非常重要，因为其他烯丙基金属试剂与芳基卤化物的交叉偶联反应通常发生在 α-碳。

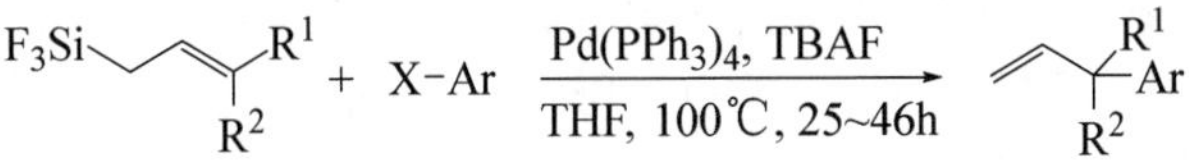

$$F_3Si\diagdown R^1/R^2 + X-Ar \xrightarrow[\text{THF, 100℃, 25~46h}]{\text{Pd(PPh}_3)_4\text{, TBAF}} \diagup R^1/Ar/R^2 \qquad (52)$$

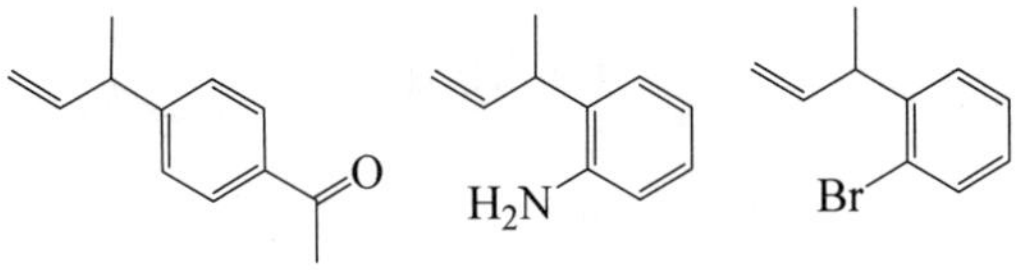

X=I, 19h, 95%   X=I, 25h, 53%   X=Br, 37h, 78%

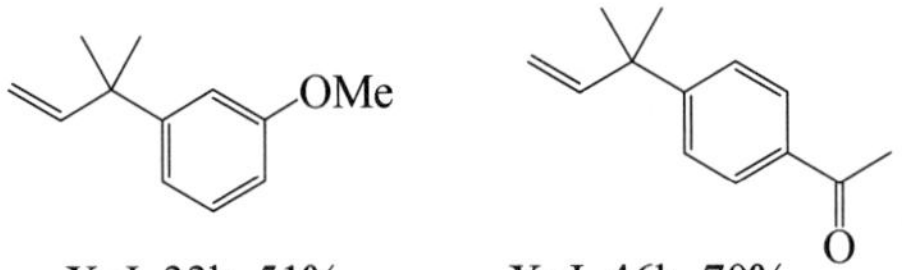

X=I, 33h, 51%   X=I, 46h, 70%

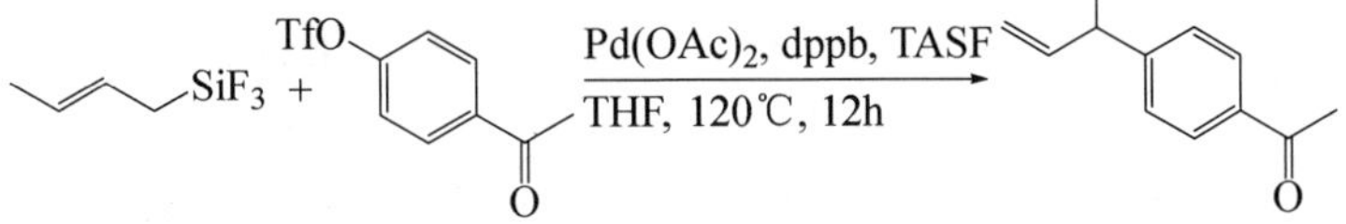

$$\diagdown SiF_3 + TfO-\text{C}_6H_4-\text{C(O)} \xrightarrow[\text{THF, 120℃, 12h}]{\text{Pd(OAc)}_2\text{, dppb, TASF}} \qquad (53)$$

芳基烯丙基化合物也可由芳基硅试剂与烯丙基卤化物间的交叉偶联生成。反应在由 PEG-600 还原 K_2PdCl_4 原位生成的钯纳米粒子催化下进行，PEG-600 同时还起稳定钯纳米粒子的作用[32]。

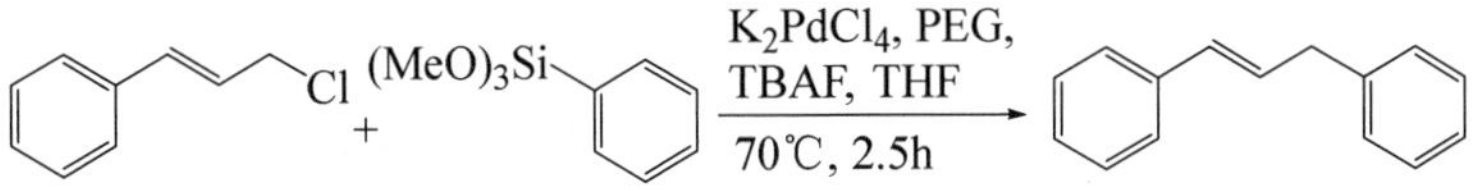

$$Ph\diagdown\diagup Cl + (MeO)_3Si-Ph \xrightarrow[\text{70℃, 2.5h}]{K_2PdCl_4\text{, PEG,}\atop \text{TBAF, THF}} Ph\diagdown\diagup Ph \qquad 95\% \qquad (54)$$

$$\text{4-MeC}_6H_4\diagdown\diagup Cl + (MeO)_3Si-\text{C}_6H_4\text{-OMe} \xrightarrow[\text{70℃, 2h}]{K_2PdCl_4\text{, PEG,}\atop \text{TBAF, THF}} \text{4-MeC}_6H_4\diagdown\diagup\text{C}_6H_4\text{-OCH}_3 \qquad 85\% \qquad (55)$$

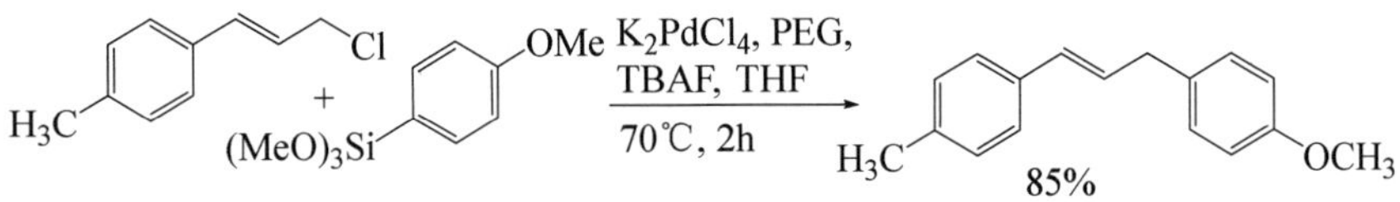

(3) 1,4-二烯化合物

烯丙基三氟化硅与三氟甲磺酸烯基酯在钯催化剂催化下交叉偶联生成 1,4-二烯类化合物,而且与前述三氟甲磺酸芳基酯作为亲电试剂类似,反应也发生在 γ-碳[31]。

$$F_3Si\diagup\diagdown\diagup\!\!\stackrel{R^1}{\underset{R^2}{=}} + TfO-R \xrightarrow[\text{THF, 120°C, 12~46h}]{\text{Pd(OAc)}_2,\text{ dppb, TASF}} \diagup\!\!\stackrel{R^1}{\underset{R^2}{=}}\diagdown\diagup R \quad (56)$$

24h, 98%   46h, 83%

与此相反,烯基硅试剂与烯丙基卤化物也能发生交叉偶联同样生成 1,4-二烯。例如,在作为活化剂的四正丁基铵二氟代三苯基硅酸盐(TBAT)存在下,不同官能团化烯基三乙氧基硅烷与烯丙基溴在 CuI 催化下顺利交叉偶联生成不同官能团化 1,4-二烯;将 CuI 换成噻吩羧酸亚铜(CuTc),γ-碳取代的烯丙基溴也能发生反应,而且反应主要发生在 γ-碳[33]。

$$R\diagup\!\!\diagdown Si(OEt)_3 + Br\diagup\!\!\diagdown \xrightarrow[\text{MeCN, rt, 16h}]{\text{CuI, TBAT}} R\diagup\!\!\diagdown\diagup\!\!\diagdown \quad (57)$$

TBAT: $n\text{-Bu}_4\text{N}^+\,[\text{Ph}_3\text{SiF}_2]^-$

98%   88%   89%

80%   85%

$$\overset{\text{OBn}}{\diagup}\!\!\diagdown Si(OEt)_3 + Br\diagup\!\!\diagdown R \xrightarrow[\text{MeCN, rt, 24h}]{\text{CuTc, TBAT}} BnO\diagup\!\!\diagdown\diagup\!\!\diagdown R \quad (58)$$

CuTc:

93%, $\gamma:\alpha$ = 84:16   93%, $\gamma:\alpha$ = 76:24   86%, $\gamma:\alpha$ = 83:17

76%, $\gamma:\alpha$ = 80:20   67%, $\gamma:\alpha$ = 88:12   79%, $\gamma:\alpha$ = 82:18

7.2.1.3.5 C(sp^3)-C(sp^3)键的构建

由于烷基卤化物与烷基硅试剂都属于惰性交叉偶联反应伙伴,而且存在自偶联倾向,因而通过 Hiyama 反应构建 C(sp^3)-C(sp^3)键比较困难,只有为数不多的反应实例。其中,烯丙基三氟化硅与乙酸烯丙酯在 TBAF 活化和钯催化剂催化下进行交叉偶联生成 1,5-二烯类化合物[31]。

$$F_3Si\diagup\diagdown + AcO\diagup\diagdown Ph \xrightarrow[\text{THF, 60°C, 5h}]{\text{Pd(PPh}_3)_4\text{, TBAF}} \underset{54\%}{\diagup\diagdown\diagup\diagdown Ph} \qquad (59)$$

过渡金属催化与光氧化还原催化使得烷基在温和条件下从硅试剂上脱离并通过单电子转移与偶联伙伴结合。正己基(双儿茶酸)硅酸钾与 18-冠-6(18-C-6)配合物在光催化剂[Ir(dF(CF$_3$)ppy)$_2$(bpy)](PF$_6$)(Ir1)、催化剂 NiCl$_2$(cod)和配体 dtbpy 共同存在下,经光照与 4-溴丁酸乙酯反应得到烷基-烷基交叉偶联产物,但伴有 4-溴丁酸乙酯自偶联产物生成[34]。

$$\text{n-C}_6\text{H}_{13}\text{-Si(catechol)}_2\text{K}^+[18\text{-C-6}] + \text{Br}\diagup\diagdown\text{CO}_2\text{Et} \xrightarrow[\text{蓝光LED, DMF, rt, 24h}]{\text{Ni(cod)}_2\text{, dtbpy, Ir1}} \underset{43\%}{\text{n-C}_6\text{H}_{13}\diagup\diagdown\text{CO}_2\text{Et}} + \underset{38\%}{\text{EtO}_2\text{C}\diagup\diagdown\text{CO}_2\text{Et}} \qquad (60)$$

Ir1: [Ir(dF(CF$_3$)ppy)$_2$(bpy)]

dtbpy: 4,4'-二叔丁基-2,2'-联吡啶

7.2.1.4 不对称 Hiyama 反应

不对称 Hiyama 反应并不普遍,往往需要特殊结构的卤代化合物。芳基上带有大空间位阻基团的 α-溴代羧酸芳基酯在手性镍催化剂催化下可以与芳基或烯基硅试剂进行不对称交叉偶联反应生成光学活性的 α-芳基或烯基取代的羧酸芳基酯[35]。配体、硅试剂以及负离子活化剂均对对映选择性起决定性作用。

$$\text{ArO-CO-CHR-Br} + (\text{MeO})_3\text{Si-R}' \xrightarrow[\text{TBAT, 1,4-二氧六环, rt}]{\text{NiCl}_2 \cdot (\text{CH}_2\text{OMe})_2\text{, (S,S)-L}} \text{BHTO-CO-CR(R')-H} \qquad (61)$$

(S,S)-L: MeHN-CHPh-CHPh-NHMe   TBAT: [F$_2$SiPh$_3$]$^-$[NBu$_4$]$^+$

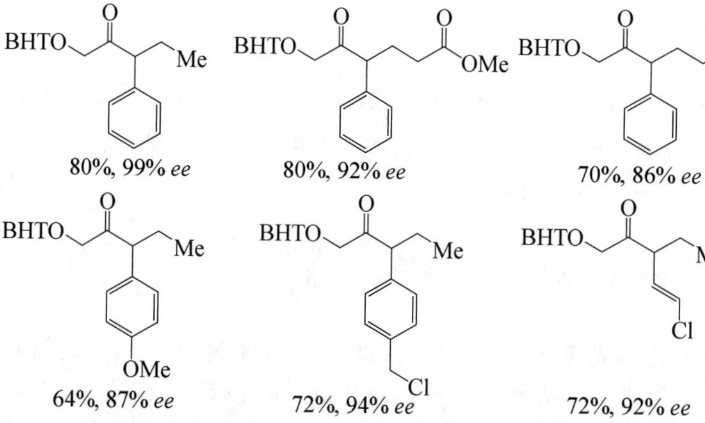

80%, 99% ee   80%, 92% ee   70%, 86% ee

64%, 87% ee   72%, 94% ee   72%, 92% ee

α-氯代-2,2,2-三氟乙基醚是另一类能够进行不对称 Hiyama 反应的底物，在手性二噁唑配体与 $NiCl_2$ 原位形成的手性配合物催化和光照下与芳基三甲氧基硅烷交叉偶联得到手性的含三氟甲基苄基醚类化合物[36]。三氟甲基广泛存在于临床药物和前药分子结构中，因而该反应为发现新药分子提供了一种有效途径。

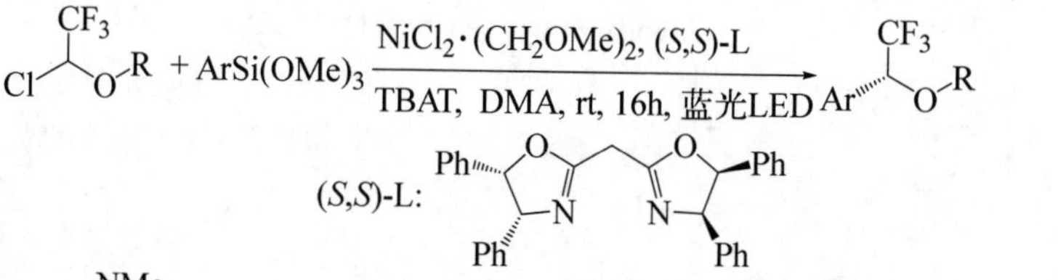

$$\text{Cl}\overset{CF_3}{\underset{}{\diagdown}}\text{O-R} + \text{ArSi(OMe)}_3 \xrightarrow[\text{TBAT, DMA, rt, 16h, 蓝光LED}]{NiCl_2 \cdot (CH_2OMe)_2, (S,S)\text{-L}} \text{Ar}\overset{CF_3}{\underset{}{\diagdown}}\text{O-R} \quad (62)$$

(S,S)-L:

93%, 97% ee   93%, 92% ee   93%, 97% ee

93%, 97% ee   80%, 92% ee   76%, 96% ee

74%, 87% ee   83%, 95% ee   84%, 95% ee

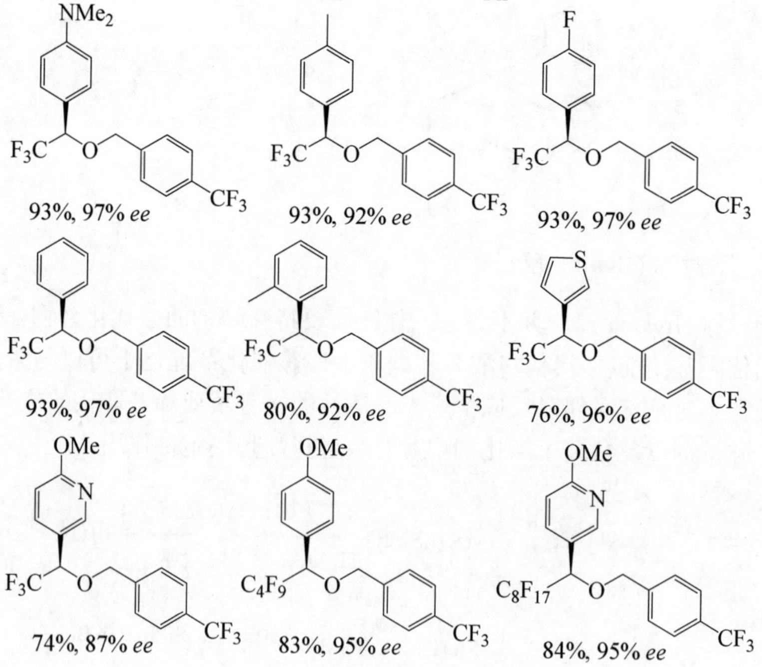

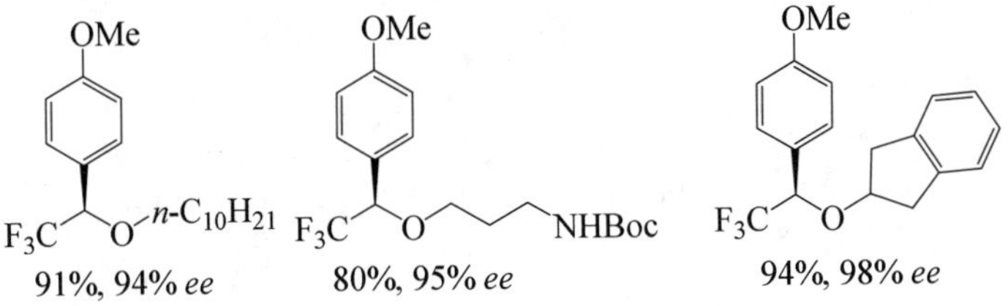

OMe
F_3C
n-$C_{10}H_{21}$
91%, 94% ee
NHBoc
80%, 95% ee
94%, 98% ee

7.2.2 分子内 Hiyama 反应

相较于分子间 Hiyama 反应，分子内 Hiyama 反应少见，关键是难以获得同时含有 Hiyama 偶联伙伴官能团的化合物。由乙烯基二甲基硅基醇醚在钼配合物催化剂 **1**（Schrock' catalyst）催化下生成同时含有碘代烯和环烯硅醚官能团的有机硅化合物。通过缓慢加料控制该有机硅化合物的浓度为 0.01mol/L，在过量的活化剂 TBAF 活化和钯催化剂$(\eta^3\text{-}C_3H_5PdCl)_2$ 催化下进行分子内交叉偶联反应生成含有 *cis,cis*-1,3-二烯单元的中环化合物。而且，通过改变碳链长度和环硅醚环的大小还能调节产物中羟基的位置[37]。

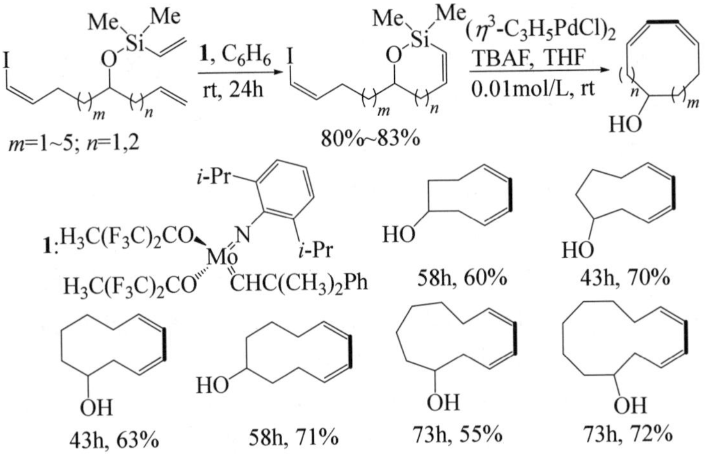

1, C_6H_6
rt, 24h
$(\eta^3\text{-}C_3H_5PdCl)_2$
TBAF, THF
0.01mol/L, rt
m=1~5; n=1,2
80%~83%
1: $H_3C(F_3C)_2CO$
$H_3C(F_3C)_2CO$
$CHC(CH_3)_2Ph$
i-Pr
58h, 60%
43h, 70%
43h, 63%
58h, 71%
73h, 55%
73h, 72%

(63)

我国科研人员在本领域也有杰出的研究成果，南开大学赵东兵研究员课题组发现邻溴芳基环丁或戊硅烷是另一类能进行分子内 Hiyama 反应的底物，其结构中的环硅烷在反应中开环，脱离硅原子的碳原子起亲核试剂作用取代芳基相邻位置的溴原子，而硅原子上生成的空位则由反应体系加入的醇的烷氧基来补充，最终生成扩环产物苯并硅氧烷[38]。

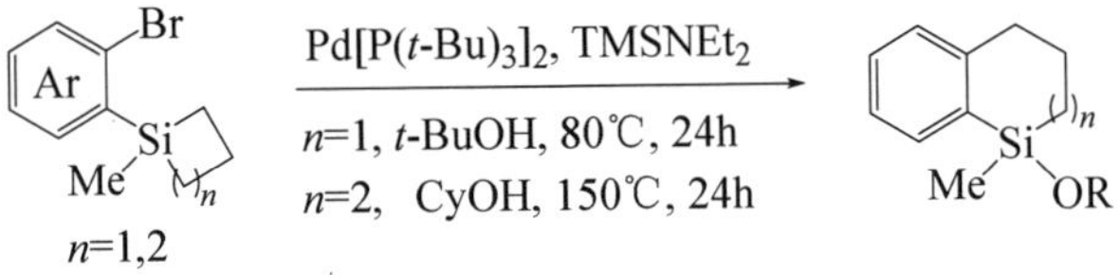

$Pd[P(t\text{-}Bu)_3]_2$, $TMSNEt_2$
n=1, t-BuOH, 80℃, 24h
n=2, CyOH, 150℃, 24h
n=1,2

(64)

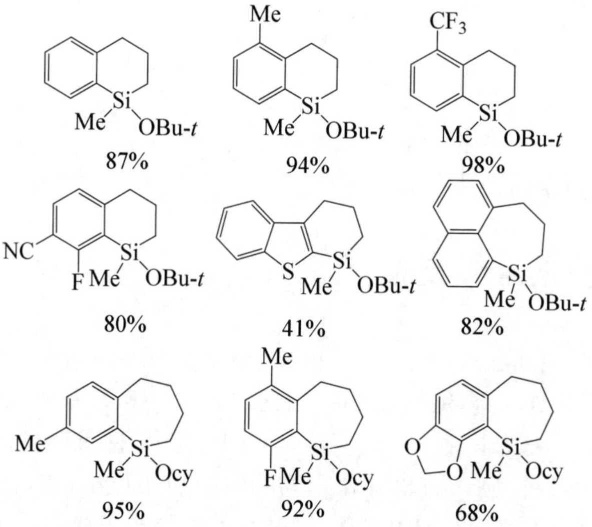

7.3 Hiyama 反应在药物合成上的应用实例

Hiyama 反应通常在中性介质中进行,条件温和,官能团耐受广泛;硅试剂种类多样,容易制备,因而与各种卤化物或其类似物进行交叉偶联反应可提供各种各样的偶联产物,为药物和药物中间体合成提供了一条有效途径。以下将介绍 Hiyama 反应在药物合成上的应用实例,为叙述方便,把 Hiyama 反应产物编号为 H-x,x 整数,代表序号。

Recifeiolide 是一种从真菌中分离出来的天然抗生素分子,结构上属于大环内酯,其合成关键中间体十二碳-(8,11)-二烯酸甲酯(H-1)可通过钯催化的烯丙基氯与烯基硅试剂的交叉偶联反应合成。合成中,8-壬炔酸甲酯先与三氯硅烷在氯铂酸催化下加成生成烯基硅试剂,进而被活化剂 KF 活化原位生成 $K_2[F_5SiCH=CH(CH_2)_6CO_2CH_3]$,再与烯丙基氯在 Pd(OAc)$_2$ 催化下反应生成 Hiyama 交叉偶联产物 H-1;后者先通过 Wacker 反应生成 11-氧代-(E)-8-十二烯酸甲酯,再经 NBH$_4$ 还原生成 recifeiolide 前体(±)-11-羟基-(E)-十二烯酸甲酯,该前体再经环合反应最终生成 recifeiolide[1]。

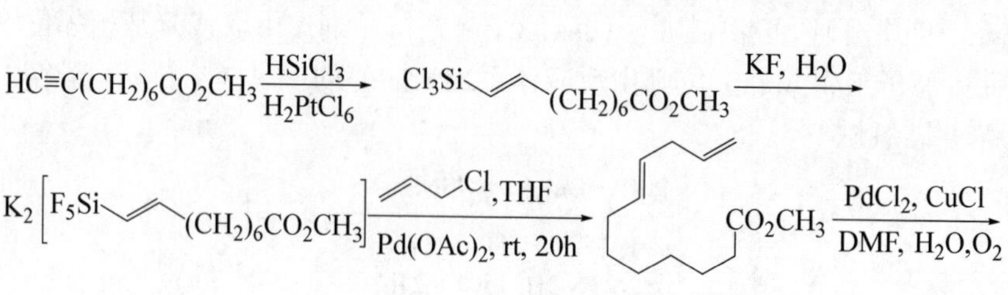

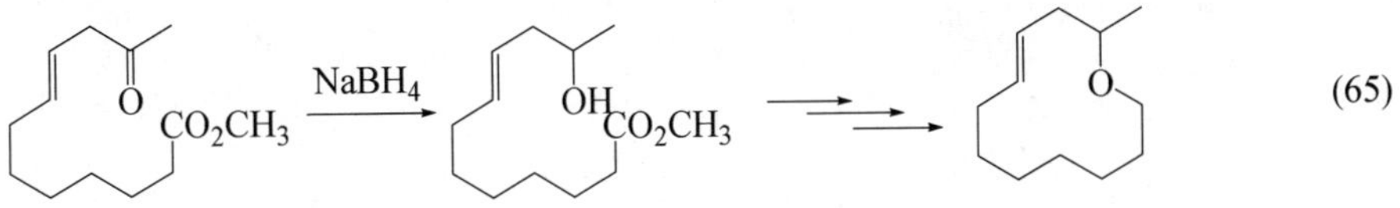

[式 (65) 反应图]

人工 HMG-CoA 还原酶抑制剂 NK-104 会降低血液中的胆固醇水平，因而被用于治疗高胆固醇血症。其合成的关键步骤之一是钯催化的烯基硅试剂与芳基碘的交叉偶联。反应如下式所示：首先，二甲基一氯硅烷与端炔在铂催化剂催化下加成生成烯基硅试剂，生成的硅试剂原位与芳基碘在常规 Hiyama 催化反应体系下进行反应生成交叉偶联产物 H-2，两步收率高达 80%；得到的 H-2 再经三氟乙酸脱叔丁基并环合生成 NK-104[39]。

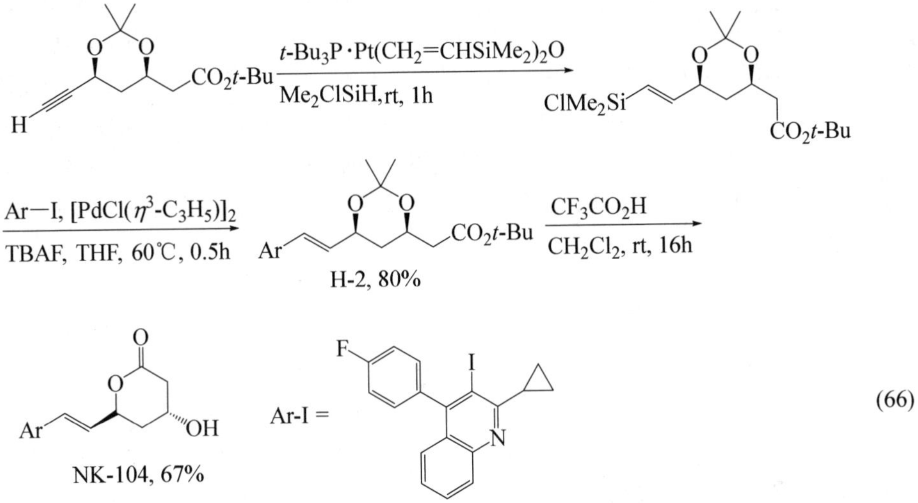

[式 (66) 反应图]

异软骨藻酸 H（isodomoic acid H）属于类胡萝卜素氨基酸家族中的一员，是神经兴奋剂。在其人工合成中，3-羧甲基脯氨酸片段与 2-甲基-4-戊烯酸酯侧链的结合是通过 Hiyama 烯-烯偶联反应实现的。如反应式(67)所示，反应前体碘代烯与烯基硅试剂在 $Pd_2(dba)_3 \cdot CHCl_3$ 和 $TBAF \cdot 8H_2O$ 存在下反应 1h 即转化完全，以 92%的分离收率得到全保护的异软骨藻酸 H(H-3)，H-3 用 LiOH 皂化、钠汞齐脱对甲苯磺酰基最终转化为异软骨藻酸 H[40]。

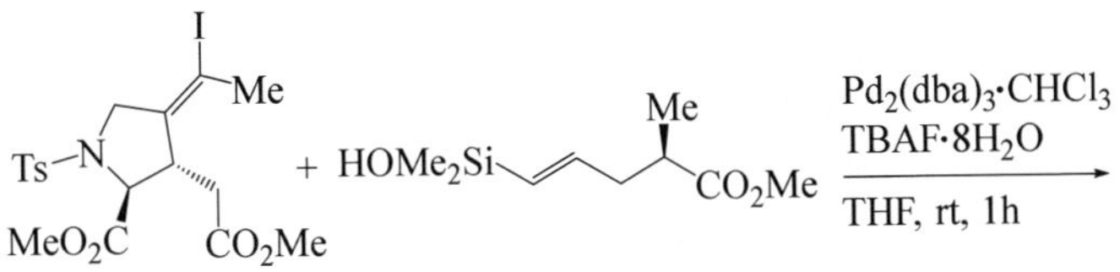

[式 (67) 反应图]

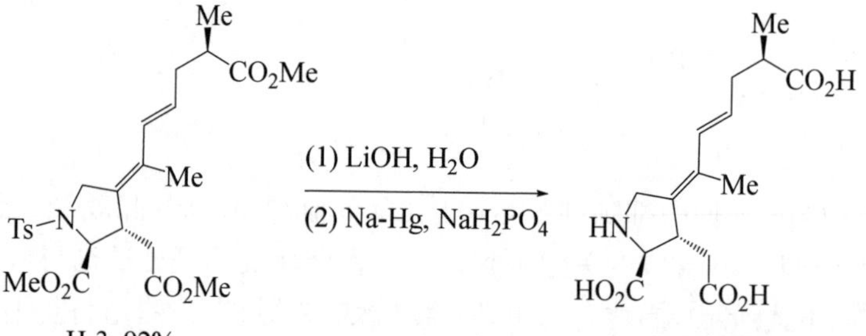

阜孢假丝菌素[(+)-papulacandin D]是从球孢巴氏菌发酵液中分离出的天然产物，结构上含 1,7-二氧螺-[5,4]癸烷骨架，该骨架连接一种衍生自 5-（羟甲基）间苯二酚的芳基-D-C-吡喃糖苷，具有抗生物和抗真菌活性。Hiyama 反应为其合成提供了一条安全可靠的途径。如反应式（68）所示，芳基碘和烯基硅醇在 $Pd_2(dba)_3 \cdot CHCl_3$ 为催化剂、NaOtBu 为活化剂的 Hiyama 催化反应体系下于 50℃反应 5h，以 82%的收率得到交叉偶联产物 H-4，H-4 用二异丁基氢化铝（DIBAL-H）脱新戊酰基(Piv)得关键中间体 KI-1，KI-1 经多步转化得关键中间体 KI-2，KI-2 再与中间体 KI-3 缩合得中间体 KI-4，KI-4 用 $HF \cdot Et_3N$ 脱三甲基硅乙氧羰基(TEOC)保护基最终得阜孢假丝菌素[41]。

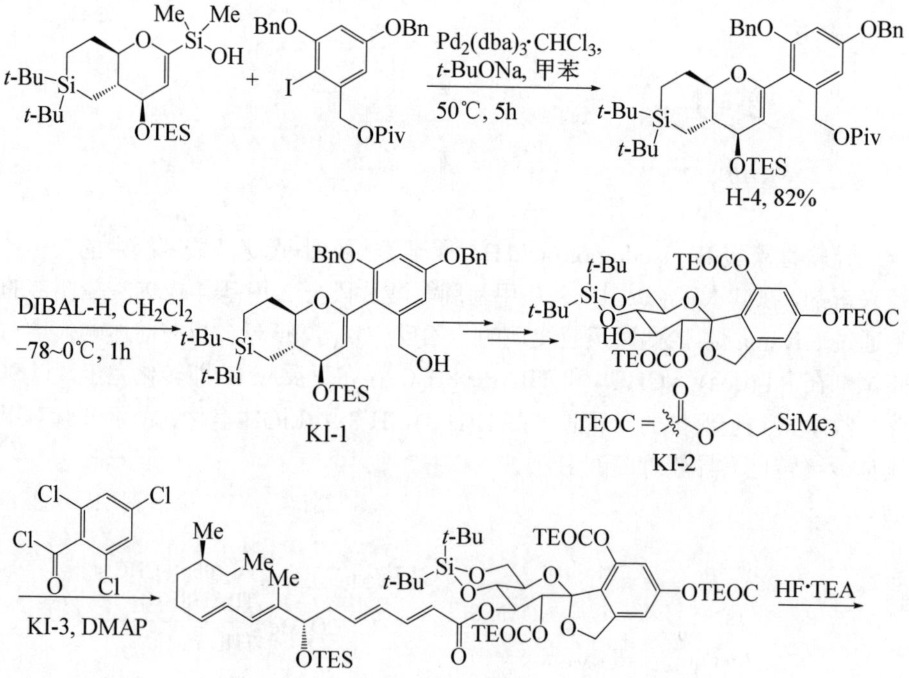

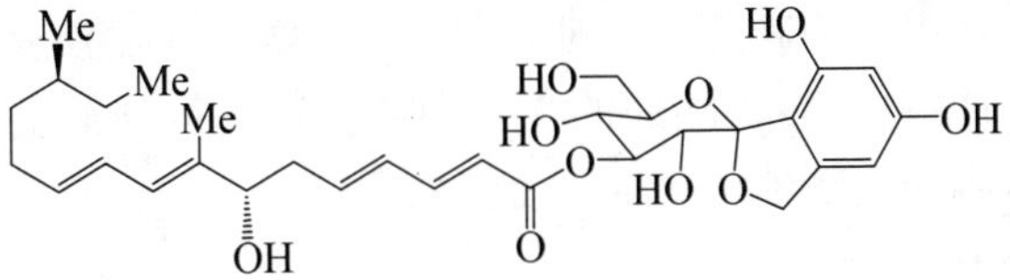

(68)

阜孢假丝菌素

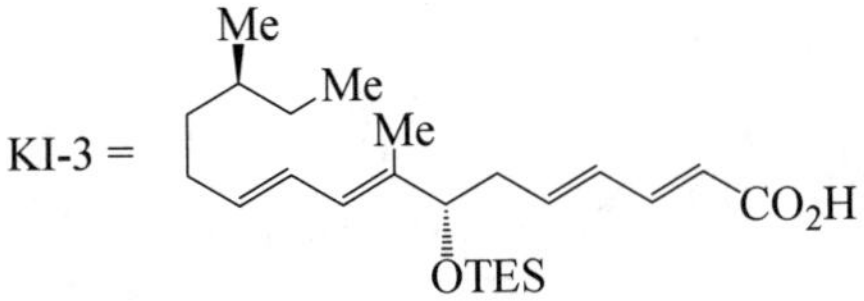

KI-3 =

　　(+)-巴西炔〔(+)-Brasilenyne〕是从海兔（*Aplysia brasiliana*）的消化腺中分离出来的一种天然产物，具有显著的拒食活性。其结构特征是含有 1,3-*cis*，*cis*-二烯结构片段的九元环醚，其合成是一个挑战性课题。利用分子内 Hiyama 反应巧妙构建了该环醚结构，从而打通了全合成(+)-巴西炔的途径。进行分子内 Hiyama 反应的前提是合成与环结构碳原子数相匹配的同时含有碘代烯和环烯硅醚的中间体。如反应式（69）所示，含有碘代烯丙基结构单元的烯醚在 Schrock 催化剂[42]催化下与乙烯基二甲基氯硅烷反应生成同时含有碘代烯和环烯硅醚结构单元的有机硅化合物，将该化合物缓慢滴加到[η^3-C_3H_5PdCl]$_2$ 为催化剂和 TBAF 为活化剂的催化反应体系中，顺利进行分子内交叉偶联得到相应的含有 1,3-*cis*，*cis*-二烯结构片段的九元环醚 H-5，H-5 经多步转化最终生成(+)-巴西炔[43]。

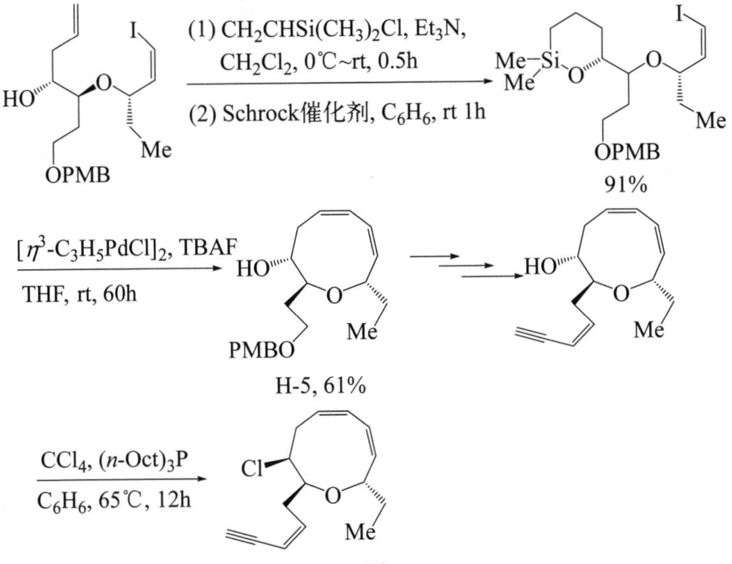

(69)

　　依泽替米贝（ezetimibe）是选择性胆固醇吸收抑制剂，临床上用于治疗高胆固醇症，对其结构修饰有望发现疗效更好的新药。中国科学技术大学王细胜教授课题组利用硅氧烷化的依泽替米贝与溴氟乙酸乙酯间的交叉偶联很容易得到氟烷基化依泽替米贝 H-6，反应在氮配体协助的镍催化剂催化下进行[44]。

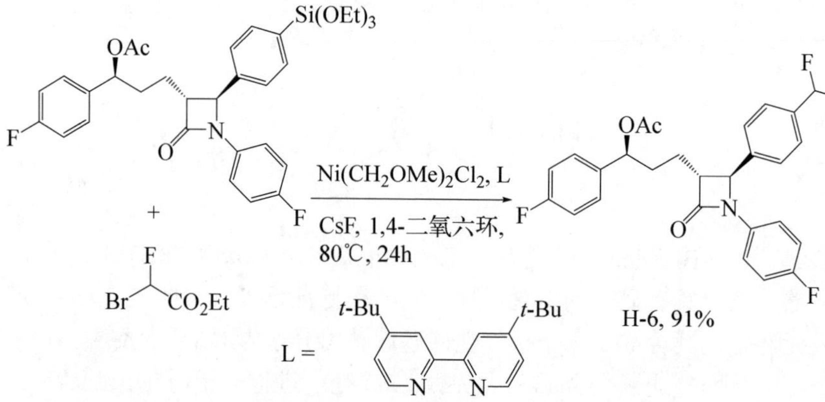

布洛芬（ibuprofen）为非甾体抗炎药，通过抑制环氧化酶，减少前列腺素的合成，产生镇痛、抗炎作用。其合成路线众多，其中之一是利用 Hiyama 交叉偶联反应进行合成。首先，在活化剂 TBAF 存在下 Pd(PPh$_3$)$_4$ 催化 4-异丁酰基溴苯与(E)-2-丁烯基三氟化硅进行 γ-位交叉偶联生成关键中间体 3-(4-异丁酰苯基)-1-丁烯（H-6），H-6 再经碳碳双键氧化、羰基还原生成布洛芬[31]。

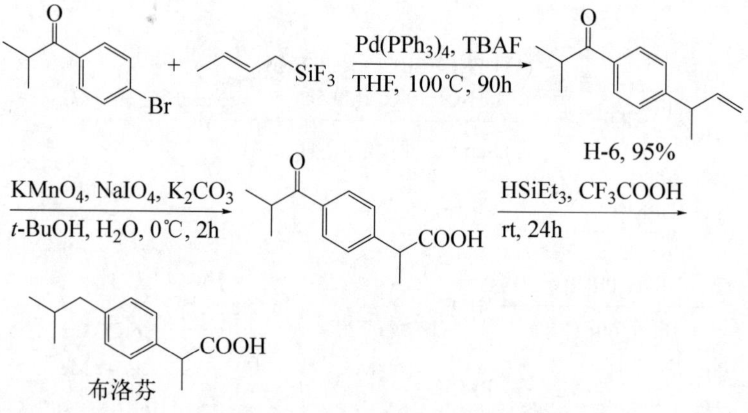

维生素 A（vitamin A）在包括视力、细胞生长和分化、生殖、胚胎发育，以及免疫反应等许多生物过程中起决定性作用，因而其合成一直受到关注。其关键中间体 H-7 可利用 Hiyama 交叉偶联反应方便合成。如下反应所示，四氢吡喃(THP)保护的 3-甲基-4-戊炔-2-烯醇与四甲基二硅氧烷加成生成相应的 1,3-二烯硅烷中间体，该中间体原位与三烯基碘在钯催化剂催化下反应生成交叉偶联产物 H-7，H-7 脱 THP 保护基得维生素 A[45]。

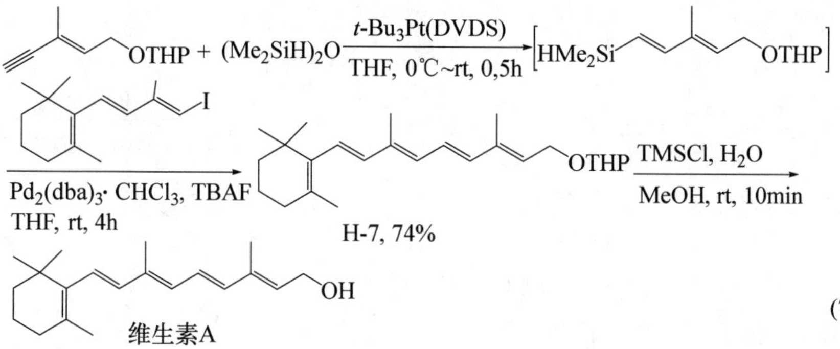

(+)-leustroducsin B 是磷霉素家族中的一员，从一种土壤细菌的培养基中分离得到，具有抗菌、抗真菌以及抗肿瘤活性。其合成路线冗长，但关键中间体由 Hiyama 交叉偶联反应得到。通过逆合成分析，要完成 Leustroducsin B 的合成，需先合成三个结构片段，即 (E)-乙烯基碘 2、炔基硅试剂 3 和 (Z)-乙烯基碘 4。首先，片段 2 由市售的 (4R)-N-丁酰基-4-苄基噁唑烷酮和丙醛硅试剂经五步反应合成；片段 3 由二乙氧基乙酸乙酯与甲酸甲酯为起始原料经六步反应得到；片段 4 以环己烯甲酸内酯和氯甲酸甲酯为起始物经六步反应生成。所合成的片段 2 与片段 3 进行偶联反应得中间体 A，A 再经两步反应生成中间体 B。B 与片段 4 进行 Hiyama 交叉偶联得关键中间体 H-8。H-8 再经五步反应最终生成目标产物 (+)-leustroducsin B[46]。

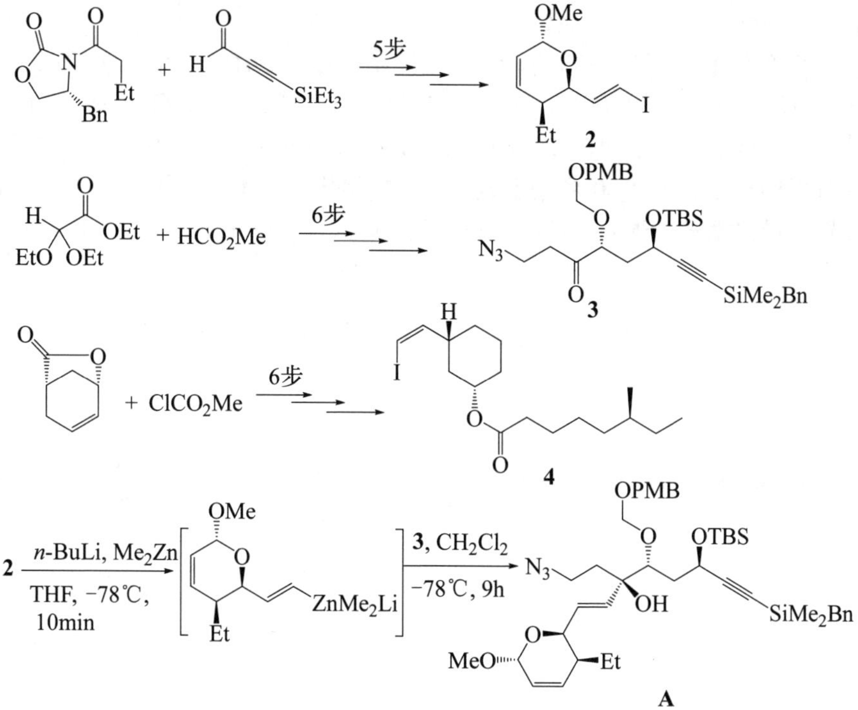

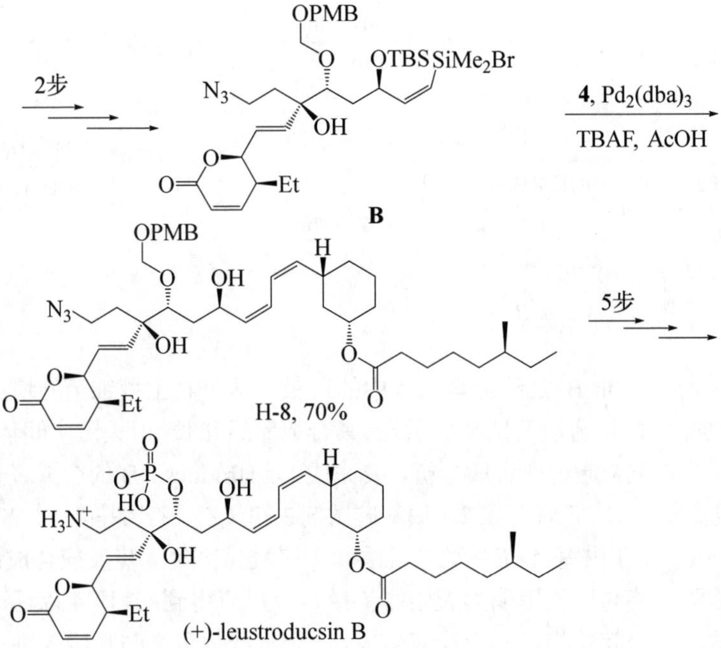

(—)-lasonolide A 是从加勒比海海洋海绵 Forcepia sp. 中提取的一种聚酮化合物,对多种癌细胞有杀伤作用,从自然界获取该化合物非常有限。采用人工合成是满足其临床需求的有效手段。其人工合成的关键步骤如反应式(74)所示,烯基硅试剂与乙酸烯丙酯在传统 Hiyama 催化体系催化下进行 sp^2-sp^3 交叉偶联生成中间体 H-9,H-9 中酯基在 LiOH 碱性介质中皂化水解生成中间体 **C**,中间体 **C** 中的乙烯基与另一含炔基的中间体 **D** 中的乙炔基在丙酮溶剂中经钌催化剂催化发生偶联,同时丙酮与 C21 和 C30 羟基生成具有缩酮结构的关键中间体 **E**。中间体 **E** 再经多步转化最终生成二十元大环内酯结构的 lasonolide A[47]。

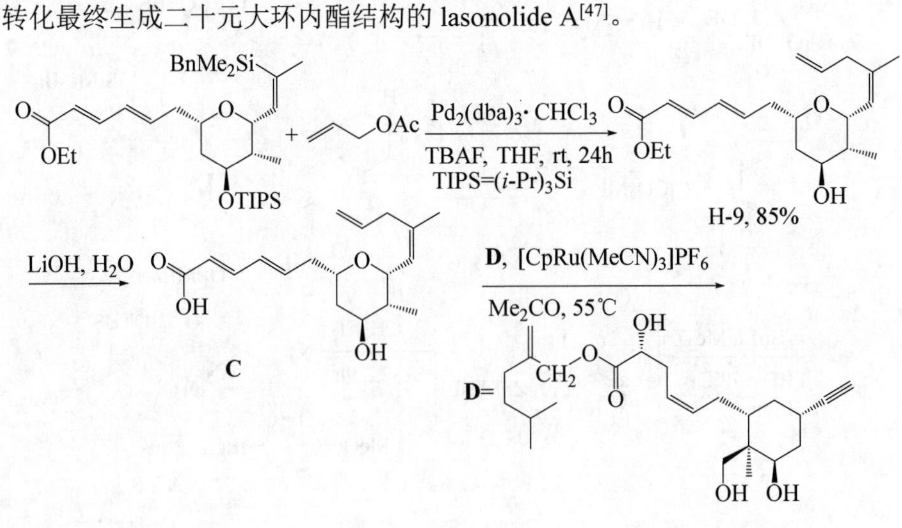

(−)-lasonolide A

参考文献

[1] Yoshida J, Tamao K, Yamamoto H, et al. Organofluorosilicates in organic synthesis. 14. Carbon-carbon bond formation promoted by palladium salts[J]. Organometallics, 1982, 1(3): 542-549.

[2] Nakao Y, Hiyama T. Silicon-based cross-coupling reaction: an environmentally benign version[J]. Chemical Society Reviews, 2011, 40(10): 4893-4901.

[3] Nishihara Y, Ikegashira K, Mori A, et al. Copper(Ⅰ)-catalyzed cross-coupling reaction of alkynylsilanes with 1-chloroalkynes[J]. Tetrahedron Letters, 1998, 39(23): 4075-4078.

[4] Hatanaka Y, Hiyama T. Cross-coupling of organosilanes with organic halides mediated by a palladium catalyst and tris(diethylamino)sulfonium difluorotrimethylsilicate[J]. The Journal of Organic Chemistry, 1988, 53(4): 918-920.

[5] Hatanaka Y, Hiyama T. Highly selective cross-coupling reactions of organosilicon compounds mediated by fluoride ion and a palladium catalyst[J]. Synlett, 1991, (12): 845-853.

[6] Nishihara Y, Noyori S, Okamoto T, et al. Copper-catalyzed Sila-Sonogashira-Hagihara cross-coupling reactions of alkynylsilanes with aryliodides under palladium-free conditions[J]. Chemistry Letters, 2011, 40(9): 972-974.

[7] Singh C, Prakasham A P, Gangwar M K, et al. One-pot tandem Hiyama alkynylation/cyclizations by palladium(Ⅱ) acyclic diaminocarbene (ADC) complexes yielding biologically relevant benzofuran scaffolds[J]. ACS Omega, 2018, 3(2): 1740-1756.

[8] Nishihara Y, Ikegashira K, Hirabayashi K, et al. Coupling reactions of alkynylsilanes mediated by a Cu(Ⅰ) salt: novel syntheses of conjugate diynes and disubstituted ethynes[J]. The Journal of Organic Chemistry, 2000, 65(6): 1780-1787.

[9] Hatanaka Y, Hiyama T. Alkenylfluorosilanes as widely applicable substrates for the palladium-catalyzed coupling of alkenylsilane/fluoride reagents with alkenyl iodides[J]. The Journal of Organic Chemistry, 1989, 54(2): 268-270.

[10] Vyvyan J R, Engles C A, Bray S L, et al. Synthesis of substituted Z-styrenes by Hiyama-type coupling of oxasilacycloalkenes: application to the synthesis of a 1-benzoxocane[J]. Beilstein Journal of Organic Chemistry, 2017, 13(1): 2122-2127.

[11] Nakao Y, Sahoo A K, Imanaka H, et al. Alkenyl-and aryl[2-(hydroxymethyl)phenyl] dimethylsilanes: tetraorganosilanes for the practical cross-coupling reaction[J]. Pure and Applied Chemistry, 2006, 78(2): 435-440.

[12] Sore H F, Galloway W R J D, Spring D R. Palladium-catalysed cross-coupling of organosilicon reagents[J]. Chemical Society Reviews, 2012, 41(5): 1845-1866.

[13] Denmark S E, Wehrli D. Highly stereospecific, palladium-catalyzed cross-coupling of alkenylsilanols[J]. Organic Letters, 2000, 2(4): 565-568.

[14] Hagiwara E, Gouda K, Hatanaka Y, et al. NaOH-Promoted cross-coupling reactions of organosilicon compounds with organic halides: practical routes to biaryls, alkenylarenes and conjugated dienes[J]. Tetrahedron Letters, 1997, 38(3): 439-442.

[15] Clarke M L. First microwave-accelerated Hiyama coupling of aryl-and vinylsiloxane derivatives: clean crosscoupling of aryl chlorides within minutes[J]. Advanced Synthesis & Catalysis, 2005, 347(2-3): 303-307.

[16] Gordillo Á, de Jesús E, López-Mardomingo C. Consecutive palladium-catalyzed Hiyama-Heck reactions in aqueous media under ligand-free conditions[J]. Chemical Communications, 2007, (39): 4056-4058.

[17] Denmark S E, Choi J Y. Highly stereospecific, cross-coupling reactions of alkenylsilacyclobutanes[J]. Journal of the American Chemical Society, 1999, 121(24): 5821-5822.

[18] Hatanaka Y, Goda K, Okahara Y, et al. Highly selective cross-coupling reactions of aryl(halo)silanes with aryl halides: a general and practical route to functionalized biaryls[J]. Tetrahedron, 1994, 50(28): 8301-8316.

[19] Mowery M E, DeShong P. Cross-coupling reactions of hypervalent siloxane derivatives: an alternative to Stille and Suzuki couplings[J]. The Journal of Organic Chemistry, 1999, 64(5): 1684-1688.

[20] Yang J, Wang L. Synthesis and characterization of dinuclear NHC–palladium complexes and their applications in the Hiyama reactions of aryltrialkyoxysilanes with aryl chlorides[J]. Dalton Transactions, 2012, 41(39): 12031-12037.

[21] Gurung S K, Thapa S, Vangala A S, et al. Copper-catalyzed Hiyama coupling of (hetero) aryltriethoxysilanes with (hetero)aryliodides[J]. Organic Letters, 2013, 15(20): 5378-5381.

[22] Hirabayashi K, Mori A, Kawashima J, et al. Palladium-catalyzed cross-coupling of silanols, silanediols, and silanetriols promoted by silver(I) oxide[J]. The Journal of Organic Chemistry, 2000, 65(17): 5342-5349.

[23] Pierrat P, Gros P, Fort Y. Hiyama cross-coupling of chloro-, fluoro-, and methoxypyridyltrimethylsilanes: room-temperature novel access to functional bi(het)aryl[J]. Organic Letters, 2005, 7(4): 697-700.

[24] Shirbin S J, Boughton B A, Zammit S C, et al. Copper-free palladium-catalyzed Sonogashira and Hiyama cross-couplings using aryl imidazol-1-ylsulfonates[J]. Tetrahedron Letters, 2010, 51(22): 2971-2974.

[25] Denmark S E, Wu Z. Synthesis of unsymmetrical biaryls from arylsilacyclobutanes[J]. Organic Letters, 1999, 1(9): 1495-1498.

[26] Hatanaka Y, Hiyama T. Pentacoordinate organosilicate as an alkylating reagent: palladium catalyzed methylation of aryl halides[J]. Tetrahedron Letters, 1988, 29(1): 97-98.

[27] Matsuhashi H, Kuroboshi M, Hatanaka Y, et al. Palladium catalyzed cross-coupling reaction of functionalized alkyltrifluorosilanes with aryl halides[J]. Tetrahedron Letters, 1994, 35(35): 6507-6510.

[28] Cornelissen L, Cirriez V, Vercruysse S, et al. Copper-catalyzed Hiyama cross-coupling using vinylsilanes and benzylic electrophiles[J]. Chemical Communications, 2014, 50(59): 8018-8020.

[29] Fier P S, Hartwig J F. Copper-mediated difluoromethylation of aryl and vinyl iodides[J]. Journal of the American Chemical Society, 2012, 134(12): 5524-5527.

[30] Strotman N A, Sommer S, Fu G C. Hiyama reactions of activated and unactivated secondary alkyl halides catalyzed by a nickel/norephedrine complex[J]. Angewandte Chemie, 2007, 119(19): 3626-3628.

[31] Hatanaka Y, Ebina Y, Hiyama T. .gamma.-Selective cross-coupling reaction of allyltrifluorosilanes: a new approach to regiochemical control in allylic systems[J]. Journal of the American Chemical Society, 1991, 113(18): 7075-7076.

[32] Srimani D, Bej A, Sarkar A. Palladium nanoparticle catalyzed Hiyama coupling reaction of benzyl halides[J]. The Journal of Organic Chemistry, 2010, 75(12): 4296-4299.

[33] Cornelissen L, Vercruysse S, Sanhadji A, et al. Copper-catalyzed vinylsilane allylation[J]. European Journal of Organic Chemistry, 2014, (1): 35-38.

[34] Lévêque C, Corcé V, Chenneberg L, et al. Photoredox/nickel dual catalysis for the $C(sp^3)$-$C(sp^3)$ cross-coupling of alkylsilicates with alkyl halides[J]. European Journal of Organic Chemistry, 2017, (15): 2118-2121.

[35] Dai X, Strotman N A, Fu G C. Catalytic asymmetric Hiyama cross-couplings of racemic α-bromo esters[J]. Journal of the American Chemical Society, 2008, 130(11): 3302-3303.

[36] Varenikov A, Gandelman M. Synthesis of chiral α-trifluoromethyl alcohols and ethers via enantioselective Hiyama cross-couplings of bisfunctionalized electrophiles[J]. Nature Communications, 2018, 9(1): 3566-3571.

[37] Denmark S E, Yang S M. Intramolecular silicon-assisted cross-coupling reactions: general synthesis of medium-sized rings containing a 1, 3-cis-cis diene unit[J]. Journal of the American Chemical Society, 2002, 124(10): 2102-2103.

[38] Qin Y, Han J L, Ju C W, et al. Ring expansion to 6-, 7-, and 8-membered benzosilacycles through strain-release silicon-based cross-coupling[J]. Angewandte Chemie International Edition, 2020, 59(22): 8481-8485.

[39] Takahashi K, Minami T, Ohara Y, et al. Synthesis of an artificial HMG-CoA reductase inhibitor NK-104 via a hydrosilylation–cross-coupling reaction[J]. Bulletin of the Chemical Society of Japan, 1995, 68(9): 2649-2656.

[40] Denmark S E, Liu J H C, Muhuhi J M. Total syntheses of isodomoic acids G and H[J]. Journal of the American Chemical Society, 2009, 131(40): 14188-14189.

[41] Denmark S E, Regens C S, Kobayashi T. Total synthesis of Papulacandin D[J]. Journal of the American Chemical Society, 2007, 129(10): 2774-2776.

[42] Fox H H, Yap K B, Robbins J, et al. Simple, high yield syntheses of molybdenum(VI) bis(imido) complexes of the type $Mo(NR)_2Cl_2$(1,2-dimethoxyethane)[J]. Inorganic Chemistry, 1992, 31(11): 2287-2289.

[43] Denmark S E, Yang S M. Intramolecular silicon-assisted cross-coupling: total synthesis of (+)-brasilenyne[J]. Journal of the American Chemical Society, 2002, 124(51): 15196-15197.

[44] Wu Y, Zhang H R, Cao Y X, et al. Nickel-catalyzed monofluoroalkylation of arylsilanes via hiyama cross-coupling[J]. Organic Letters, 2016, 18(21): 5564-5567.

[45] Montenegro J, Bergueiro J, Saa C, et al. Hiyama cross-coupling reaction in the stereospecific synthesis of retinoids[J]. Organic Letters, 2009, 11(1): 141-144.

[46] Trost B M, Biannic B, Brindle C S, et al. A Highly convergent total synthesis of leustroducsin B[J]. Journal of the American Chemical Society, 2015, 137(36): 11594-11597.

[47] Trost B M, Stivala C E, Hull K L, et al. A concise synthesis of (−)-lasonolide A[J]. Journal of the American Chemical Society, 2014, 136(1): 88-91.

第 8 章
Tsuji-Trost 反应及其在药物合成上的应用

8.1 Tsuji-Trost 反应及其机理

8.1.1 Tsuji-Trost 反应

经典的 Tsuji-Trost 反应，是指在钯催化剂催化下亲核试剂对烯丙基化合物进行亲核取代得到烯丙基化产物的反应。此类反应是通过 π-烯丙基中间体进行的，其反应如图 8-1 所示。

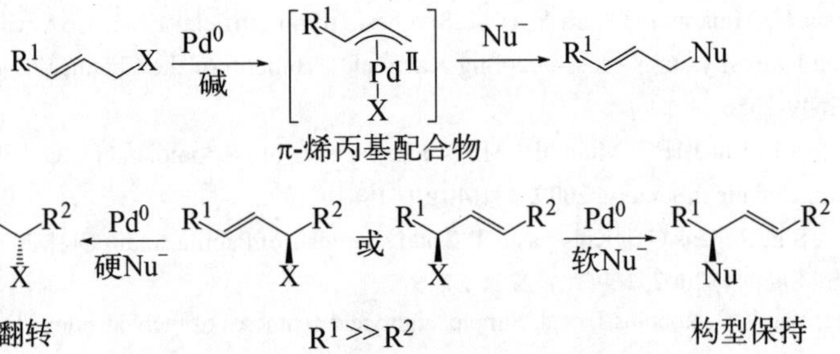

X=OCOR, OCO_2R, OCONHR, $OP(O)(OR)_2$, OPh, Cl, NO_2, SO_2Ph, NR_3X, SR_2X, OH

HNu= 活泼亚甲基类配合物、烯醇类、胺、醇等

图 8-1 Tsuji-Trost 反应及范围

烯丙基底物很广泛，除了氯、磺酰基等传统的离去基团外，其他吸电子基也可作为离去基团，甚至在经典亲核取代反应中不易离去的羟基都可被亲核试剂取代。图 8-2 是经典 Tsuji-Trost 反应常见的烯丙基底物及相应的离去基团。这些离去基团的反应活性顺序不同，氯作为离去基团反应活性最高，甚至在无催化剂的情况下反应都能进行；羟基直接作为离去基团反应较难进行，烯丙醇酯，通常为乙酸烯丙酯，

则能很顺利地进行反应,而烯丙醇磷酸酯比乙酸酯更活泼。然而,这些烯丙基底物与软亲核试剂的反应需要在碱存在下才能进行。

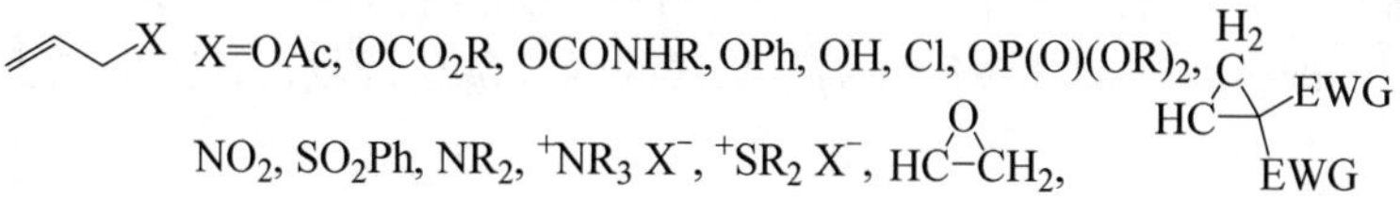

$$\text{烯丙基-X} \quad X=OAc, OCO_2R, OCONHR, OPh, OH, Cl, OP(O)(OR)_2, \text{C(EWG)(EWG)H}, NO_2, SO_2Ph, NR_2, {}^+NR_3\ X^-, {}^+SR_2\ X^-, \text{环氧乙烷}$$

图 8-2　常见烯丙基底物及离去基团

相比而言,烯丙基甲基碳酸酯、烯丙基氨基甲酸酯、烯丙基芳基醚、乙烯基环氧化物反应活性更高,反应在无碱存在的中性条件下就能进行,这在涉及对酸、碱敏感反应物的有机合成上是非常重要的。之所以烯丙基甲基碳酸酯能够在无碱存在的中性条件下进行反应,是因为烯丙基甲基碳酸酯与钯氧化加成过程中伴随脱羧生成烯丙基钯甲氧化物,此时生成的甲氧基负离子从亲核试剂(NuH)上夺取氢离子,起到外加碱的作用(图 8-3)。而且,这种氧化加成脱羧反应是不可逆的,这也是反应得以顺利进行的原因之一;相反,烯丙基醋酸酯与钯加成生成烯丙基钯醋酸盐是可逆的,反应不容易进行。

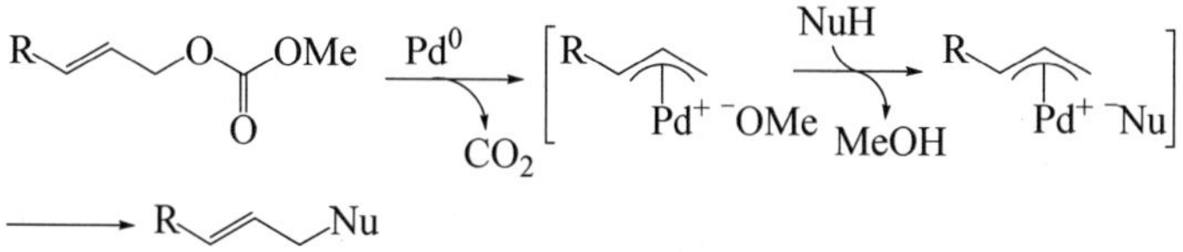

图 8-3　钯催化的烯丙基甲基碳酸酯与亲核试剂的烯丙基化反应

乙烯基环氧化物作为活泼烯丙基化试剂则另有原因。如图 8-4 所示,乙烯基环氧化物氧化加成到 Pd^0 伴随环氧环的开环生成结构上含烷氧基负离子的 π-烯丙基钯配合物,配合物上的烷氧基负离子起碱的作用从亲核试剂(NuH)攫取一个质子,生成的 Nu^- 再进攻 π-烯丙基远离烷氧基的碳原子,最终主要生成 1,4-加成产物,而非 1,2-加成产物[1]。

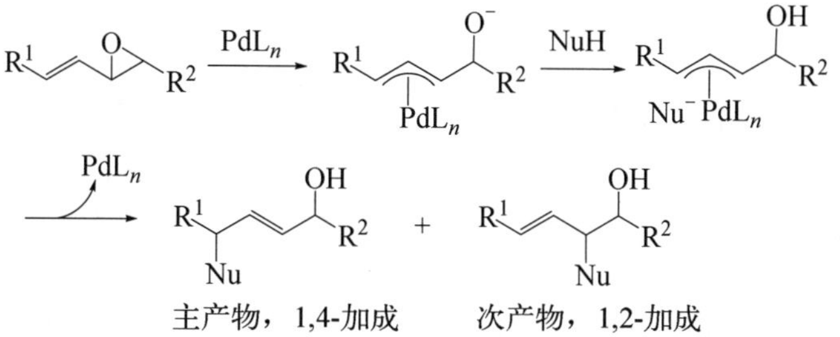

主产物,1,4-加成　　　次产物,1,2-加成

图 8-4　钯催化乙烯基环氧化物对亲核试剂的烯丙基化反应

在 Tsuji-Trost 反应中，亲核试剂分为硬亲核试剂和软亲核试剂。硬亲核试剂通常为格氏试剂等主族金属有机化合物，而软亲核试剂包括含活泼亚甲基的化合物、烯醇盐、烯胺等化合物（图 8-5）。当不对称的烯丙基化合物（$R^1 \gg R^2$）作为烯丙基化试剂时，硬亲核试剂进攻得到构型翻转的烯丙基化产物，而软亲核试剂进攻得到构型保持的产物。

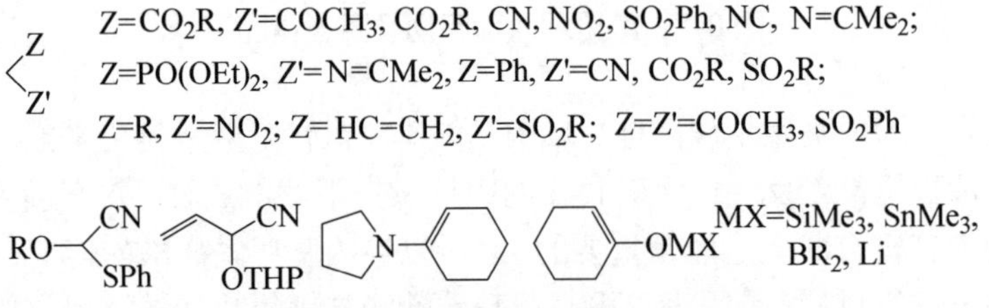
Z
Z'
Z=CO2R, Z'=COCH3, CO2R, CN, NO2, SO2Ph, NC, N=CMe2;
Z=PO(OEt)2, Z'=N=CMe2, Z=Ph, Z'=CN, CO2R, SO2R;
Z=R, Z'=NO2; Z= HC=CH2, Z'=SO2R; Z=Z'=COCH3, SO2Ph
RO
CN
SPh
CN
OTHP
N
OMX
MX=SiMe3, SnMe3,
BR2, Li

图 8-5　软亲核试剂的范围

钯最早被发现可用以催化 Tsuji-Trost 反应，而且至今仍是最主要的催化剂，后来发现其他一些金属包括银、镍、钌、铑、铱、铁以及钼等过渡金属也可作为 Tsuji-Trost 反应的催化剂[2-5]。

8.1.2　反应机理

钯催化的 Tsuji-Trost 反应的机理可由图 8-6 所示的催化循环来描述[6]。首先，催化剂前体在反应体系中转化为零价钯活性物种 **A**；一旦 **A** 生成，其与烯丙基底物配位生成 π-配合物 **B**；**B** 中配位的烯丙基发生C-X键断裂并与钯原子进行氧化加成生成 η^3-烯丙基配合物 **C**，体系中的配体与离去基团交换生成另一 η^3-烯丙基配合物 **D**，**C** 与 **D** 处于动态平衡状态；此时，脱质子的亲核试剂进攻空间位阻小的烯丙基碳与烯丙基偶联生成 π-配合物 **E**；**E** 中配位的偶联产物烯烃从金属钯上脱离生成烯丙基化产物并再生活性物种 **A**，完成催化循环。从催化循环图中可以看出，配体对催化剂的催化性能具有重要的影响。

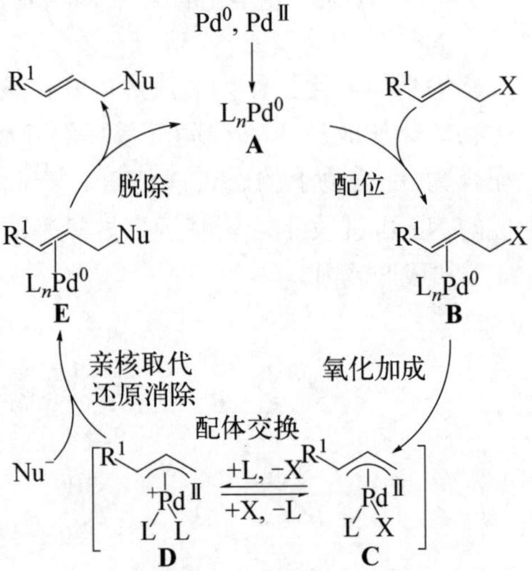
Pd0, PdII
LnPd0
A
R1
Nu
R1
X
脱除
配位
R1
Nu
LnPd0
E
R1
X
LnPd0
B
亲核取代
还原消除
氧化加成
配体交换
Nu−
R1
+Pd II
L L
D
+L, −X
+X, −L
R1
Pd II
L X
C

图 8-6　Tsuji-Trost 反应的催化循环

8.2 基本的 Tsuji-Trost 反应

由上节可以看出，烯丙基底物和亲核试剂的范围广泛，两者之间的组合可以产生庞大数目的 Tsuji-Trost 反应；而且，当同一分子结构中同时存在烯丙基和亲核试剂单元，并且二者处于分子结构的适当位置时，还可以发生分子内 Tsuji-Trost 反应。因此，关于 Tsuji-Trost 反应不可能面面俱到，只介绍碳亲核试剂进攻的一些最基本的反应。

8.2.1 分子间 Tsuji-Trost 反应

8.2.1.1 含活泼亚甲基亲核试剂的反应

当亚甲基连有两个强吸电子基时，亚甲基能够与碱作用生成亚甲基负离子，亲核性增强，因而可作为亲核试剂与带离去基团的烯丙基衍生物在钯等过渡金属催化剂催化下发生亲核取代反应，反应结果生成烯丙基化的亚甲基化合物。

β-羰基羧酸酯、丙二酸二甲酯、β-二酮等是最经典的活泼亚甲基化合物，它们与烯丙基醋酸酯类化合物在钯配合物-膦配体催化体系催化下发生亚甲基烯丙基化反应，反应一般在极性溶剂中进行，而且需要强碱将亚甲基转化为亚甲基负离子[7]。当烯丙基不对称时，如反应式（3）～式（5）所示，反应发生在空间位阻小的烯丙基碳上[8-10]。

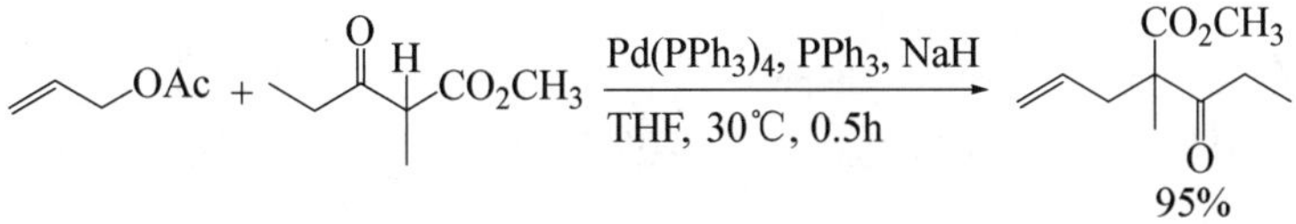

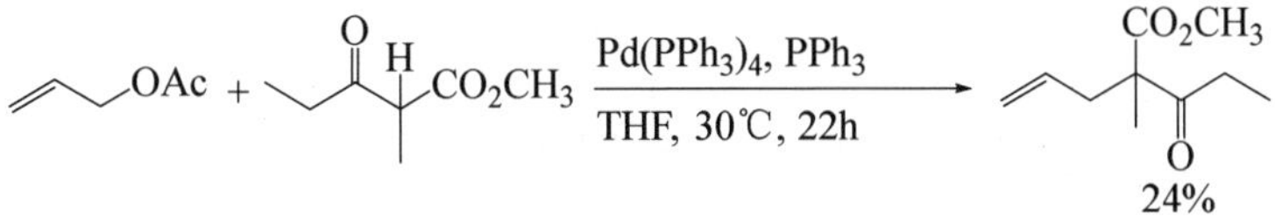

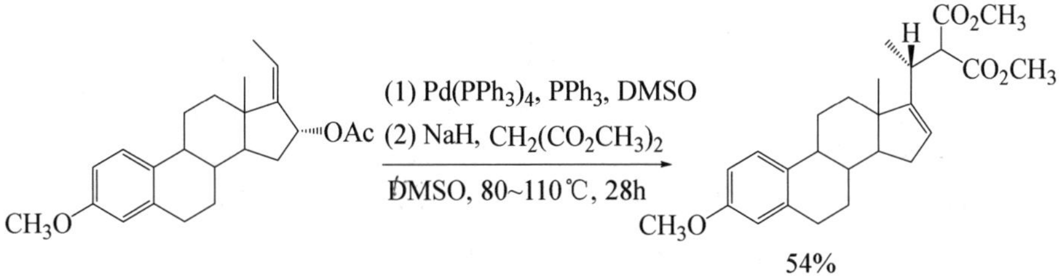

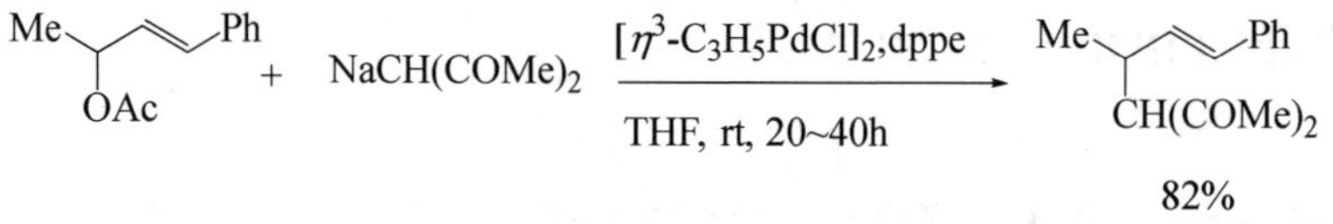

$$\text{Me-CH(OAc)-CH=CH-Ph} + \text{NaCH(COMe)}_2 \xrightarrow[\text{THF, rt, 20~40h}]{[\eta^3\text{-C}_3\text{H}_5\text{PdCl}]_2,\text{dppe}} \text{Me-CH(CH(COMe)}_2)\text{-CH=CH-Ph} \quad 82\% \quad (4)$$

$$\text{AcO-CMe}_2\text{-CH=CH-CH=CH}_2 + \text{NaCMe(CO}_2\text{Et)}_2 \xrightarrow[\text{THF, 40℃, 3h}]{\text{Pd(OAc)}_2, \text{PBu}_3} \text{产物} \quad 83\% \quad (5)$$

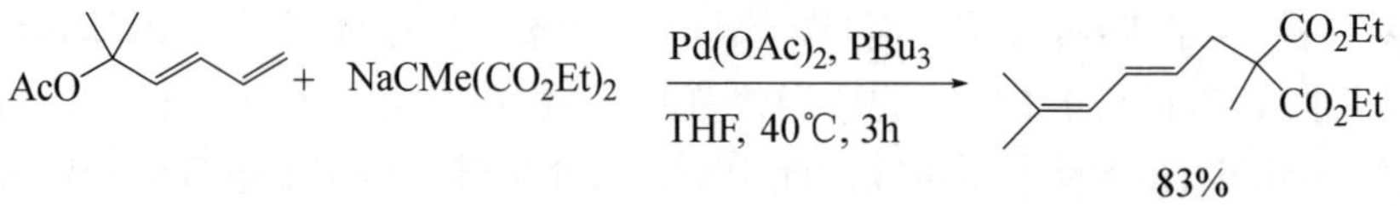

另外，如反应式（6）和式（7），E-构型的烯丙基醋酸酯与丙二酸二甲酯负离子反应后，与乙酰氧基相连手性碳的构型在反应后得到保持，而 Z-构型的异构体，相同的手性中心的构型反应后发生翻转；对比反应式（7）与式（8），Z-构型的烯丙基醋酸酯异构体与丙二酸二甲酯负离子反应得到相同的反应结果[11]。

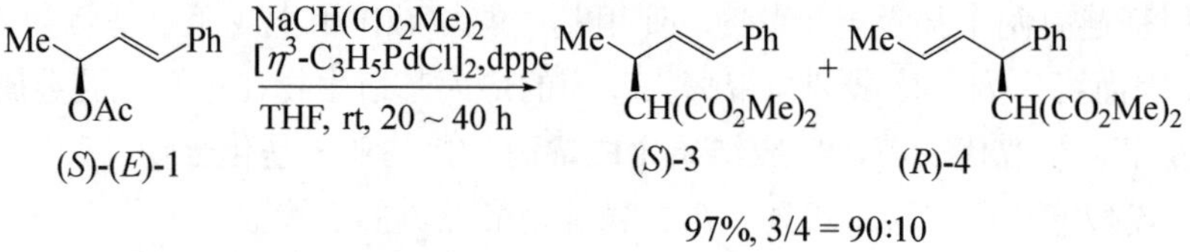

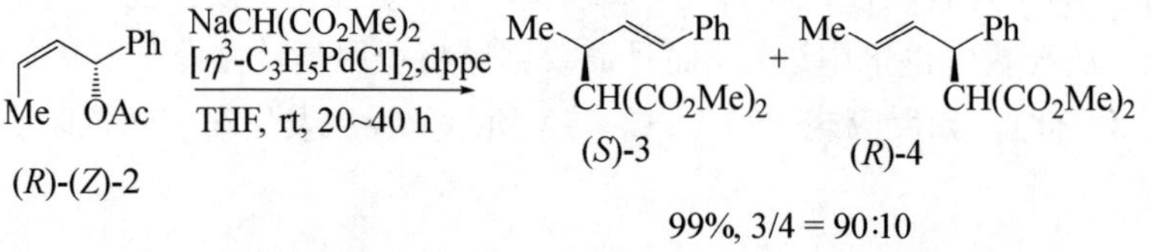

$$(S)\text{-}(E)\text{-}1 \xrightarrow[\text{THF, rt, 20~40 h}]{\text{NaCH(CO}_2\text{Me)}_2,\ [\eta^3\text{-C}_3\text{H}_5\text{PdCl}]_2,\text{dppe}} (S)\text{-}3 + (R)\text{-}4 \quad 97\%,\ 3/4=90:10 \quad (6)$$

$$(R)\text{-}(Z)\text{-}2 \xrightarrow[\text{THF, rt, 20~40 h}]{\text{NaCH(CO}_2\text{Me)}_2,\ [\eta^3\text{-C}_3\text{H}_5\text{PdCl}]_2,\text{dppe}} (S)\text{-}3 + (R)\text{-}4 \quad 99\%,\ 3/4=90:10 \quad (7)$$

$$(R)\text{-}(Z)\text{-}1 \xrightarrow[\text{THF, rt, 20~40 h}]{\text{NaCH(CO}_2\text{Me)}_2,\ [\eta^3\text{-C}_3\text{H}_5\text{PdCl}]_2,\text{dppe}} (S)\text{-}3 + (R)\text{-}4 \quad 97\%,\ 3/4=90:10 \quad (8)$$

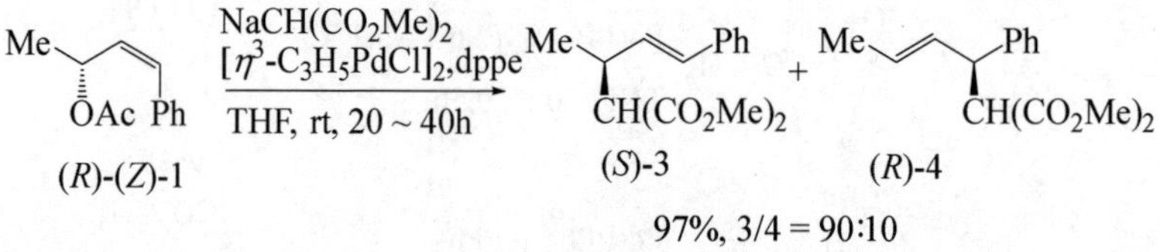

区域异构体相同的反应结果可以归因于图 8-7 所示的反应经由共同的中间体，而且手性中心相反且为 E-构型的烯丙基醋酸酯也按此机理进行反应，这也是反应式（6）、式（7）以及式（8）结果完全相同的原因[8]。

需要明确的是，多数情况下烯丙醇醋酸酯与软亲核试剂的 Tsuji-Trost 反应手性碳原子的构型得到保持。以相反构型的乙酰氧基环己烯羧酸甲酯为例，与丙二酸二甲酯负离子反应后手性中心的构型不发生变化，原来的 cis-反应物得到 cis-产物，而 trans-反应物得到 trans-产物。这一立体化学结果可解释为两次相继的构型反转。如图 8-8 所示，首先钯取代离去基团发生一次构型反转，紧接着亲核试剂从 exo-面进攻再发生一次构型反转，总的结果是构型得到保持。

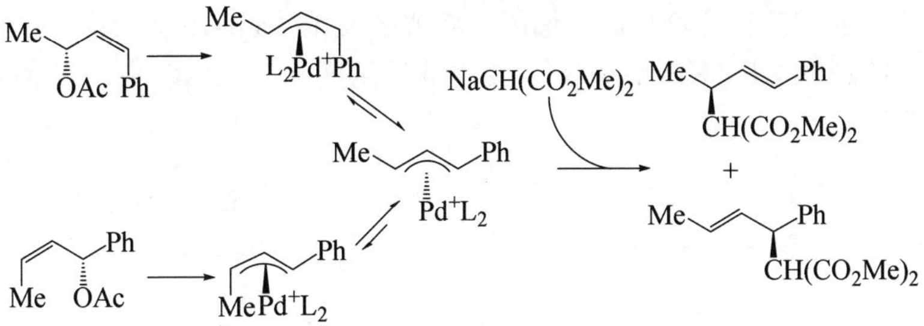

图 8-7　区域异构体与亲核试剂反应结果相同的机理

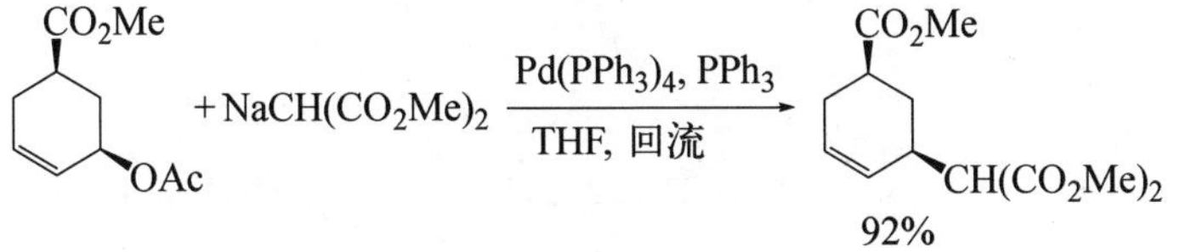

(9)

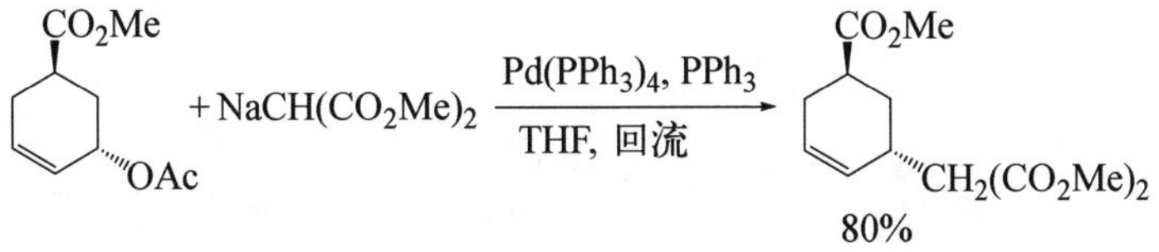

(10)

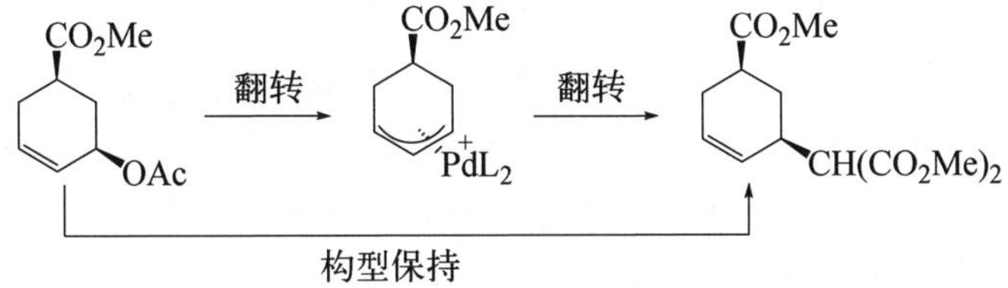

图 8-8　构型保持原理

当有手性配体存在并严格选择溶剂和控制反应条件时，上述 β-羰基酸酯能够实现不对称烯丙基化，如下反应以优异的收率和 86%的对映体过量值得到光学活性的烯丙基化产物[12]。

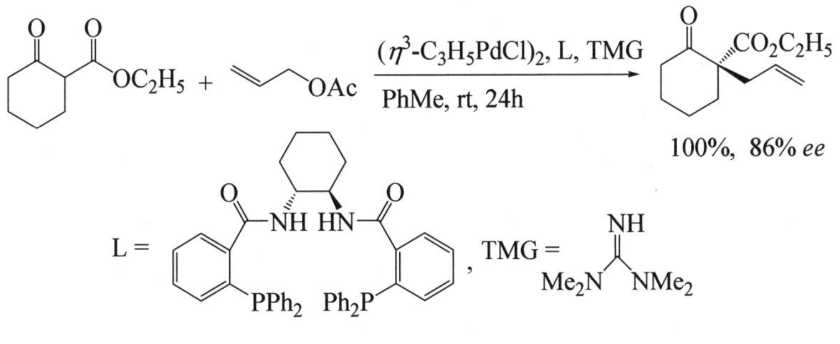

(11)

其他活泼亚甲基化合物也能与烯丙基醋酸酯发生 Tsuji-Trost 反应，反应以负载于 Al_2O_3 上的碱性 KF（$KF-Al_2O_3$）为亚甲基活化剂[13]，或以路易斯酸钛酸四异丙酯为亚甲基活化剂[14]。

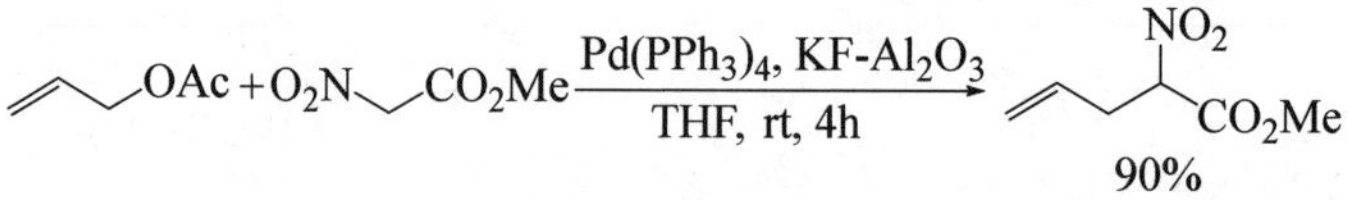

$$\text{allyl-OAc} + O_2N\text{—CH}_2\text{—CO}_2Me \xrightarrow[\text{THF, rt, 4h}]{Pd(PPh_3)_4, KF-Al_2O_3} \underset{90\%}{\text{CH}_2\text{=CHCH}_2\text{CH(NO}_2\text{)CO}_2Me} \quad (12)$$

$$\text{allyl-OAc} + \underset{\text{CN}}{\text{Me—C}_6H_4\text{—SO}_2\text{—CH}_2} \xrightarrow[\text{THF, rt, 1.5h}]{Pd(dppe)_2, KF-Al_2O_3} \underset{75\%}{\text{产物}} \quad (13)$$

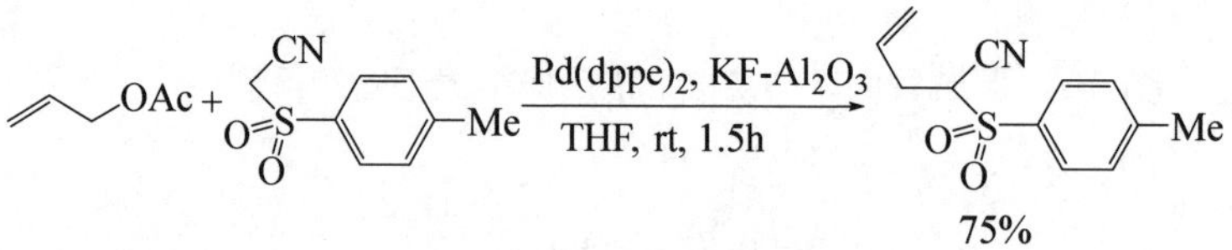

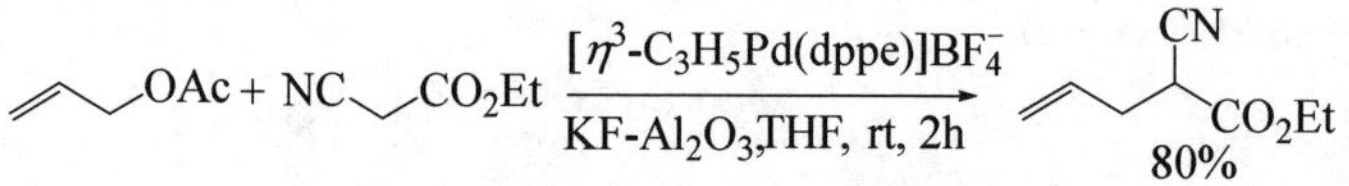

$$\text{allyl-OAc} + NC\text{—CH}_2\text{—CO}_2Et \xrightarrow[\text{KF-Al}_2O_3,\text{THF, rt, 2h}]{[\eta^3-C_3H_5Pd(dppe)]BF_4^-} \underset{80\%}{\text{产物}} \quad (14)$$

$$\text{geranyl-OAc} + \underset{SO_2Ph}{\overset{SO_2Ph}{\text{CH}_2}} \xrightarrow[\text{CH}_2Cl_2, 85℃, 12h]{Pd_2(dba)_3, PPh_3, Ti(OPr-i)_4} \underset{88\%}{\text{产物}} \quad (15)$$

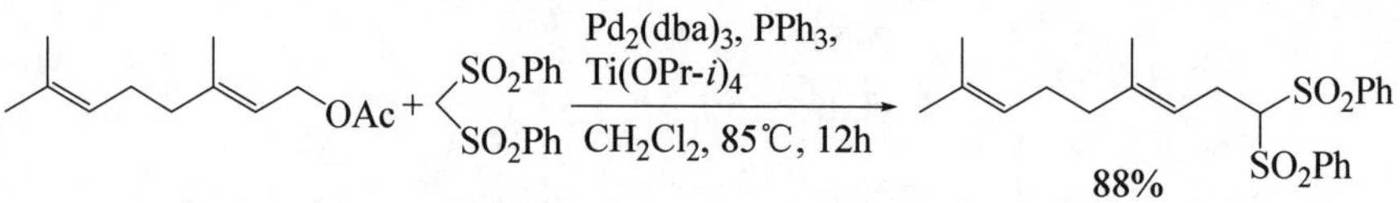

烯丙基碳酸酯和烯丙基氨基甲酸酯是比烯丙基醋酸酯更活泼的烯丙基化试剂，与活泼亚甲基等碳为亲核原子的亲核试剂能够在无碱存在的室温下进行反应，也就是说反应在中性条件下进行。烯丙基碳酸酯能够在中性条件下与活泼亚甲基化合物反应的原因已在前面介绍，如图 8-3 所示。表 8-1 是一些丙二酸酯和 β-羰基羧酸酯与烯丙基碳酸酯或烯丙基氨基甲酸酯的烯丙基化反应的实例[7]。

表 8-1 丙二酸酯和 β-羰基羧酸酯与烯丙基碳酸酯或烯丙基氨基甲酸酯的烯丙基化反应

烯丙基化合物	亲核试剂	配体	温度 /℃	时间 /h	产物	收率 /%
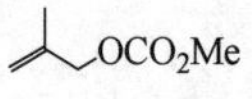 异戊烯基-OCO₂Me	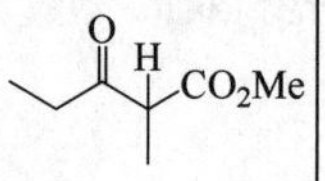 CH₃COCH(CO₂Me)	PPh₃	30	0.2	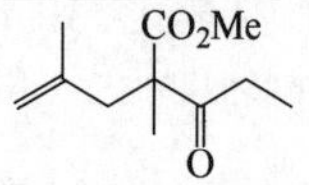 产物	92
Ph-CH=CH-CH₂-OCO₂Me	CH₃COCH₂CO₂Me	PPh₃	25	1	产物	90
O=C-CH₂CH₂CH=CHCH₂-OCO₂Et	CH₂(CO₂Et)₂	dppe	30	0.5	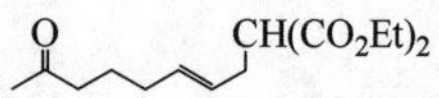 产物	91
CH₃CH=CHCH₂-OCO₂Et	CH₂(CO₂Et)₂	PPh₃	30	h	产物	86

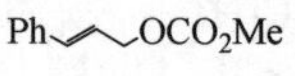

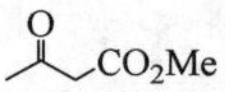

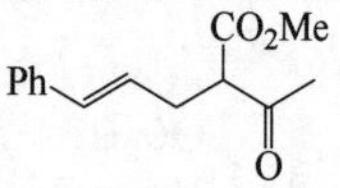

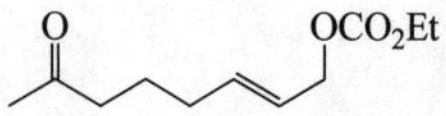

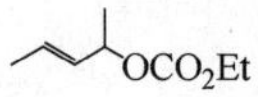

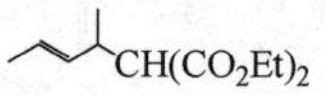

续表

烯丙基化合物	亲核试剂	配体	温度/°C	时间/h	产物	收率/%
(结构式: 含CN和OCO₂Me的化合物)	(结构式: 酮酯)	PPh₃	65	5	(结构式: 含NC和CO₂Me的产物)	93
(结构式: 含酮和OCO₂Me的化合物)	(结构式: 酮酯)	dppe	65	2	(结构式: 含酮和CO₂Me的产物)	99
(结构式: 烯丙基氨基甲酸酯)	(结构式: 酮酯)	PPh₃	30	0.17	(结构式: 含CO₂Me的产物)	100

注：本表中催化剂为 $Pd_2(dba)_3 \cdot CHCl_3$，溶剂为 THF。

其他烯丙基衍生物与活泼亚甲基化合物的反应也有所见。丙二酸二甲酯负离子与烯丙基磷酸酯在钯催化下能够顺利发生烯丙基化反应，而且反应活性比醋酸烯酯的活性要高，因此同时含有膦酸酯基和乙酸酯基的烯丙基化合物与丙二酸二甲酯负离子或乙酰乙酸乙酯负离子反应，相同条件下只有膦酸酯基一端反应，而乙酸酯基一端保持不变，并且主要生成构型保持的产物[15]。

$$\text{CH}_2=\text{CHCH}_2\text{OP(OEt)}_2 + \text{NaCH(CO}_2\text{CH}_3)_2 \xrightarrow[\text{THF, rt, 1h}]{\text{Pd(PPh}_3)_4} \text{CH}_2=\text{CHCH}_2\text{CH(CO}_2\text{CH}_3)_2 \quad 81\% \tag{16}$$

$$\text{AcO-CH}_2\text{CH=CHCH}_2\text{OP(OEt)}_2 + \text{NaCH(CO}_2\text{CH}_3)_2 \xrightarrow[\text{THF, rt, 1h}]{\text{Pd(PPh}_3)_4} \text{AcO-CH}_2\text{CH=CHCH}_2\text{CH(CO}_2\text{CH}_3)_2 \quad 83\%, Z/E=87:13 \tag{17}$$

$$\text{AcO-CH}_2\text{CH=CHCH}_2\text{OP(OEt)}_2 + \text{NaCH(COCH}_3)(\text{CO}_2\text{CH}_3) \xrightarrow[\text{THF, rt, 1h}]{\text{Pd(PPh}_3)_4} \text{AcO-CH}_2\text{CH=CHCH}_2\text{CH(COCH}_3)(\text{CO}_2\text{CH}_3) \quad 68\%, Z/E=93:7 \tag{18}$$

N-烯丙基酰胺在 $FeCl_3$ 催化下能够与活泼亚甲基化合物在没有碱存在下于硝基甲烷中直接发生偶联反应，酰胺基是离去基团，反应条件温和，而且 $FeCl_3$ 廉价无毒[16]。

$$\text{(结构式: N-烯丙基苯甲酰胺)} + \text{(乙酰丙酮)} \xrightarrow[\text{rt, 24h}]{\text{FeCl}_3, \text{MeNO}_2} \text{(产物)} \quad 100\% \tag{19}$$

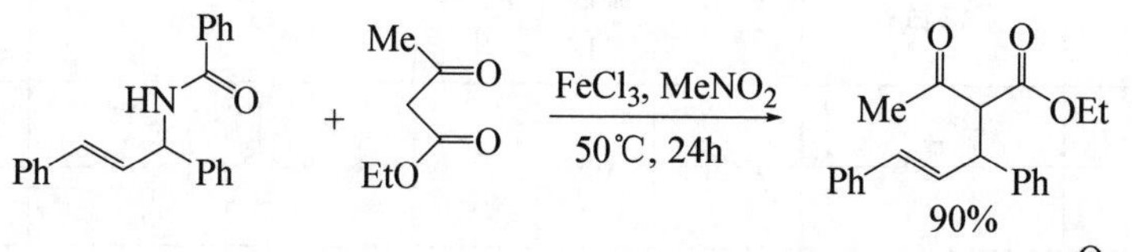

$$\text{(20)}$$

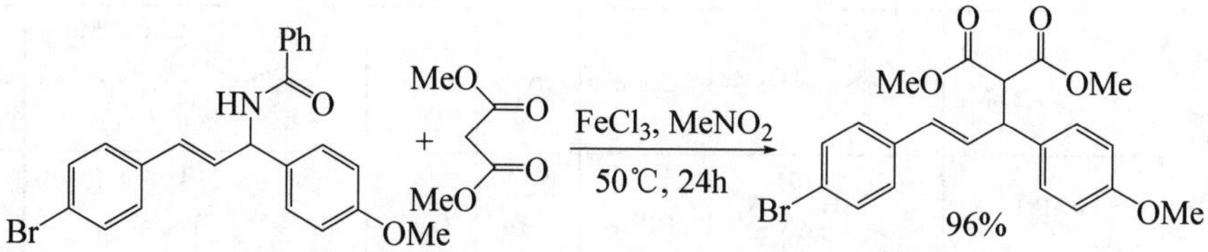

$$\text{(21)}$$

烯丙醇本身并非好的烯丙基化试剂,因为羟基不是好的离去基团。然而,当烯丙醇被取代的苯基活化使得烯丙醇容易发生离子化时,此类烯丙醇则可在钯催化下作为烯丙基化试剂与活化亚甲基化合物反应,高收率得到烯丙基化产物[17]。

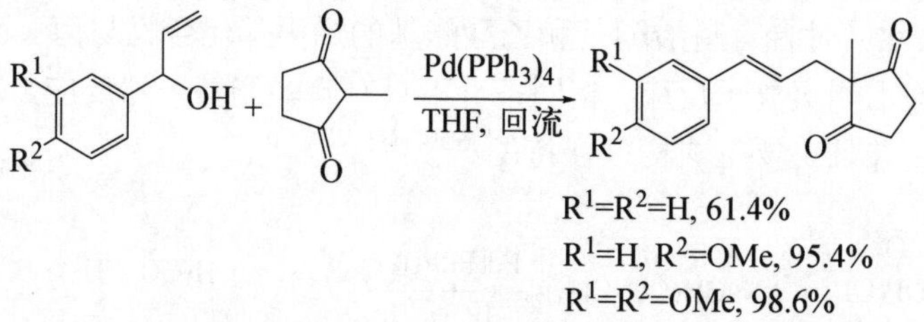

$$\text{(22)}$$

$R^1=R^2=H, 61.4\%$
$R^1=H, R^2=OMe, 95.4\%$
$R^1=R^2=OMe, 98.6\%$

苯基烯丙基醚以及类似结构的丙烯醛缩二甲醇也可作为烯丙基化试剂进行烯丙基化反应,并且可以生成二烯丙基化的产物[18,19]。

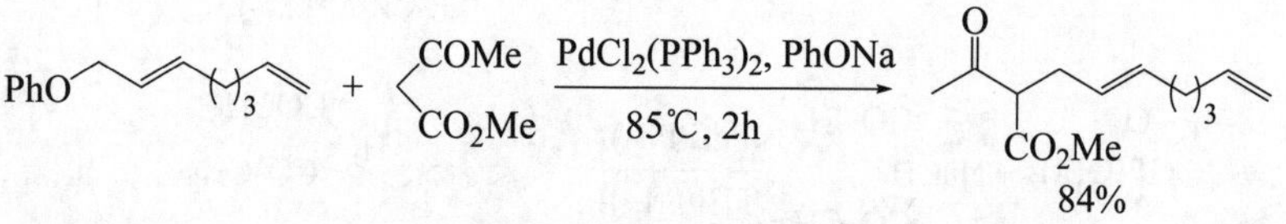

$$\text{(23)}$$

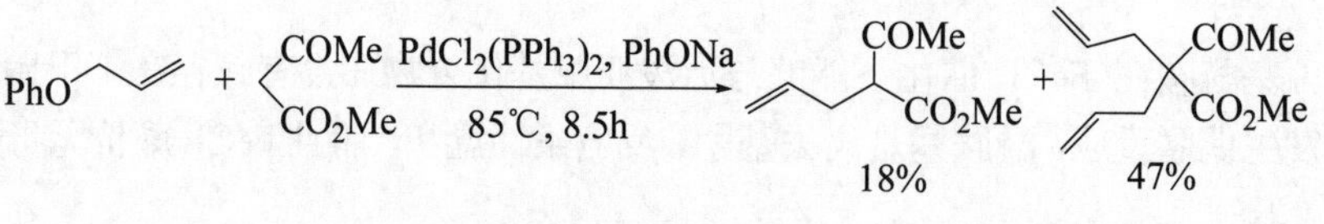

$$\text{(24)}$$

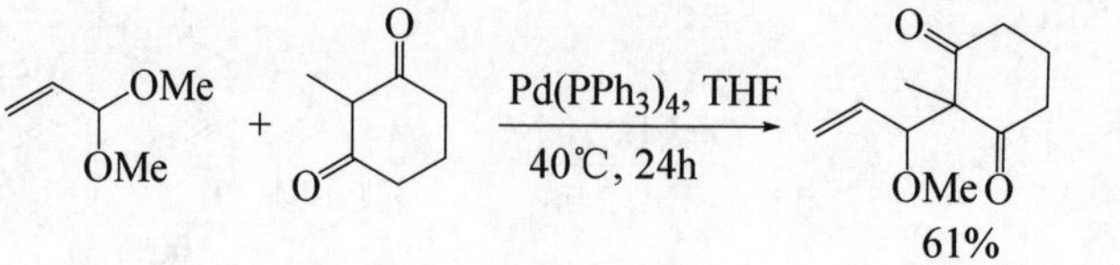

$$\text{(25)}$$

烯基环氧化物是另一种无须碱对亲核试剂活化就能在钯催化剂催化下对亲核试剂进行烯丙基化的试剂。以乙烯基环氧乙烷为例，反应机理如图 8-9 所示：乙烯基环氧乙烷与钯配合物相遇时，环氧化 C2 位的 C-O 键发生异裂，与钯配合物形成含烷氧基负离子的 η^3-烯丙基配合物中间体 **A**；**A** 上的烷氧基负离子从亲核试剂上夺取质子生成中间体 **B**，并使亲核试剂生成负离子得到活化；亲核试剂负离子一旦生成，则分别进攻中间体 **B** 中的 η^3-烯丙基的两个碳即 C2 和 C4，分别生成 1,2-和 1,4-加成产物。亲核试剂选择性地进攻空间位阻小的 C4，主要生成 1,4-加成产物。

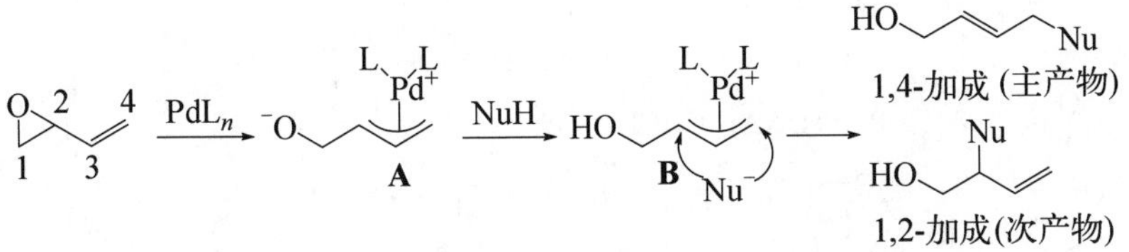

图 8-9 钯催化的乙烯基环氧化物与亲核试剂的加成反应

例如，乙烯基环氧乙烷在钯催化剂催化下与丙二酸二乙酯生成 1,4-加成产物；相同反应条件下环状的乙烯基环氧化物也生成 1,4-加成产物，并且是从氧原子的同侧进攻，而在标准的碱催化下则得到构型翻转的产物[20]。

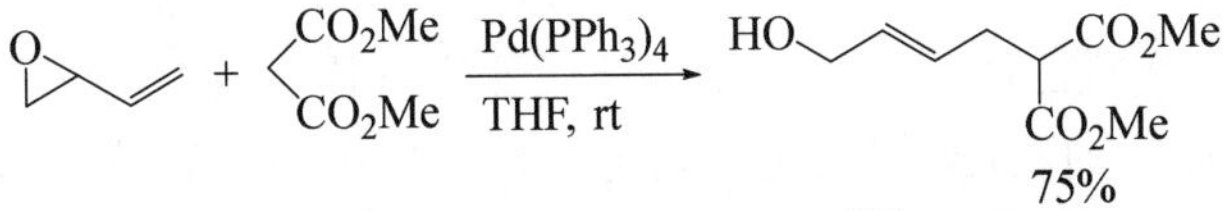

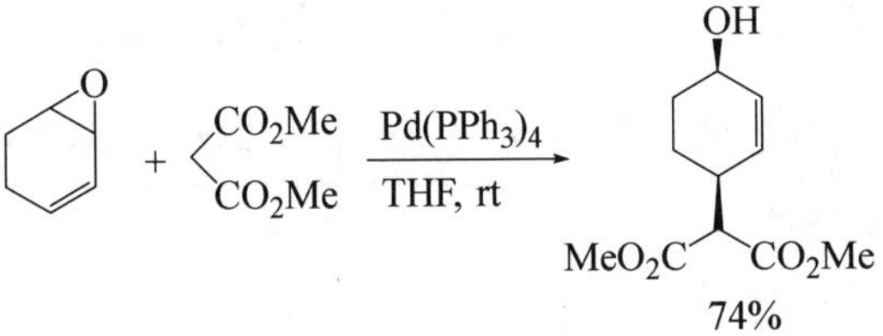

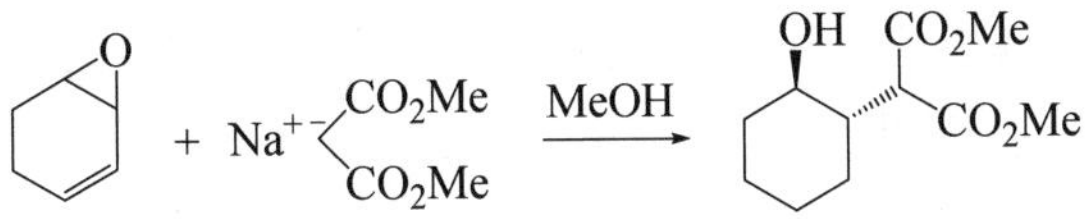

对于取代的乙烯基环氧化物，取代基的电子性能和空间位阻对亲核试剂的进攻位置有很大的影响，往往决定生成 1,2-还是 1,4-加成产物。下面反应是取代基对反应区域选择性影响的一些实例[21,22]。

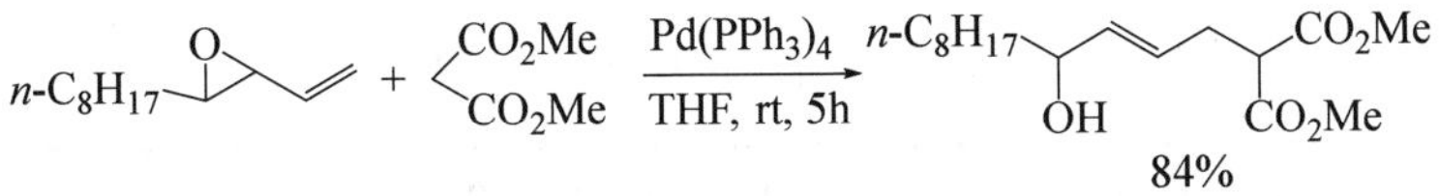

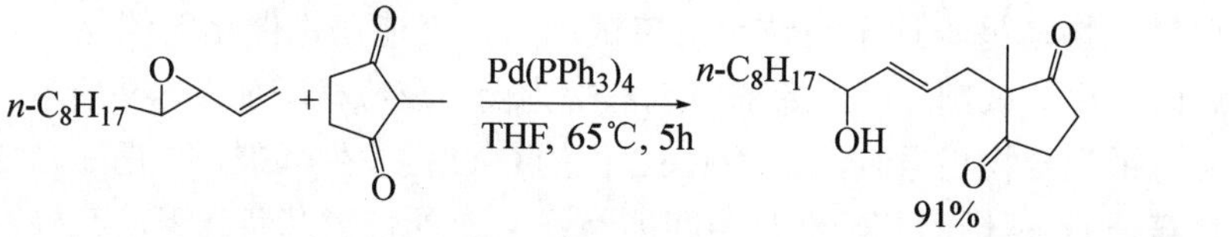

如表 8-2 所示，环状乙烯基环氧化物与活泼亚甲基化合物的 Tsuji-Trost 反应通常是区域和立体选择性的，1,4-加成是主要过程，一个例外是邻苯二酚衍生的环氧化物与丙二酸二甲酯在钯催化下的 Tsuji-Trost 反应是非立体选择性的[20,23]。

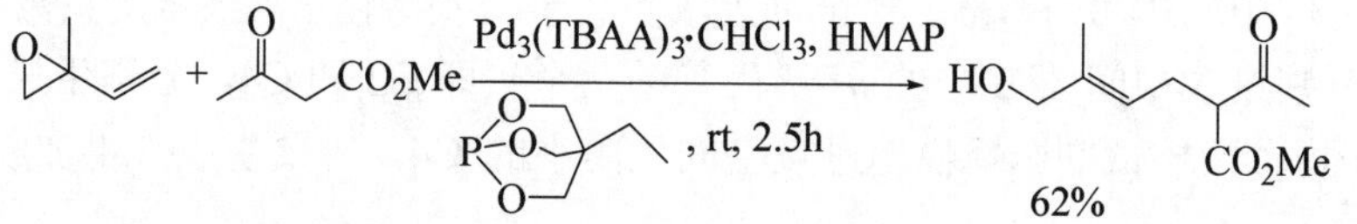

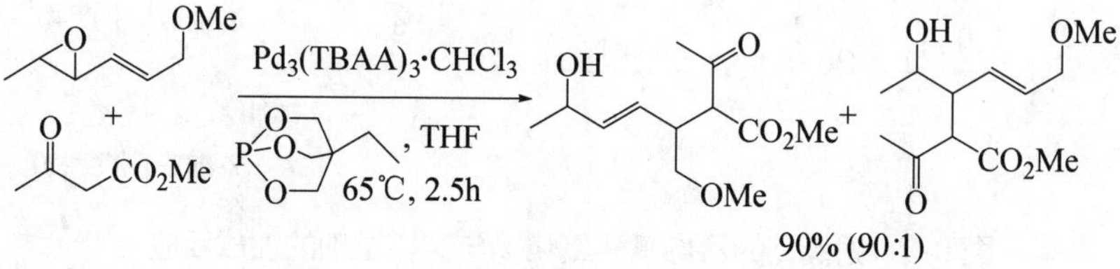

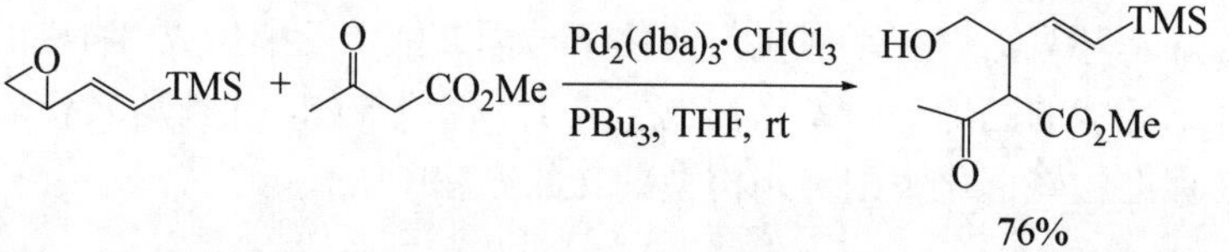

表 8-2 钯催化的环状乙烯基环氧化物与活泼亚甲基化合物的反应

乙烯基环氧化物	亲核试剂	产物	产率/%
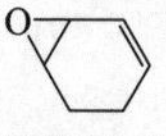	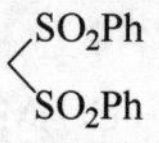 SO$_2$Ph / SO$_2$Ph	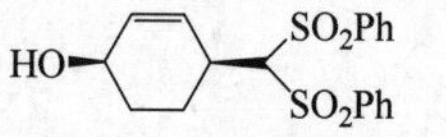	85
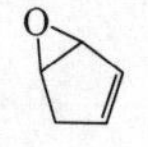	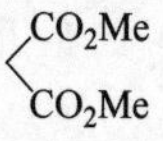 CO$_2$Me / CO$_2$Me	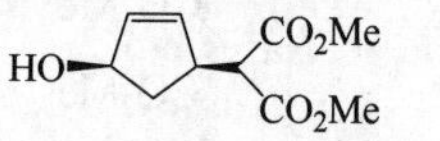	57
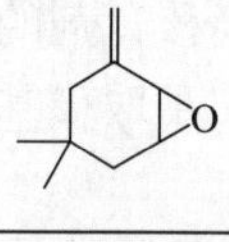	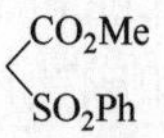 CO$_2$Me / SO$_2$Ph	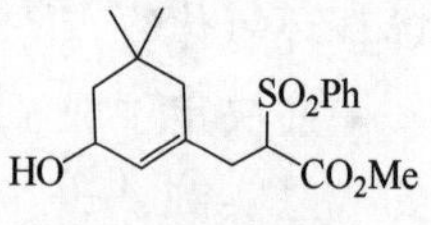	81
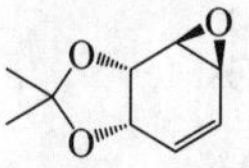	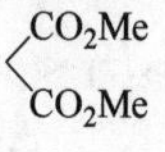 CO$_2$Me / CO$_2$Me	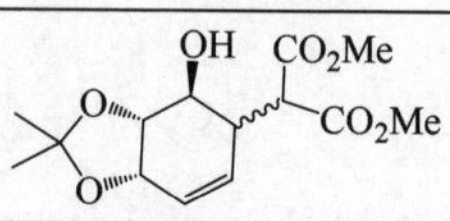	65

注：反应条件为催化剂 Pd(PPh$_3$)$_4$，溶剂 THF，反应温度 40℃。

以上反应均可归属为含氧烯丙基衍生物对活泼亚甲基软亲核试剂的烯丙基化反应。含氮烯丙基衍生物包括烯丙基胺、铵盐、甲苯磺酰胺硝基化合物等也可作为烯丙基化试剂对活泼亚甲基化合物进行烯丙基化反应。对于烯丙基胺、铵盐以及甲苯磺酰胺，以单烯丙基化产物为主，但往往伴生少量的二烯丙基化产物[24-26]。

$$\diagup\!\!\!\diagup\!\!\!\diagdown NEt_2 + CH_2(COMe)_2 \xrightarrow[85℃, 24h]{Pd(acac)_2, PPh_3} \diagup\!\!\!\diagup\!\!\!\diagdown CH(COMe)_2 + C(COMe)_2 \quad (34)$$
$$70\% \quad 20\%$$

$$\diagup\!\!\!\diagup\!\!\!\diagdown \overset{+}{N}Et_3Br^- + NaCH(CO_2Et)_2 \xrightarrow[THF, rt, 24h]{Pd(PPh_3)_3} \diagup\!\!\!\diagup\!\!\!\diagdown CH(CO_2Et)_2 + C(CO_2Et)_2 \quad (35)$$
$$62\% \quad 13\%$$

$$\diagup\!\!\!\diagup\!\!\!\diagdown NTs_2 + NaCH(CO_2Me)_2 \xrightarrow[23℃, 10min]{Pd(OAc)_2, P(OPr\text{-}i)_3} \diagup\!\!\!\diagup\!\!\!\diagdown CH(CO_2Me)_2 + C(CO_2Me)_2 \quad (36)$$
$$63\% \quad 17\%$$

环状烯丙基硝基化合物与丙二酸二甲酯钠盐反应因环的结构不同而异，但相同条件下环戊烯基、环庚烯基硝基甲烷与丙二酸二甲酯钠盐的反应结果相当，而且都生成环外取代产物，而环己烯基硝基甲烷则几乎不反应[27]。非环状的烯丙基硝基化合物与稳定的亚甲基负离子反应生成区域异构体混合物，产物异构体的比例受烯丙基单元取代基、亲核试剂和配体空间位阻的共同影响[28]。见表8-3。

$$\underset{n=1,2,3}{\text{环}^{NO_2}} + \underset{CO_2Me}{\overset{CO_2Me}{|}} \xrightarrow[DMF, 70℃, 3\sim18h]{Pd(PPh_3)_4, PPh_3} \underset{n}{\text{环}^{CH(CO_2Me)_2}} \quad (37)$$

$n=1, 70\%$ (3h)
$n=2, 0\%$
$n=3, 69\%$ (18h)

8.2.1.2 醛、酮等羰基化合物作为亲核试剂的反应

醛、酮、酯以及相关的羰基化合物的烯丙醇化通常通过它们的碱金属烯醇盐与烯丙基卤化物间的亲核取代反应实现。Tsuji-Trost 反应拓展了这类羰基化合物通过其烯醇盐进行烯丙基化及相关反应的底物范围。烯丙基化试剂从传统的烯丙基氯或

表 8-3 非环状烯丙基硝基化物与稳定亚甲基负离子的反应

$$\underset{1}{O_2N-\underset{R}{\overset{Me}{C}}-\overset{E}{\underset{H}{C}}=CH_2} + NaCH\overset{E}{\underset{E'}{<}} \xrightarrow[THF, 回流]{Pd(PPh_3)_4, PPh_3} \underset{2}{\overset{Me}{R-\underset{Nu}{C}}-C=CH_2} + \underset{3}{\overset{Me}{\underset{R}{>}}C=C\overset{H}{\underset{Nu}{<}}}$$

R	1	E,E'	时间/h	产物	收率/%	2:3	E:Z (3)
Me	1a	CO$_2$Me, CO$_2$Me	4	2a+3a	60	73:27	—
n-C$_6$H$_{13}$	1b	CO$_2$Me, CO$_2$Me	6	2b+3b	78	42:58	71:29
CH$_2$CH$_2$CO$_2$Me	1c	CO$_2$Me, CO$_2$Me	6	2c+3c	66	36:64	75:25
CH$_2$Ph	1d	CO$_2$Me, CO$_2$Me	8	2d+3d	77	18:82	82:18
CH$_2$OAc	1e	CO$_2$Me, CO$_2$Me	3	3e	91	0:100	85:15
Me	1a	CN,CO$_2$Me	6	2a'+3a'	75	99:1	—
Me	1a	NaCMeEE'	15	3a″	90	0:100	—

注：NaCMeEE'=NaC(Me)(SO$_2$C$_6$H$_4$Me-p)(CO$_2$Et)。

溴拓展到含有氧、硫、氮及其他杂原子的烯丙基衍生物。以下将介绍一些基本的醛、酮等羰基化合物经其烯醇盐或硅醚等与各种烯丙醇化试剂的反应。

（1）醛酮烯醇盐的烯丙基化

多种金属离子或离子基团作为反离子的烯醇盐在钯催化剂催化下可以与烯丙基衍生物发生 Tsuji-Trost 反应，最终生成 α-烯丙基化的醛或酮。这里的金属可以是 Li、Mg、Zn、B、Al、Si 和 Sn，而这些金属的烯醇盐主要由烯醇锂制备。

最早发现苯乙酮的烯醇锂盐在 Pd(PPh$_3$)$_4$ 催化下与醋酸烯丙酯发生烯丙基化反应，但以二取代产物为主，二取代产物与单取代产物的比例接近 2:1，改变催化剂、溶剂等常规反应条件并不能对反应有所改进［反应式（38）］。以烯基硅醚替代锂盐能够将取代的产物产率提高到 59%，但反应不能拓展到取代的烯丙基［反应式（39）］。当将酮如 α-甲基环己酮转化为烯基有机锡醚时，底物范围得到拓展，以高收率得到各种单烯丙基化的酮［反应式（40）～式（44）］[29]。

$$\underset{65\%}{\text{PhC(OLi)=CH}_2} + AcO\text{-allyl} \xrightarrow{Pd(PPh_3)_4, THF, rt} \text{PhCOCH(allyl)(allyl)} + \underset{34\%}{\text{PhCOCH}_2\text{CH}_2\text{CH=CH}_2} \quad (38)$$

$$\underset{59\%}{\text{PhC(OSiMe}_3\text{)=CH}_2} + AcO\text{-allyl} \xrightarrow{Pd(PPh_3)_4, THF, rt} \text{PhCOCH}_2\text{CH}_2\text{CH=CH}_2 \quad (39)$$

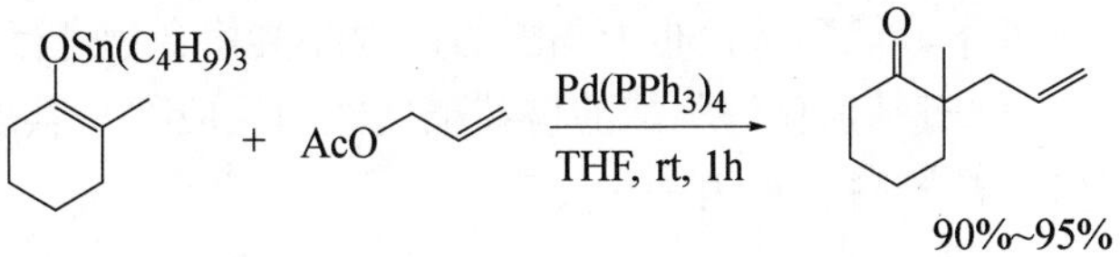

$$\text{(40)} \quad 90\%\sim95\%$$

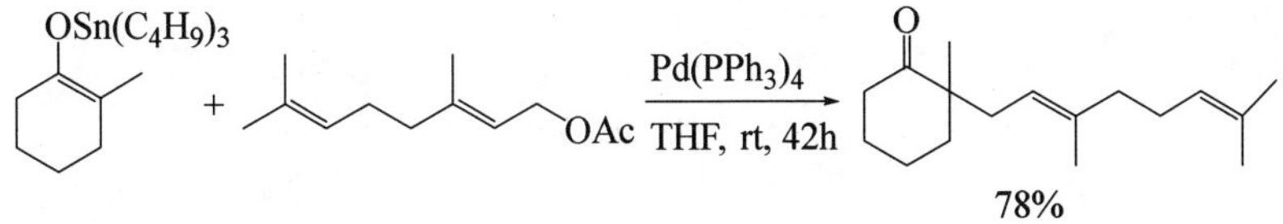

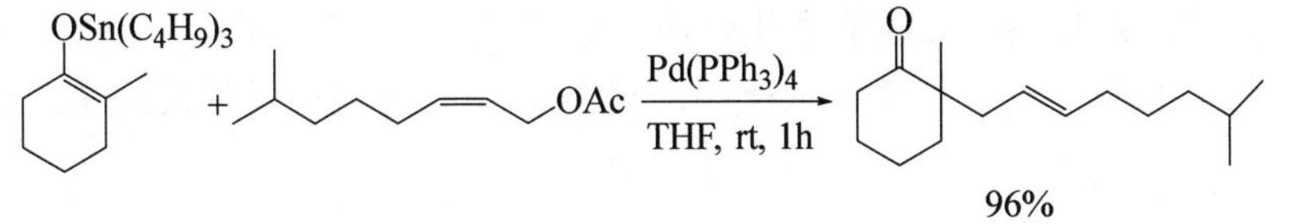

$$\text{(41)} \quad 78\%$$

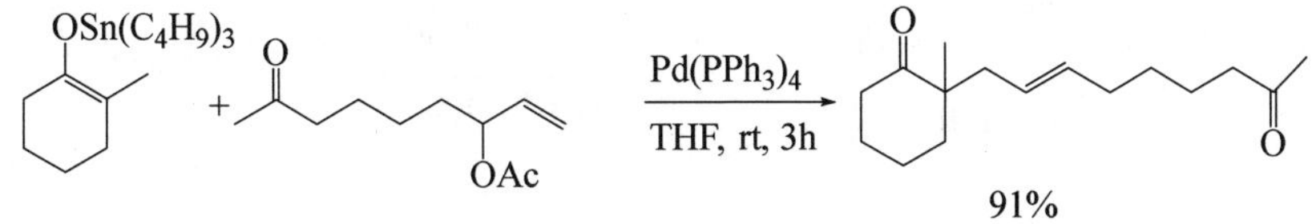

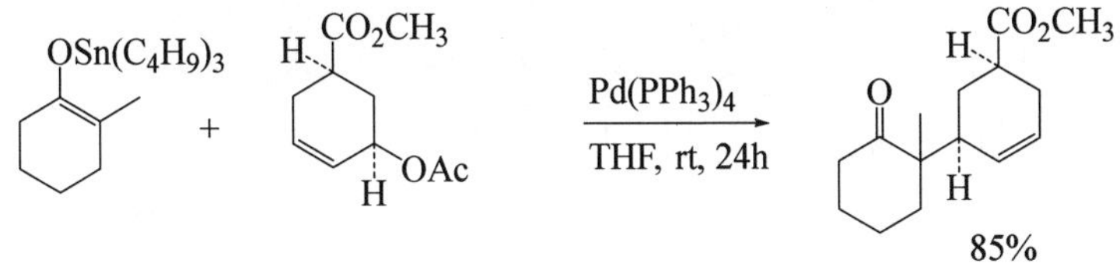

$$\text{(42)} \quad 96\%$$

$$\text{(43)} \quad 91\%$$

$$\text{(44)} \quad 85\%$$

三乙基（烯氧基）硼酸钾，即由烯醇钾与三乙基硼络合生成，与醋酸烯丙酯顺利发生烯丙基化反应。以 α-甲基环己酮的烯丙基化反应为例，无论是热力学优先还是动力学优先的三乙基（甲基环己烯氧基）硼酸钾均能与醋酸香叶酯在钯催化剂催化下反应生成相应的香叶基化产物。α-甲基环己酮分别用 KN(SiMe$_3$)$_2$ 和 KH 处理生成相应的动力学和热力学优先的烯醇钾，然后与 BEt$_3$ 配位生成对应的三乙基（甲基环己烯氧基）硼酸钾，它们再在 Pd(PPh$_3$)$_4$ 催化下与醋酸香叶酯缩合生成甲基香叶基环己酮。

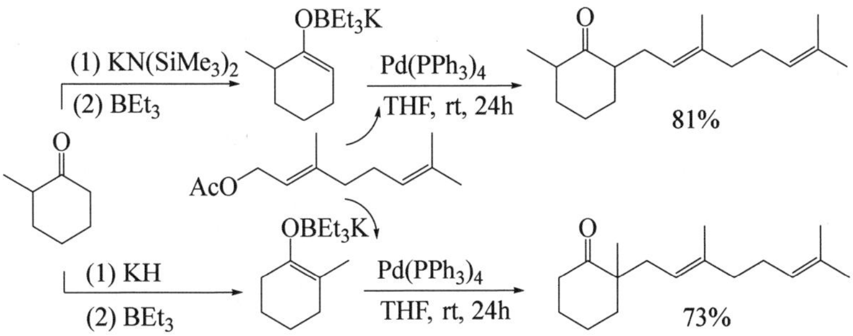

$$\text{(45)} \quad 81\%$$

$$\text{(46)} \quad 73\%$$

在与上相同催化反应条件下，烯醇锌盐也能与醋酸烯丙酯类化合物进行 Tsuji-Trost 反应，例如 α-甲基环己烯氧基氯化锌与醋酸橙花酯之间的反应[30]。其他的金属烯醇盐则不适合与醋酸烯丙酯类进行烯丙基化反应。

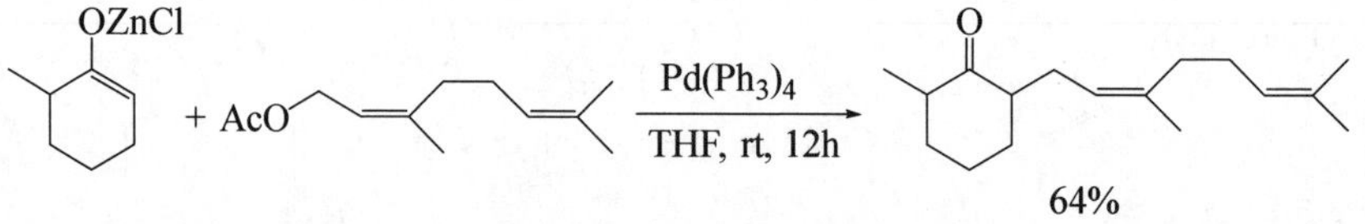

$$\text{(47)}$$

尽管烯丙醇锂与醋酸烯丙酯类化合物不是 Tsuji-Trost 反应的良好偶联试剂，但烯丙醇锂能与烯丙基季铵盐等顺利反应，其中 α-甲基环己醇锂的反应是区域选择性的[25]。

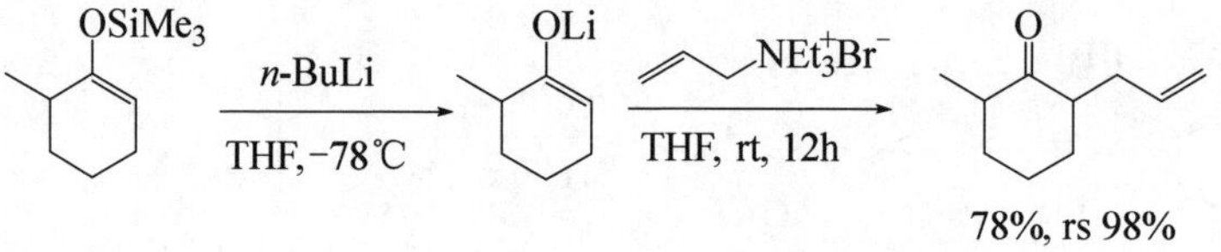

$$\text{(48)}$$

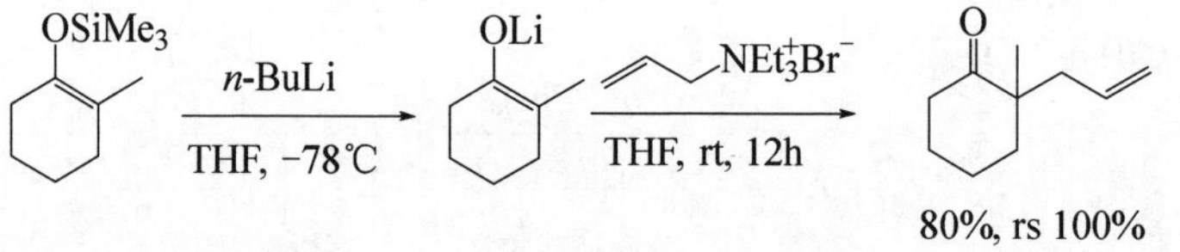

$$\text{(49)}$$

与活泼亚甲基的烯丙基化反应类似，碳酸烯丙酯作为烯丙基化试剂比醋酸烯丙酯更活泼，底物适用范围更广。例如，烯醇硅醚不与醋酸烯丙酯进行反应，但各种结构的烯醇硅醚却能与碳酸烯丙酯在 $Pd_2(dba)_3 \cdot CHCl_3$/dppe 催化下反应生成烯丙基化的醛或酮[31]。

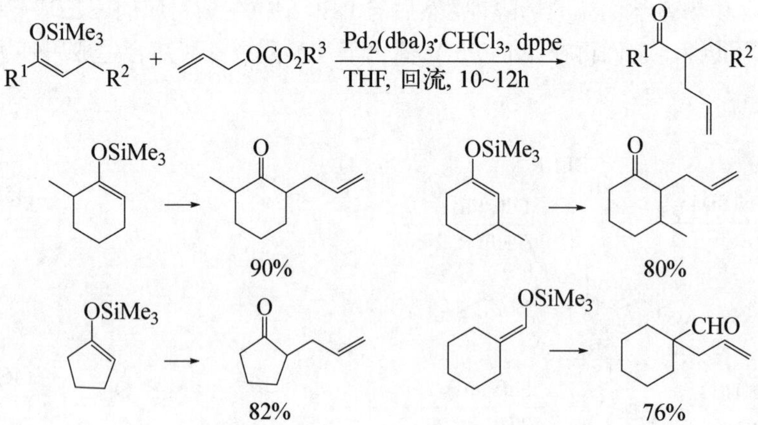

$$\text{(50)}$$

$$\text{Ph-C(OSiMe}_3\text{)=CHCH}_2\text{CH}_3 \longrightarrow \text{PhCOCH(Et)CH}_2\text{CH=CH}_2 \quad 64\%$$

$$\text{PhCH}_2\text{CH=CHOSiMe}_3 \longrightarrow \text{PhCH}_2\text{CH(CHO)CH}_2\text{CH=CH}_2 \quad 67\%$$

与烯醇硅醚类似，烯醇锡醚也能与碳酸烯丙酯反应生成烯丙基化酮，其中烯醇锡醚由醋酸烯醇酯与甲氧基三丁基锡反应原位生成[32]。

环己烯-OAc + Bu$_3$SnOCH$_3$ → [环己烯-OSnBu$_3$] + 烯丙基-OCO$_2$CH$_3$, Pd$_2$(dba)$_3$·CHCl$_3$, 1,4-二氧六环, 回流, 12h → 2-甲基-2-烯丙基环己酮 82% (51)

6-甲基环己烯-OAc + Bu$_3$SnOCH$_3$ → [6-甲基环己烯-OSnBu$_3$] + 烯丙基-OCO$_2$CH$_3$, Pd$_2$(dba)$_3$·CHCl$_3$, 1,4-二氧六环, 回流, 12h → 2-甲基-6-烯丙基环己酮 78% (52)

（2）酯烯醇盐或相关亲核试剂的烯丙基化

酯不像醛或酮在碱性或酸性条件下容易发生烯醇化，但在碱存在下与三甲基氯硅烷反应生成三甲硅基缩烯醛，后者在 Pd$_2$(dba)$_3$·CHCl$_3$/dppe 催化下与碳酸烯丙酯类化合物在二氧六环溶剂中反应得到单烯丙基化酯 [反应式（53）～式（55）]。溶剂对该反应的影响很大，如果是乙腈或苯甲腈作溶剂，并在没有配体存在下会发生消除反应得到 α,β-不饱和酯，而非 α-烯丙酯 [反应式（56）、式（57）]。

环己烯-C(OMe)=OTMS + CH$_2$=CHCH$_2$CH$_2$CH=CHCH$_2$OCO$_2$Me $\xrightarrow{\text{Pd}_2\text{(dba)}_3\cdot\text{CHCl}_3,\ \text{dppe}}_{1,4\text{-二氧六环, 回流}}$ 产物 74% (53)

CH$_3$(CH$_2$)$_2$CH=C(OSiMe$_3$)(OEt) + 烯丙基-OCO$_2$Et $\xrightarrow{\text{Pd}_2\text{(dba)}_3\cdot\text{CHCl}_3,\ \text{dppe}}_{1,4\text{-二氧六环, 回流}}$ 产物-CO$_2$Et 79% (54)

二氢吡喃-OSiMe$_3$ + 烯丙基-OCO$_2$Me $\xrightarrow{\text{Pd}_2\text{(dba)}_3\cdot\text{CHCl}_3,\ \text{dppe}}_{1,4\text{-二氧六环, 回流}}$ δ-戊内酯-α-烯丙基 48% (55)

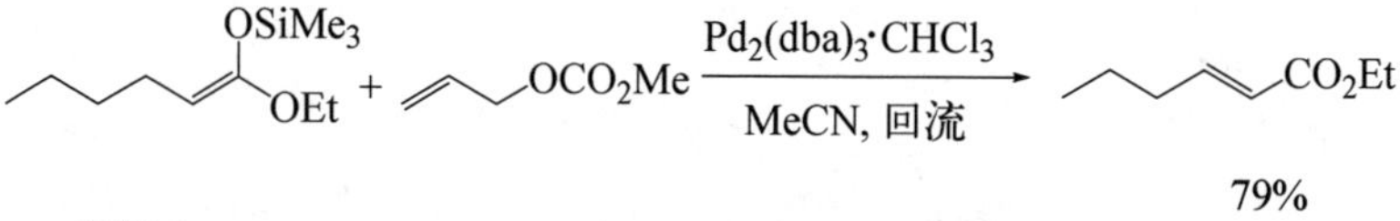

$$\text{OSiMe}_3\text{-OEt} + \diagup\!\!\!\diagdown\text{OCO}_2\text{Me} \xrightarrow[\text{MeCN, 回流}]{\text{Pd}_2(\text{dba})_3\cdot\text{CHCl}_3} \diagup\!\!\!\diagdown\text{CO}_2\text{Et} \quad (56)$$

79%

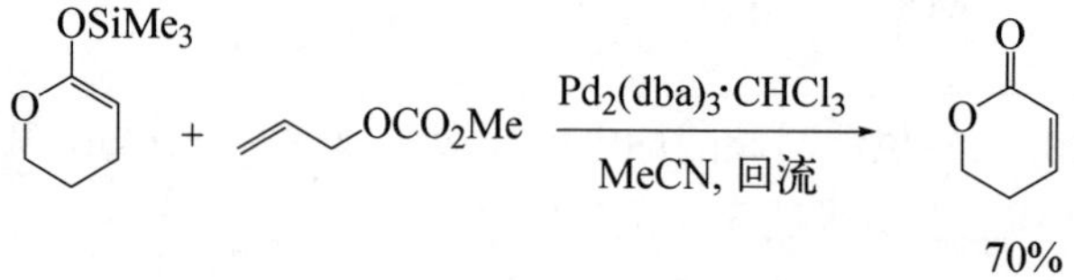

$$(57)$$

70%

雷夫马斯基(Reformatsky)试剂也能与醋酸烯丙酯类化合物发生酯的烯丙基化反应，反应的区域选择性由立体效应控制，具体地说雷夫马斯基试剂主要是从烯丙基取代基少的碳进攻；而且不管烯丙醇基的构型是 Z-式还是 E-式，都生成 E-式产物[33]。

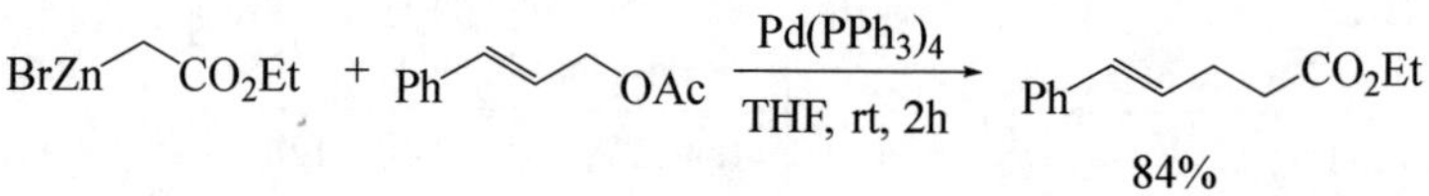

$$\text{BrZnCH}_2\text{CO}_2\text{Et} + \text{Ph-CH=CH-CH}_2\text{OAc} \xrightarrow[\text{THF, rt, 2h}]{\text{Pd(PPh}_3)_4} \text{Ph-CH=CH-CH}_2\text{-CH}_2\text{CO}_2\text{Et} \quad (58)$$

84%

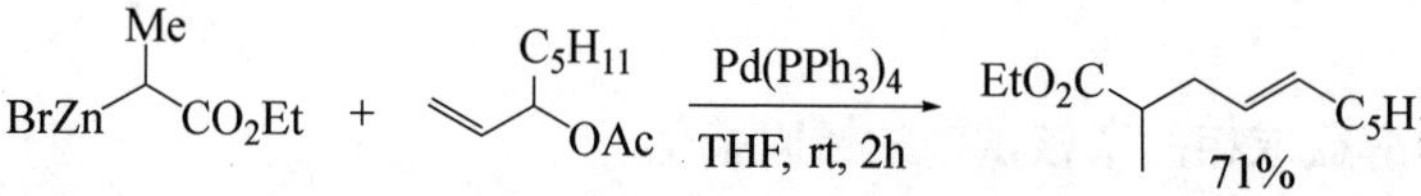

$$(59)$$

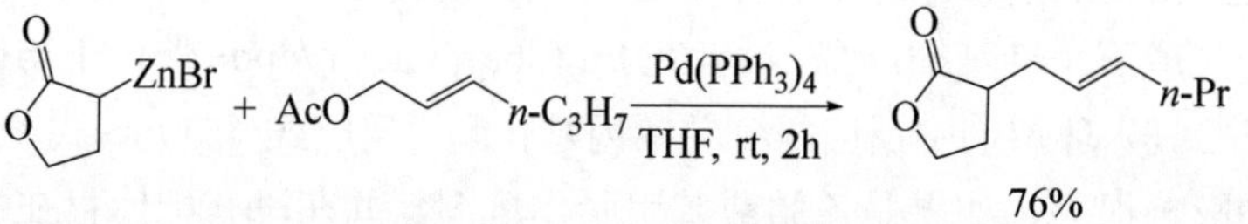

71%

$$(60)$$

76%

一些近似酯烯醇锂结构的化合物与醋酸烯丙酯在钯催化下也能发生烯丙基化反应。例如，2-羟基苯乙酸与丙酮缩二甲醇进行交换反应生成二氧杂环戊酮结构单元，其与酯结构相似，用二异丙基氨基锂（LDA）处理得到相应的烯醇锂，烯醇锂与醋酸烯丙酯类化合物得到烯丙基化二氧杂环戊酮，最后用氢氧化钾甲醇溶液处理再用稀盐酸酸化得到烯丙基化的 2-羟基苯乙酸[34]。

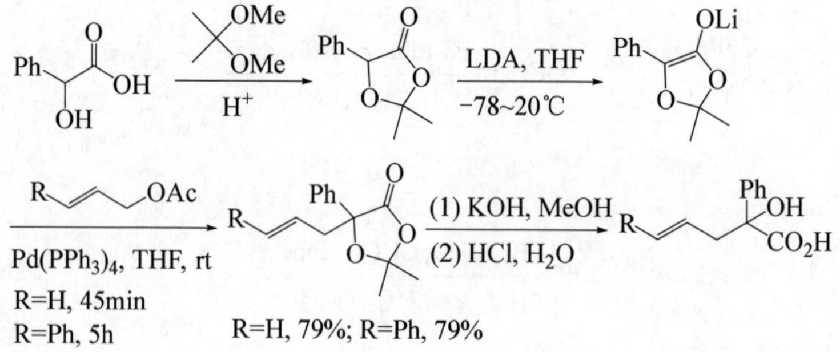

$$(61)$$

R=H, 79%; R=Ph, 79%

（3）烯胺、亚胺、烯醇醚以及硝基化合物的烯丙基化

烯胺由酮、醛与仲胺缩合而得，其在钯催化剂催化下与烯丙基醚或醋酸烯丙酯缩合得到烯丙基酮。由于烯丙基化过程中经历了亚胺盐中间体，而亚胺盐极易转化为新的烯胺，所以往往生成一部分双烯丙基化产物。例如，N-环己烯-1-基哌啶与苯基烯丙基醚的反应除生成 2-烯丙基环己酮外，还生成相当数量的 2,6-二烯丙基环己酮[35]。也正是因为酰胺烯丙基化这一特征，因而可利用同一底物的双烯丙基化反应合成桥环化合物[36]。

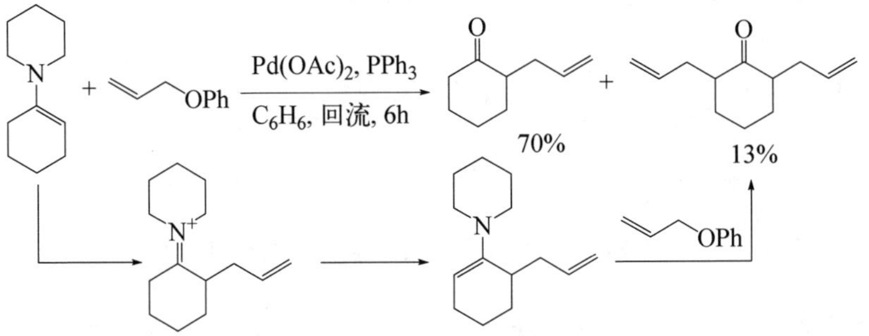

（62）

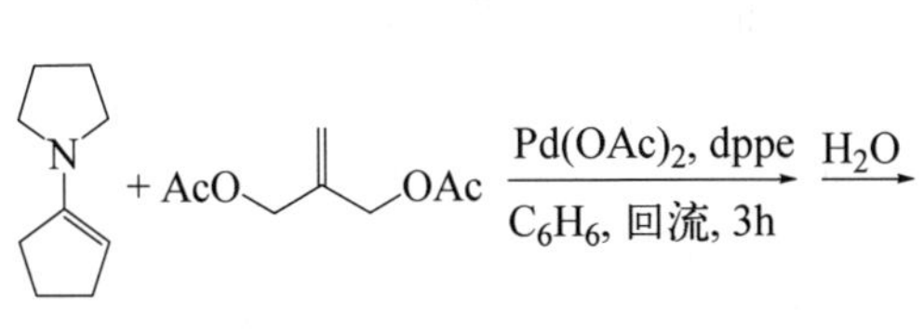

（63）

由于烯胺很容易由醛、酮与氨基缩合而得，与手性氨基酸烯丙酯或类似物缩合则得到光学活性的烯胺，进而实现对映选择性烯丙基化，得到光学活性的 α-烯丙基醛、酮。以下是一些代表性的实例[37,38]。

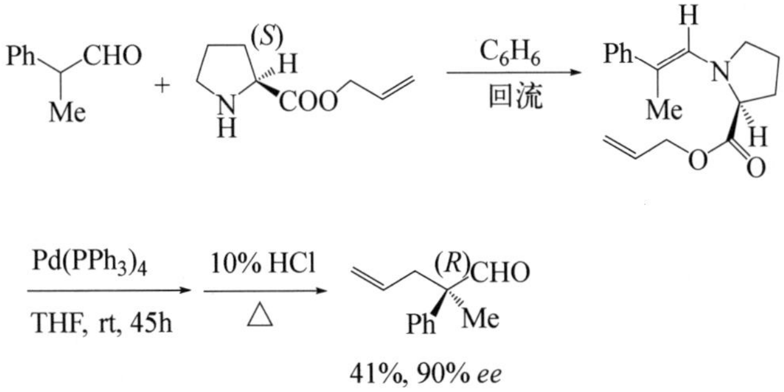

（64）

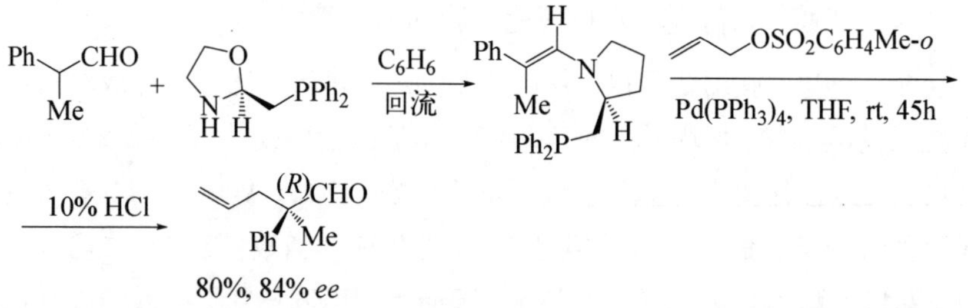

$$\text{Ph}\underset{\text{Me}}{\overset{\text{CHO}}{\diagup}} + \underset{\overset{|}{\text{H}}}{\overset{\text{O}}{\diagdown}}\text{N}\text{—}\text{CH}_2\text{PPh}_2 \xrightarrow[\text{回流}]{\text{C}_6\text{H}_6} \underset{\text{Me}}{\overset{\text{Ph}}{\diagup}}\overset{\text{H}}{=}\underset{\overset{|}{\text{Ph}_2\text{MP}}}{\text{N}\text{—}\overset{\text{H}}{\text{C}}} \xrightarrow[\text{Pd(PPh}_3)_4,\text{ THF, rt, 45h}]{\diagup\diagup\diagdown\text{OSO}_2\text{C}_6\text{H}_4\text{Me-}o}$$

$$\xrightarrow{10\% \text{ HCl}} \underset{\underset{80\%,\ 84\%\ ee}{\text{Ph } \text{Me}}}{\diagup\diagup\diagdown\overset{(R)\text{CHO}}{\diagup}} \tag{65}$$

α-甲基苯乙醛与（S）-脯氨酸烯丙酯反应最终生成（R）-α-甲基-α-烯丙酯苯乙醛是通过如下所示的过渡态进行的[37]。如图 8-10 所示。

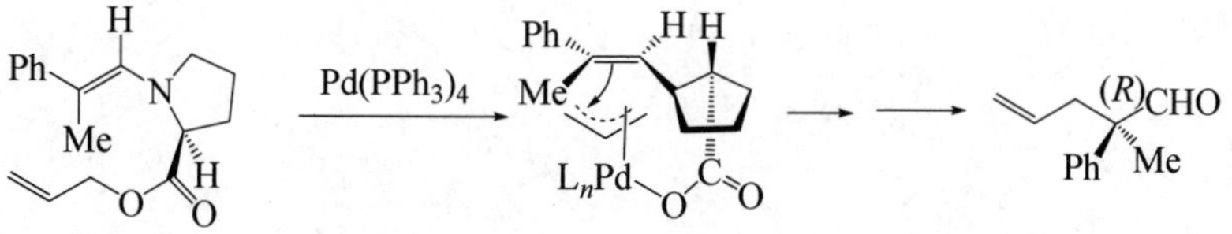

图 8-10　α-甲基苯乙醛与（S）-脯氨酸烯丙酯不对称烯丙基化

烯醇醚只能与一些取代的烯丙醇在钯催化下发生 Tsuji-Trost 反应，生成 α-烯丙基酮。实际上烯醇烯丙基醚在无钯催化下加热即可发生克莱森（Claisen）重排也得到 α-烯丙基酮，但与钯催化的反应产物的构型不同。例如，6-甲氧基二氢萘-1-甲醚与（E）-1-异丁基-2-丁烯醇在钯催化下得到反式（anti-）产物，而在无钯存在的 2,6-二甲基苯酚催化下得到顺式（syn-）的产物[39]。

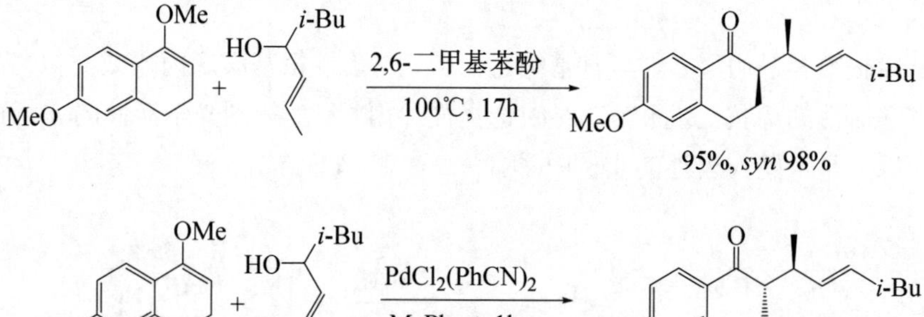

$$\text{(66)} \quad 95\%,\ syn\ 98\%$$

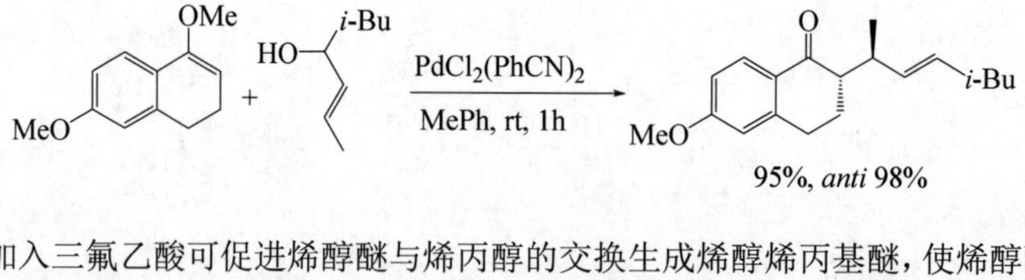

$$\text{(67)} \quad 95\%,\ anti\ 98\%$$

加入三氟乙酸可促进烯醇醚与烯丙醇的交换生成烯醇烯丙基醚，使烯醇醚底物范围更宽［反应式（68）和表 8-4］。从烯丙醇结构方面看有以下结论：（E）-烯丙醇，特别是 α-取代的烯丙醇，给出好的收率和高的 anti-选择性；相反，（Z）-烯丙醇反应的收率低，但主要还是以 anti-产物为主，尽管选择性略低；反应生成的碳碳双键主要以（E）-构型为主；α,α-二取代的烯丙醇反应很差，收率很低[40]。

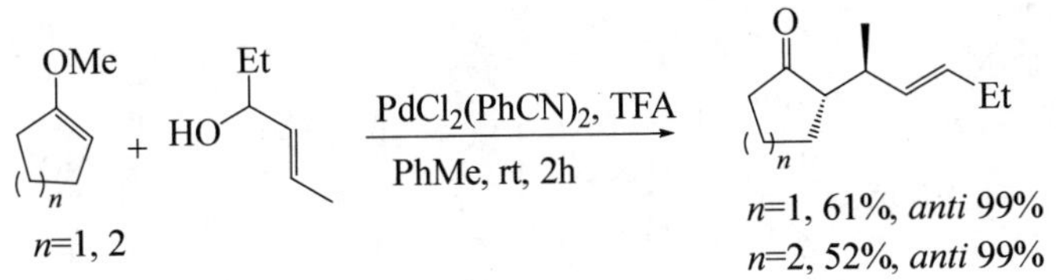

$$\text{(68)}$$

n=1, 61%, anti 99%
n=2, 52%, anti 99%

表 8-4 钯催化的二氢环甲醚与各种烯丙醇的反应

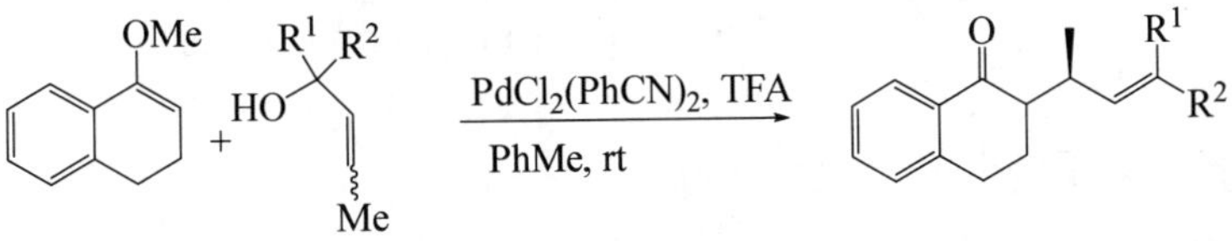

烯丙醇	产物	时间/h	产率/%
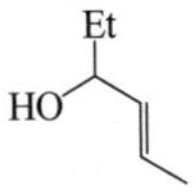	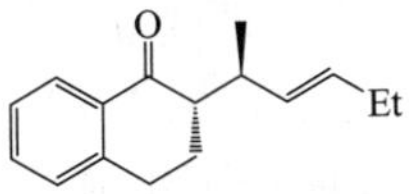	1	68 anti 96% E 100%
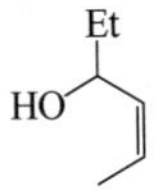	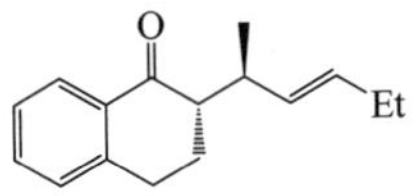	1	12 anti 88% E 100%
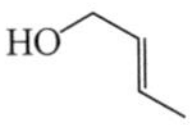	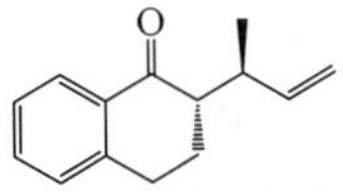	1	59 anti 90%
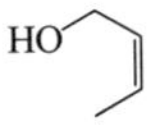	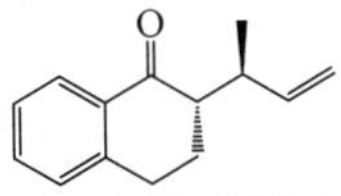	3	18 anti 82%
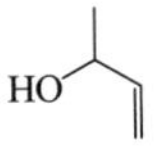	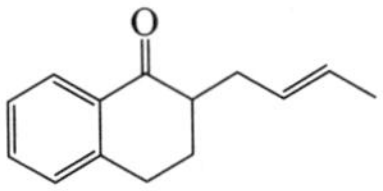	0.5	69 E 100%
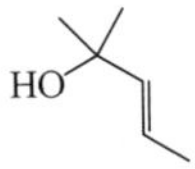	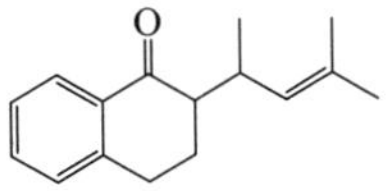	3	16 anti 44%

烷基硝基化合物由于硝基的强吸电子作用很容易生成一价负离子，该负离子与卤代烷发生亲核取代反应生成烷基化的硝基化合物，但烷基化往往发生在氧原子而非碳原子上。钯催化的烷基硝基化合物的一价负离子与醋酸烯丙酯的反应可选择性地生成碳原子烯丙基化的产物，但烯丙基的 C1 和 C3 均能与烷基硝基化合物结合，因而生成两个烯丙基化合物的混合物，空间位阻小的碳进攻生成的产物为主[41]。

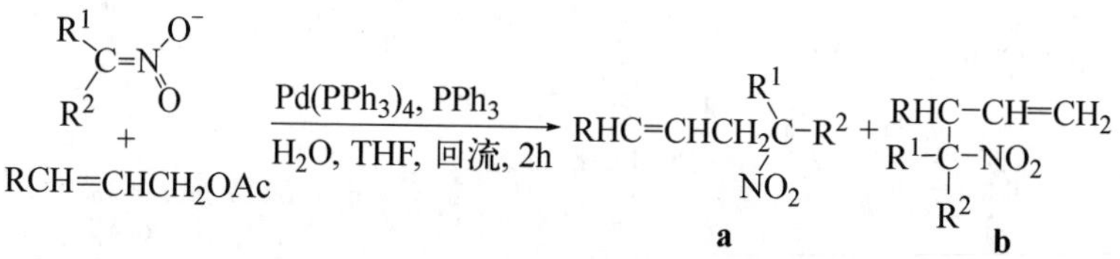

$$\begin{array}{c}\text{R}^1\\\text{C=N}^+\text{O}^-\\\text{R}^2\end{array} \quad \xrightarrow[\text{H}_2\text{O, THF, 回流, 2h}]{\text{Pd(PPh}_3)_4\text{, PPh}_3} \quad \begin{array}{c}\text{R}^1\\\text{RHC=CHCH}_2\text{C-R}^2\\\text{NO}_2\\\textbf{a}\end{array} + \begin{array}{c}\text{RHC-CH=CH}_2\\\text{R}^1\text{-C-NO}_2\\\text{R}^2\\\textbf{b}\end{array} \quad (69)$$
+ RCH=CHCH$_2$OAc

R=Ph, R^1=CH$_3$, R^2=H, 54% (**a:b**=87:13)
R=Ph, R^1=*n*-Bu, R^2=H, 57% (**a:b**=86:14)
R=CH$_3$, R^1=Ph, R^2=H, 81% (**a:b**=49:51)
R=Ph, R^1=EtO$_2$C, R^2=CH$_3$, 89% (**a:b**=97:3)

8.2.2 分子内 Tsuji-Trost 反应

当同一分子中同时存在 Tsuji-Trost 反应的两个结构反应单元时，在钯催化剂催化下这两个反应单元发生分子内反应。亲电试剂结构单元通常为烯丙基醋酸酯、烯丙基芳基醚和乙烯基环氧乙烷，而亲核试剂通常为含各种强吸电子基的活泼亚甲基。这里的分子内反应存在两种情况：其一是反应后不生成环，类似于简单的重排反应；其二是反应后形成环状化合物。

8.2.2.1 分子内重排反应

β-羰基羧酸烯丙酯、丙二酸烯丙酯等化合物结构中同时含有活泼亚甲基和烯丙基酯两个 Tsuji-Trost 反应的结构单元，而此类化合物可以通过其他酯与烯丙醇酯交换获得，因而在有机合成中具有应用价值。实际上，一些此类化合物在无催化剂、

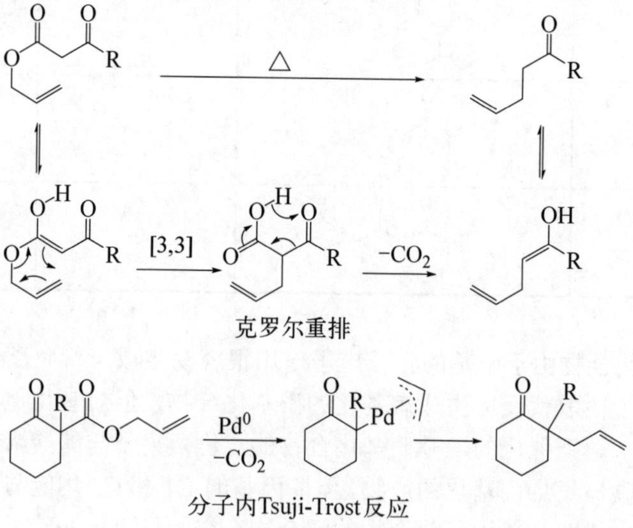

图 8-11 克罗尔重排与分子内 Tsuji-Trost 反应机理

加热到 200℃ 的条件下也能发生反应，称作克罗尔重排（Carroll rearrangement）反应。克罗尔重排反应条件苛刻，反应底物适应性窄。引入钯催化剂使反应条件变得温和，并拓宽了底物范围。尽管有时两个反应生成相同的反应产物，但二者的反应机理完全不同（图 8-11）[42]。克罗尔重排反应被认为是烯醇酯的[3,3]σ-重排反应，当亚甲基上没有活泼氢时不能形成烯醇式，反应不能进行；而钯催化的分子内 Tsuji-Trost 反应即使亚甲基上没有活泼氢反应也能进行。

下面的反应充分说明，钯催化的分子内烯丙基化反应条件温和、底物适用范围广泛[43, 44]。

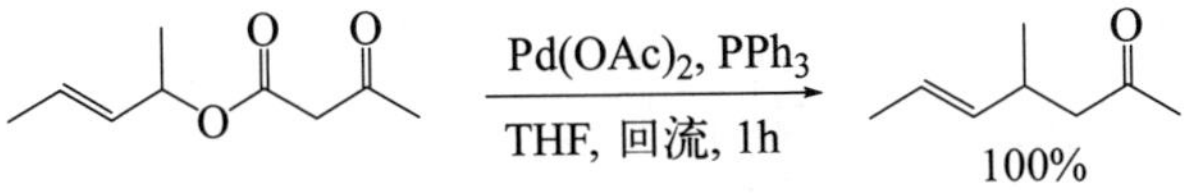

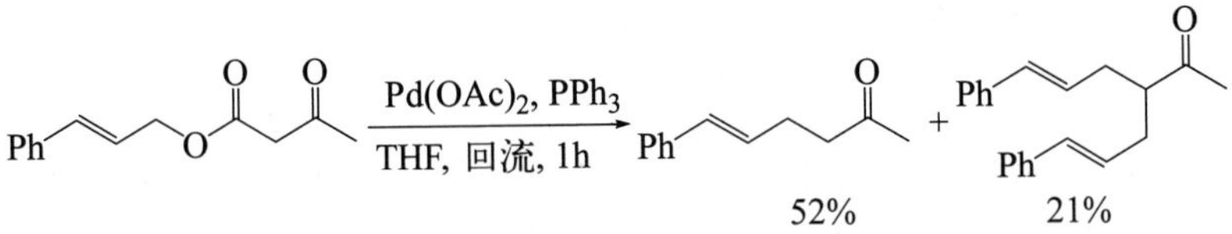

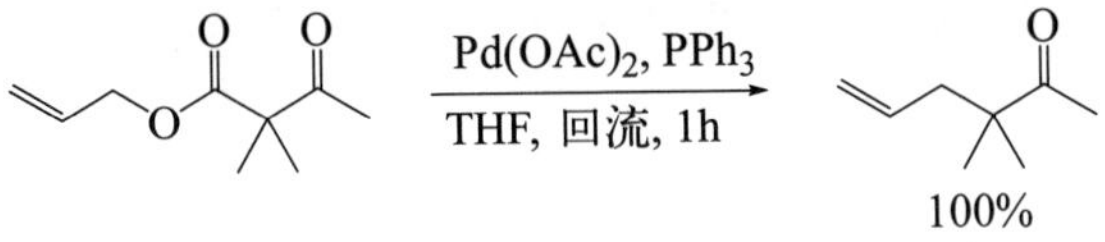

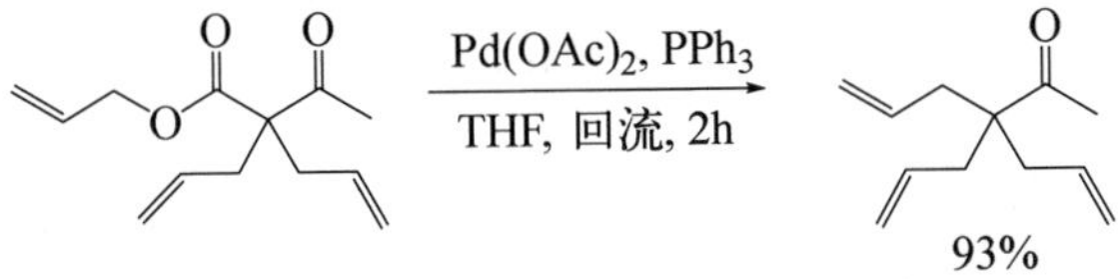

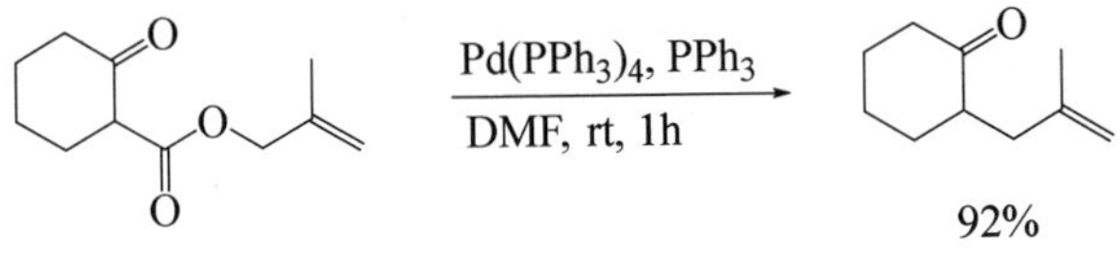

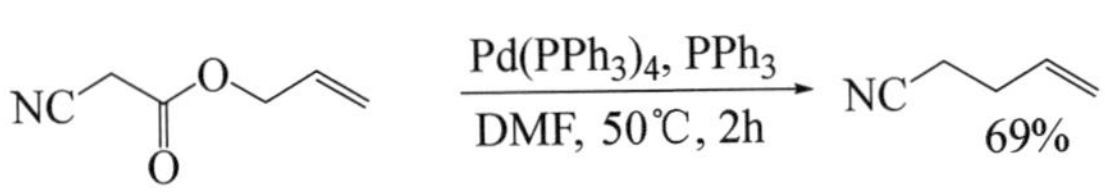

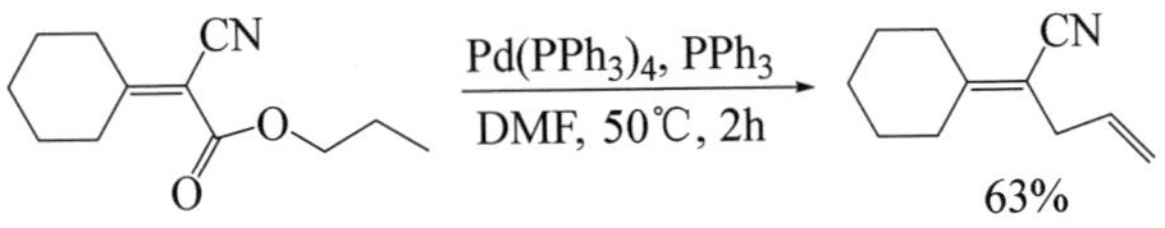

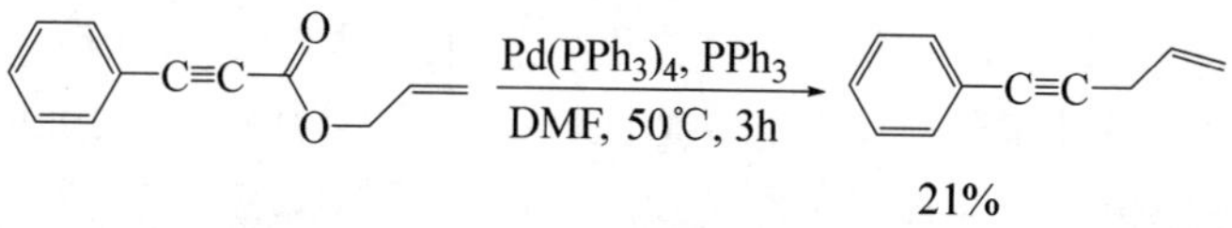

$$\text{Ph-C≡C-COO-CH}_2\text{CH=CH}_2 \xrightarrow[\text{DMF, 50℃, 3h}]{\text{Pd(PPh}_3)_4, \text{PPh}_3} \text{Ph-C≡C-CH}_2\text{CH=CH}_2 \quad 21\% \tag{77}$$

带有烯丙基碳酸酯单元的 β-羰基酸烯丙酯的分子内烯丙基化生成中间体 β-羰基酸烯丙酯，紧接着该中间体发生脱羧基-烯丙基化反应生成 2-烯丙基-3-乙烯基环己酮。之所以能够发生该二烯丙基化反应是基于这样一个事实，即 β-羰基酸酯与烯丙基碳酸酯单元间的分子内烯丙基化比 β-羰基酸烯丙酯的脱羧基烯丙基化反应快[45]。

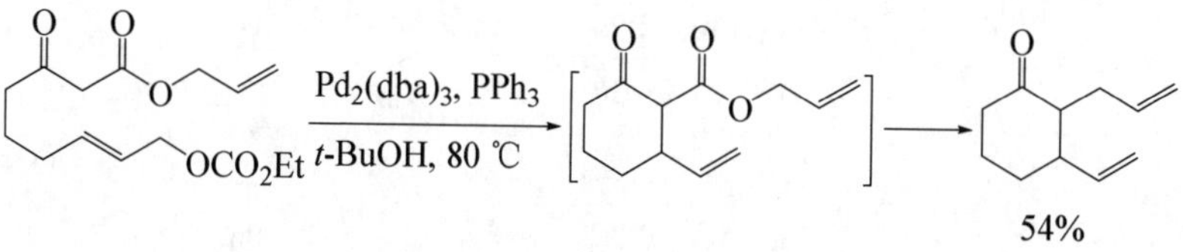

$$\tag{78}\ 54\%$$

烯丙基烯基碳酸酯在钯催化下很容易发生重排以高收率生成烯丙基酮，而烯丙基烯基碳酸酯很容易由氯甲酸烯丙酯与醛或酮的烯醇盐反应制得。烯丙基烯基碳酸酯比 β-羰基酸烯丙酯更容易发生分子内烯丙基化反应，甚至反应在 0℃ 就能进行[46]。

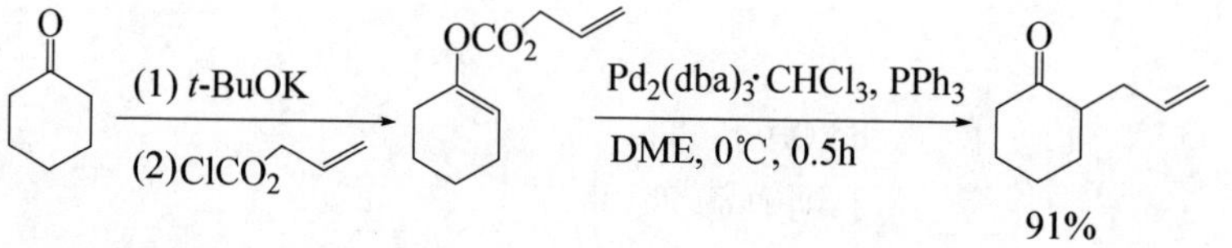

$$\tag{79}\ 91\%$$

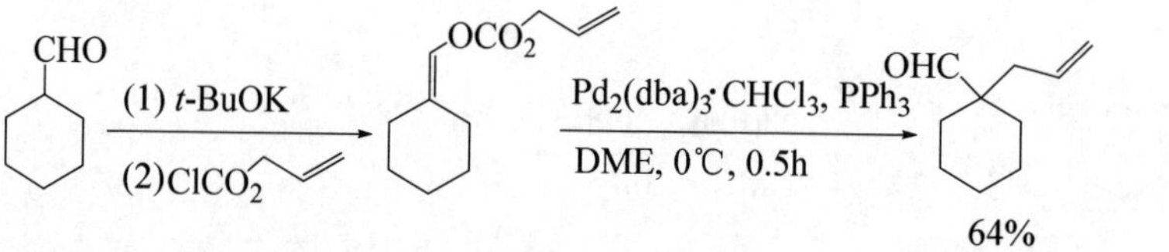

$$\tag{80}\ 64\%$$

8.2.2.2 成环反应

（1）小环化合物的生成

第一个分子内 Tsuji-Trost 反应是生成五元环的反应。如下反应所示，反应物结构中存在一个醋酸烯丙酯单元，而活泼亚甲基同时被甲酯基和砜基活化，两个结构单元通过 3 个碳原子（包括一个六元环上的碳）的碳链连接。反应时加入氢化钠使活泼亚甲基生成碳负离子，进而在钯催化剂 Pd(PPh$_3$)$_4$ 催化下进攻烯丙基发生环合反应生成五元环[47]。

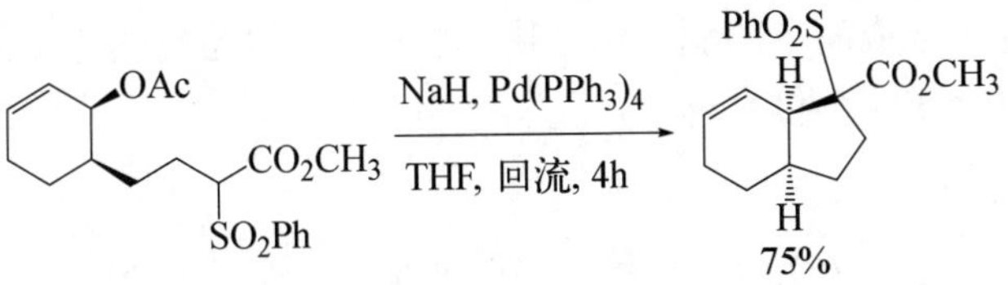

$$\text{(81)} \quad 75\%$$

当碳链少一个碳原子时,相同反应条件下生成 2∶1 的四元环与六元环的混合物。其原因是反应的过渡态不同。其中生成四元环过渡态中的支链处于热力学稳定的假平伏键位置,而生成六元环的过渡态中的支链处于热力学不稳定的假直立键位置。很显然,此分子内环合反应不是由环的大小决定的。

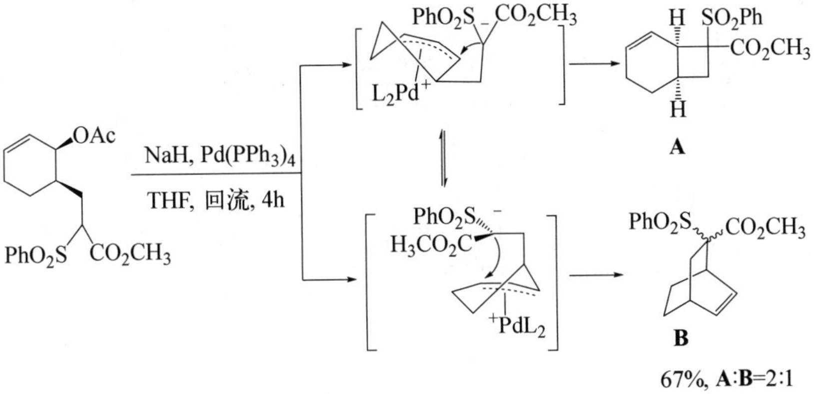

$$\text{(82)} \quad 67\%,\ \mathbf{A}\mathbf{:}\mathbf{B}=2\mathbf{:}1$$

另外,分子内 Tsuji-Trost 环合反应的构型保持不变,也就是环合产物的构型与反应物的构型一致,而传统的通过亲核取代进行的环合反应通常发生构型翻转。因而,此环合反应与传统环合反应互为补充。例如,在八氢化萘衍生物的合成中,顺式的原料生成顺式的环合产物,而反式原料则生成相应的反式环合产物[48]。

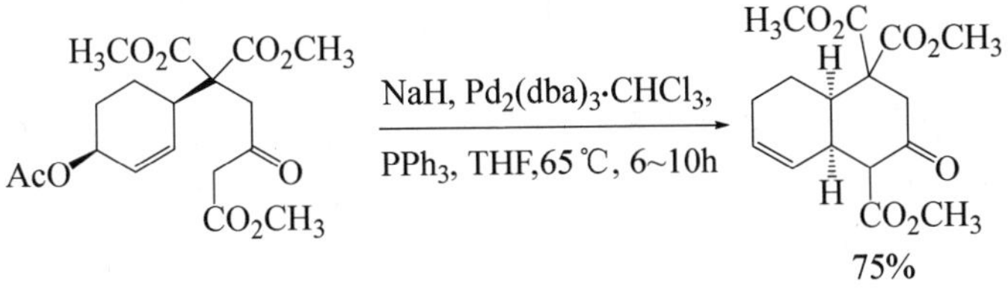

$$\text{(83)} \quad 75\%$$

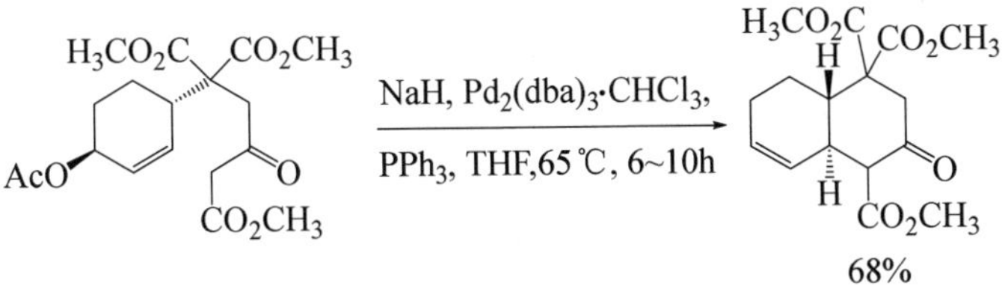

$$\text{(84)} \quad 68\%$$

该反应除了用于合成简单的稠环化合物，改变烯丙醇与侧链连接位置，在相同反应条件下还可以合成螺环化合物[49]。

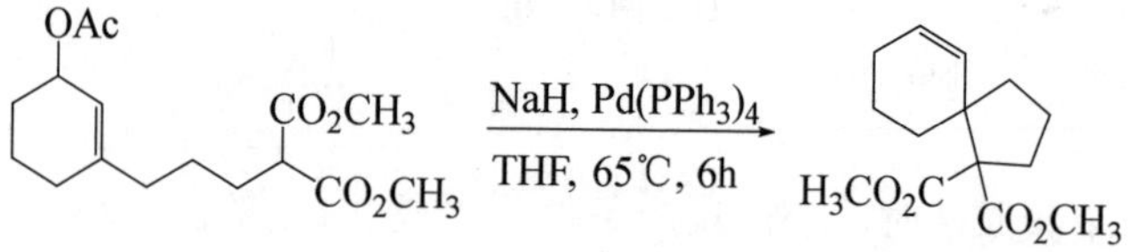

$$\text{（反应式 85）} \tag{85}$$

当亲核的活泼亚基单元不是丙二酸二甲酯，而是 β-羰基酸甲酯时则存在氧负离子与碳负离子的亲核竞争问题，如果能够形成五元杂环，则生成四氢呋喃环。然而，将酯基中的甲基替换为大空间位阻的烷基，并且碳链长度合适时，则得到碳负离子进攻的环合产物。

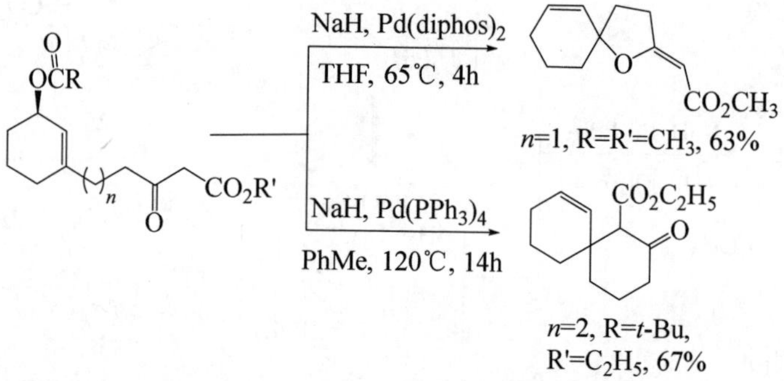

$$\text{（反应式 86）} \tag{86}$$

与 β-羰基酸酯相反，相同催化反应条件下，β-羰基砜则更容易发生碳亲核环化反应，并且立体构型得到保持[50]。

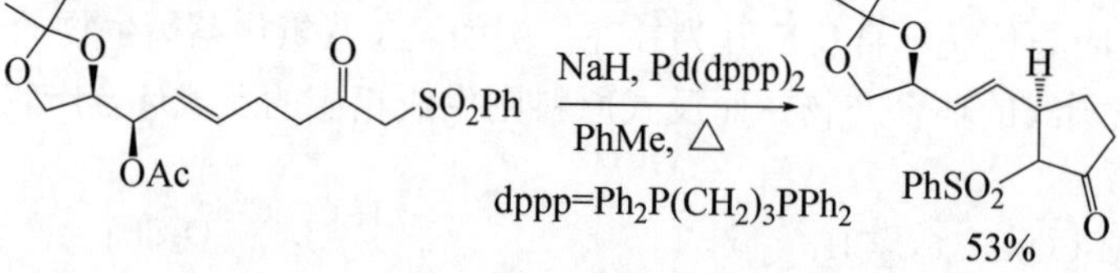

$$\text{（反应式 87）} \tag{87}$$

苯基烯丙基醚也可替代醋酸烯丙酯与活泼亚甲基发生分子内 Tsuji-Trost 环合反应，构成了合成甾体类化合物 D 环的基础，这里苯氧基作为离去基团[51]。用容易操作的钯盐与膦配体原位生成 Pd^0 活性物种进行催化反应是钯催化反应中常用的方法，钯盐可以是氯化钯和醋酸钯，其中氯化钯更容易还原所以更常用。

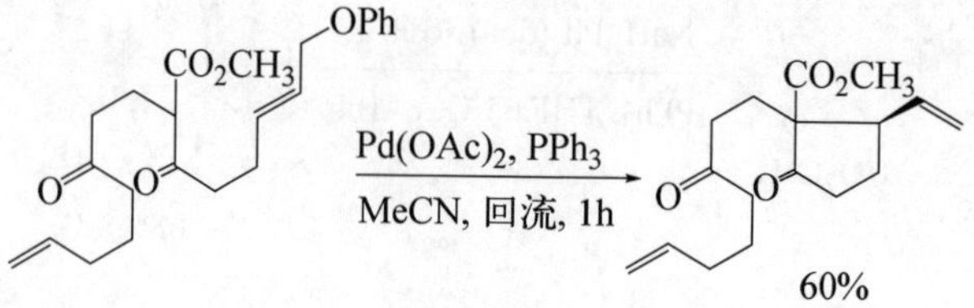

$$\text{（反应式 88）} \tag{88}$$

利用过渡金属催化获得光学活性化合物在有机和药物合成上占有重要地位,因此对映选择性烯丙基化反应自然受到关注。然而,高对映选择性烯丙基化并不容易,因为在烯丙基化反应中碳碳键的构建位点远离手性中心。尽管如此,一些精心设计的不对称催化反应体系获得了中等的对映选择性,其代表性实例是反应式(89)所示的麦角生物碱的成功合成[52,53]。

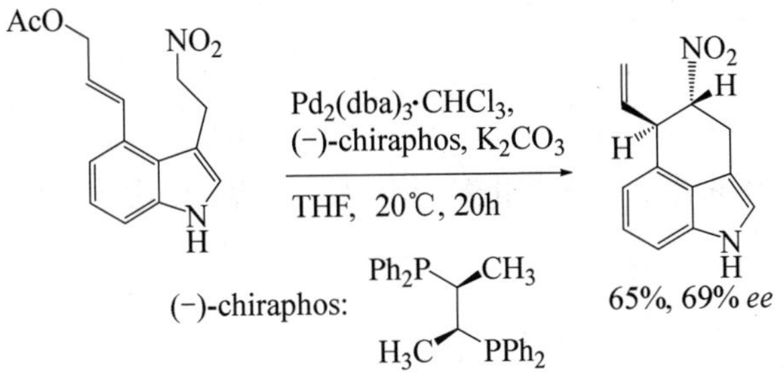

(89)

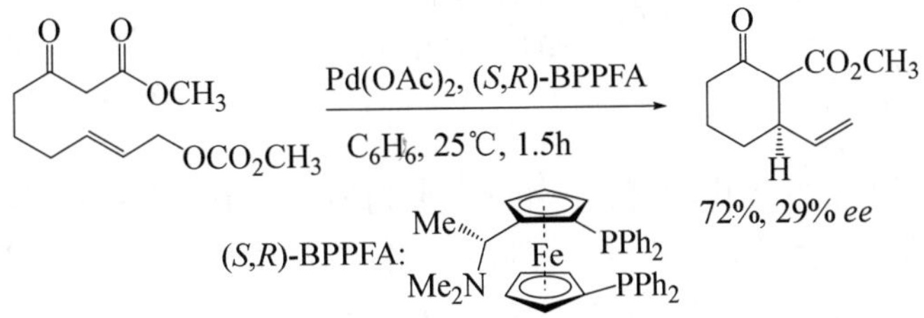

(90)

钯催化的分子内烯丙基化反应还可用来合成更小的环丙烷环。在下面的反应式中,双碳酸甲酯化的烯丙醇衍生物与丙二酸二甲酯钠盐在钯催化下先以端碳进行分子间烯丙基化反应,生成的烯丙基化的丙二酸二甲酯再原位发生分子内烯丙基化反应生成环丙烷环[54]。

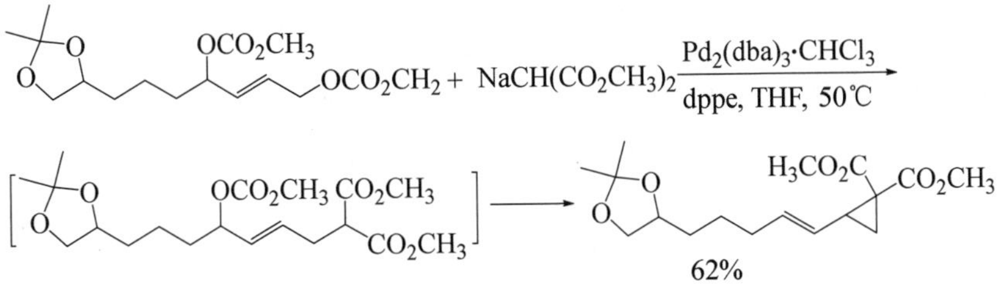

(91)

无论是分子间烯丙基化还是分子内烯丙基化,反应后手性中心的构型得到保持,因此,采用上述串联反应策略,可实现非对映选择性的环丙烷化反应。下面的串联反应就是非对映选择性的环丙烷化反应实例,由于烯丙基氯的反应活性比醋酸烯丙酯的高,烯丙基氯优先与丙酸二甲酯钠发生构型保持的分子间烯丙基化,然后再进行乙酰氧基为离去基团的分子内环丙烷化反应，手性中心的构型得到保持[55]。

$$\text{(92)}$$

以下是类似的反应，采用二氯苯甲酰氧基作为离去基团能更好地保持手性中心的构型[56]。

$$\text{(93)}$$

关于亲核碳上的立体化学研究得很少，因为该碳原子要么是非手性的，要么是在反应后手性消失。然而，在下面的反应中，钯催化的环丙烷化引导空间位阻小的 CN 基绝大部分指向环丙环空间位阻大的一面，而在没有催化剂存在下的反应没有立体选择性[57]。

$$\text{(94)}$$

70% (a:b=19:1)

（2）中和大环化合物的生成

通常，区域选择性决定钯催化的环合反应所生成环的大小。在如图 8-12 所示的钯配合物结构中，若环合反应存在三元环与六元环、八元环或是六元环与八元环的竞争，则主要生成小的环；当存在四元环与六元环或是五元环与七元环的竞争生成问题时，两个环都能生成。

图 8-12　钯催化环合反应的两个途径

在全碳体系环合反应中，若是存在七元环和九元环的竞争生成问题，通过调节环合反应前体的结构，既可以生成七元环也可以生成九元环，而在非过渡金属催化的竞争环合反应中，九元环总是作为次要产物生成。例如，下面两个反应，前者生成七元环，而后者由于提高了亲核基团的空间位阻，则生成九元环[58]。

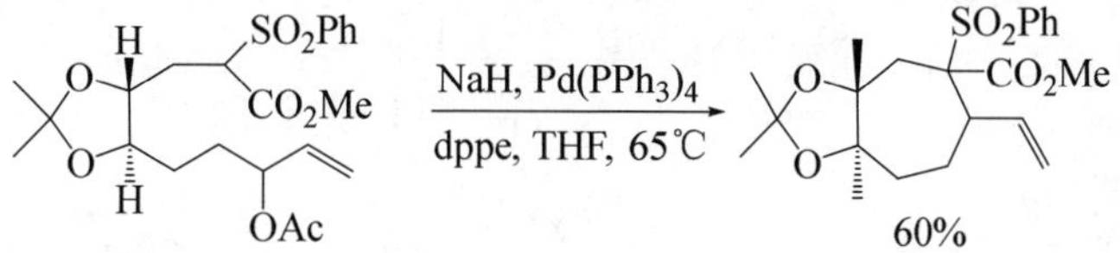

$$\text{(95)}$$

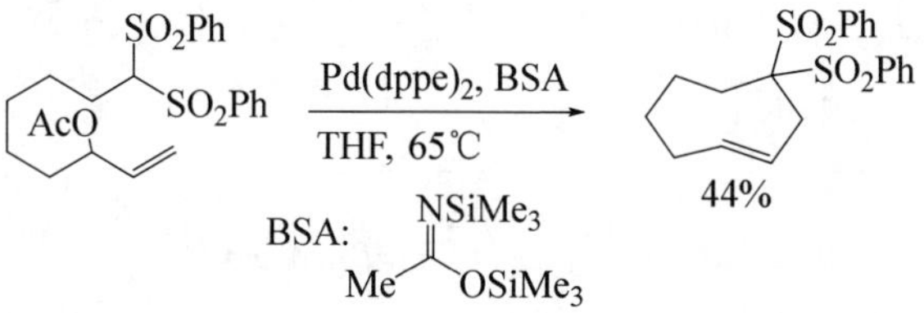

$$\text{(96)}$$

尝试更大环的合成也取得了成功，以下是十一元环[59]和十四元环合成的实例[60]。而且，十四元环的生成是非对映选择性的，因为亲核中心碳原子反应后只生成了如反应式所示的一种非对映异构体。

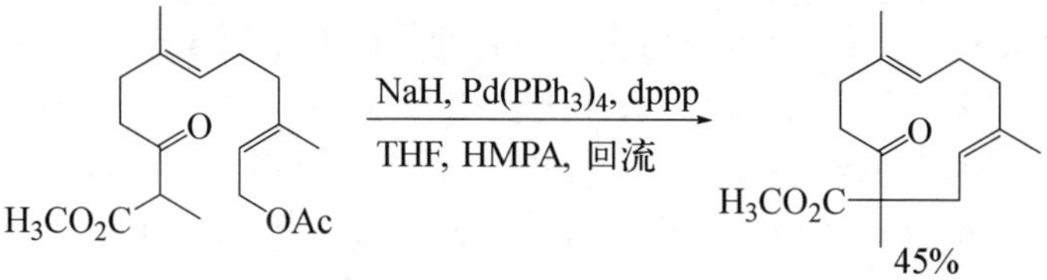

$$\text{(97)}$$

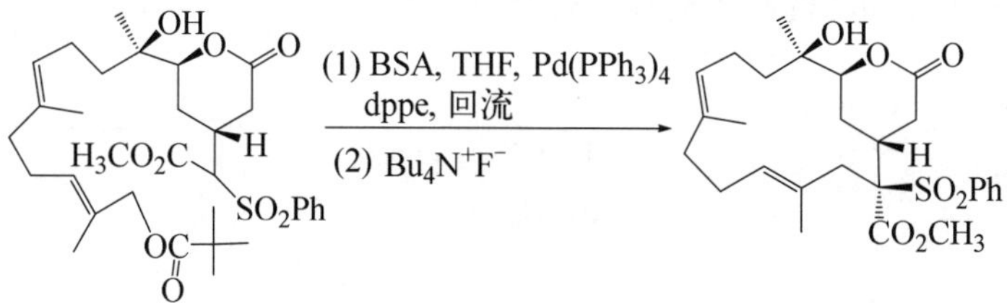

$$\text{(98)}$$

将亲电基团由羧酸酯变成乙烯基环氧化物使环合反应更加容易，而且可用来合成更大的环。需要注意的是，为了避免高浓度下底物发生分子间烯丙基化反应，分子内反应往往需要在低底物浓度下进行。利用聚苯乙烯负载钯催化剂可以产生"伪低浓度效应（pseudodilution effect）"，使得反应在高底物浓度下进行。这是因为在无催化剂存在下惰性的底物间不发生反应，必须扩散到聚苯乙烯并与活性中心相遇才被活化，而聚合物上稀疏的活性位点使得被活化的分子间不能相互接触，仍不能发生分子间反应，只能进行分子内烯丙基化反应从而形成大环。采用这种策略成功合成了十元环和十五元环[61]。此策略还可用于构建大于十五元环的环和小于十元环的环，但只能以很低的收率生成九元环。

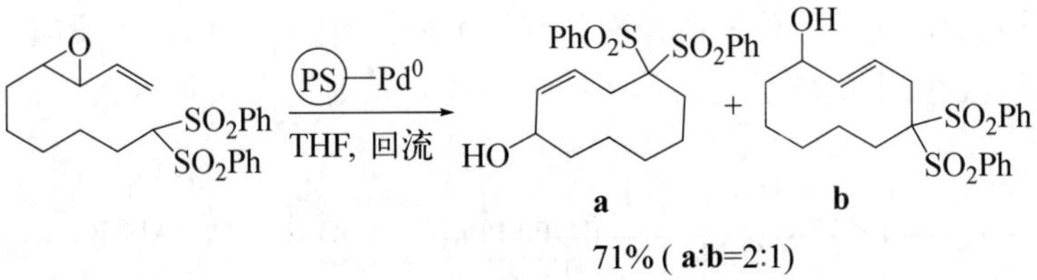

（3）杂环化合物的构建

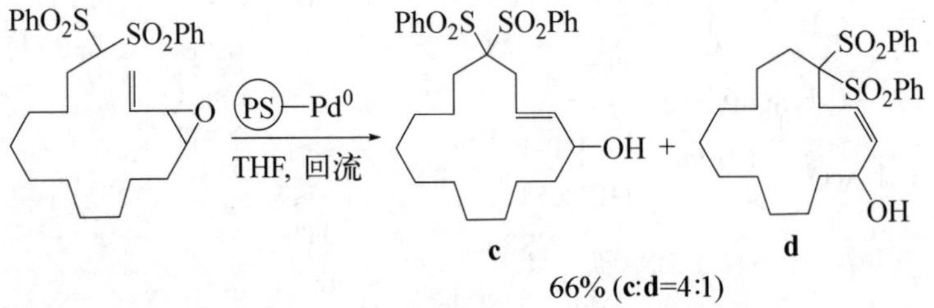

以上反应全部涉及碳环化合物的合成。将杂原子引入连接两个 Tsuji-Trost 反应官能团的连接链中，可以用来合成杂环化合物，但由于杂原子对底物施加的构象作用以及杂原子与催化剂中过渡金属的配位作用，环合反应会受到杂原子的显著影响。在碳环化合物合成中，八元、九元环难以合成，而下面两个连接链上含氧的底物则更倾向几乎完全生成八元、九元含氧杂环化合物[58,62]。其中反应式（101）倾向于生成八元环的部分原因是亲核基团更容易进攻烯丙基取代基少的碳原子一端。因此，反应式（103）所示的底物的反应则生成六元环而非八元环[63]。这种过渡金属促进的反应还具有手性中心的高效转移的优势。

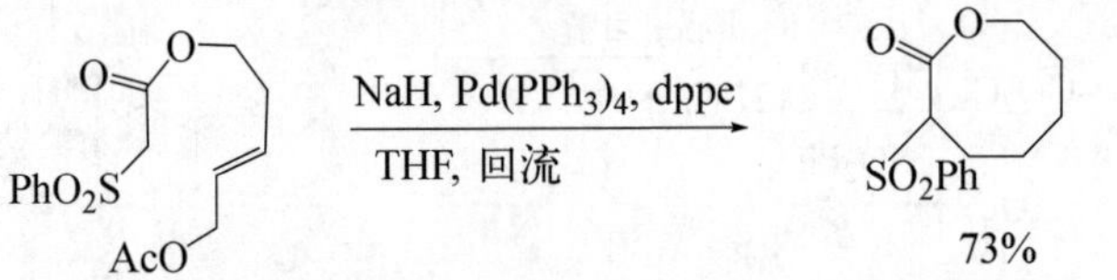

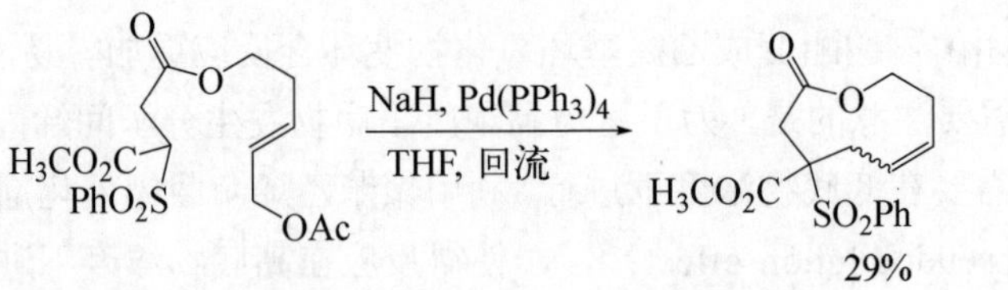

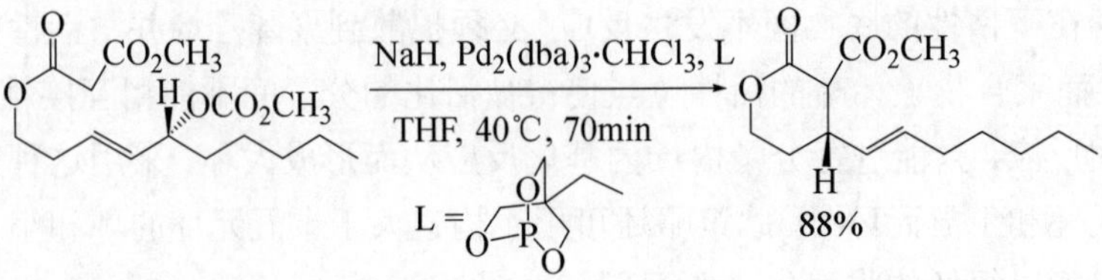

既然难以合成的中环化合物都能通过此类反应构建，其用于容易合成的大环化合物则顺理成章。的确，采用此反应成功合成了十元[64]、十二元[65]以及十六元氧杂环化合物[47]。

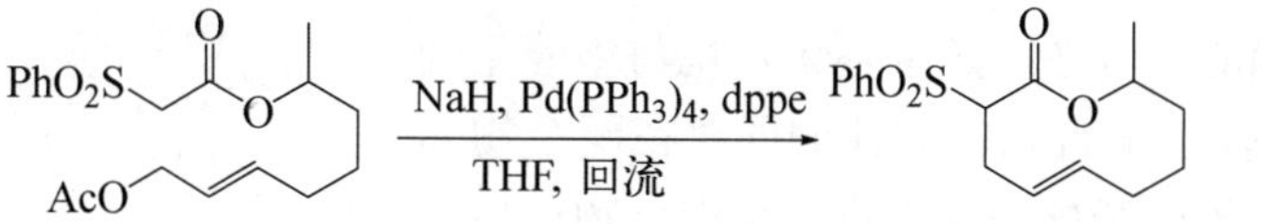

$$\text{(104)} \quad 88\%$$

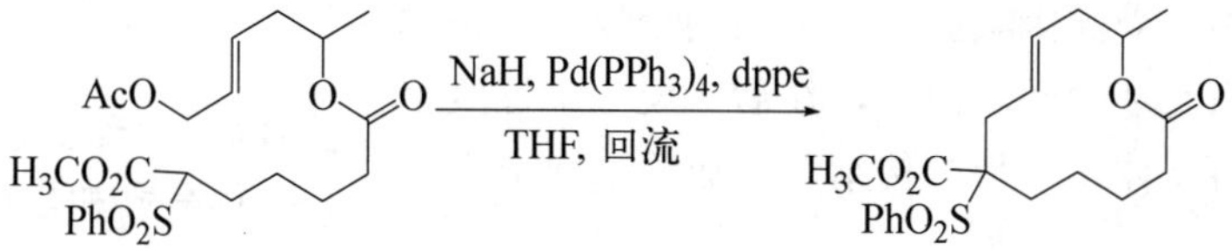

$$\text{(105)} \quad 78\%$$

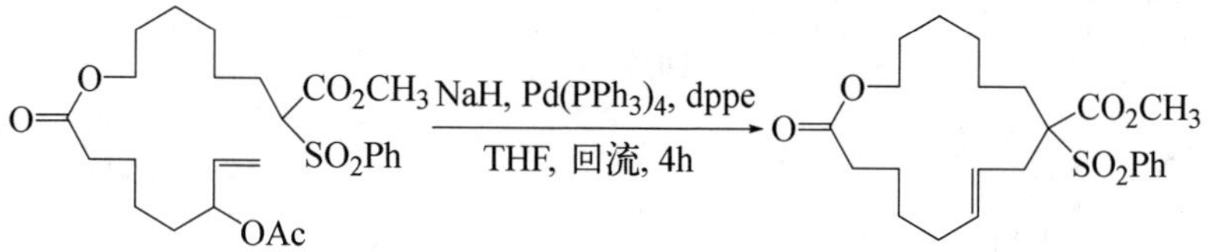

$$\text{(106)} \quad 69\%$$

采用与前述碳环合成中"伪低浓度效应"相同的策略，从乙烯基环氧乙烷出发也能合成大环内酯，这在有机与药物合成上非常有用[66]。例如，下面在 0.1～0.5 mol/L 的较高的底物浓度下以 74%～87%的收率得到十七元大环内酯。底物中活泼次甲基的动力学酸性至关重要，例如当以高聚物负载钯为催化剂时，动力学上酸性弱的砜酯底物不能发生环合反应，而采用均相的标准钯催化剂时则反应顺利进行（动力学酸性与热力学酸性不同，热力学酸性与 pK_a 值关联）。此外，将活泼次甲基换成活泼亚甲基作为预亲核基团成功进行聚合物负载钯（伪低浓度效应）催化的高浓度乙烯基环氧化物的大环构建，进一步说明底物分子动力学酸性的重要性。

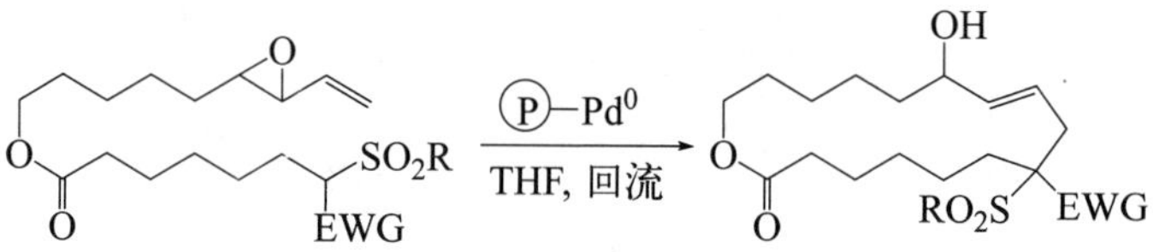

$$\text{(107)}$$

R=Ph, EWG=SO$_2$Ph, 3h, 74%
R=CH$_3$, EWG=SO$_2$CH$_3$, 0.4h, 87%
R=Ph, EWG=CO$_2$CH$_3$, 3h, 0%

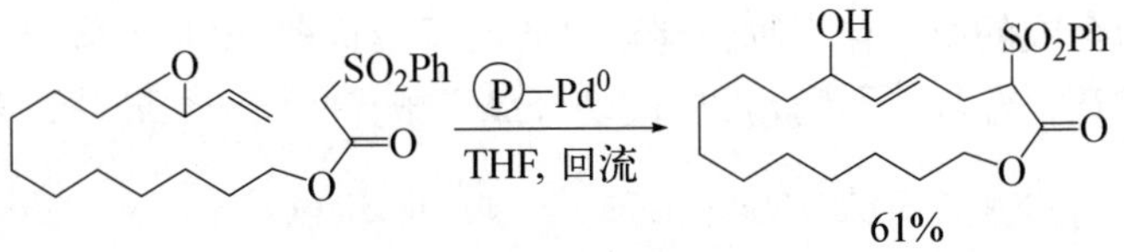

$$(108)$$

需要明确的是，在乙烯基环氧乙烷与钯催化剂形成 η^3-烯丙基钯配合物中间体的同时在配合物邻位生成羟基，随后亲核基团将进攻钯配合物远离羟基的一端环合（图 8-13）。这种选择性进攻将决定生成环的大小。受此因素影响，下面两个反应分别生成十四和六元环[58,67]。而且，在后一反应中过渡金属配合物将立体化学特征传给新生成的手性中心。

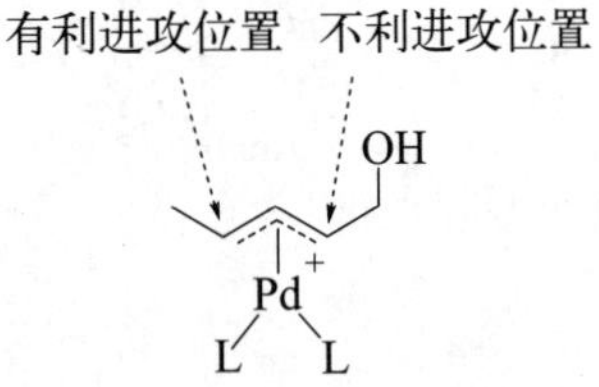

图 8-13 亲核试剂进攻 η^3-烯丙基钯中间体的方向

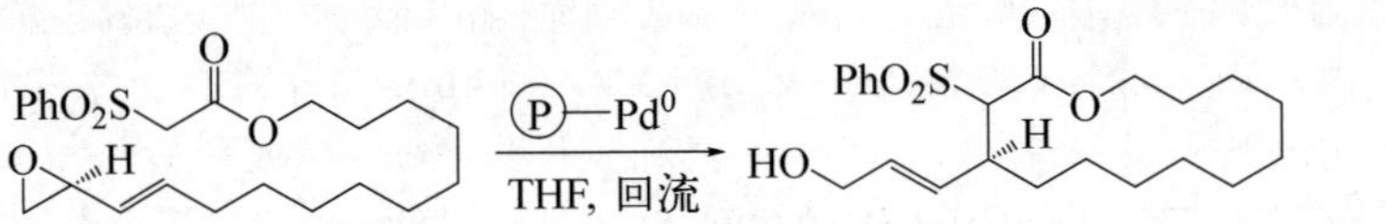

$$(109)$$

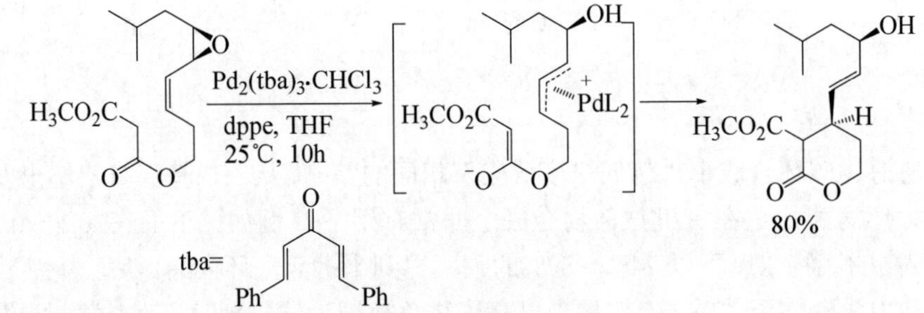

$$(110)$$

tba=

最后，用酰胺连接链替代酯连接链环合反应仍可进行，得到氮杂环化合物。而且，钯催化的化合反应（路线 **a**）与无金属催化的反应（路线 **b**）得到手性中心构型相反的产物，二者在有机和药物合成上互为补充[68]。

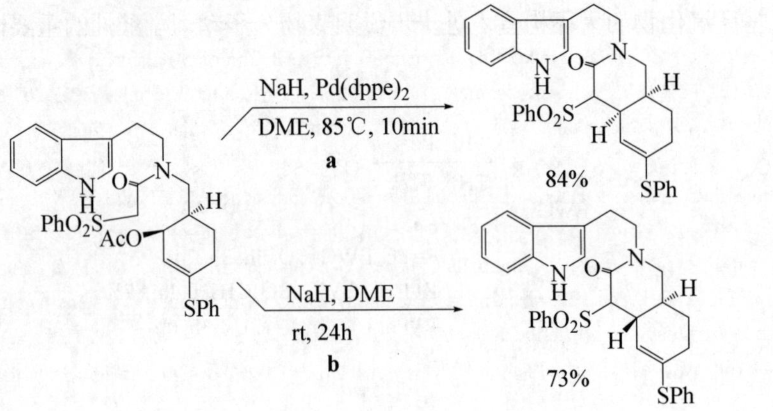

$$(111)$$

8.3 Tsuji-Trost 反应在药物和天然产物合成上的应用实例

由基本反应可知，Tsuji-Trost 反应的亲核底物与烯丙基化试剂种类多样，通过不同底物间的耦合可实现种类繁多的烯丙基化产物的合成；而且，反应在温和条件下进行，具有很高且可预测的化学、区域和立体选择性，因而在药物和天然产物的合成上应用广泛。以下按底物的不同结构类型介绍 Tsuji-Trost 反应在药物或生物活性化合物合成上的一些应用实例。为叙述方便，把 Tsuji-Trost 反应产物编号为 TT-x，x 为整数,代表序号。

茉莉酮酸及其甲酯是一类植物激素，能够促进植物蛋白质的储存。茉莉酮酸和其甲酯可以顺利地以乙烯基环戊烯醇锂为起始原料经 Tsuji-Trost 反应合成。首先乙烯基环戊烯醇锂与 BEt_3 生成复盐，复盐原位与 (Z)-2-戊烯基醋酸酯在钯催化下进行 Tsuji-Trost 反应生成中间体 TT-1，TT-1 再经多步转化生成茉莉酮酸甲酯 [methyl (Z)-jasmonate] [69]。

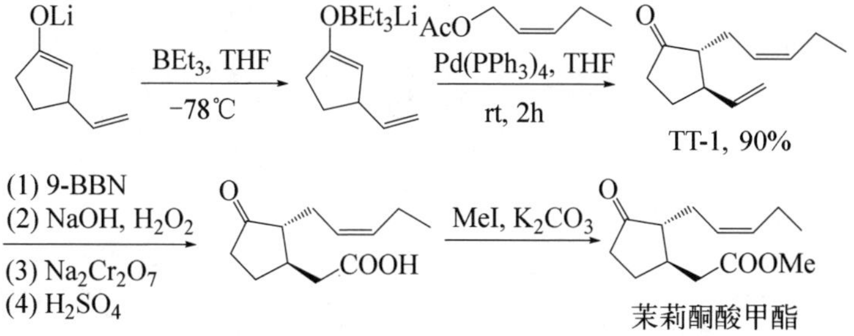

$$(112)$$

Kujounin A_2 是从日本一种葱属植物种分离出的天然产物，具有抗癌和抗炎活性。其结构和合成路线如反应式（113）所示。合成的起始步骤是抗坏血酸与一种碳酸烯丙酯，即乙氧基甲酸（4-苯基丁烯-2-醇）酯，以 $Pd(acac)_2/PPh_3$ 为催化剂在温和条件下进行烯丙基化反应，能以 6∶1 的比例生成直接环合的中间体 TT-2 和开环的异构体 TT-2′，二者不易分离，故不经分离直接进行臭氧化以 86% 总收率生成非对映异构体 **A** 和 **B**，二者比例 63∶37。**A** 再经多步反应生成 Kujounin A_2 [70]。

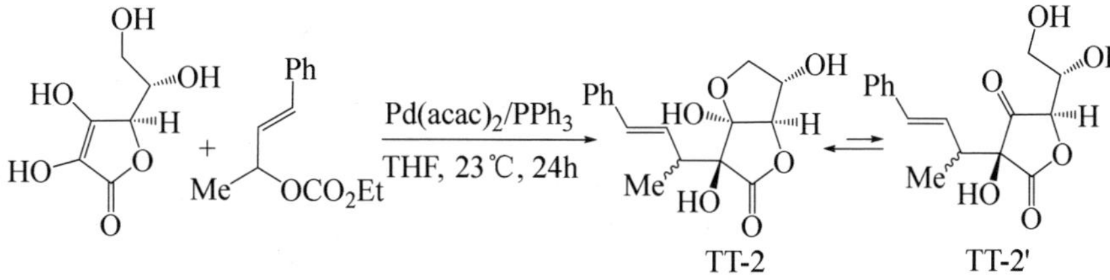

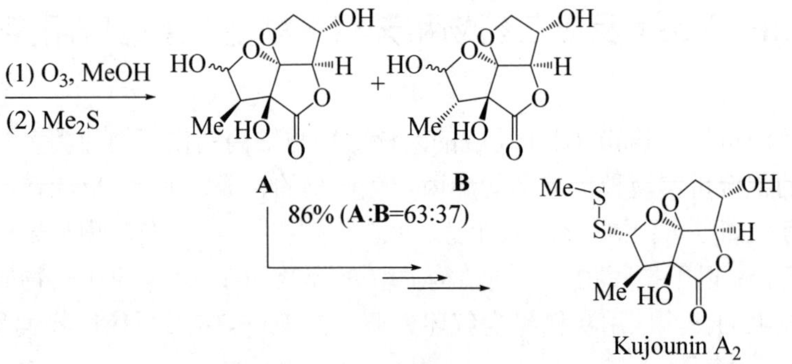

吲嗪生物碱（indolizidine alkaloids）是一类生物活性天然产物，*allo*-pumiliotoxin 339B 是其中最复杂的一种。利用乙烯基环氧化物与砜类化合物前体在温和的 Tsuji-Trost 催化反应体系下顺利反应生成关键中间体 TT-3，反应过程中乙烯基环氧化物重排与砜基活化亚甲基偶联，同时环氧基开环衍生一个羟基，手性中心的构型得到传递，TT-3 用 Na-Hg 齐脱砜基，两步总收率 24%；最后用 LiAlH$_4$ 还原羰基并同时脱 TBDPS（叔丁基二苯基硅基）以 68% 的收率得到 *allo*-pumiliotoxin 339B[71]。

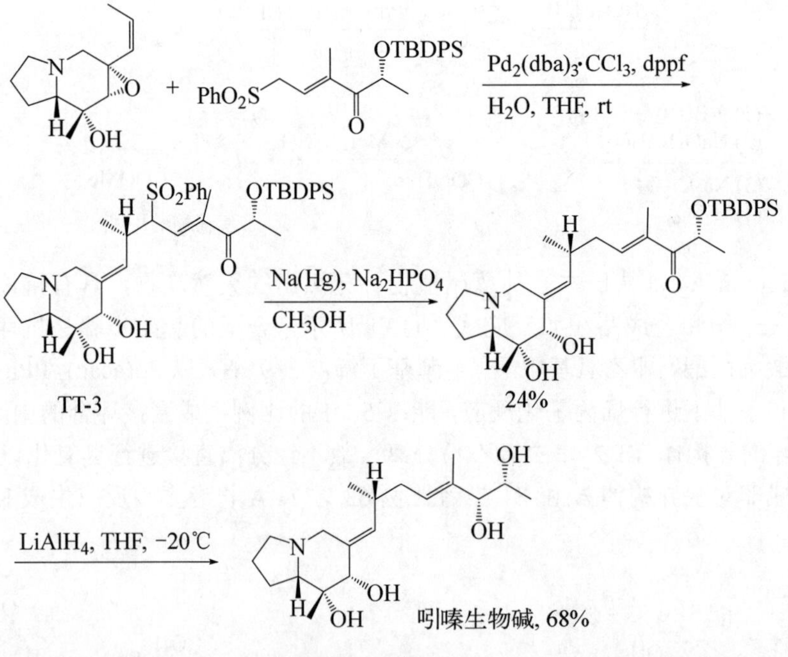

河鲀毒素（tetrodotoxin）最初是由河豚中分离出的，能够特异性抑制电压依赖性钠离子通道，被广泛应用于神经生理学和神经科学，其合成极具挑战性。其合成方法之一如反应式(115)所示。关键步骤之一是利用 Tsuji-Trost 反应在氧桥双环化合物中引入烯丙基合成中间体 TT-4。TT-4 再经多步转化生成河鲀毒素。

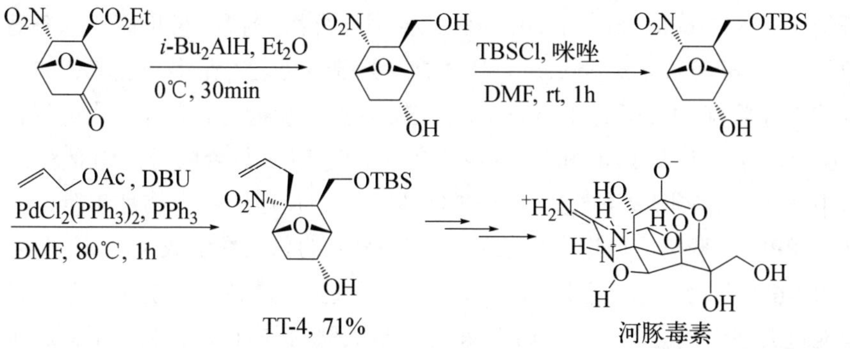

$$\text{(115)}$$

士的宁(strychnine)又名番木鳖碱，是由马钱子中提取的一种生物碱，能选择性兴奋脊髓，增强骨骼肌的紧张度，临床用于轻瘫或弱视的治疗。其合成的第一个关键步骤是在经典 Tsuji-Trost 催化反应体系下乙酰乙酸乙酯衍生物被对映体纯的碳酸烯丙酯烯丙基化，生成关键中间体 TT-5。在此步反应中以下几点值得注意：首先，连接离去基团手性碳的构型反应后完全得到保持；其次，烯丙基化具有区域选择性，尽管环戊烯衍生物含有碳酸烯丙酯和乙酰氧基两个离去基团，但反应只发生在碳酸酯一端，因此只生成 TT-5；最后，TT-5 是与羰基相邻手性中心的对映体的混合物，但二者经后续反应得到相同的中间体 **a**。中间体 **a** 与芳基碘进行羰基化 Stille 反应得到另一关键中间体 **b**，中间体 **b** 再多步转化最终完成（−）-士的宁的全合成[72,73]。

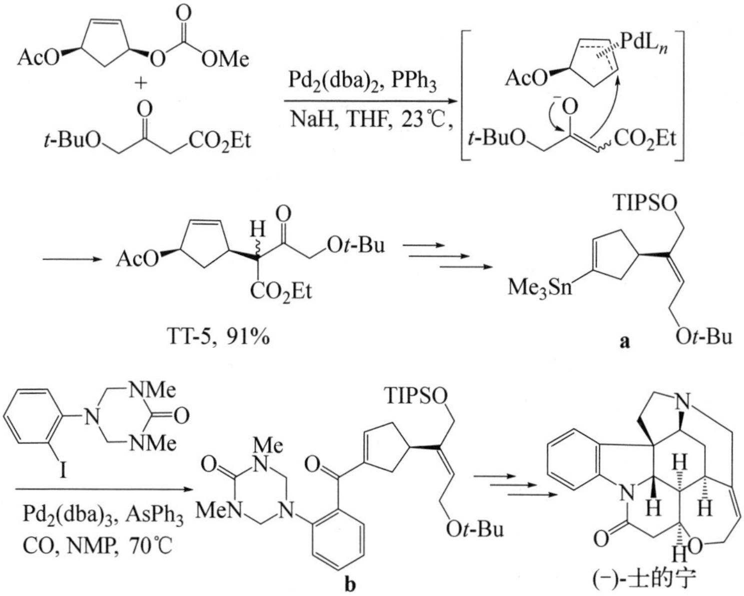

$$\text{(116)}$$

哥伦比亚素 A(colombiasin A)是一种二萜类化合物,分离自哥伦比亚一海岛的戈氏八角珊瑚（*Gorgonian octocoral*）。其合成的关键步骤之一是多基团取代的二氢萘酚基碳酸-2-丁烯酯的分子内 Tsuji-Trost 反应。在此步反应中,烯丙基化反应有 **a** 和 **b** 两种途径,分别生成 TT-6 和 TT-6'两个产物,而 TT-6 是合成哥伦比亚素 A 所需的关键中间体。正常情况下,受空间位阻效应影响,反应的主产物是 TT-6'而非 TT-6 。加入 PPh_3 改变了区域选择性的方向,以 88%的总收率生成了以 TT-6 为主的 TT-6 与 TT-6'混合物,二者之比为 2.4∶1。区域选择性的改变可能是三苯基膦配体使得反应过渡态中巴豆基（甲基烯丙基）上显示部分正电荷,有利于亲核反应发生在多取代的一端。用 σ-供电子的配体如 dppe、$P(Oi\text{-}Pr)_3$ 等替换 PPh_3 以降低过渡态的正电荷密度,将生成正常区域选择的产物（TT-6'）为主的产物[74]。TT-6 经后续多步反应生成哥伦比亚素 A。

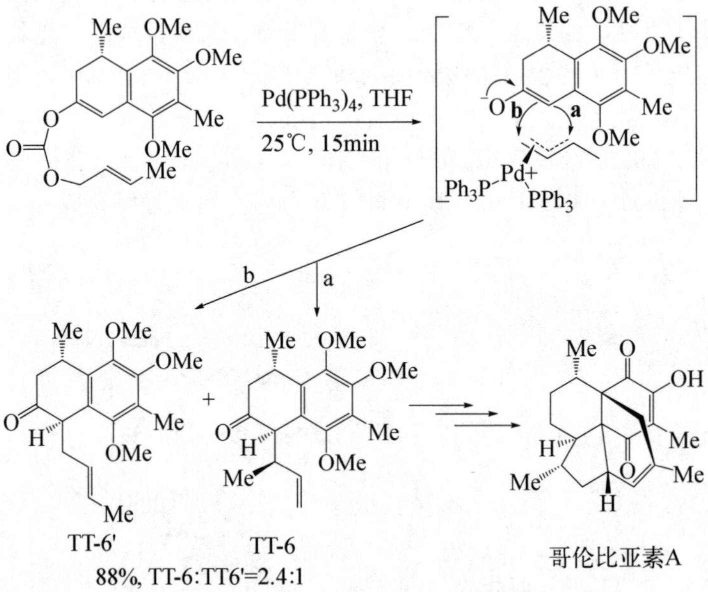

$$\text{(117)}$$

Tsuji-Trost 反应最精妙的应用实例之一是结构独特的生物碱 roseophilin 的全合成。其合成存在两个挑战,即氮杂富烯发色团和大张力的十三元碳环的构建,这两个挑战都通过 Tsuji-Trost 反应得到解决。如反应式（118）所示,将烯丙基环氧化物的稀溶液缓慢加入加热回流的催化量的 $Pd(PPh_3)_3$-dppe THF 溶液中,反应 6h,以 85%的收率得到大环化合物 TT-7;TT-7 经两步反应得到另一中间体 A, A 与苄胺在钯催化剂催化下发生分子间氮原子为亲核试剂的 Tsuji-Trost 反应,生成间位双取代吡咯环化合物 TT-7'。这种有趣的钯催化生成吡咯环的反应在温和的中性条件下就能顺利地进行，成为传统上酸催化的由非环状的二酮或其等同物缩合合成吡咯环的有力替代方法。TT-7'再经后续转化最终生成 roseophilin[74]。

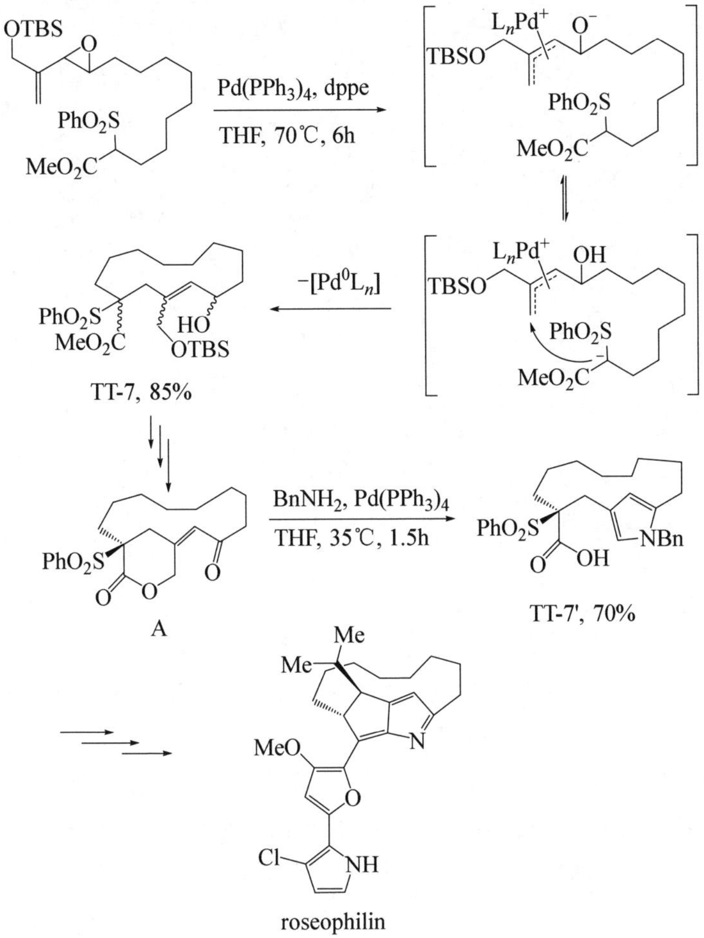

(118)

不对称 Tsuji-Trost 反应被成功应用于天然产物 hamigeran B 合成中的手性中心的构建。该天然产物分离自海洋海绵 *Hamigera tarangaensis*，具有抗病毒活性。如下反应所示，起始原料 2-甲基-5-(叔丁氧基)亚甲基环戊酮与醋酸烯丙酯在钯催化剂和手性配体原位生成的手性催化剂催化下进行不对称烯丙基化反应以 77%的收率和 93%的对映选择性生成烯丙基化产物 TT-8，TT-8 再经包括分子内 Heck 反应在内的多步反应生成 hamigeran B。由不对称 Tsuji-Trost 反应先期构建的手性中心诱导，在分子内 Heck 反应中又非对映选择性地生成第二个手性中心，最终合成的 hamigeran B 具有光学活性[74,75]。

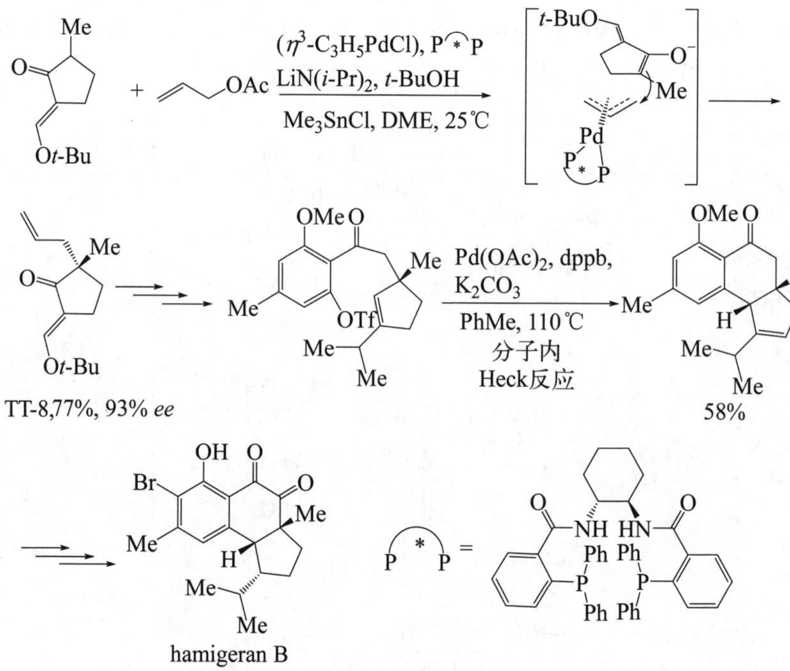

钯催化的不对称 Tsuji-Trost 反应最广泛的应用之一是内消旋底物的去对称化。容易得到的常处于环状结构中的内消旋 2-烯-1,4-二醇衍生物是最常用的亲电底物。该策略被成功用于对映选择性全合成生物碱 γ-lycorane[74]。后者是存在于石蒜科植物中的重要天然产物,具有抗肿瘤、抗乙酰胆碱酯酶、抗炎等功效。合成的关键一步是内消旋 1,4-环己烯二醇二苯甲酸酯与由酰胺与二异丙基氨基锂反应原位生成的酰胺碳负离子间的偶联去内消旋化,反应在催化量的 Pd(OAc)$_2$ 和手性膦配体（L）存在下进行反应生成光学活性的关键中间体 TT-9,但对映体选择性不高,只有 54%。得到关键中间体 TT-9 后,剩下的主要问题是目标产物分子结构中与环己烯骨架稠合的两个环的构建。从结构上分析可以预期 TT-9 结构中尚存烯丙基酯,能够进行钯催化的对相邻酰胺氮的分子内烯丙基化,即杂原子为亲核试剂的 Tsuji-Trost 反应,进而与分子内 Heck 反应串联从而连续构建两个环。因此,将 TT-9 在 DMF 中用 Pd(OAc)$_2$、dppb 和 NaH 在 50℃处理引发分子内烯丙基化生成预期的中间体 TT-9',然后加入 i-Pr$_2$NEt 并升温到 100℃,顺理成章地进行分子内 Heck 环合反应,以唯一的非对映异构体和 58%的收率生成五环稠合的中间体 TH-1。值得注意的是分子内 Heck 反应和氮杂原子烯丙基化反应的区域和立体化选择性全部由 TT-9 的立体效应和结构特征所决定,无须额外手性配体的引入。这些实例说明 Tsuji-Trost 反应是构建碳-碳键的有效方法,而且在不对称合成上有潜在的应用前景。

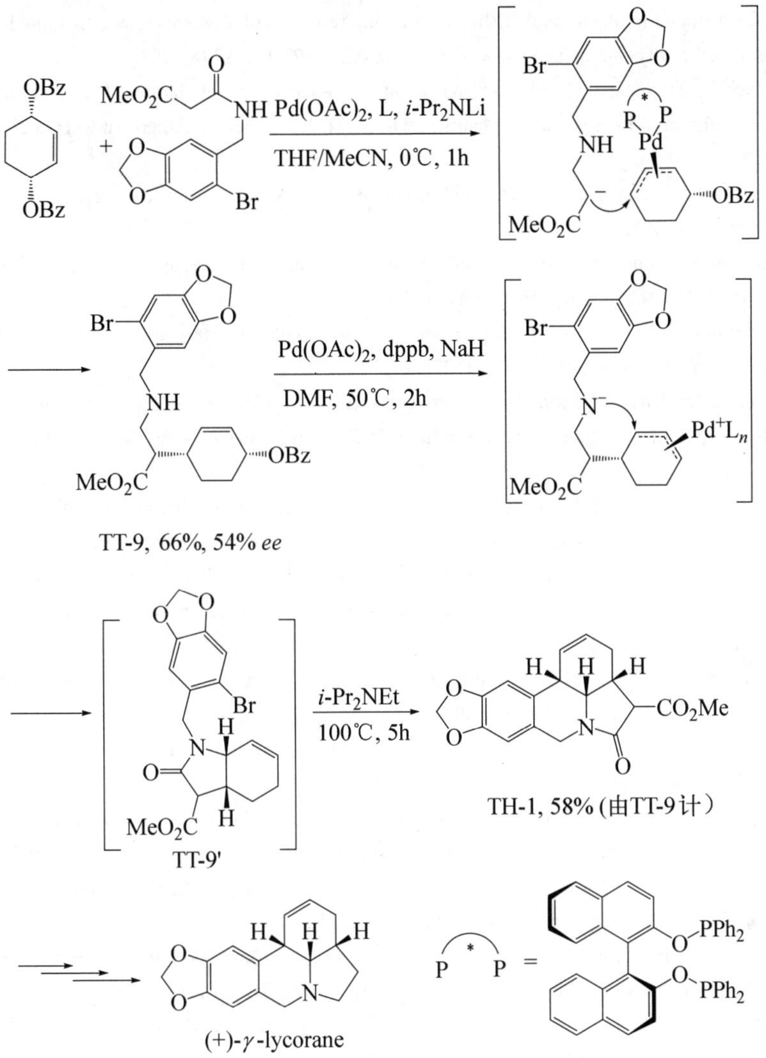

参考文献

[1] Tsuji J. The Tsuji–Trost reaction and related carbon–carbon bond formation reactions palladium-catalyzed nucleophilic substitution involving allylpalladium, propargylpalladium, and related derivatives[M]// Negishi E I. Handbook of organopalladium chemistry for organic synthesis. New York: John Wiley & Sons Inc, 2002: 1669-1675.

[2] Zhang M M, Wang Y N, Lu L Q, et al. Light up the transition metal-catalyzed single-electron allylation[J]. Trends in Chemistry, 2020, 2(8): 764-775.

[3] Yamamoto Y, Nakagai Y, Itoh K. Ruthenium-catalyzed one-pot double allylation/cycloisomerization of 1,3-dicarbonyl compounds leading to *exo*-methylenecyclopentanes[J]. Chemistry: A European Journal, 2004, 10(1): 231-236.

[4] Bricout H, Carpentier J F, Mortreux A. Efficient coupling reactions of allylamines with soft nucleophiles using nickel-based catalysts[J]. Chemical Communications, 1997(15): 1393-1394.

[5] Suzuki Y, Seki T, Tanaka S, et al. Intramolecular Tsuji-Trost-type allylation of carboxylic acids: Asymmetric synthesis of highly π-allyl donative lactones[J]. Journal of the American Chemical Society, 2015, 137(30): 9539-9542.

[6] Li J J. Name reactions: a collection of detailed mechanisms and synthetic applications[M]. 5th ed. New York: Springer, 2014: 605.

[7] Tsuji J. New general synthetic methods involving π-allylpalladium complexes as intermediates and neutral reaction conditions[J]. Tetrahedron, 1986, 42(16): 4361-4401.

[8] Frost C G, Howarth J, Williams J M J. Selectivity in palladium catalysed allylic substitution[J]. Tetrahedron: Asymmetry, 1992, 3(9): 1089-1122.

[9] Trost B M, Verhoeven T R. Stereocontrolled approach to steroid side chain via organopalladium chemistry. Partial synthesis of 5. alpha.-cholestanone[J]. Journal of The American Chemical Society, 1978, 100(11): 3435-3443.

[10] Andersson P G, Baeckvall J E. Synthesis of furanoid terpenes via an efficient palladium-catalyzed cyclization of 4, 6-dienols[J]. The Journal of Organic Chemistry, 1991, 56(18): 5349-5353.

[11] Hayashi T, Yamamoto A, Hagihara T. Stereo-and regiochemistry in palladium-catalyzed nucleophilic substitution of optically active (E)-and (Z)-allyl acetates[J]. The Journal of Organic Chemistry, 1986, 51(5): 723-727.

[12] Trost B M, Radinov R, Grenzer E M. Asymmetric alkylation of β-ketoesters[J]. Journal of the American Chemical Society, 1997, 119(33): 7879-7880.

[13] Ferroud D, Genet J P, Muzart J. Allylic alkylations catalyzed by the couple palladium complexes-alumina[J]. Tetrahedron Letters, 1984, 25(39): 4379-4382.

[14] Poli G, Giambastiani G, Mordini A. Palladium-catalyzed allylic alkylations via titanated nucleophiles: a new early-late heterobimetallic system[J]. The Journal of Organic Chemistry, 1999, 64(8): 2962-2965.

[15] Tanigawa Y, Nishimura K, Kawasaki A, et al. Palladium(0)-catalyzed allylic alkylation and amination of allylic phosphates[J]. Tetrahedron Letters, 1982, 23(52): 5549-5552.

[16] Jiang Z Y, Zhang C H, Gu F L, et al. Efficient iron-catalyzed tsuji-trost coupling reaction of aromatic allylic amides through a sp^3-carbon-nitrogen breaking[J]. Synlett, 2010(8): 1251-1254.

[17] Sakakibara M, Ogawa A. Pd-catalyzed allylic alkylation of phenylvinylcarbinols with some nucleophiles[J]. Tetrahedron Letters, 1994, 35(43): 8013-8014.

[18] Takahashi K, Miyake A, Hata G. Palladium-catalyzed exchange of allylic groups of ethers and esters with active hydrogen compounds. II [J]. Bulletin of the Chemical Society of Japan, 1972, 45(1): 230-236.

[19] Vicart N, Goré J, Cazes B. Palladium-catalyzed substitution of acrolein acetals by β-dicarbonyl nucleophiles[J]. Tetrahedron, 1998, 54(37): 11063-11078.

[20] Trost B M, Molander G A. Neutral alkylations via palladium(0) catalysis[J]. Journal of the American Chemical Society, 1981, 103(19): 5969-5972.

[21] Tsuji J, Kataoka H, Kobayashi Y. Regioselective 1, 4-addition of nucleophiles to 1, 3-diene monoepoxides catalyzed by palladium complex[J]. Tetrahedron Letters, 1981, 22(27): 2575-2578.

[22] Tsuji J, Yuhara M, Minato M, et al. Palladium-catalyzed regioselective reactions of silyl-substituted allylic carbonates and vinyl epoxide[J]. Tetrahedron Letters, 1988, 29(3): 343-346.

[23] Banwell M G, Nugent T C, Mackay M F, et al. Regio-and stereo-chemical outcomes in the nucleophilic

ring cleavage reactions of mono-epoxides derived from *cis*-1, 2-dihydrocatechols[J]. Journal of the Chemical Society, Perkin Transactions 1, 1997,(12): 1779-1792.

[24] Atkins K E, Walker W E, Manyik R M. Palladium catalyzed transfer of allylic groups[J]. Tetrahedron Letters, 1970, 11(43): 3821-3824.

[25] Hirao T, Yamada N, Ohshiro Y, et al. Palladium-catalyzed reaction of allylic ammonium bromides with nucleophiles[J]. Journal of Organometallic Chemistry, 1982, 236(3): 409-414.

[26] Rhee H, Yoon D, Kim S. Pd(0)-catalyzed coupling reaction of allyl *N*, *N*-ditosylimide and allyl *N*-tosylamides as new substrates for the formation of π-allylpalladium complexes[J]. Bulletin of the Korean Chemical Society, 1998, 19(1): 25-27.

[27] Tamura R, Hegedus L S. Palladium(0)-catalyzed allylic alkylation and amination of allylnitroalkanes[J]. Journal of the American Chemical Society, 1982, 104(13): 3727-3729.

[28] Ono N, Hamamoto I, Kaji A. Palladium-catalysed allylic alkylation of allylic nitro compounds with stabilized carbanions[J]. Journal of the Chemical Society, Perkin Transactions 1, 1986: 1439-1443.

[29] Trost B M, Keinan E. Enolstannanes as electrofugal groups in allylic alkylation[J]. Tetrahedron Letters, 1980, 21(27): 2591-2594.

[30] Negishi E, John R A. Selective carbon-carbon bond formation via transition-metal catalysis. Countercation effects on the palladium-catalyzed allylation of enolates[J]. The Journal of Organic Chemistry, 1983, 48(22): 4098-4102.

[31] Tsuji J, Minami I, Shimizu I. Palladium-catalyzed allylation of ketones and aldehydes with allylic carbonates via silyl enol ethers under neutral conditions[J]. Chemistry Letters, 1983, 12(8): 1325-1326.

[32] Tsuji J, Minami I, Shimizu I. Allylation of ketones via their enol acetates catalyzed by palladium-phosphine complexes and organotin compounds[J]. Tetrahedron Letters, 1983, 24(43): 4713-4714.

[33] Boldrini G P, Mengoli M, Tagliavini E, et al. Palladium catalyzed allylation of reformatsky reagents. Synthesis of γ, δ-unsaturated esters[J]. Tetrahedron Letters, 1986, 27(35): 4223-4226.

[34] Moorlag H, De Vries J G, Kaptein B, et al. Palladium-catalyzed allylation of α-hydroxy acids[J]. Recueil Des Travaux Chimiques Des Pays-Bas, 1992, 111(3): 129-137.

[35] Onoue H, Moritani I, Murahashi S I. Reaction of cycloalkanone enamines with allylic compounds in the presence of palladium complexes[J]. Tetrahedron Letters, 1973, 14(2): 121-124.

[36] Huang Y, Lu X. Palladium catalyzed annulation reaction using a bifunctional allylic alkylating agent[J]. Tetrahedron Letters, 1988, 29(44): 5663-5664.

[37] Hiroi K, Abe J, Suya K, et al. Palladium-catalyzed asymmetric allylations of aldehydes via (*S*)-proline allyl ester enamines[J]. Tetrahedron Letters, 1989, 30(12): 1543-1546.

[38] Hiroi K, Abe J. Palladium catalyzed asymmetric allylations of chiral enamines bearing phosphine functionality. Effects of anionic counterparts of allylating reagents on asymmetric induction[J]. Tetrahedron Letters, 1990, 31(25): 3623-3626.

[39] Mikami K, Takahashi K, Nakai T. Diastereocontrol via the phenol-and palladium(Ⅱ)-catalyzed claisen rearrangement with cyclic enol ethers[J]. Tetrahedron Letters, 1987, 28(47): 5879-5882.

[40] Sugiura M, Yanagisawa M, Nakai T. An improved procedure for the Pd(Ⅱ)-catalyzed claisen rearrangement via in situ enol ether exchange[J]. Synlett, 1995(5): 447-448.

[41] Wade P A, Morrow S D, Hardinger S A. Palladium catalysis as a means for promoting the allylic C-alkylation of nitro compounds[J]. The Journal of Organic Chemistry, 1982, 47(2): 365-367.

[42] Hatcher M A, Posner G H. The carroll rearrangement[M]// Hiersemann M, Nubbemeyer U. The Claisen rearrangement: methods and applications. Weinheim: Wiley-VCH Verlag GmbH & Co. KGaA, 2007:

397-430.

[43] Shimizu I, Yamada T, Tsuji J. Palladium-catalyzed rearrangement of allylic esters of acetoacetic acid to give γ, δ-unsaturated methyl ketones[J]. Tetrahedron Letters, 1980, 21(33): 3199-3202.

[44] Tsuda T, Chujo Y, Nishi S, et al. Facile generation of a reactive palladium(Ⅱ) enolate intermediate by the decarboxylation of palladium(Ⅱ). beta.-ketocarboxylate and its utilization in allylic acylation[J]. Journal of the American Chemical Society, 1980, 102(20): 6381-6384.

[45] Shimizu I, Ohashi Y, Tsuji J. A new one-pot method for α, α-diallylation of ketones based on the palladium-catalyzed reaction of allylic carbonates and allyl β-keto carboxylates under neutral conditions[J]. Tetrahedron Letters, 1983, 24(36): 3865-3868.

[46] Tsuji J, Minami I, Shimizu I. Palladium-catalyzed allylation of ketones and aldehydes via allyl enol carbonates[J]. Tetrahedron Letters, 1983, 24(17): 1793-1796.

[47] Trost B M, Verhoeven T R. Cyclizations via organopalladium intermediates. Macrolide formation[J]. Journal of the American Chemical Society, 1977, 99(11): 3867-3868.

[48] Bäckvall J E, Granberg K L. Palladium-catalyzed cis-and trans-annulations to 1, 3-cyclohexadiene and 1, 3-cycloheptadiene1[J]. Tetrahedron Letters, 1989, 30(5): 617-620.

[49] Godleski S A, Valpey R S. Palladium-catalyzed preparation of carbon and oxygen spirocycles[J]. The Journal of Organic Chemistry, 1982, 47(2): 381-383.

[50] Trost B M. Cyclizations via palladium-catalyzed allylic alkylations[J]. Angewandte Chemie International Edition, 1989, 28(9): 1173-1192.

[51] Tsuji J, Kobayashi Y, Kataoka H, et al. New bis-and tris-annulation reagents for the syntheses of CD rings of steroids, bearing a functionalized 18-methyl group, by the palladium- catalyzed cyclization[J]. Tetrahedron Letters, 1980, 21(35): 3393- 3394.

[52] Genet J P, Grisoni S. Asymmetric synthesis. An entry into tricyclic nitro ergoline synthon[J]. Tetrahedron Letters, 1988, 29(36): 4543-4546.

[53] Yamamoto K, Deguchi R, Ogimura Y, et al. Complete retention of chirality in the palladium catalyzed cyclization of methyl (R)-3-oxo-7-(methoxycarbonyloxy)-8-nonenoate[J]. Chemistry Letters, 1984, 13(10): 1657-1660.

[54] Trost B M, Tometzki G B, Hung M H. Unusual chemoselectivity using difunctional allylic alkylating agents[J]. Journal of the American Chemical Society, 1987, 109(7): 2176-2177.

[55] Baeckvall J E, Vågberg J O, Zercher C, et al. Stereoselective synthesis of vinylcyclopropanes via palladium-catalyzed reactions[J]. The Journal of Organic Chemistry, 1987, 52(24): 5430-5435.

[56] Colobert F, Genet J P. Synthesis of (+)-dictyopterene a constituent of marine brown algae and (+)-dictyopterene C′ by chirality transfer of optically active allylic benzoate with palladium(0) catalyst[J]. Tetrahedron Letters, 1985, 26(23): 2779-2782.

[57] Genet J P, Piau F. Palladium-catalyzed reactions: stereoselective synthesis of substituted cyclopropanes related to chrysanthemic acid. A simple route to cis-chrysanthemonitrile[J]. The Journal of Organic Chemistry, 1981, 46(11): 2414-2417.

[58] Trost B M, Verhoeven T R. Influence of a transition metal on the regiochemistry of ring closures. An approach to medium-ring compounds[J]. Journal of the American chemical Society, 1979, 101(6):1595-1597.

[59] Kitagawa Y, Itoh A, Hashimoto S, et al. Total synthesis of humulene. A stereoselective approach[J]. Journal of the American Chemical Society, 1977, 99(11): 3864-3867.

[60] Marshall J A, Andrews R C, Lebioda L. Synthetic studies on cembranolides. Stereoselective total synthesis of isolobophytolide[J]. The Journal of Organic Chemistry, 1987, 52(12): 2378-2388.

[61] Trost B M, Warner R W. Macrocyclization via an isomerization reaction at high concentrations promoted by palladium templates[J]. Journal of the American Chemical Society, 1982, 104(22): 6112-6114.

[62] Trost B M, Verhoeven T R. Influence of a transition metal on the regiochemistry of ring closures. An approach to medium-ring compounds[J]. Journal of the American Chemical Society, 1979, 101(6): 1595-1597.

[63] Takahashi T, Jinbo Y, Kitamura K, et al. Chirality transfer from CO to CC in the palladium catalyzed ScN' reaction of (E)-and (Z)-allylic carbonates with carbonucleophile[J]. Tetrahedron Letters, 1984, 25(51): 5921-5924.

[64] Schultz A J, Williams J M, Schrock R R, et al. Interaction of hydrogen and hydrocarbons with transition metals. Neutron diffraction evidence for an activated carbon-hydrogen bond in an electron-deficient tantalum-neopentylidene complex[J]. Journal of the American Chemical Society, 1979, 101(6): 1593-1595.

[65] Trost B M, Verhoeven T R. Stereocontrolled synthesis of (±)-recifeiolide via organopalladium chemistry[J]. Tetrahedron Letters, 1978, 19(26): 2275-2278.

[66] Trost B M, Warner R W. Macrolide formation via an isomerization reaction. An unusual dependence on nucleophile[J]. Journal of the American Chemical Society, 1983, 105(18): 5940-5942.

[67] Takahashi T, Miyazawa M, Ueno H, et al. Palladium-catalyzed stereocontrolled cyclization of 1, 3-diene monoepoxide: a route to a new synthetic intermediate for de-ab-cholestane derivative[J]. Tetrahedron Letters, 1986, 27(33): 3881-3884.

[68] Godleski S A, Villhauer E B. Application of a transition-metal-mediated stereospecific Michael reaction equivalent to the synthesis of alloyohimbone[J]. The Journal of Organic Chemistry, 1986, 51(4): 486-491.

[69] Luo F T, Negishi E. Palladium-catalyzed allylation of lithium 3-alkenyl-1-cyclopentenolates-triethylborane and its application to a selective synthesis of methyl (Z)-Jasmonate1[J]. Tetrahedron Letters, 1985, 26(18): 2177-2180.

[70] Burtea A, Rychnovsky S D. Biosynthesis-inspired approach to Kujounin A_2 using a stereoselective Tsuji-Trost alkylation[J]. Organic Letters, 2018, 20(18): 5849-5852.

[71] Trost B M, Scanlan T S. Stereoelectronic requirements of palladium(0)-catalyzed cyclization. A synthesis of *allo*-pumiliotoxin 339B[J]. Journal of the American Chemical Society, 1989, 111(13): 4988-4990.

[72] Knight S D, Overman L E, Pairaudeau G. Synthesis applications of cationic aza-Cope rearrangements. 26. Enantioselective total synthesis of (−)-strychnine[J]. Journal of the American Chemical Society, 1993, 115(20): 9293-9294.

[73] Knight S D, Overman L E, Pairaudeau G. Asymmetric total syntheses of (−)-and (+)-strychnine and the Wieland-Gumlich aldehyde[J]. Journal of the American Chemical Society, 1995, 117(21): 5776-5788.

[74] Nicolaou K C, Bulger P G, Sarlah D. Palladium-catalyzed cross-coupling reactions in total synthesis/metathesis reactions in total synthesis [J]. Angewandte Chemie International Edition, 2005, 44(29): 4442-4489.

[75] Trost B M, Pissot-Soldermann C, Chen I, et al. An asymmetric synthesis of hamigeran B via a Pd asymmetric allylic alkylation for enantiodiscrimination[J]. Journal of the American Chemical Society, 2004, 126(14): 4480-4481.

第 9 章
过渡金属催化串联反应及其在药物合成上的应用

前面各章分别介绍了几种重要的过渡金属催化的偶联反应，这些反应各有自身优势，为药物合成提供新的途径，使一些复杂药物分子的合成得以实现。有机合成化学家并未满足于此，而是不断探索以求获得进一步突破。他们发现，通过精准设计，可以将上述一些反应完美串联，使得复杂结构的有机和药物分子合成更加高效、简捷，原子经济性更高。以下将介绍部分与前述过渡金属催化偶联反应相关的串联反应及在药物合成上的应用。

9.1 Heck/Tsuji-Trost 串联反应及机理

9.1.1 Heck/Tsuji-Trost 串联反应

Heck/Tsuji-Trost 串联反应以炔丙基化合物、联烯化合物、1,3-二烯化合物、1,4-二烯化合物为起始底物，先与零价钯发生氧化加成得到 π-烯丙基钯中间体；该中间体一旦生成，即可与各种 Tsuji-Trost 反应的亲核试剂进行交叉偶联得到相应烯丙基化产物[1]。如图 9-1 所示。

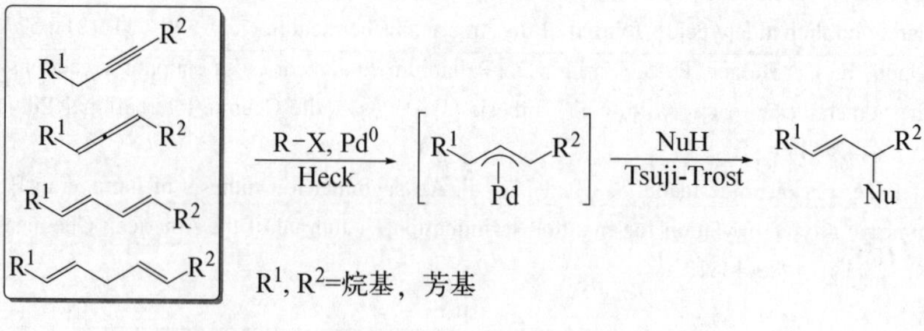

图 9-1 钯催化 Heck/Tsuji-Trost 串联反应

炔丙基化合物、联烯化合物及二烯化合物因其反应活性高、廉价易得、来源广泛等优点，广泛应用于药物、染料和聚合物的合成，是一类重要的有机合成子。以联烯和 1,3-二烯类化合物为底物，通过钯催化 Heck/Tsuji-Trost 串联反应可以构建吡咯、吡啶、环戊烷、环己烷、呋喃、吡喃等化合物。

9.1.2 反应机理

钯催化 1,3-二烯类化合物的 Heck/Tsuji-Trost 串联反应机理如图 9-2 所示[2]。首先，零价钯络合物催化剂与 RX（R 为芳基或烯基，X 为卤素、四氟重氮硼酸盐或三氟甲磺酸酯等）氧化加成得到二价钯中间体 **A**，随后与 1,3-二烯类化合物发生迁移插入生成烯丙基钯中间体 **B**，进一步异构化得到共振式烯丙基钯中间体 **C**。然后有两种可能的反应路径：路径 a，软亲核试剂（活泼亚甲基的化合物、烯醇盐、烯胺等化合物）进攻烯丙基钯中间体 **C**，再经过还原消除过程生成 1,2-或者 1,4-加成产物，同时伴随零价钯催化剂的再生从而完成催化循环；路径 b，硬亲核试剂（格氏试剂等主族金属有机化合物）进攻共振式烯丙基钯中间体 **C** 发生转金属化生成烯丙基钯中间体 **D**，进一步发生还原消除过程得到 1,2-或者 1,4-加成产物。

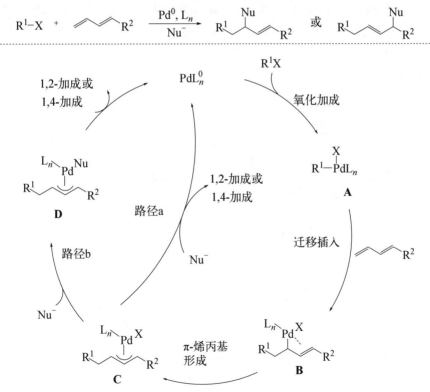

图 9-2 钯催化 1,3-二烯类化合物的 Heck/Tsuji-Trost 串联反应机理

钯催化联烯的 Heck/Tsuji-Trost 串联反应是联烯化合物双官能团化的重要反应，得到了有机化学家的广泛关注，该反应的一般过程如图 9-3 所示。首先，零价钯与卤代烃 R^1X（R^1 为烯基或芳基）经过氧化加成生成二价钯中间体 **A**，随后联烯迁移插入生成烯丙基钯中间体 **B**，进而亲核试剂进攻中间体 **B** 并通过还原消除反应生成相应的加成产物。

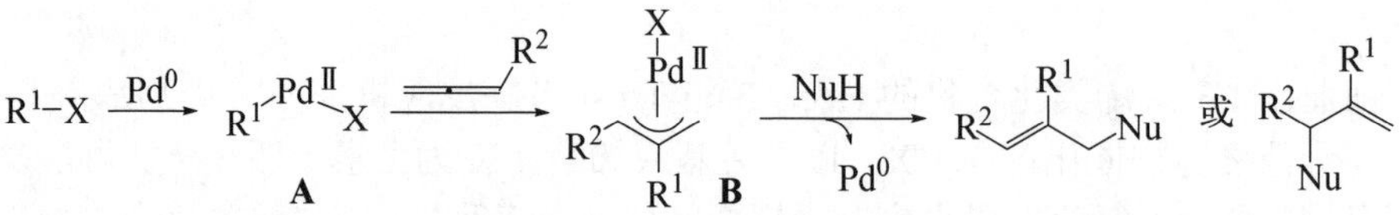

图 9-3　钯催化联烯的 Heck/Tsuji-Trost 串联反应机理

与 1,3-二烯类化合物和联烯化合物相比，钯催化 1,4-二烯类化合物的 Heck/Tsuji-Trost 串联反应更为复杂。由于 1,4-二烯类化合物中没有共轭体系，属非活化的烯烃，因此需要发生多次与钯催化剂的配位以及迁移插入才能转化为烯丙基钯中间体。具体的反应机理如图 9-4 所示。首先，芳基卤化物与零价钯发生氧化加成生成二价钯中间体 **A**，**A** 随之与 1,4-二烯类化合物发生迁移插入得到中间体 **B**，并进一步异构化生成二价钯中间体 **C**；**C** 发生钯迁移转化为中间体 **D**，进而通过烯丙基化生成烯丙基钯中间体 **E**；最后，中间体 **E** 被亲核试剂捕获，并通过还原消除反应得到烯丙基化产物。

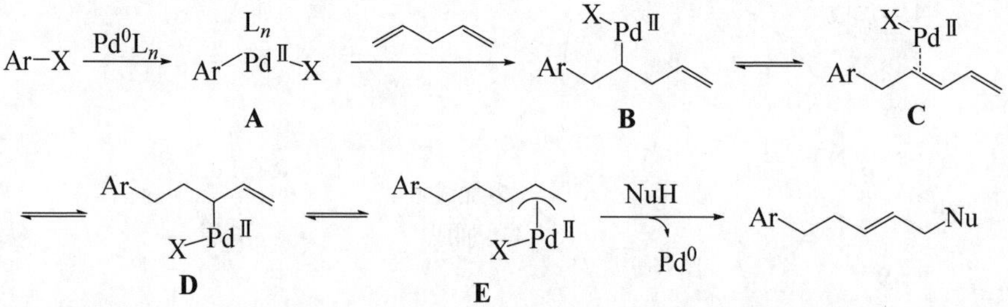

图 9-4　钯催化 1,4-二烯类化合物的 Heck/Tsuji-Trost 串联反应机理

9.1.3　基本的 Heck/Tsuji-Trost 反应

由上节可以看出，炔丙基化合物、联烯化合物、1,3-二烯化合物、1,4-二烯化合物和各种不同的亲核试剂在钯催化下都能够发生 Heck/Tsuji-Trost 串联反应，得到一系列烯丙基化产物。亲核试剂可以是氮亲核试剂、碳亲核试剂和氧亲核试剂。

9.1.3.1 氮亲核试剂

（1）1,3-二烯化合物为底物

在以共轭二烯烃代替简单烯烃与芳基卤代物进行钯催化的 Heck 反应时意外生成了 1,3-二烯胺化/芳基化产物[3]。当以溴苯、异戊二烯为起始底物，哌啶作为氮亲核试剂时，以 57%的收率得到共轭二烯烃双官能团胺化产物。该反应的总体收率不佳，原因是以 35%的收率直接生成了 Tsuji-Trost 偶联反应产物苯基异戊二烯。

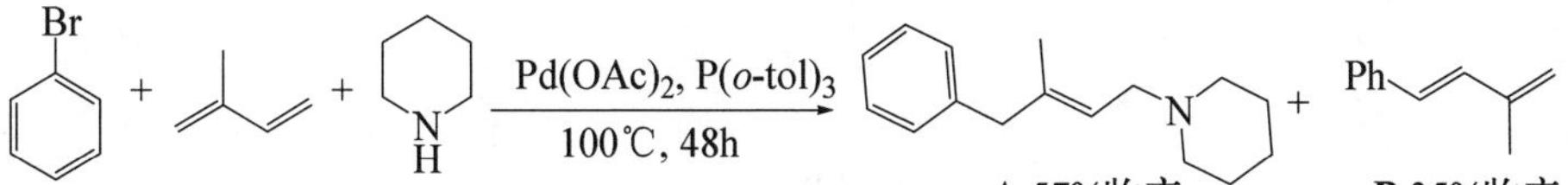

$$\text{PhBr} + \text{异戊二烯} + \text{哌啶} \xrightarrow[100℃, 48h]{Pd(OAc)_2, P(o\text{-tol})_3} \mathbf{A}\ 57\%收率 + \mathbf{B}\ 35\%收率 \tag{1}$$

利用三苯基膦代替三(邻甲基苯基)膦作为配体，并经条件优化，反应的产率得以提高，并拓宽了仲胺亲核试剂的范围，而且兼容各种伯胺亲核试剂[4]。值得注意的是，该反应得到的是 1,2-加成产物而非 1,4-加成产物，并且亲核试剂进攻位点在双键的内侧，可能是由于配体和亲核试剂之间存在空间位阻所致。此外，在标准条件下邻碘苯胺也可作为氮亲核试剂分别与 2-甲基-1,3-丁二烯和 1,3-环己二烯顺利地进行反应，分别以 72%和 70%的收率得到相应的二氢吲哚衍生物。

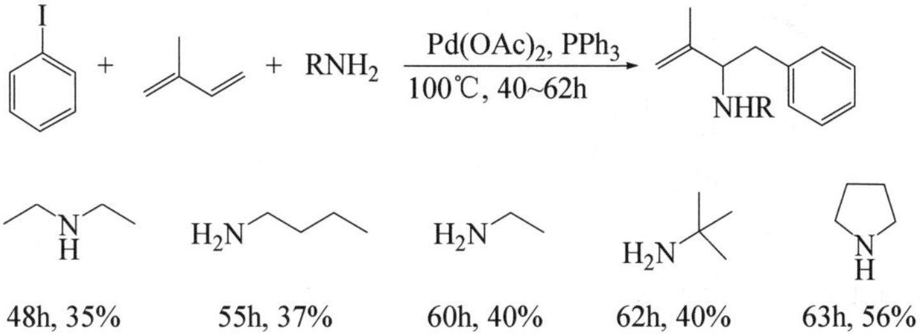

$$\text{PhI} + \text{异戊二烯} + \text{RNH}_2 \xrightarrow[100℃, 40\sim62h]{Pd(OAc)_2, PPh_3} \tag{2}$$

Et_2NH 48h, 35%　　$H_2N{-}n{-}Bu$ 55h, 37%　　H_2NEt 60h, 40%　　$H_2N{-}t{-}Bu$ 62h, 40%　　吡咯烷 63h, 56%

$$\text{2-碘苯胺} + \text{异戊二烯} \xrightarrow[100℃, 48h]{Pd(OAc)_2, PPh_3} \text{72\%} \tag{3}$$

$$\text{2-碘苯胺} + \text{1,3-环己二烯} \xrightarrow[100℃, 48h]{Pd(OAc)_2, PPh_3} \text{70\%} \tag{4}$$

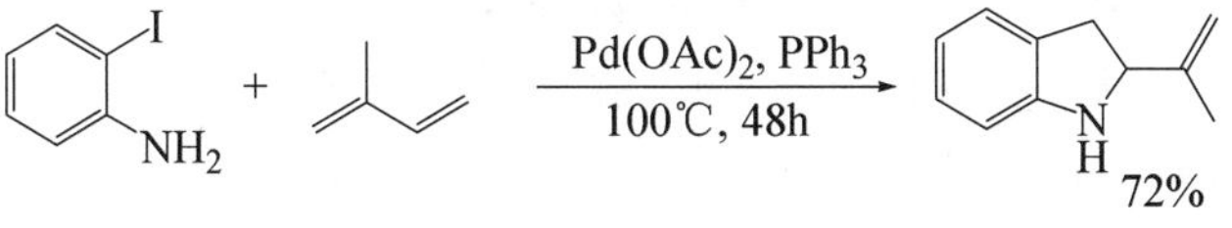

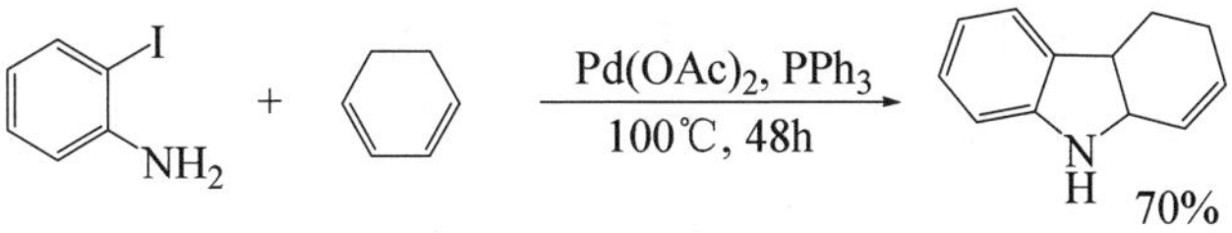

光学活性的二氢吲哚化合物是众多具有生物活性的天然产物、药物和农药分子的基本结构单元。发展高效合成光学活性的二氢吲哚化合物的方法一直是化学家追求的目标之一，其中钯催化的邻碘苯胺和 1,3-二烯化合物的对映选择性 Heck/Tsuji-Trost 串联反应是实现手性二氢吲哚的合成的有效方法[5]。以 Pd(OAc)$_2$ 和光学活性的手性磷酰胺配体 BINOL 构成的催化体系催化邻碘苯胺与 1,3-二烯化合物的对映选择性 Heck/Tsuji-Trost 串联反应以优良的产率和对映选择性得到光学活性的二氢吲哚产物。其中，BINOL 上的吸电子基是反应取得高对映选择性的关键。此外，以邻碘苄醇为起始原料进行反应，也能以较高的产率和对映选择性生成光学活性的苯并异色满衍生物。

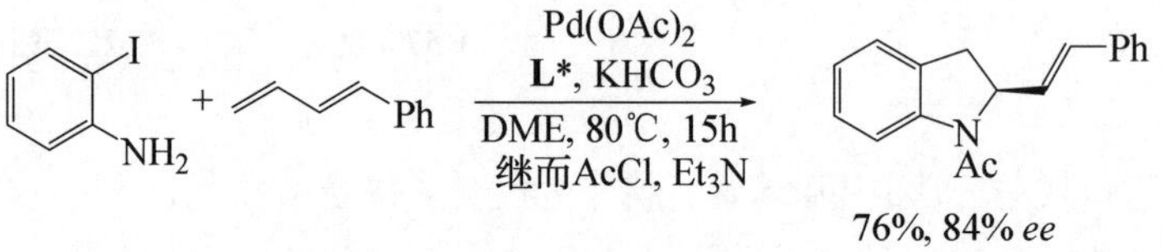

$$\text{邻碘苯胺} + \text{1,3-二烯-Ph} \xrightarrow[\text{DME, 80℃, 15h}]{\text{Pd(OAc)}_2, \text{L*, KHCO}_3} \text{继而AcCl, Et}_3\text{N} \text{二氢吲哚-Ph} \quad (5)$$

76%, 84% ee

$$\text{邻碘苄醇} + \text{1,3-二烯-Ph} \xrightarrow[\text{DME, 80℃, 48h}]{\text{Pd(OAc)}_2, \text{L*, KHCO}_3} \text{苯并异色满-Ph} \quad (6)$$

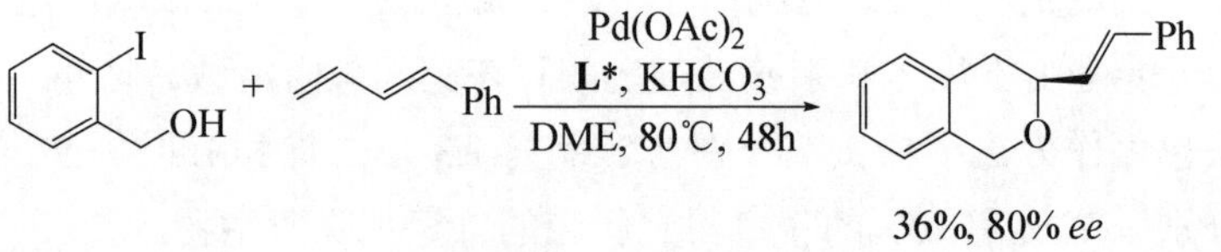

36%, 80% ee

L* = (BINOL-磷酰胺配体)，R = 3,5-(CF$_3$)$_2$C$_6$H$_3$

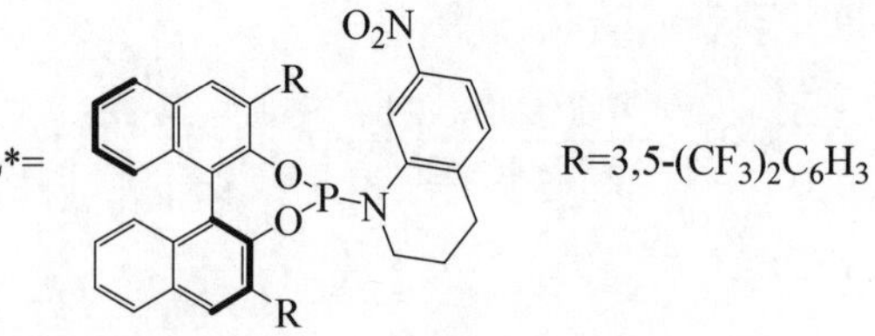

（2）联烯化合物为底物

联烯类化合物是含有 1,2-丙二烯官能团结构单元的一类不饱和化合物，它的两个 π-轨道互相垂直，两个双键上的电子云密度不同，因此具有独特的反应活性与选择性，在药物化学、材料化学和天然产物合成等领域具有广泛的应用。

其中以 2,3-累积二烯基肼作为氮亲核试剂，在 Pd(PPh$_3$)$_4$ 催化下和芳基卤化物 Heck/Tsuji-Trost 串联反应，能够非对映选择性合成一类含有两个氮原子的四元环二氮杂丁[6]。此外，使用光学活性的 2,3-累积二烯基肼作为底物，也能顺利进行反应并以良好的产率和优异的对映选择性得到手性 1,2-二氮杂环丁烷产物。

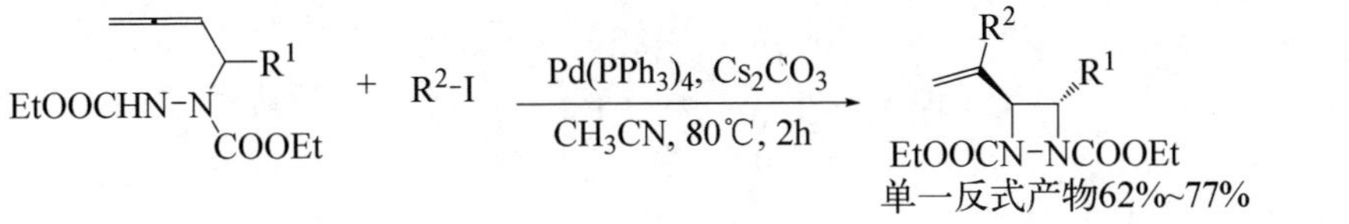

关于联烯化合物的不对称 Heck/Tsuji-Trost 串联反应的实例较少，原因是此类反应难以受手性因素控制。在为数不多的实例中，Pd(OAc)$_2$ 与手性双噁唑啉配体（Box）构成的催化体系可有效催化此类反应，以优异的产率和良好的对映选择性（82% ee）得到二氢吲哚产物[7]。

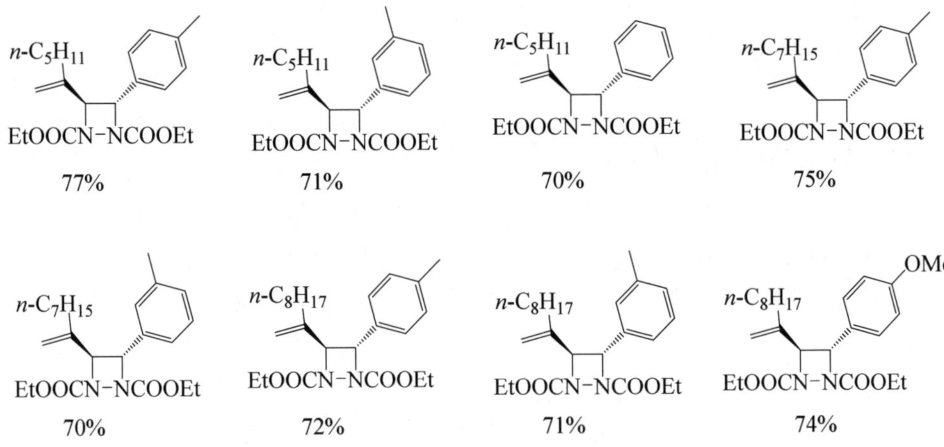

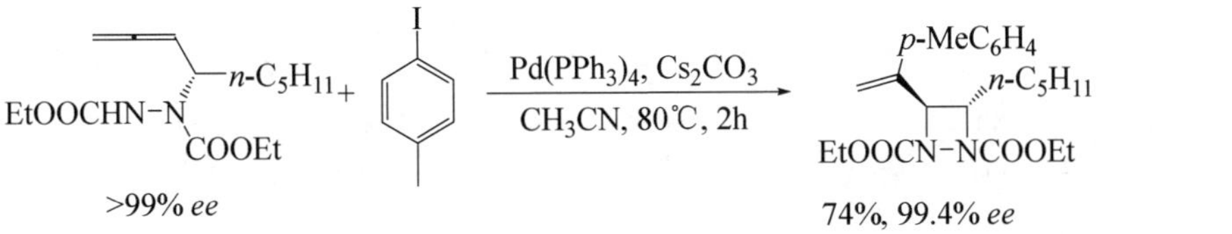

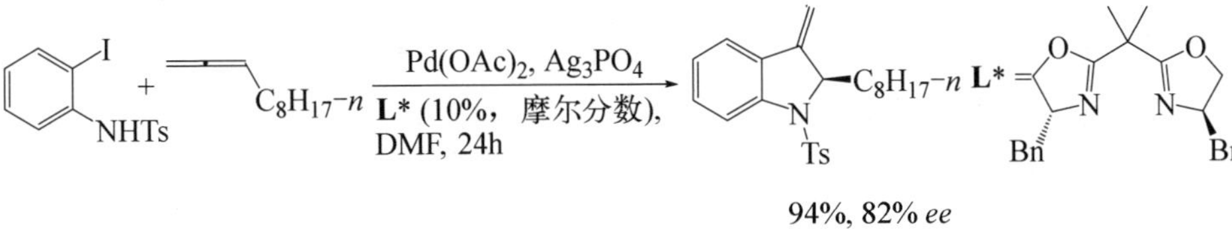

为了提高反应的对映选择性，对催化体系进行了进一步探索，发现(R_a, S, S)型萘甲基取代的手性螺环双噁唑啉配体与 Pd(dba)$_2$ 构成的催化体系显示良好的催化性能，能以较高的产率和优异的对映选择性得到二氢吲哚产物[8]。

$$\text{(reaction 10)} \quad \text{ArI(NHTs)} + \text{CH}_2=\text{CH-C}_8\text{H}_{17}\text{-}n \xrightarrow[\text{110℃, 48h}]{\text{Pd(dba)}_2, \text{Ag}_3\text{PO}_4 \atop \text{L*}(R_a,S,S), \text{THF}} \text{产物} \quad (10)$$

76%, 95% ee

L* = (SPINOL-derived bisoxazoline), R = 萘甲基

（3）1,4-二烯化合物为底物

1,4-二烯类化合物中没有共轭体系，为非活化的二烯烃，因此以 1,4-二烯作为底物的 Heck/Tsuji-Trost 串联反应并不多见。下面的反应以 1,4-戊二烯作为起始底物，邻碘苯胺作为氮亲核试剂，能以中等收率得到含氮六元杂环化合物[9]。

$$\text{邻碘苯胺} + \text{1,4-戊二烯} \xrightarrow[\text{Na}_2\text{CO}_3, \text{DMF}, 100℃, 3.5d]{\text{Pd(OAc)}_2, n\text{-Bu}_4\text{NCl}} \text{四氢喹啉衍生物} \quad (11)$$

65%

除了直链 1,4-二烯类化合物，脂环 1,4-二烯类化合物同样能够参与 Heck/Tsuji-Trost 串联反应。钯催化芳基碘化物、1,4-环己二烯和吗啉发生偶联反应，能够以较高的收率得到单一的烯丙胺类化合物[10]。

$$\text{PhI} + \text{1,4-环己二烯} + \text{吗啉} \xrightarrow[\text{DMA, 100℃, 24h}]{\text{Pd(dba)}_2, n\text{-Bu}_4\text{NCl}} \text{产物} \quad (12)$$

70%

在上述 Pd(dba)$_2$/n-Bu$_4$NCl 催化条件下，2,5-环己二烯取代的芳基碘化物可以和不同氮亲核试剂反应，以中等至良好的收率得到环化的偶联产物。此外，2,5-环己二烯取代的烯基碘化物同样适用于该反应条件，使用环状的 1,4-二烯类化合物可以避免迁移插入发生在烯基卤代物双键上[11]。

$$\text{(邻碘苯氧甲基环己二烯)} + \text{吗啉} \xrightarrow[\text{Na}_2\text{CO}_3, \text{DMSO}, 100℃, 36h]{\text{Pd(OAc)}_2, n\text{-Bu}_4\text{NCl}} \text{三环产物} \quad (13)$$

91%

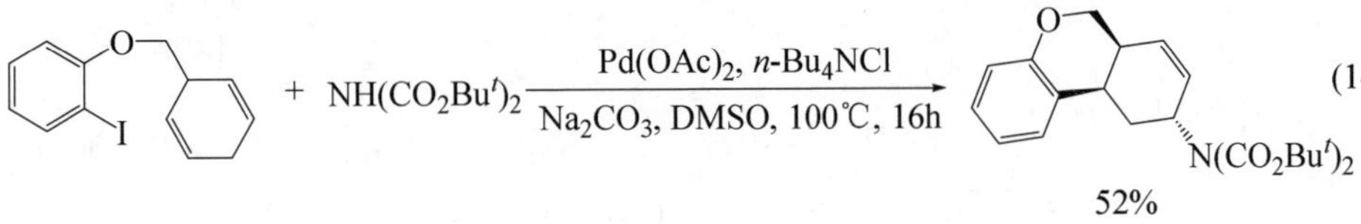

$$\text{(14)} \quad 52\%$$

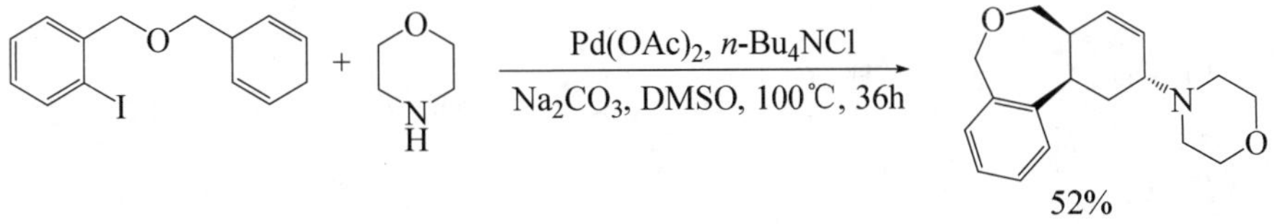

$$\text{(15)} \quad 52\%$$

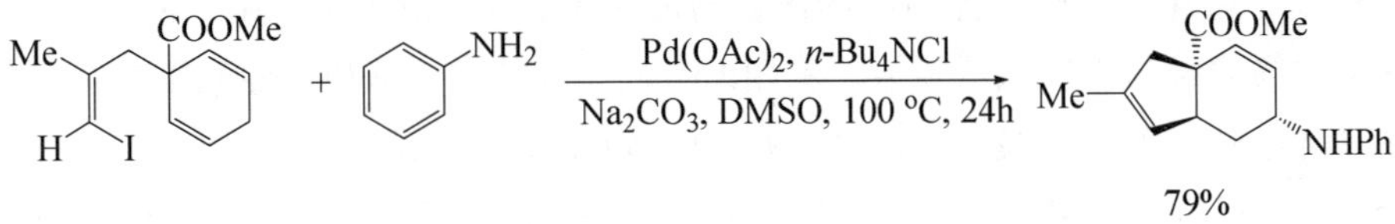

$$\text{(16)} \quad 79\%$$

9.1.3.2 碳亲核试剂

联烯、1,3-二烯和 1,4-二烯化合物除了能与氮亲核试剂发生 Heck/Tsuji-Trost 串联反应以外，还能够和多种碳亲核试剂（活性亚甲基化合物、芳基硼酸、四氟重氮硼酸盐、对苯基苯酚等）发生 Heck/Tsuji-Trost 串联反应。

（1）活性亚甲基化合物为碳亲核试剂

活性亚甲基化合物是指一个饱和碳原子上连有两个强吸电子基如硝基、羰基、磺酰基、酯基、氰基、苯基等的化合物。亚甲基氢原子具有一定的酸性，在碱作用下亚甲基生成亚甲基负离子，作为碳亲核试剂参与 Heck/Tsuji-Trost 串联反应。

如反应式（17）所示，丙二酸二乙酯钠盐与联烯、2-溴丙烯在 Pd(dba)$_2$ 催化下进行三组分偶联反应，化学选择性地生成 E 式 1,3-丁二烯衍生物[12]。参考图 9-3，反应中零价钯首先与 2-溴丙烯进行氧化加成，随后与联烯迁移插入得到 π-烯丙基钯中间体，然后碳亲核试剂进攻该中间体空间位阻小的一侧，再经还原消除生成 E 式 1,3-丁二烯产物。

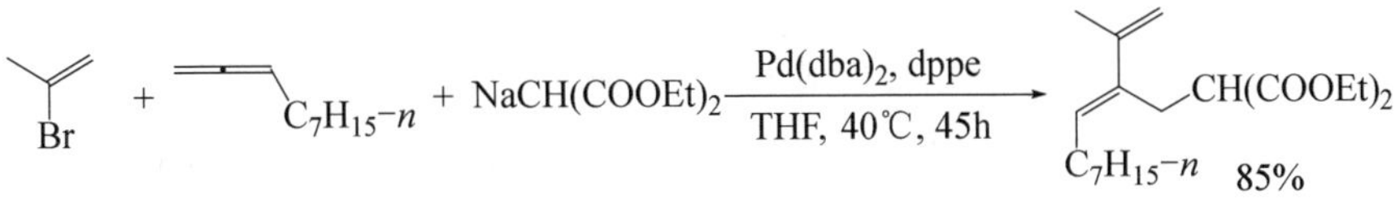

$$\text{(17)} \quad 85\%$$

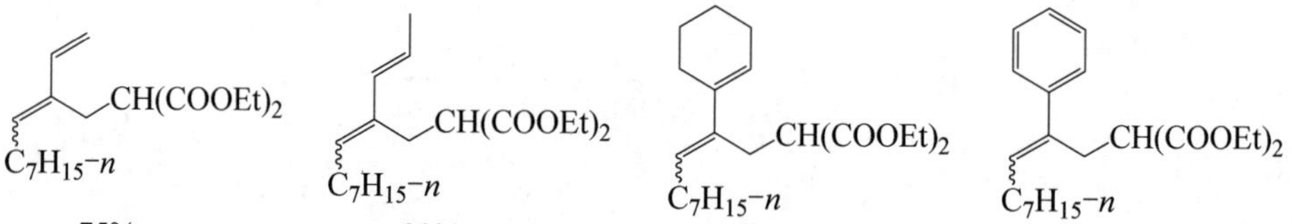

产率分别为 75%、80%、63%、71%、60%、80%、70%、85%（见结构式）。

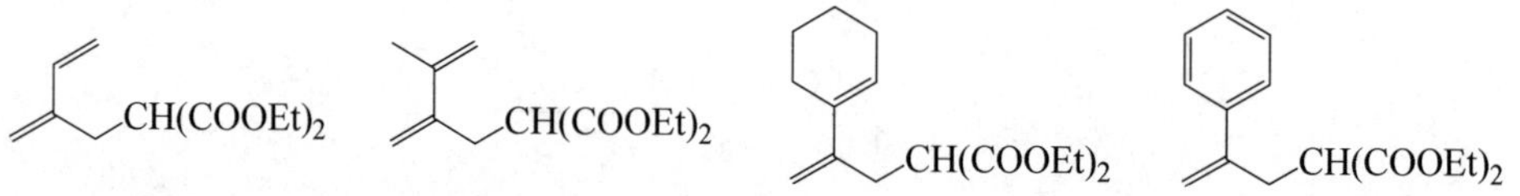

活性亚甲基化合物除了能与联烯化合物发生 Heck/Tsuji-Trost 串联反应外，还能与 1,3-二烯化合物进行偶联反应。例如，同时含有碘代芳基和活性亚甲基结构单元的邻碘苯基丙二酸二甲酯在 Na_2CO_3 协助和 $Pd(OAc)_2$ 催化下与 1,3-环己二烯进行偶联反应，以优良的收率得到二氢吲哚三环产物[13]。需要强调的是反应无须三苯基膦配体即可顺利进行，由于苯基对亚甲基也有活化作用，邻碘苯基乙酸乙酯也能作为底物与 1,3-环己二烯顺利地进行 Heck/Tsuji-Trost 串联反应以 85%的收率得到相应的三环化合物。

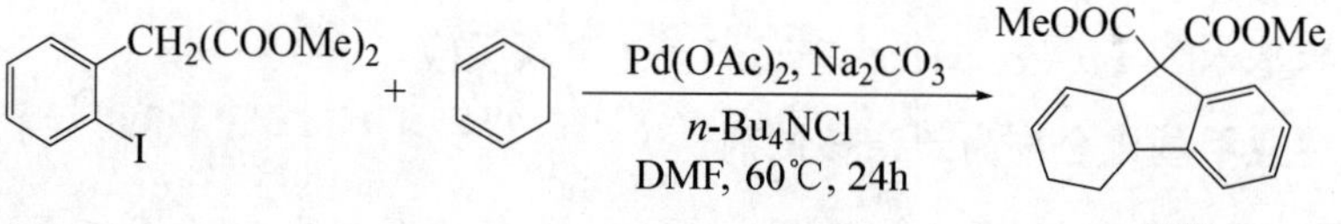

$$\text{o-I-C}_6\text{H}_4\text{-CH}_2(\text{COOMe})_2 + \text{环己二烯} \xrightarrow[\text{DMF, 60℃, 24h}]{Pd(OAc)_2, Na_2CO_3, n\text{-Bu}_4NCl} \text{三环产物}\ 87\% \quad (18)$$

$$\text{o-I-C}_6\text{H}_4\text{-CHCOOEt} + \text{环己二烯} \xrightarrow[\text{DMF, 80℃, 48h}]{Pd(OAc)_2, KOAc, n\text{-Bu}_4NCl} \text{三环产物}\ 85\% \quad (19)$$

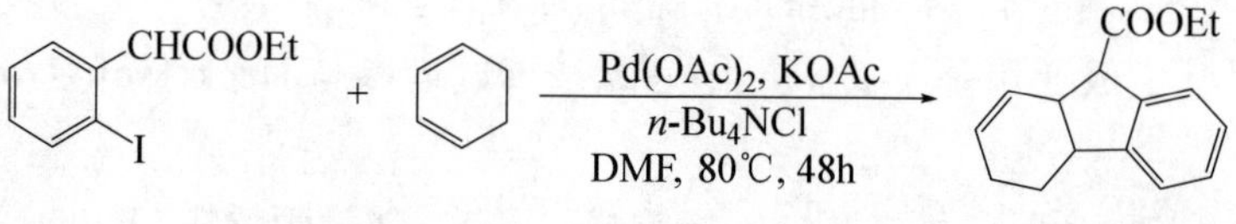

活性亚甲基化合物、芳基碘和 1,3-二烯化合物在适宜的手性钯催化剂催化下可进行对映选择性 Heck/Tsuji-Trost 三组分串联反应[14]。其中，手性配体是关键的影响因素，因为手性配体与整个反应过程中形成的钯中间体配位，不仅能促进钯与芳基碘的氧化加成和插入反应，而且能有效地控制不对称烯丙基烷基化的立体选择性。$Pd(OAc)_2$ 与 BINOL-型手性磷酰胺形成的手性催化剂的催化性能最佳，在其催化下 1,3-二烯化合物、芳基碘化物和丙二酸二乙酯钠盐能顺利地进行三组分偶联，以优异的产率和对映选择性生成光学活性的 1,2-双官能团化产物。

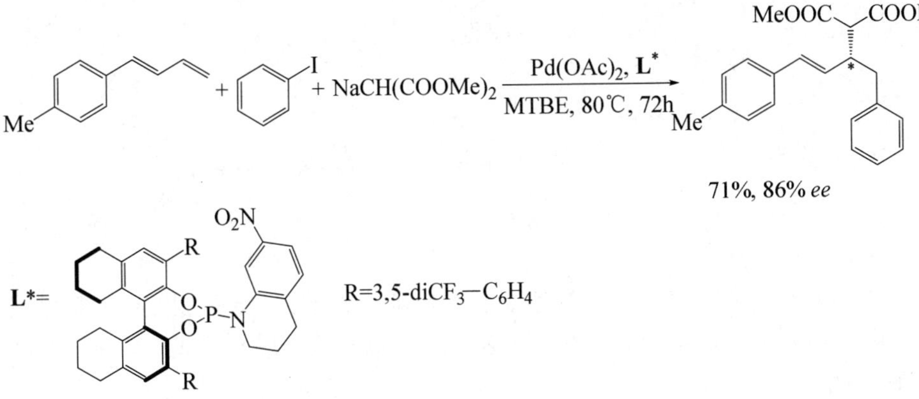

$$\text{(20)}$$

（图：Me-苯基-丁二烯 + 碘苯 + NaCH(COOMe)$_2$ $\xrightarrow{\text{Pd(OAc)}_2, \text{L}^*}{\text{MTBE, 80℃, 72h}}$ 产物，71%, 86% ee）

L* = （联萘氧膦-7-硝基四氢喹啉结构），R = 3,5-diCF$_3$-C$_6$H$_4$

活性亚甲基化合物与非活化的 1,4-二烯化合物也能发生 Heck/Tsuji-Trost 串联反应。邻位官能团化的芳基碘化物与直链 1,4-二烯化合物在钯催化下能够顺利发生烯丙基化反应，得到相应的环状化合物[15]。

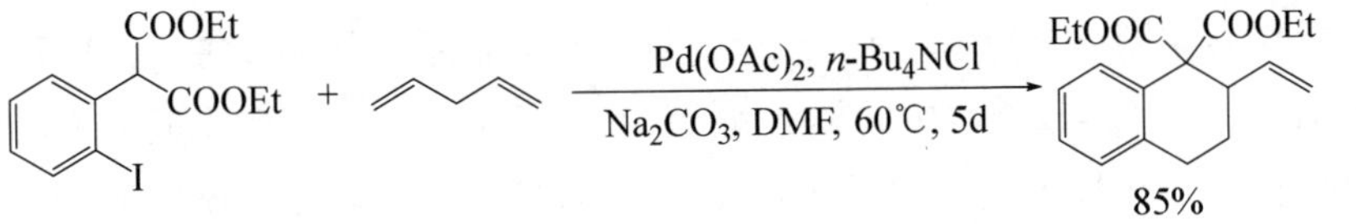

$$\text{(21)}$$

（图：邻碘苯基丙二酸二乙酯 + 1,4-戊二烯 $\xrightarrow{\text{Pd(OAc)}_2, n\text{-Bu}_4\text{NCl}}{\text{Na}_2\text{CO}_3, \text{DMF, 60℃, 5d}}$ 四氢萘产物，85%）

（2）有机硼酸试剂为碳亲核试剂

有机硼酸试剂具有性质温和、易于制备、无毒、耐热、空气中不分解、能够兼容各种不同官能团等优点。因此，有机硼酸作为碳亲核试剂广泛应用于钯催化的交叉偶联反应中，为碳碳键的构建提供了一种简单高效的途径。

其中，在温和条件下钯催化联烯、碘化物和硼酸的 Heck/Tsuji-Trost 三组分串联反应，同时构建了 C(sp^3)-C(sp^2)键和 C(sp^3)-C(sp^3)键，以高区域和立体选择性生成联烯双官能团化产物，产物构型为 Z 式，底物适用范围广泛[16]。

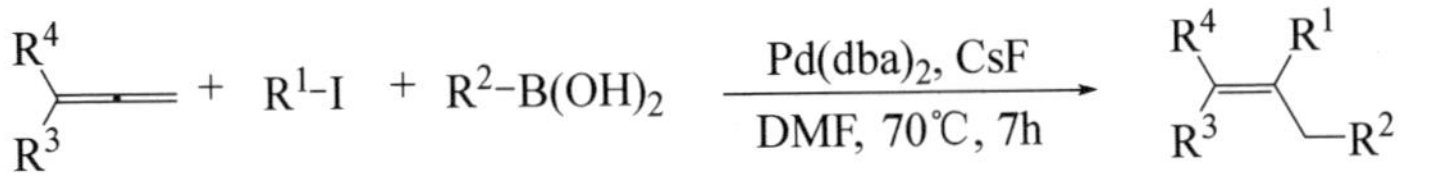

$$\text{(22)}$$

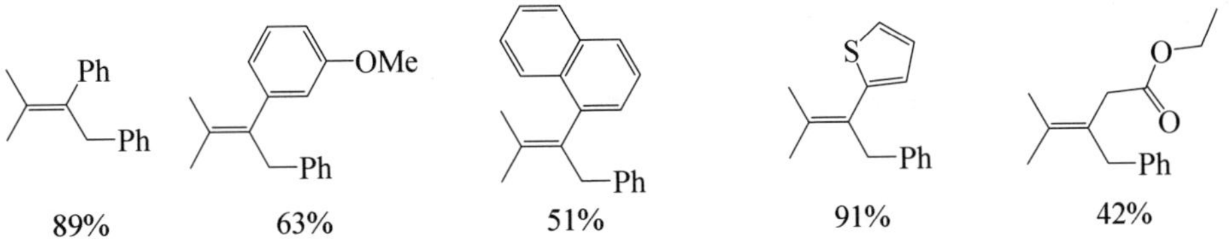

（图：R^{4}R^{3}C=C=CH$_2$ + R^1-I + R^2-B(OH)$_2$ $\xrightarrow{\text{Pd(dba)}_2, \text{CsF}}{\text{DMF, 70℃, 7h}}$ R^{4}R^{3}C=C(R^1)-CH$_2$R^2）

产物示例：
- 2-甲基-3-苯基-4-苯基-2-丁烯，89%
- (3-甲氧基苯基)衍生物，63%
- 萘基衍生物，51%
- 噻吩基衍生物，91%
- 乙酯基衍生物，42%

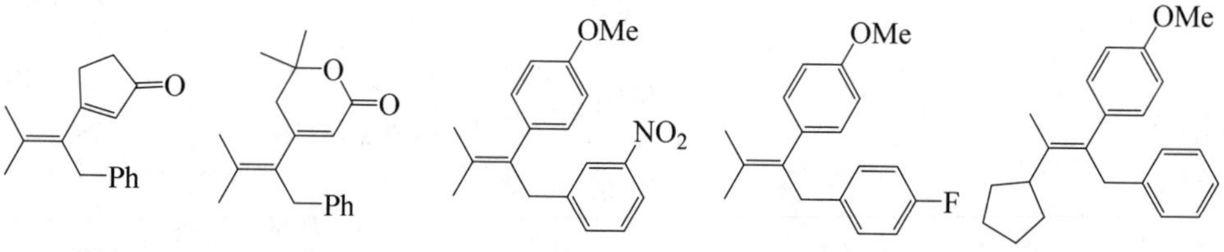

1,3-二烯化合物作为底物、三氟乙烯酯作为亲电试剂以及芳基硼酸作为亲核试剂的三组分反应物在 $Pd_2(dba)_3$ 催化下发生交叉偶联反应可顺利生成二烯化合物的 1,2-双官能团化产物[17]。

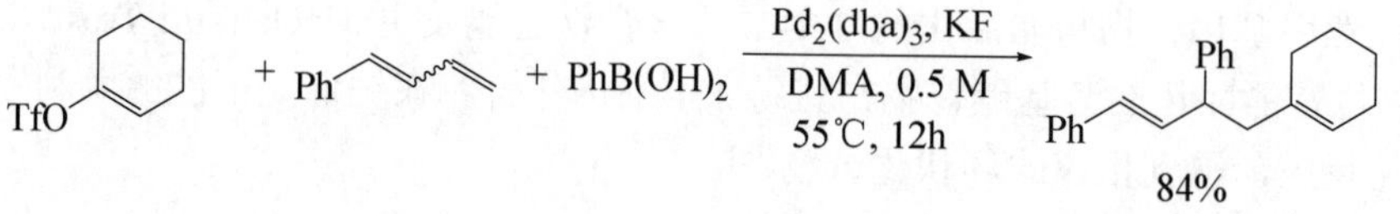

$$\text{TfO-cyclohexenyl} + Ph\text{-diene} + PhB(OH)_2 \xrightarrow[55℃, 12h]{Pd_2(dba)_3, KF \atop DMA, 0.5 M} \text{product, 84\%} \quad (23)$$

该反应机理如图 9-5 所示：首先三氟乙烯酯对零价钯氧化加成生成乙烯基二价钯中间体 A，随后 1,3-二烯对中间体 A 通过 Heck 插入得到烷基钯中间体 B，进一步经过异构化得到 π-烯丙基钯中间体 C，随后通过硼酸转金属化和还原消除过程得到反式的 1,5-二烯化合物。应该注意的是，通过三氟甲磺酸乙烯酯来启动催化循环，是因为高度亲电的乙烯基二价钯中间体更倾向于发生烯烃插入而不是硼酸插入的 Suzuki 反应。

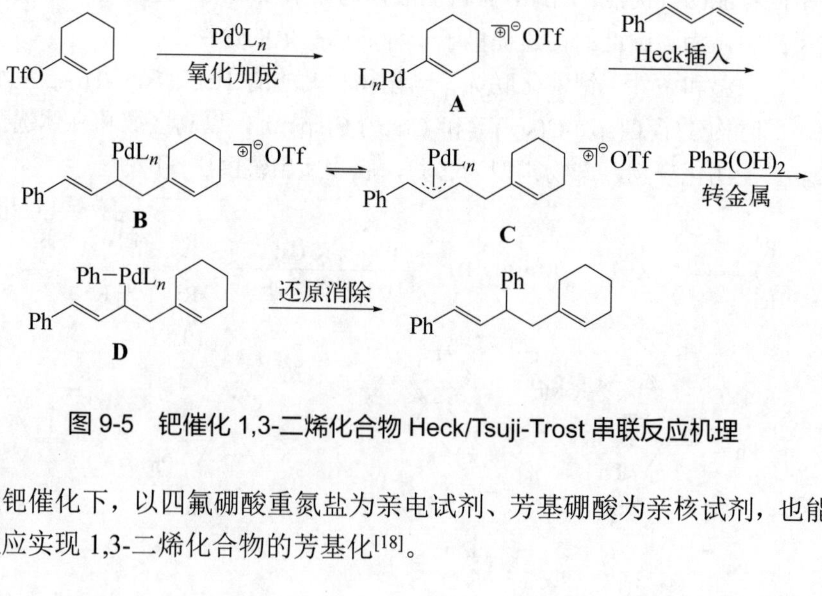

图 9-5　钯催化 1,3-二烯化合物 Heck/Tsuji-Trost 串联反应机理

在钯催化下，以四氟硼酸重氮盐为亲电试剂、芳基硼酸为亲核试剂，也能通过串联反应实现 1,3-二烯化合物的芳基化[18]。

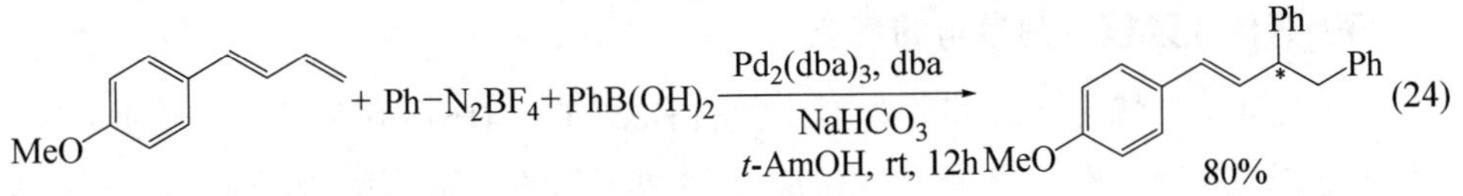

$$\text{MeO-C}_6\text{H}_4\text{-CH=CH-CH=CH}_2 + \text{Ph-N}_2\text{BF}_4 + \text{PhB(OH)}_2 \xrightarrow[\text{t-AmOH, rt, 12h}]{\text{Pd}_2(\text{dba})_3, \text{dba}, \text{NaHCO}_3} \text{产物} \quad 80\% \quad (24)$$

（3）对苯基苯酚为碳亲核试剂

钯催化 1,3-二烯化合物与对苯基苯酚串联脱芳构化反应，以很高的化学选择性和区域选择性生成含有三级碳和四级碳两个手性中心相邻的螺环化合物[19]。反应经过 C—I 键氧化加成、1,3-二烯区域选择性插入、π-烯丙基钯中间体脱芳构化等过程。烷基钯中间体 A 是整个反应的一个关键中间体，相比于直接发生 β-氢消除反应，中间体 A 倾向于生成更稳定的 π-烯丙基钯中间体，随后发生脱芳构化、还原消除得到相应螺环产物。此外，在手性亚磷酰胺配体存在下，反应以良好的收率和对映选择性生成光学活性的脱芳构化产物。

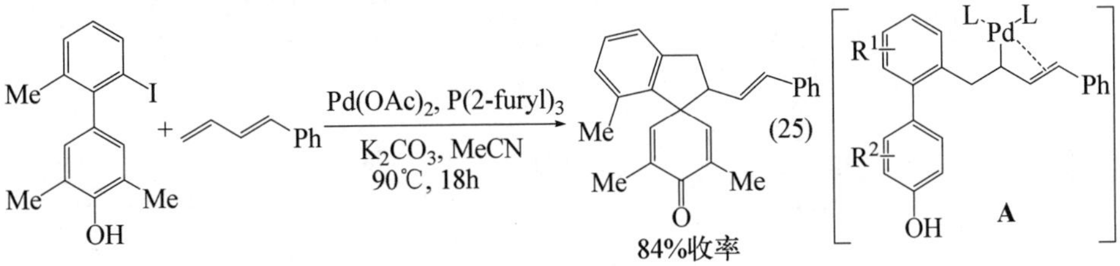

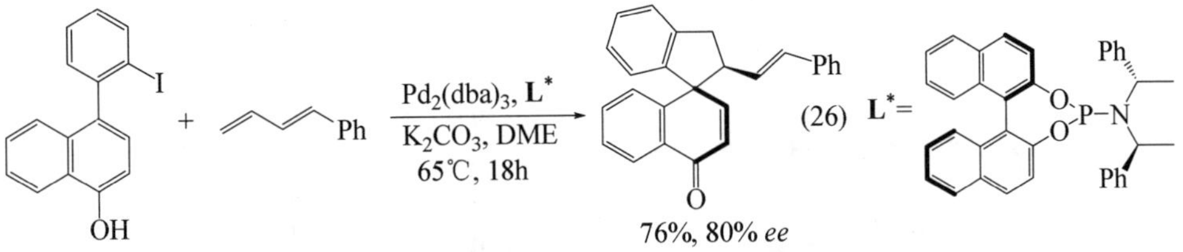

[反应式 (25)：使用 Pd(OAc)$_2$, P(2-furyl)$_3$, K$_2$CO$_3$, MeCN, 90℃, 18h，84%收率；中间体 A]

[反应式 (26)：使用 Pd$_2$(dba)$_3$, L*, K$_2$CO$_3$, DME, 65℃, 18h，76%, 80% ee]

（4）四氟硼酸重氮盐为碳亲核试剂

在醋酸钯和手性螺环磷酸构成的催化体系催化下，1,3-二烯化合物、四氟硼酸重氮盐和芳香醛可对映选择性地进行三组分羰基烯丙化反应，以良好的产率和高对映选择性生成光学活性的 Z-构型高烯丙醇产物[20]。

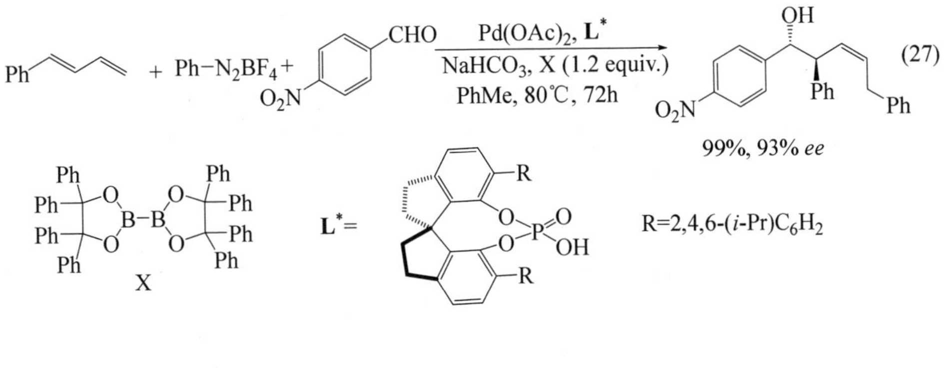

$$\text{Ph-CH=CH-CH=CH}_2 + \text{Ph-N}_2\text{BF}_4 + \text{O}_2\text{N-C}_6\text{H}_4\text{-CHO} \xrightarrow[\text{PhMe, 80℃, 72h}]{\text{Pd(OAc)}_2, \text{L}^*, \text{NaHCO}_3, \text{X (1.2 equiv.)}} \text{产物} \quad 99\%, 93\%\ ee \quad (27)$$

X = 四苯基联硼酸酯；L* = 螺环磷酸，R = 2,4,6-(i-Pr)C$_6$H$_2$

（5）其他碳亲核试剂参与的反应

钯催化 1,3-二烯环丁醇和芳基碘化物的 Heck/Tsuji-Trost 串联反应，可使 1,3-二烯环丁醇发生扩环，化学选择性地生成 2-烯基-环戊酮产物[21]。其反应的机理如图 9-6 所示：首先，零价钯与芳基碘化物氧化加成生成芳基二价钯中间体 **A**，随后 **A** 与 1,3-二烯发生迁移插入得到烷基钯中间体 **B**，**B** 在 Ag_2CO_3 的作用下通过脱去一分子二氧化碳和水得到 π-烯丙基钯中间体 **C**，**C** 再经过分子内片呐醇重排以及还原消除生成 2-烯基-环戊酮产物。

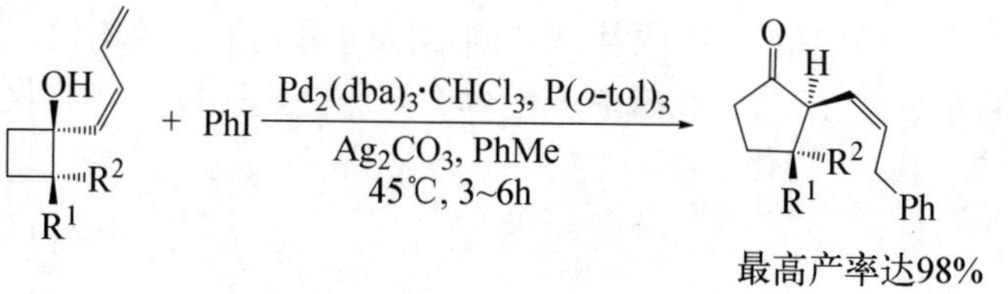

$$(28)$$

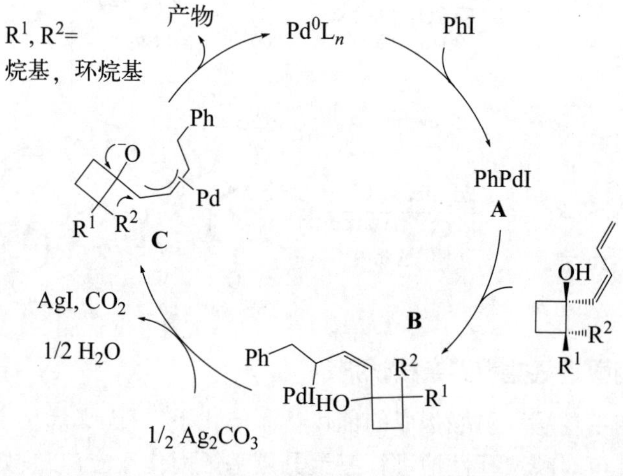

最高产率达98%

图 9-6 钯催化 1,3-二烯环丁醇 Heck/Tsuji-Trost 串联反应机理

在适宜的手性配体存在下，钯催化剂可催化 3-(2'-碘苯甲基)-3-烷氧甲酰基-2,4-环己二烯与各种亲核试剂进行对映选择性分子内 Heck/Tsuji-Trost 反应，快速构建光学活性的四氢芴衍生物[22]。例如，在 $Pd(dba)_2$/手性亚磷酰胺配体 H8-BINOL 催化体系催化下，上述 2,4-环己二烯衍生物作为反应底物经过连续的氧化加成、迁移插入、β-氢消除、烯烃插入、π-烯丙基的形成、碳原子或杂原子亲核试剂的进攻等步骤，以良好的产率和对映选择性生成与 C、N、O 各种亲核试剂偶联的手性四氢芴衍生物。

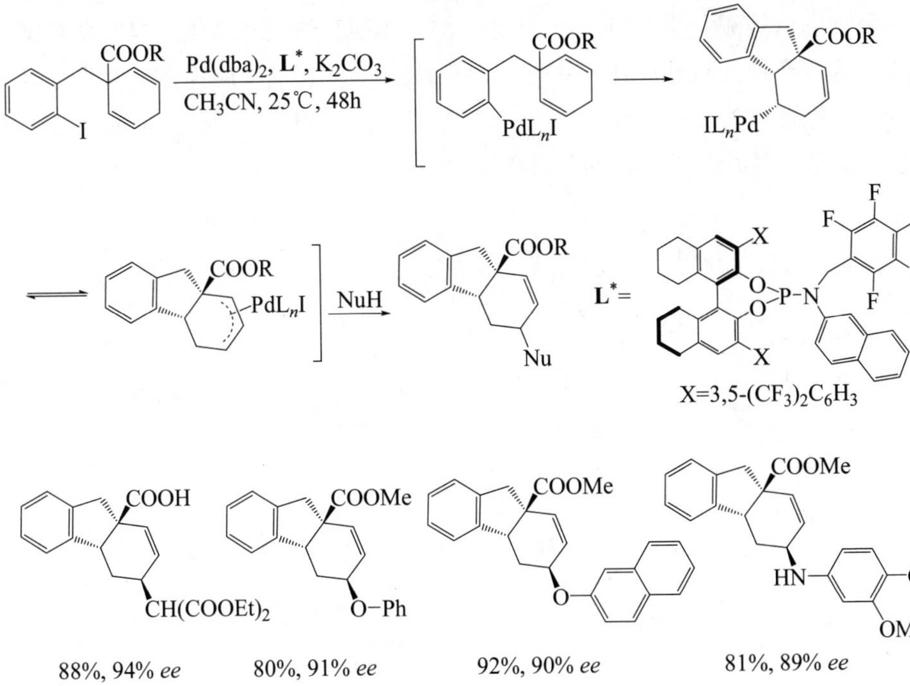
COOR
Pd(dba)2, L*, K2CO3
CH3CN, 25℃, 48h
I
PdLnI
ILnPd
NuH
Nu
L*=
X
X=3,5-(CF3)2C6H3
COOH
CH(COOEt)2
COOMe
O−Ph
HN
OMe
88%, 94% ee
80%, 91% ee
92%, 90% ee
81%, 89% ee

(29)

醋酸钯与手性磷酸构成的催化体系可催化 1,3-二烯、炔基溴和醛类化合物三组分烯丙基化反应，生成具有结构多样性和对映选择性的功能化高烯丙醇产物[23]。

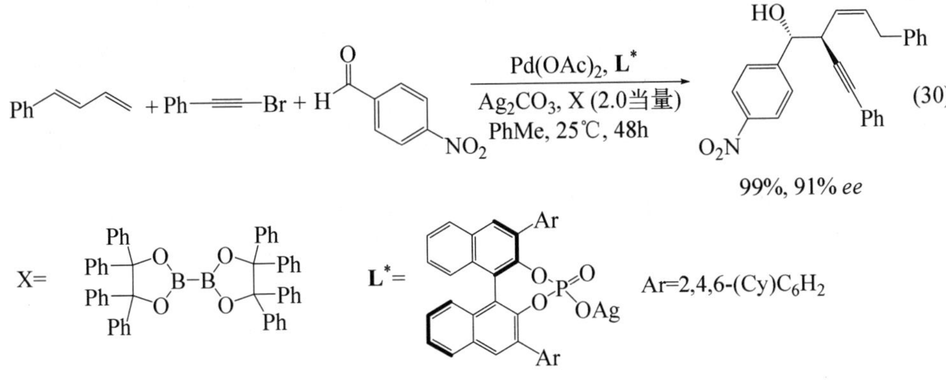
Ph
+ Ph———Br + H
NO2
Pd(OAc)2, L*
Ag2CO3, X (2.0当量)
PhMe, 25℃, 48h
HO
O2N
99%, 91% ee
X=
L*=
OAg
Ar
Ar=2,4,6-(Cy)C6H2

(30)

反应机理如图 9-7 所示：炔基溴化物在温和条件下与钯催化剂进行氧化加成得到炔基溴化钯中间体 **A**，**A** 随后与手性磷酸银发生复分解反应生成手性磷酸钯中间体 **B**，其进一步与 1,3-二烯经过 Heck 插入反应生成手性烯丙基钯中间体 **C**，随后发生烯丙基化得到 π-烯丙基钯中间体 **D**。中间体 **D** 与联硼酸酯经过不对称硼化反应生成手性烯丙基硼酸盐 **E**，**E** 进一步通过醛的不对称烯丙基硼化反应生成光学活性的高烯丙醇产物，同时释放出磷酸硼中间体 **F**，**F** 与银盐反应再生手性磷酸银从而完成催化循环。在该催化循环中，手性磷酸银与炔基溴化钯原位生成的炔基磷酸钯是实现高效立体选择性控制的关键中间体。

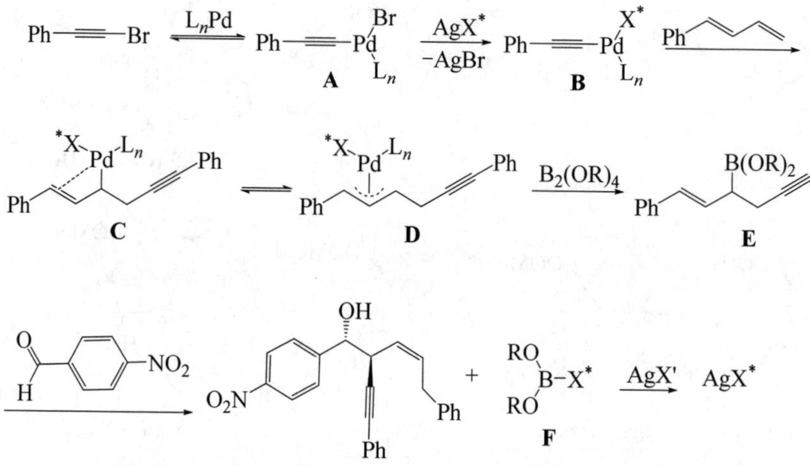

图 9-7　钯催化 1,3-二烯 Heck/Tsuji-Trost 串联反应机理

此外，烯丙基异氰化物也能作为碳亲核试剂与 *N*-芳甲酰基吲哚发生 Heck/Tsuji-Trost 串联反应，化学选择性地生成含 C3 亚胺杂环的吲哚啉骨架结构[24]。

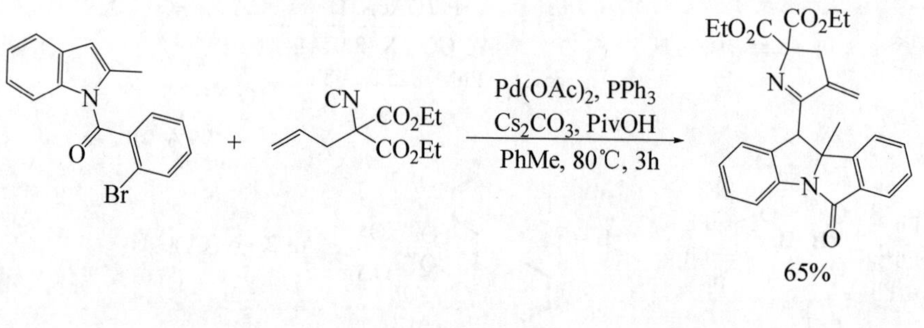

(31)

三甲基乙酸负离子在该反应中起了关键作用，因为不加三甲基乙酸反应几乎不能进行。反应机理如图 9-8 所示：首先，芳基溴化物氧化加成到 Pd^0 生成苄基钯中间体 **A**，随后吲哚双键迁移插入生成五元环钯中间体 **B**，**B** 进一步发生异氰插入生成亚胺钯中间体 **C**，然后通过第二次双键迁移插入生成烷基钯中间体 **D**，最后，**D** 经过还原消除得到含亚胺杂环的吲哚啉产物。

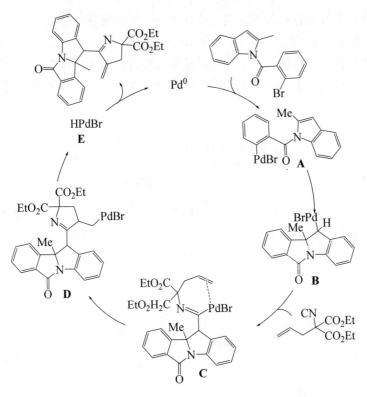

图 9-8 钯催化 1,3-二烯 Heck/Tsuji-Trost 串联反应机理

9.1.3.3 氧亲核试剂

除了氮亲核试剂和碳亲核试剂，氧亲核试剂（醇、酚和羧酸）也能够参与 Heck/Tsuji-Trost 串联反应。例如，邻碘苯酚与联烯在 $Pd(OAc)_2$ 催化下进行反应，以良好的产率生成苯并呋喃衍生物[25]。该催化反应体系不仅兼容链状联烯，环己基联烯也能顺利发生反应，以 63% 的产率得到苯并异色满衍生物。

在适宜的手性配体诱导下联烯化合物可发生对映选择性 Heck/Tsuji-Trost 反应，生成光学活性的化合物。其中，以 $Pd_2(dba)_3 \cdot CHCl_3$ 和手性双噁唑啉配体构成的催化体系可催化丙二烯羧酸类化合物和芳基碘化物进行对映选择性环加成反应，在低温条件下以中等的收率和对映选择性生成光学活性的 β-芳基丁烯内酯[26]。

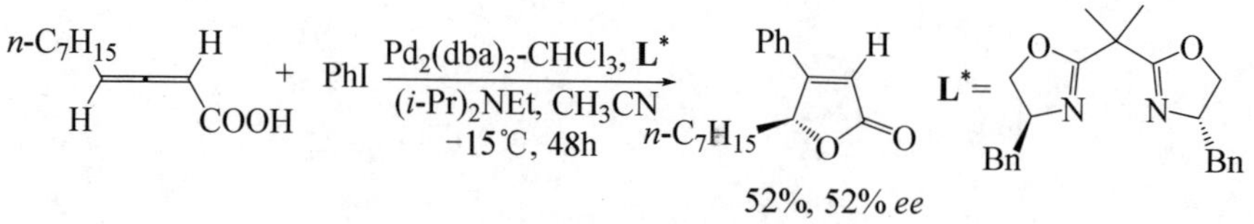

$$\text{反应式 (34)}\qquad 52\%,\ 52\%\ ee$$

除了苯酚和丙二烯羧酸化合物，γ-烯醇也能作为氧亲核试剂，在手性二苯基膦苯甲酰胺作为配体的条件下发生不对称 Heck/Tsuji-Trost 反应，以良好的收率和对映选择性生成光学活性的四氢呋喃衍生物[27]。

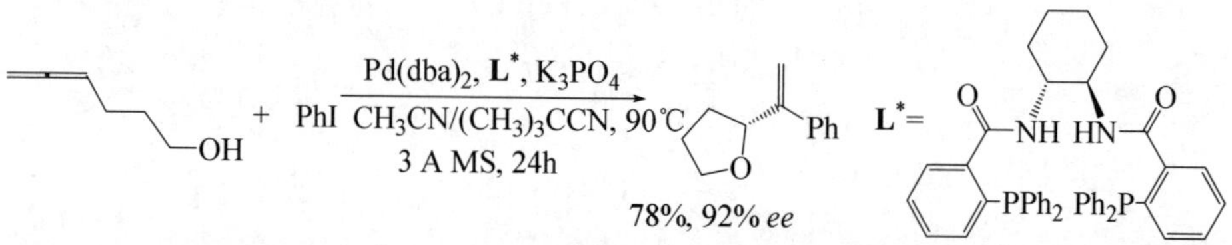

$$\text{反应式 (35)}\qquad 78\%,\ 92\%\ ee$$

非活化的 1,4-二烯化合物同样能与邻碘苯酚发生 Heck/Tsuji-Trost 串联反应，实现 1,4-二烯的 1,2-加成，以较高的收率生成苯并吡喃衍生物[28]。

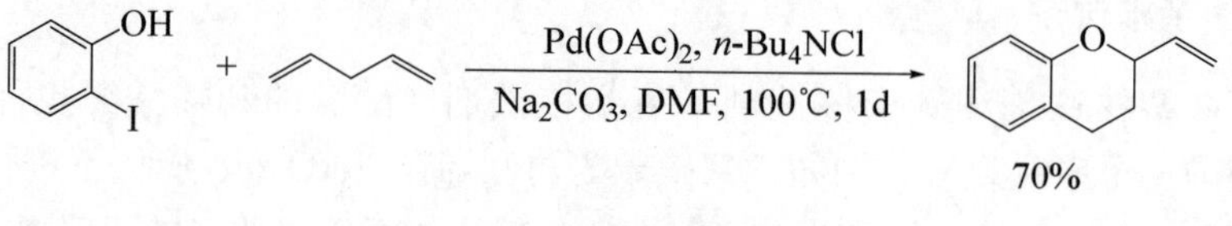

$$\text{反应式 (36)}\qquad 70\%$$

9.1.4 Heck/Tsuji-Trost 反应在药物和天然产物合成上的应用

在现代合成化学领域，从简单易得的起始材料出发，通过原子经济性和步骤经济性的方法，合成结构复杂多样的药物和天然产物分子一直是合成化学家们不断追求的目标。其中，多组分多米诺串联反应是实现上述"理想合成"最重要的策略之一。近年来，通过钯催化 Heck/Tsuji-Trost 策略合成结构多样化、功能化的手性分子受到化学家广泛的关注。

麦角酸及其衍生物（lysergic acid、lysergol 和 isolysergol）是非常重要的吲哚类生物碱，由寄生在黑麦和其他谷物上的真菌 Claviceps purpurea 分泌产生，具有广泛的生物活性，用作抗催乳素和抗帕金森病的药物。其结构和代表性的化合物

isolysergol 的合成路线如反应式（37）所示。首先，以 4-溴吲哚为起始原料，经过溴甲基化和 Ts 保护得到 3-(溴甲基)吲哚，然后通过硫缩醛水解、还原、脱硅和共轭加成等一系列步骤得到联烯化合物 TT-1，其进一步发生分子内钯催化 Heck/Tsuji-Trost 反应生成 TT-2，再经多步转化生成麦角酸衍生物 isolysergol[29]。

lysergic acid　　lysergol　　isolysergol

(37)

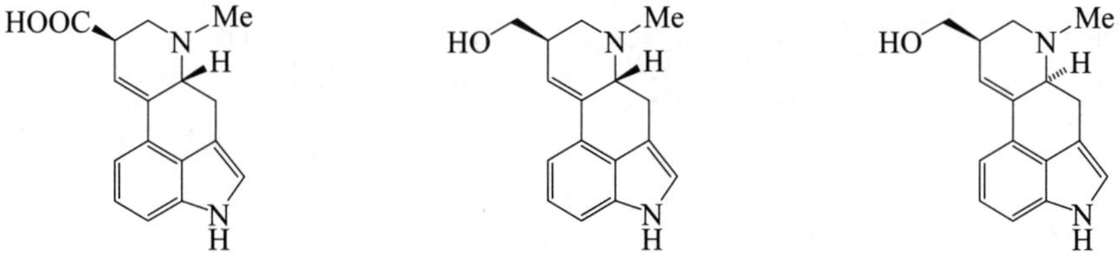

螺旋前列腺素(spirotryprostatins)是一种从真菌 *Aspergillus fumigatus* 发酵液中分离出来的二酮哌嗪生物碱，可以在微摩尔浓度下抑制哺乳动物细胞 G2/M 生长周期。以炔丙醇为原料，经过碘代、氧化、Witting 反应得到 1,3-二烯化合物 TT-3，随后通过羰基化、氨解反应、必要的官能团保护和脱保护、氧化反应等步骤得到关键的中间体 TT-4，最后在 Pd/(*R*)-BINAP 催化体系下发生 Heck/Tsuji-Trost 反应得到螺旋前列腺素前体，再脱除 N 保护基得到螺旋前列腺素[30]。

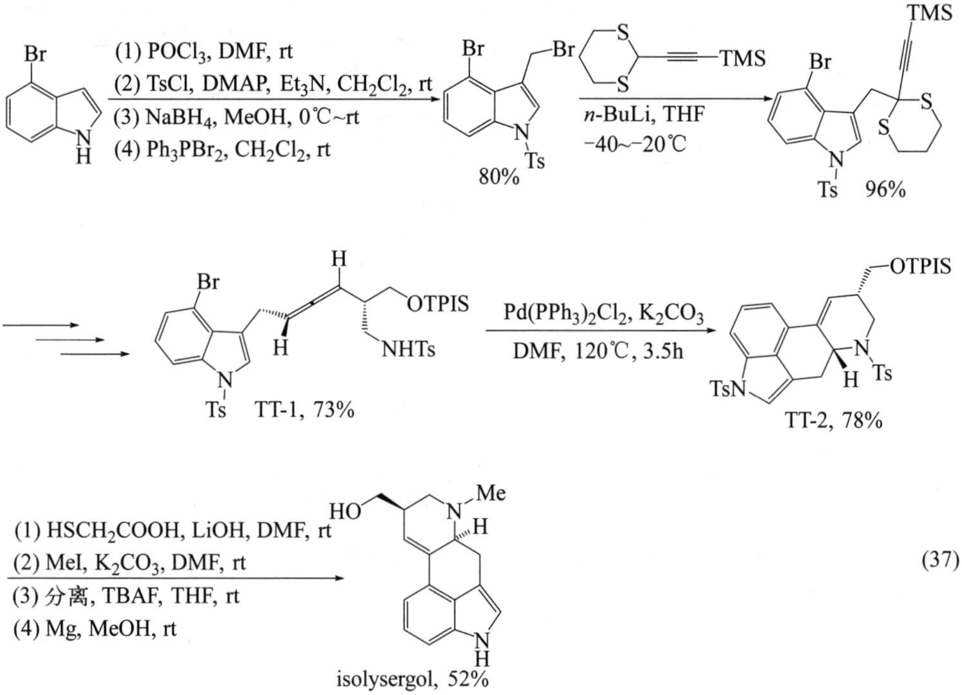

(38)

螺旋前列腺素，95%

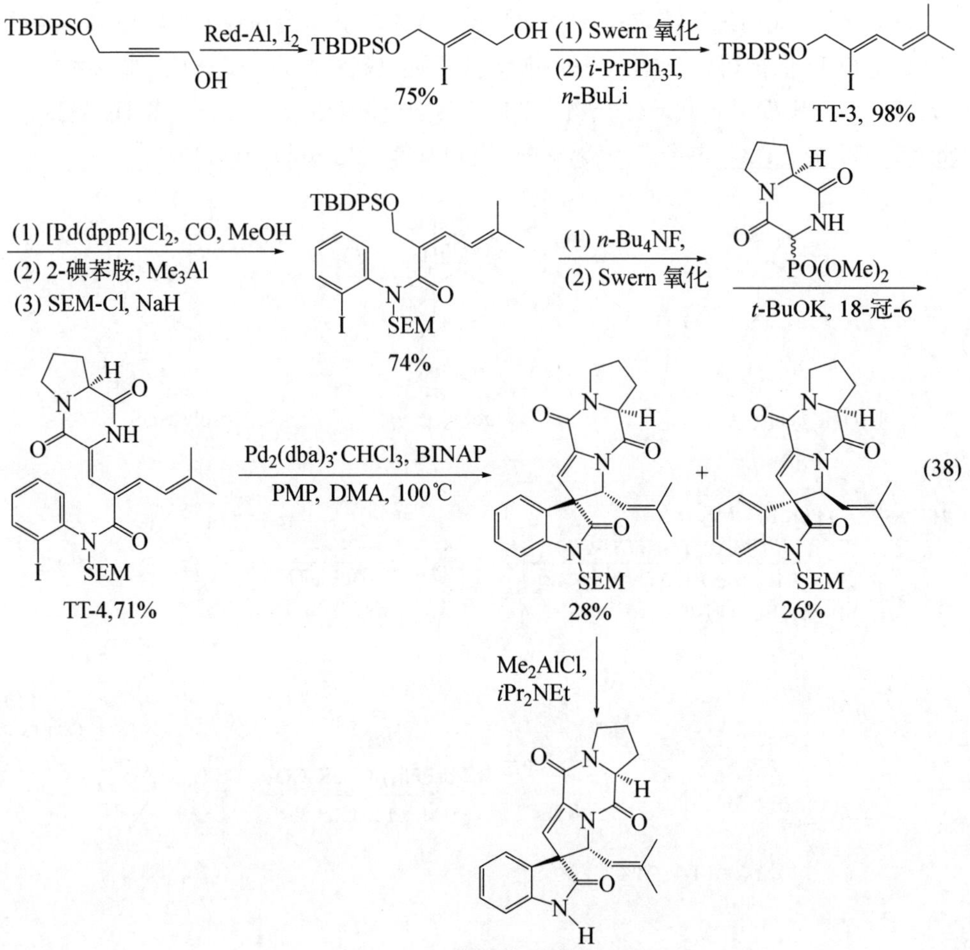

昆虫信息素(insect pheromone)是从雄性“干豆甲虫”中分离出来的一种丙二烯化合物，其在农作物保护领域有着重要的作用。它的合成路线如反应式（39）所示。

(39)

昆虫信息素,17%, 76% *ee*

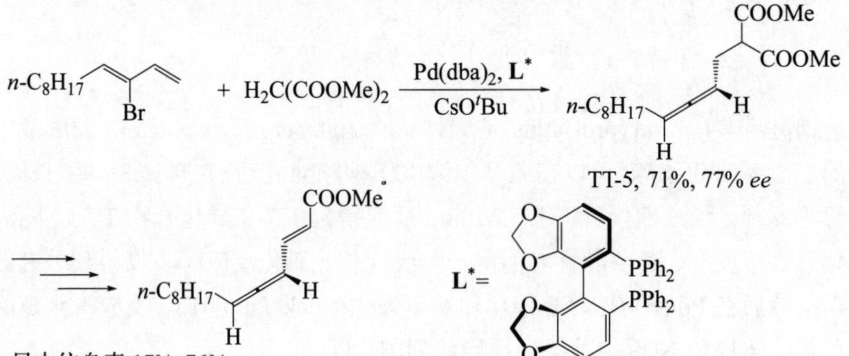

首先，在钯催化下 1,3-二烯化合物和丙二酸二甲酯经过 S_N2 取代反应生成联烯化合物 TT-5，进而通过脱羧、硒氧化物促进的脱氢等步骤转化为目标分子[31]。

9.2 Heck/Heck 串联反应

9.2.1 Heck/Heck 串联反应及机理

Heck/Heck 串联反应如图 9-9 所示：首先，芳基卤化物或类芳基卤化物（如芳基三氟甲磺酸酯）氧化加成到 Pd^0 生成二价钯引发反应，然后与第一个烯烃配位、插入生成 σ-烷基钯配合物中间体 **A**；**A** 不直接发生 β-消除完成 Heck 反应，而是与分子内或分子间的第二个烯烃单元配位、插入生成另一个 σ-烷基钯配合物中间体 **B**；最后，**B** 通过 β-消除生成三组分偶联产物。如果是分子内 Heck/Heck 串联反应则一步生成多环骨架结构的化合物。

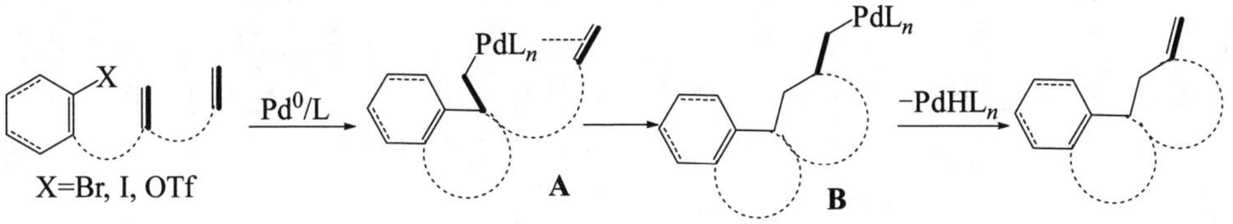

图 9-9 钯催化 Heck/Heck 串联反应机理

根据反应底物的不同，Heck/Heck 串联反应可分为分子内双 Heck 反应、分子间-分子内双 Heck 反应、分子内-分子间双 Heck 反应。在有机和药物合成中，通过底物的精准设计进行 Heck/Heck 串联反应，无须中间体分离，"一锅"反应合成结构复杂的化合物，具有高原子经济性、绿色高效的特征。

9.2.2 基本的 Heck/Heck 反应

9.2.2.1 分子内双 Heck 反应

分子内双 Heck 反应是一种高效构建含有多个季碳中心稠环化合物的方法。其中含 1,4-二烯结构片段的烷基邻溴苯基醚在钯催化下进行双 Heck 反应，能够实现复杂多元螺环化合物的高效合成[32]。需要强调的是，由于反应所采用的膦配体不同，得到两种结构完全不同的产物。当采用大位阻的三环己基膦单齿配体时，双 Heck 反应经过连续两次烯烃的迁移插入生成[4,5]-螺环化合物。而采用双齿膦配体 DPE-Phos 时，该配体能够促进远程 β-碳消除得到相应的苯并呋喃衍生物。

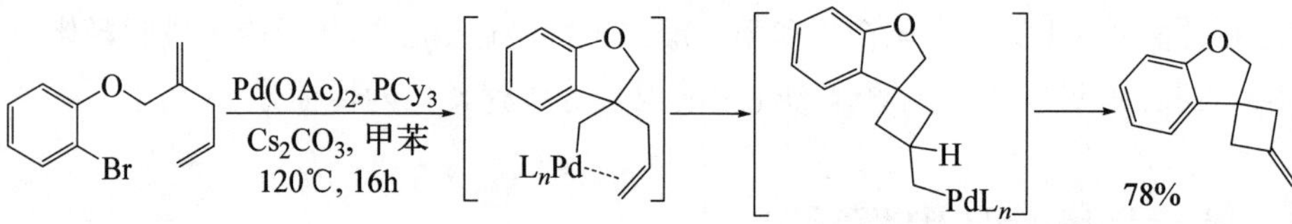

钯催化 1,5-二烯芳基碘化物的分子内双 Heck 反应，以良好的收率得到苯并螺环化合物[33]。反应是非对映选择性的，当 R^1 为甲基，R^2 为氢原子时，以 81%的收率和 13/1 的非对映选择性生成相应的苯并螺环化合物；当 R^1 为氢原子，R^2 为甲基时，则以 73%的收率和 6.5/1 的非对映选择性得到相应的产物。该现象可以由双键迁移插入得到芳基-钯(II)中间体的优势构象来解释。

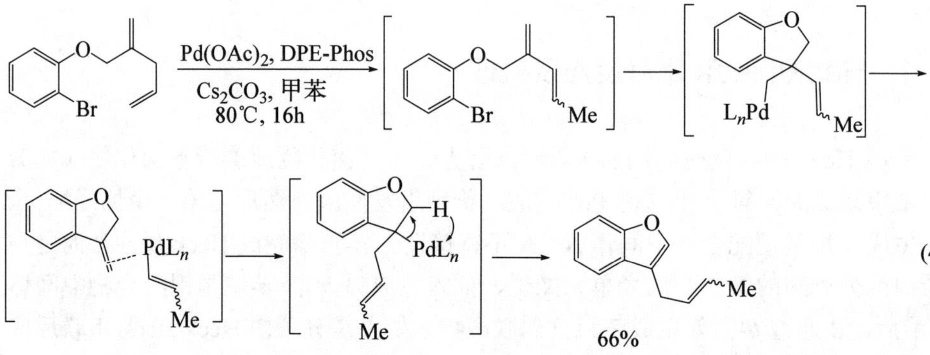

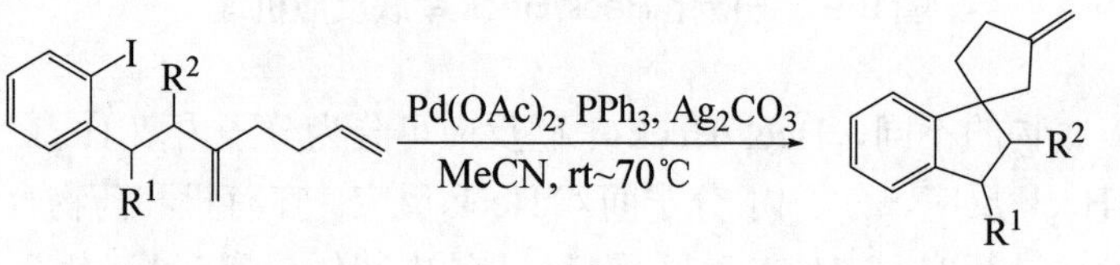

R^1=H, R^2=H   86%
R^1=Me, R^2=H   81% (d.r. 13/1)
R^1=H, R^2=Me   73% (d.r. 6.5/1)

1,6-二烯化合物同样能够发生分子内双 Heck 反应，经过 5-exo-exo（发生在双键内侧的环加成）化学选择性生成吡咯并环丙烷衍生物[34]。在 $Pd(OAc)_2/PPh_3$ 催化体系下，1,6-二烯化合物首先得到关键的四氢吡咯甲基二价钯中间体，进而通过分子内立体选择性环加成生成 6/5/3/-稠环产物。

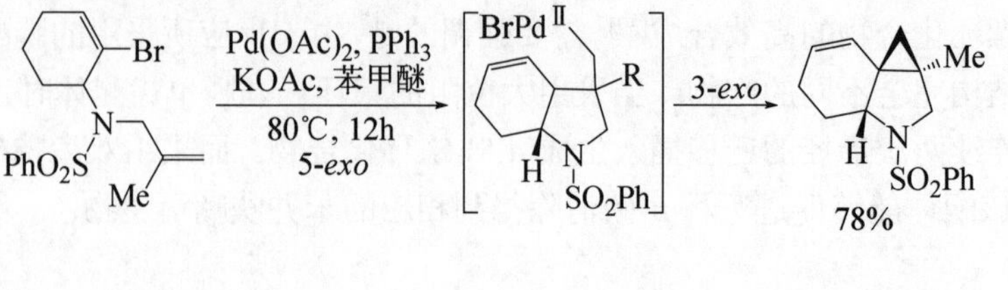

此外，以 1,5-二烯碘苯化合物为起始原料，在钯催化下能够高效合成多元螺环化合物[35]。反应首先生成 σ-烷基钯中间体 **A**，随后该二价钯物种与分子内的吲哚环发生迁移插入得到环化的中间体 **B**，**B** 再进一步进行还原消除生成苯并螺环产物。反应经过分子内碳碳双键对二价钯中间体迁移插入，然后进一步发生还原消除，得到 5/5 螺环产物。

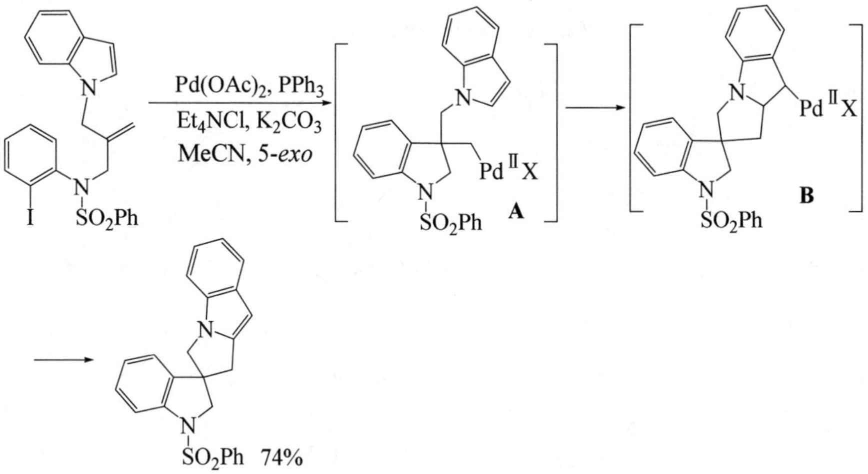

(44)

除了二烯类化合物，1,3,5-三烯化合物同样能够发生分子内双 Heck 反应。以 Pd(OAc)$_2$/(R)-BINAP 构成的不对称催化体系，可以催化 1,3,5-三烯碘苯类化合物对映选择性生成苯并四环产物[36]。该反应经历碘苯氧化加成到 Pd^0 生成二价钯中间体，再进行分子内双键迁移插入生成烷基钯中间体，随后发生 6-endo 双键插入得到相应的产物。

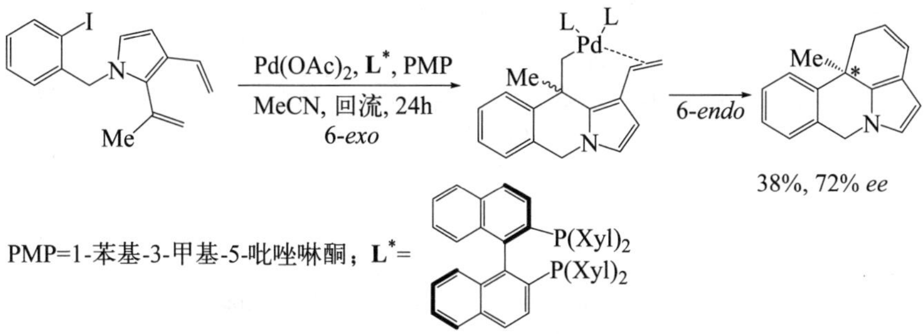

(45)

PMP=1-苯基-3-甲基-5-吡唑啉酮；L* =

1,6-烯炔化合物同时含有双键和三键官能团，其反应活性高，是药物化学和有机合成中广泛应用的合成子。以 1,6-烯炔化合物为底物，含有手性亚砜结构的环戊二烯基钌(CpRu)作为催化剂，能以中等至良好的产率及良好的对映选择性得到光学活性的[3.1.0]氮杂环产物[37]。

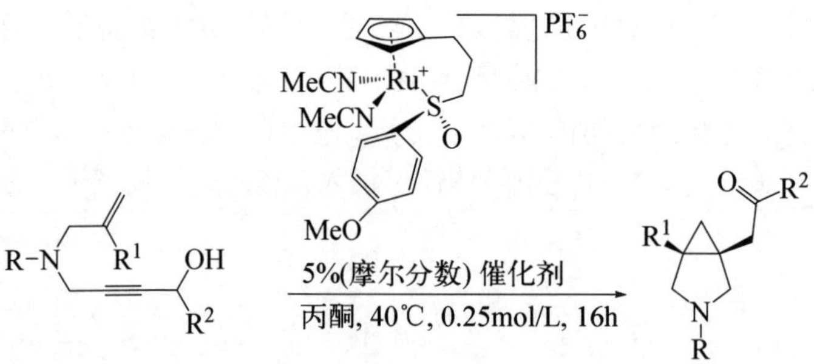

$$\text{(46)}$$

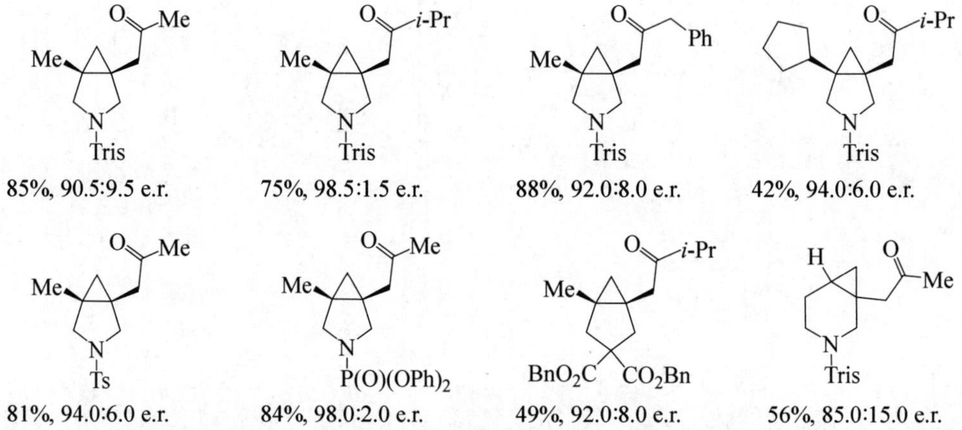

催化剂: Ru复合物 PF$_6^-$，含MeCN、S(O)-对甲氧苯基配体

反应条件：5%(摩尔分数) 催化剂，丙酮, 40℃, 0.25 mol/L, 16 h

产物示例：

- 85%, 90.5:9.5 e.r. （Me, Tris, 乙酰基）
- 75%, 98.5:1.5 e.r. （Me, Tris, i-Pr酮）
- 88%, 92.0:8.0 e.r. （Me, Tris, CH$_2$Ph酮）
- 42%, 94.0:6.0 e.r. （环戊基, Tris, i-Pr酮）
- 81%, 94.0:6.0 e.r. （Me, Ts, 乙酰基）
- 84%, 98.0:2.0 e.r. （Me, P(O)(OPh)$_2$, 乙酰基）
- 49%, 92.0:8.0 e.r. （Me, BnO$_2$C/CO$_2$Bn, i-Pr酮）
- 56%, 85.0:15.0 e.r. （Tris氮杂, 乙酰基）

以 1,6-烯炔化合物为起始原料，二苯基硅烷为还原剂，在钯催化剂作用下发生不对称还原烷基化双 Heck 串联反应，能以良好的产率和对映选择性得到一系列光学活性的 5/3-双氮杂环产物[38]。此外，以苯乙炔作为还原剂，也能以 75% 的收率和 88% ee 的对映选择性得到氮杂[3.1.0]双环产物。

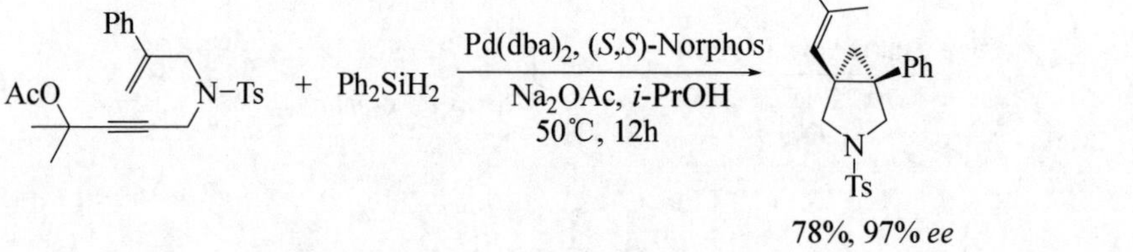

$$\text{(47)}$$

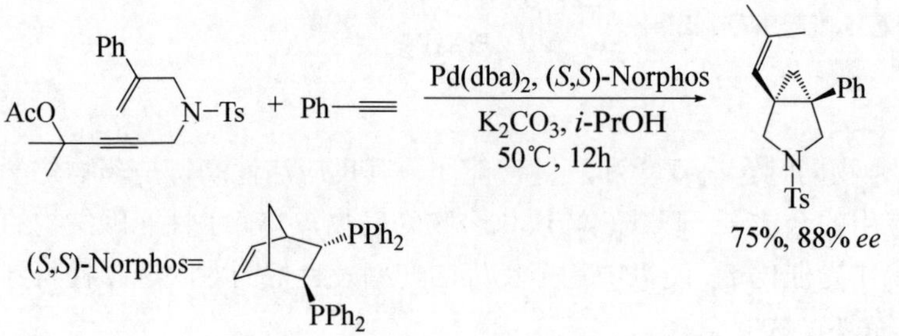

AcO-C(Me)$_2$-C≡C-CH$_2$-N(Ts)-CH$_2$-C(Ph)=CH$_2$ + Ph$_2$SiH$_2$ →（Pd(dba)$_2$, (S,S)-Norphos, Na$_2$OAc, i-PrOH, 50℃, 12 h）→ 产物

78%, 97% ee

$$\text{(48)}$$

AcO-C(Me)$_2$-C≡C-CH$_2$-N(Ts)-CH$_2$-C(Ph)=CH$_2$ + Ph-C≡CH →（Pd(dba)$_2$, (S,S)-Norphos, K$_2$CO$_3$, i-PrOH, 50℃, 12 h）→ 产物

75%, 88% ee

(S,S)-Norphos = 降冰片烯-2,3-双(二苯基膦)

9.2.2.2 分子间-分子内双 Heck 反应

过渡金属催化的烯烃多次迁移插入反应大多数是分子内迁移插入引发的,而分子间迁移插入引发的反应相对较少。下面的实例则属于后者,先由分子间 Heck 反应引发,然后再进行分子内 Heck 偶联。反应底物苄基卤化物和 4-氮杂-1,6-二烯在钯催化下反应生成二氢吡咯衍生物[39]。

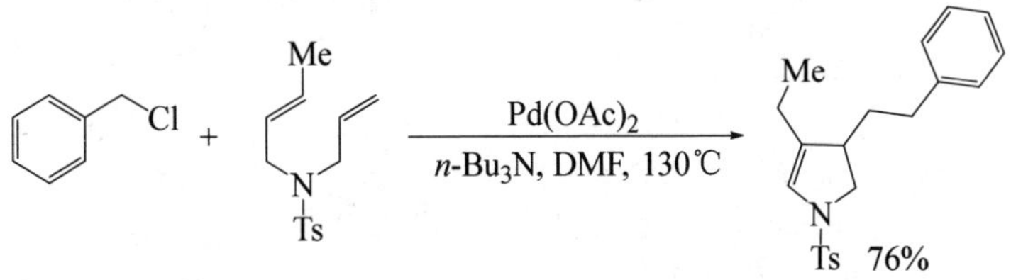

(49)

反应机理如图 9-10 所示:首先,苄基卤化物氧化加成到 Pd^0 生成苄基二价钯中间体 **A**,然后 4-氮杂-1,6-二烯的末端双键与中间体 **A** 发生分子间迁移插入生成烷基钯中间体 **B**,随后经过分子内 5-*exo-trig*(对双键的环外加成)环加成/异构化得到二氢吡咯环产物。

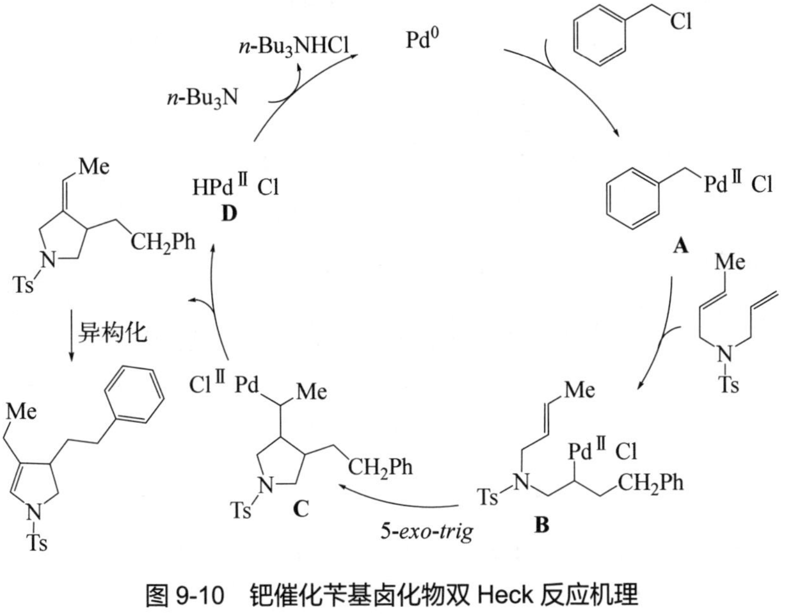

图 9-10 钯催化苄基卤化物双 Heck 反应机理

以邻乙烯基苯基三氟甲磺酸酯与 2,3-二氢呋喃为底物,在 $Pd(dba)_2$/(*R*)-BINAP(O)或(*R*)-Xyl-SDP(O)催化体系催化下进行不对称分子间-分子内双 Heck 反应,以中等至较高的收率和良好的对映选择性生成一系列苯并亚甲基环戊烷并呋喃衍生物[40]。此外，其他一些环烯类化合物如环戊烯、环庚烯、环辛烯、二氢吡咯等同样可以作为亲核试剂参与分子间-分子内双 Heck 反应，生成相应的稠合三环化合物。

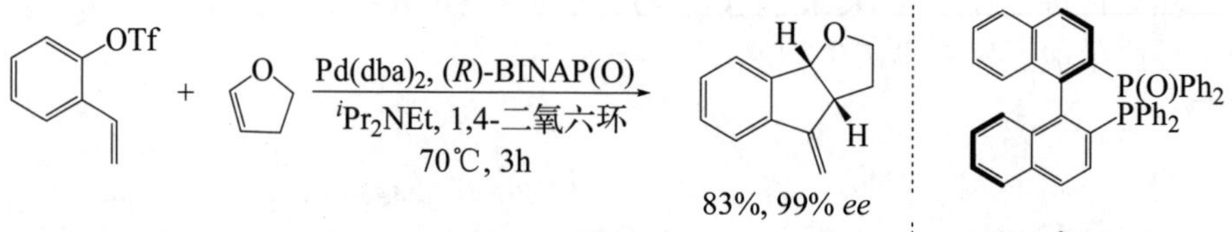

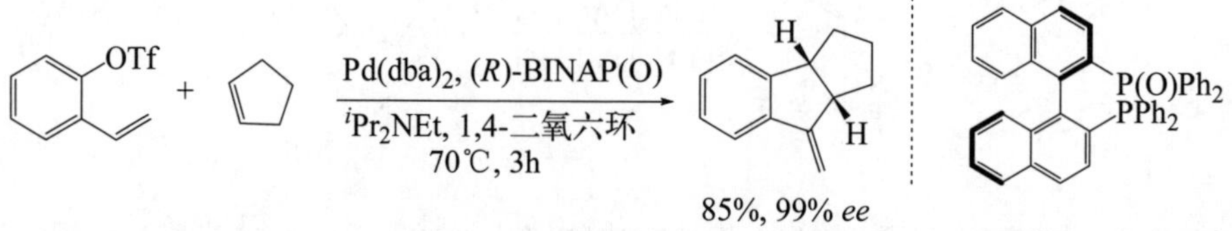

9.2.2.3 分子内-分子间双 Heck 反应

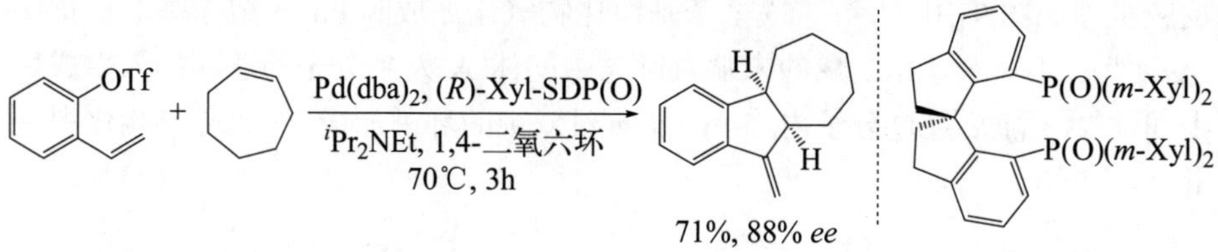

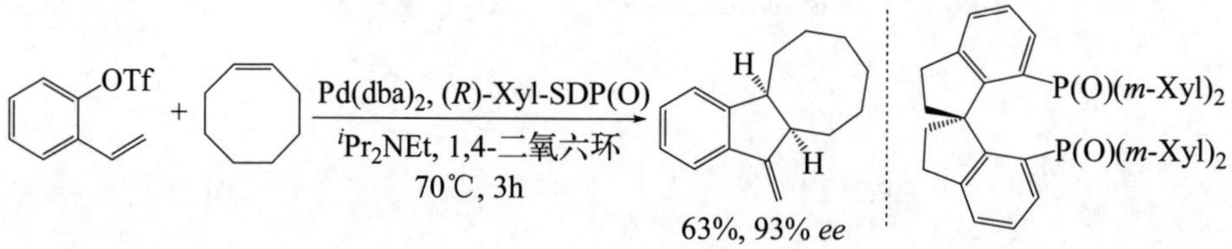

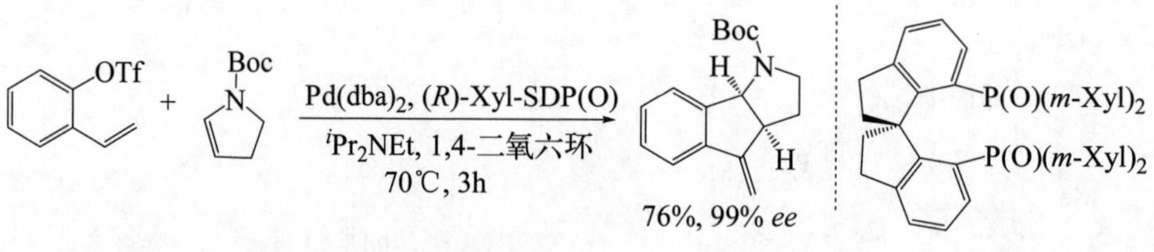

除了上述涉及的分子内双 Heck 反应和分子间-分子内双 Heck 反应，分子内-分子间双 Heck 反应同样是一种重要的双 Heck 反应，为构建各种碳环和杂环化合物提供了一种非常简捷高效的途径。

以烯丙基芳基醚和丙烯酸甲酯为起始原料，通过分子内-分子间双 Heck 反应，能以中等的产率生成苯并呋喃衍生物，伴随直接分子内 Heck 偶联副产物以及 Heck 氢化副产物[41]。

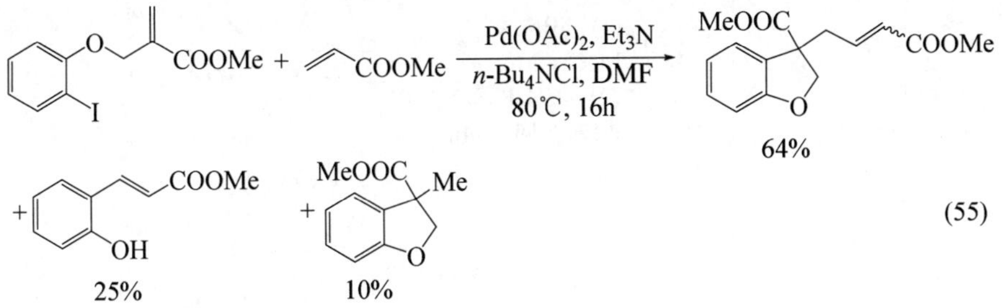

(55)

以 N-邻溴苯基丙烯酰胺衍生物和 β-芳基溴乙烯类化合物为起始原料,在镍催化剂催化下于二甲基亚砜等极性溶剂中进行分子内-分子间双 Heck 反应,以良好的产率生成苯并吡咯烷酮衍生物[42]。当以手性双噁唑啉替代联吡啶为配体时,以高达 74%的产率和 82% ee 的对映选择性生成光学活性的苯并吡咯烷酮衍生物。

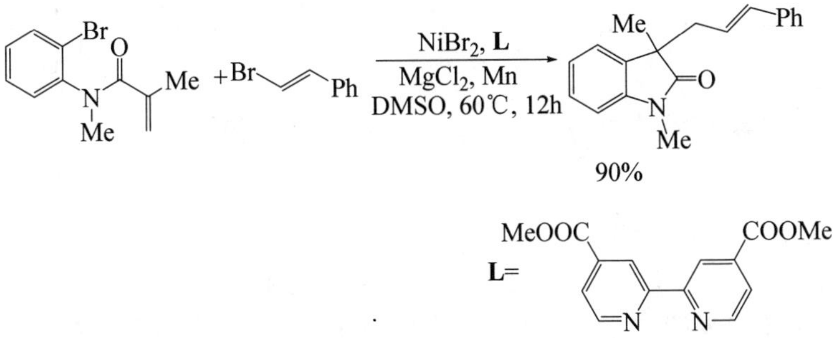

(56)

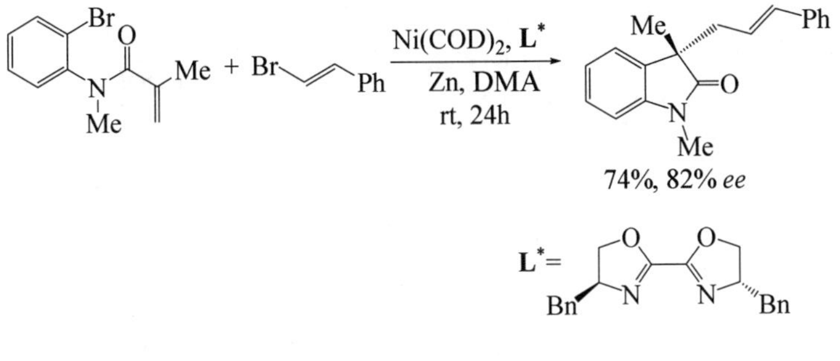

(57)

偕二氟乙烯类化合物同样可以作为偶联试剂参与分子内-分子间双 Heck 反应。N-邻溴苯基丙烯酰胺衍生物和芳基偕二氟乙烯在不对称镍催化剂催化下进行不对称还原芳化-单氟烯基化反应,以优良的对映选择性得到光学活性的苯并吡咯酮衍生物[43]。此方法成功用于复杂生物活性分子如生育酚、雌素酮和果糖衍生物的后期单氟烯化反应,并耐受其他敏感官能团如酯、缩醛和酮等。

(58)

生育酚衍生物

雌素酮衍生物

果糖衍生物

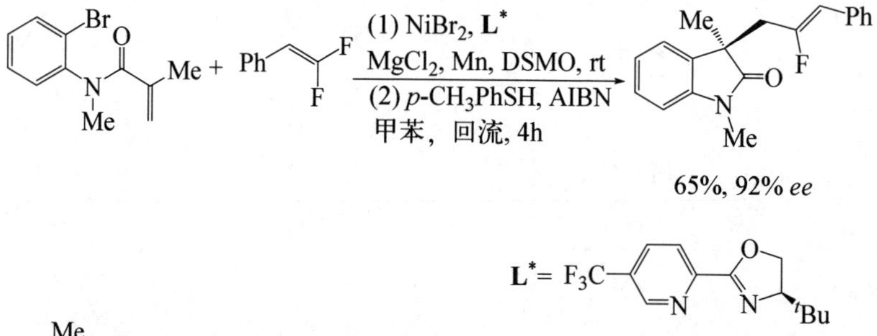

在手性钯催化剂催化下，芳基四氟硼酸重氮盐邻烯丙基醚与简单烯烃进行不对称双 Heck 串联反应，能够构建含有季碳中心的光学活性的苯并二氢呋喃衍生物[44]。Ag_2CO_3 是该反应一个重要的添加剂，不仅作为路易斯酸促进 $PdCl_2$ 和底物芳基四氟硼酸重氮盐的氧化加成，还能消除反应体系中的氯离子。

(59)

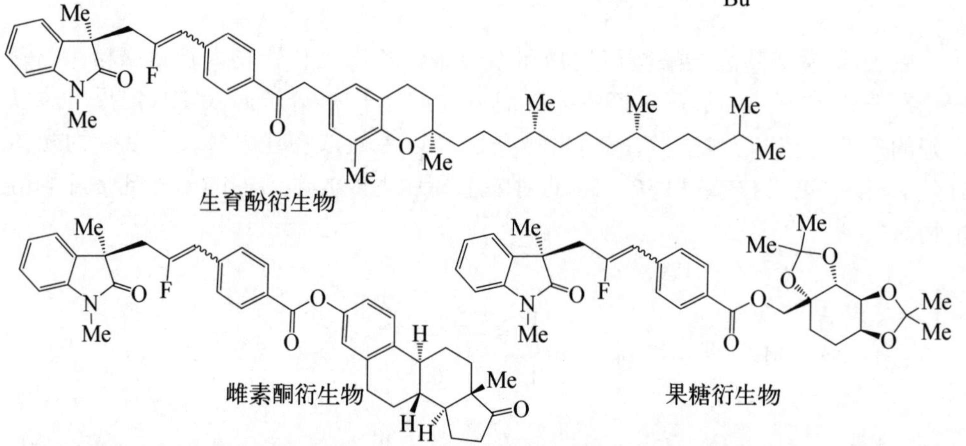

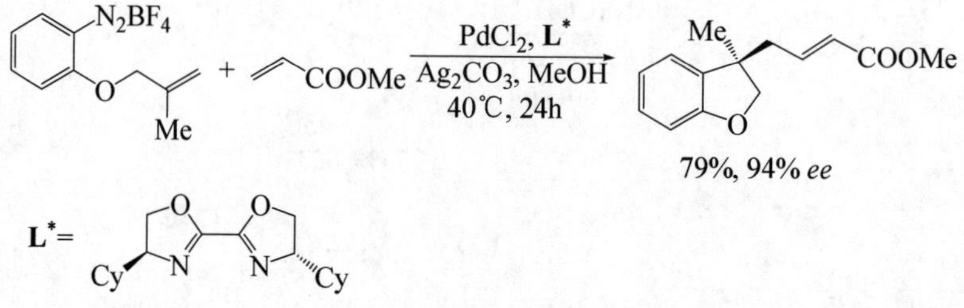

9.2.3 Heck/Heck 串联反应在药物和天然产物合成上的应用

双 Heck 串联反应具有绿色高效、原子经济、操作方便等优点，为构建各种碳环和杂环化合物提供了一种非常简捷高效的方法。因而，该串联反应被应用于一些含螺环、稠环以及桥环结构单元的复杂天然产物和药物分子的合成。

四环二萜茛菪酸 B（scopadulcic acid B）是从药用植物茛菪碱中分离得到的一种天然产物，具有很强的生物活性，能够抑制哺乳动物胃中氢离子、钾离子和腺苷三磷酸酶。该天然产物可以邻碘苯甲醛为原料经 31 步反应合成，其中最关键的一步是含 1,5-二烯基结构单元的烷基邻碘苯基酮 TT-6 在钯催化下发生分子内 Heck/Heck 串联反应生成多环中间体 TT-7[45]。得到的关键中间体 TT-7 再经多步反应最终生成 scopadulcic acid B。该合成路线过于冗长，总体收率不高。

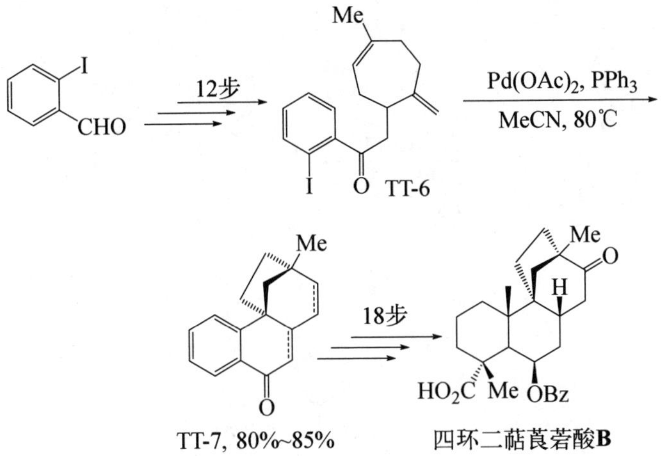

(60)

基于四环二萜茛菪酸 B 的合成经验，将分子内 Heck/Heck 串联反应成功应用于四环二萜茛菪酸 A 的全合成。反应以同时含有 1,5-二烯和碘乙烯结构单元的化合物为原料，在 Pd(OAc)$_2$/PPh$_3$ 催化体系下进行反应，以良好的收率得到关键的桥接三环化合物中间体 TT-8。TT-8 再经 15 步反应最终对映选择性地生成具有光学活性的四环二萜茛菪酸 A[46]。通过路线的重新设计，相较于四环二萜茛菪酸 B 的合成，四环二萜茛菪酸 A 的合成路线大大缩短，整体收率明显提高。

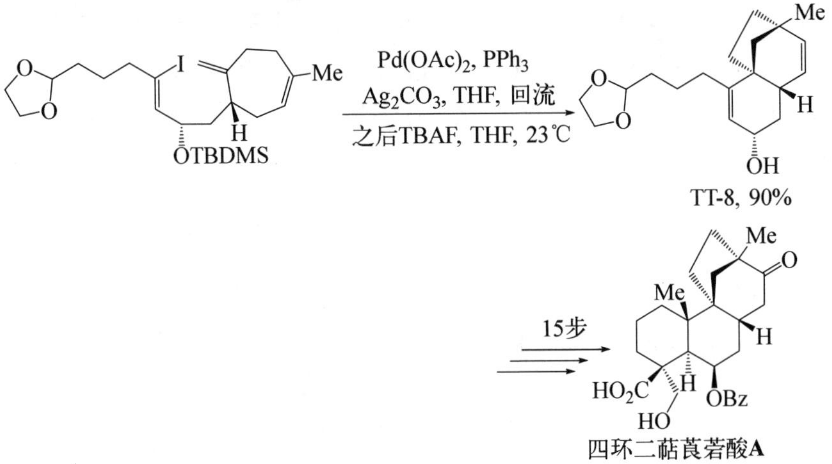

(61)

(+)-齐斯托醌［(+)-xestoquinone］是从太平洋西斯托海绵体中提取的生物活性化合物，对心肌具有独特的正性肌力作用，是一种强效的心肌强化剂。在手性钯催化剂催化下，三氟甲磺酸萘酯-二乙烯基呋喃甲酮衍生物进行不对称双 Heck 反应，以良好的收率和中等的对映选择性生成关键的五环中间体 TT-9。随后，该中间体再经 Pd/C 催化还原和硝酸铈铵(CAN)氧化得到光学活性的(+)-齐斯托醌[47]。

1

TT-9, 82%, 68% *ee*

(+)-齐斯托醌　(62)　L^*=

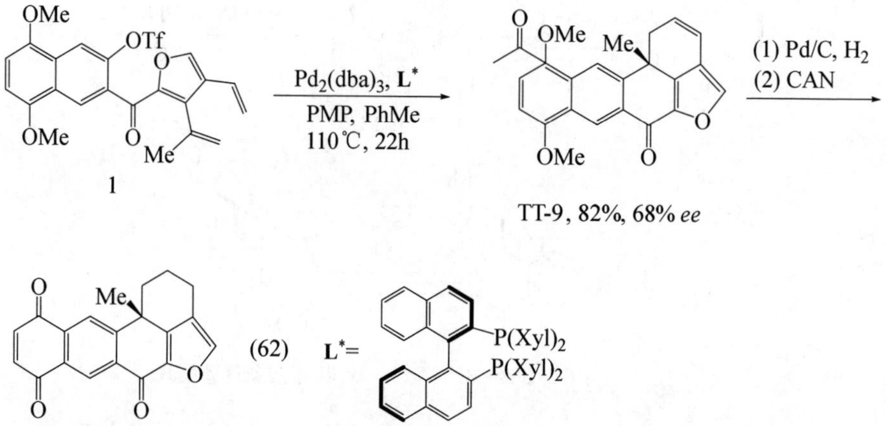

(+)-毒扁豆碱[(+)-physostigmine]和(+)-毒扁豆次碱[(+)-physovenine]是从非洲出产的毒扁豆种子中提取得到的生物碱，为抗胆碱酯酶的药物，临床上用来抑制青光眼眼压偏高的症状。以 2-溴-4-甲氧基苯胺为起始原料，经过与 *α*-甲基丙烯酰氯缩合、碘甲烷甲基化生成 *N*-（2-溴-4-甲氧基苯基）-*α*-甲基丙烯酰胺 TT-10，TT-10 进一步与 *β*-芳基乙烯基溴进行分子内-分子间双 Heck 反应得到环化的二氢吲哚产物 TT-11。TT-11 经过氧化反应、还原反应等多步转化生成(+)-毒扁豆碱和(+)-毒扁豆次碱[42]。

TT-10, 92%

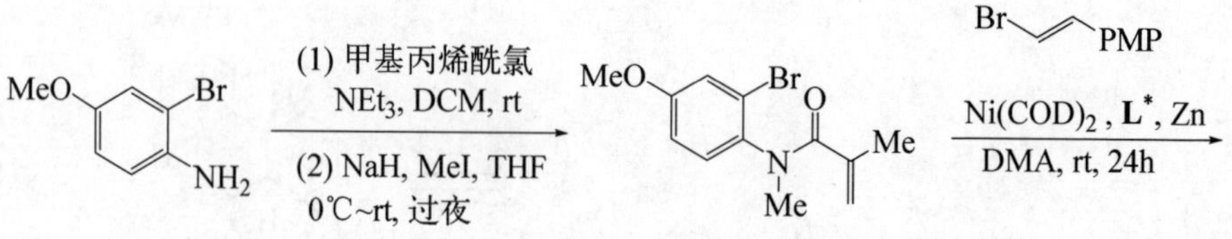

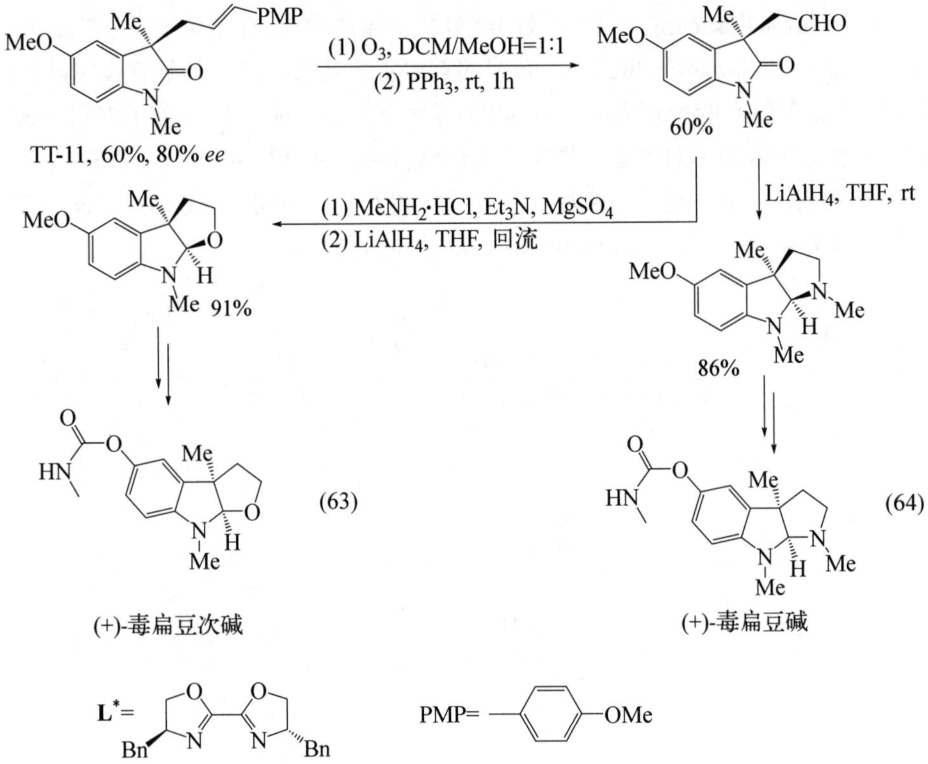

isoabietenin A 是一种从细叶香茶菜中提取得到的天然产物,对部分癌细胞有较好的细胞毒性。以 2-硝基苯酚为原料,经过与 2-甲基烯丙基氯的亲核取代成醚、硝基还原、重氮化得到四氟硼酸重氮盐 TT-12,随后通过钯催化的不对称双 Heck 串联反应生成苯并二氢呋喃衍生物 TT-13。TT-13 再进一步经过环氧化、开环-亲电取代串联反应生成 isoabietenin A[48]。

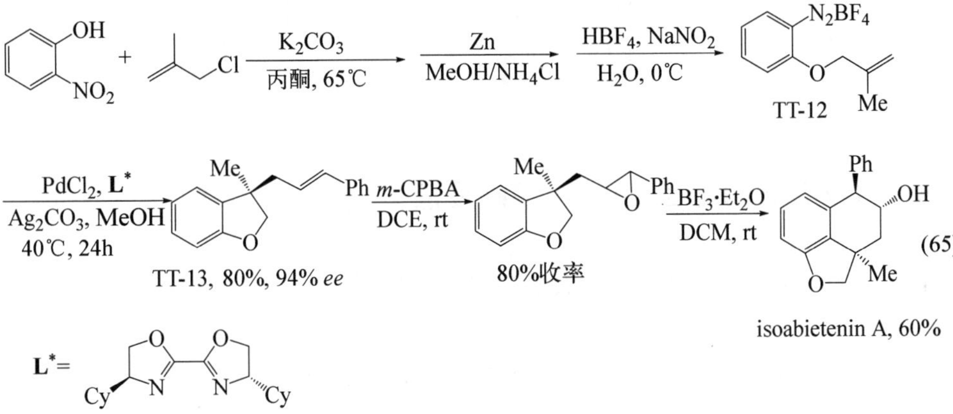

(−)-martinellic acid 是一种有效的缓激肽受体拮抗剂，在南美国家被广泛用于治疗眼部感染。(−)-martinellic acid 合成的首要步骤是邻乙烯基苯基三氟甲磺酸酯与 2,3-二氢吡咯在 Pd(dba)$_2$/(R)-Xyl-SDP(O)催化下的不对称分子间-分子内双 Heck 反应，以良好的收率和对映选择性生成光学活性中间体 TT-14。随后，TT-14 经臭氧化、烷氧胺化、亚胺还原、扩环反应等反应得到关键中间体 TT-15；中间体 TT-15 再经多步转化最终完成(−)-martinellic acid 的全合成[49]。

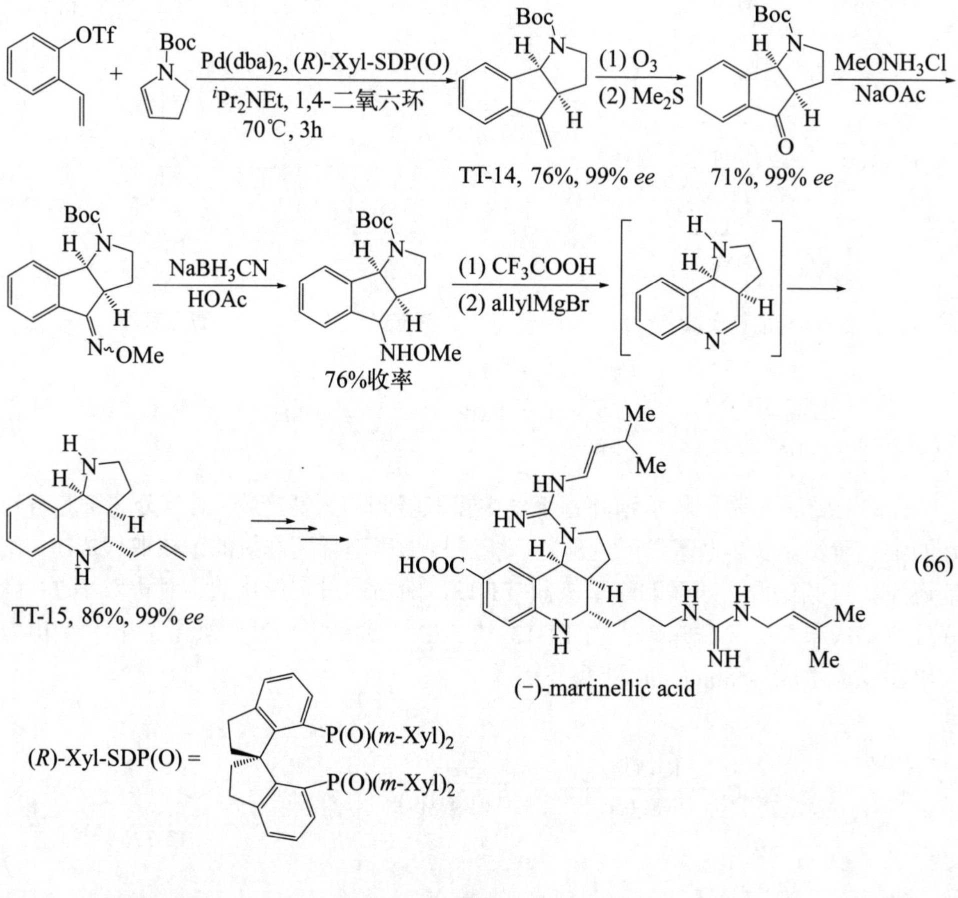

9.3 Heck/Sonogashira 串联反应

9.3.1 Heck/Sonogashira 串联反应及机理

钯催化的 Heck 反应和钯/铜催化的 Sonogashira 反应的引发步骤都是烯基卤代物或芳基卤代物与 Pd^0 的氧化加成，通常 Heck 反应的活性较高，而 Sonogashira 反应的活性较低。利用两个反应的活性差，可以将二者巧妙串联，同时构建多个碳-碳键，将大大提高反应的效率，使复杂结构的药物和生物活性化合物的合成更简捷高效。Heck/Sonogashira 串联反应主要有两类。其一是钯催化的邻二卤代烯烃衍生物与末端烯烃、末端炔烃三组分 Heck/Sonogashira 串联反应；其二是 Pd/Cu 双金属催化的带有碳碳双键结构片段的芳基卤化物与末端炔烃化合物的双组分 Heck/Sonogashira 串联反应。

第一种 Heck/Sonogashira 串联反应过程如图 9-11 所示。首先，在钯催化剂催化下邻二卤代烯烃活性高的一端与末端烯烃经过氧化加成、还原消除等步骤完成单独的 Heck 反应，生成偶联中间体 **A**；**A** 一旦生成即刻与末端炔烃进行单独的 Sonogashira 反应，最终实现三组分的偶联，生成目标产物[50]。

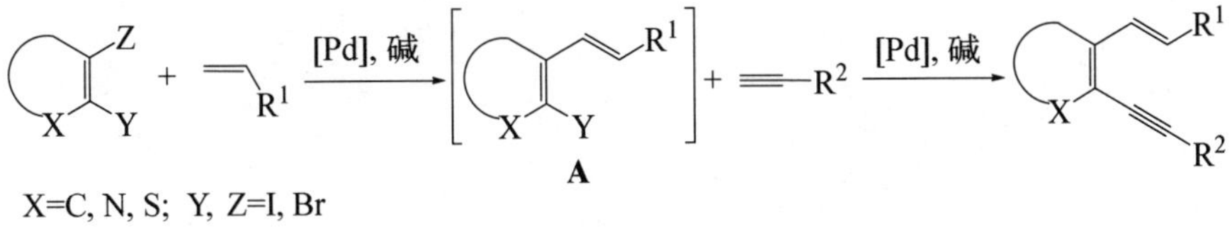

[Pd], 碱
[Pd], 碱
A
X=C, N, S; Y, Z=I, Br

图 9-11　钯催化 Heck/Sonogashira 三组分反应机理

第二种 Heck/Sonogashira 串联反应机理如图 9-12 所示。首先，芳基碘化物氧化加成得到苄基二价钯中间体 **A**，随后发生分子内烯烃迁移插入得到五元环烷基钯中间体 **B**；由于 **B** 中 R^2 的存在不能发生 β-消除完成 Heck 反应，其进一步与炔基铜配合物发生转金属得到炔基二价钯中间体 **C**；最后，**C** 经过还原消除生成炔丙基取代的二氢吲哚酮产物[51]。

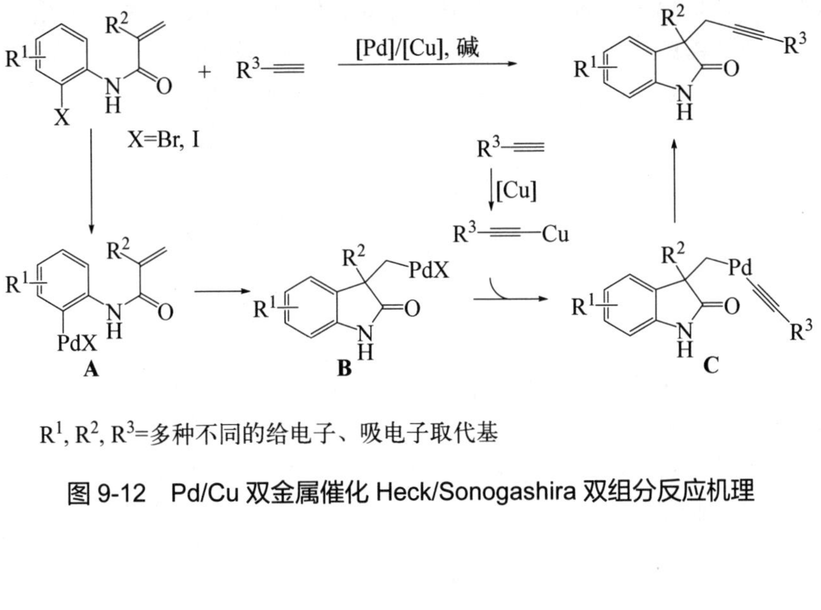

[Pd]/[Cu], 碱
X=Br, I
[Cu]
A
B
C
R1, R2, R3=多种不同的给电子、吸电子取代基

图 9-12　Pd/Cu 双金属催化 Heck/Sonogashira 双组分反应机理

9.3.2 基本的 Heck/Sonogashira 反应

芳基碘化物、活化烯烃、末端芳香炔烃化合物在醋酸钯催化下进行 Heck/Sonogashira 三组分串联反应，"一锅"生成相应的烯炔衍生物[52]。

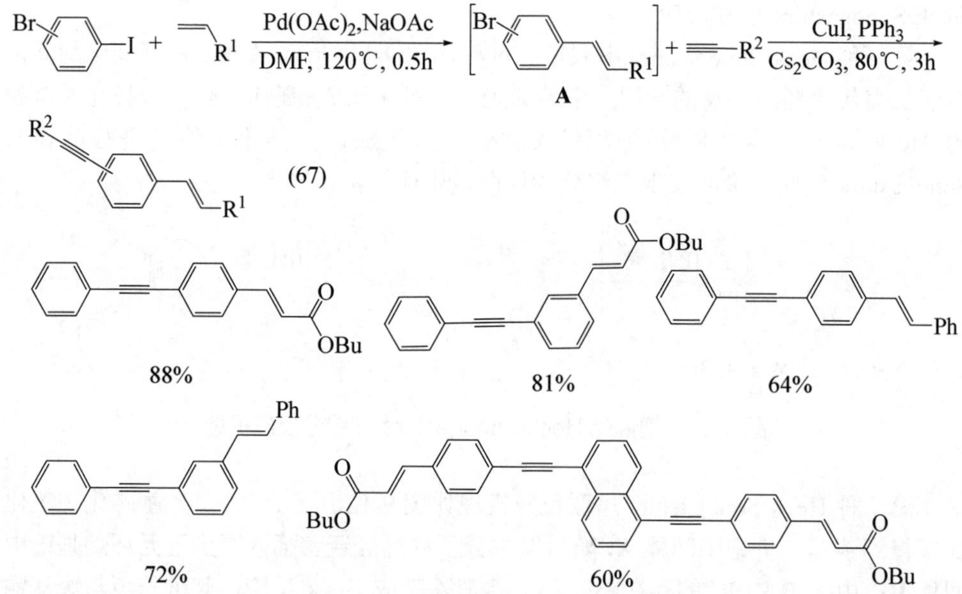

$$\text{Br}\diagdown\text{C}_6\text{H}_4\diagdown\text{I} + \diagdown\text{R}^1 \xrightarrow[\text{DMF, 120℃, 0.5h}]{\text{Pd(OAc)}_2,\text{NaOAc}} [\text{A}] + \equiv\text{—R}^2 \xrightarrow[\text{Cs}_2\text{CO}_3, 80℃, 3h]{\text{CuI, PPh}_3}$$

(67)

88%　　81%　　64%

72%　　60%

含有乙烯基片段的芳基卤化物和炔类化合物在钯催化下无须铜催化剂就能顺利地进行 Heck/Sonogashira 串联反应，以良好的产率生成一系列炔丙基取代的含季碳中心的二氢吲哚酮衍生物[53]。反应结果与卤素的种类相关，卤素负离子的离去能力越强，反应越容易进行，如反应式（68）所示，芳基碘化物、芳基溴化物分别以95%和59%的产率得到相应的偶联产物，而芳基氯化物不进行反应。

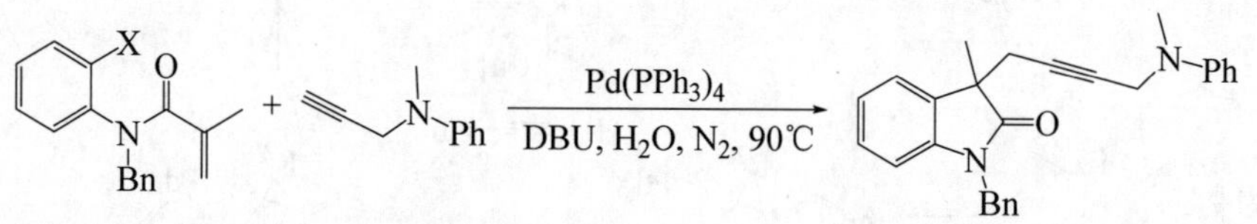

(68)

X=I, 95%; X=Br, 59%; X=Cl, 0%

采用 Pd(MeCN)$_2$Cl$_2$/CuI 共催化体系，以 *N*-(2-溴苯甲酰基)-吲哚和末端炔烃为底物，经 Heck 环化/Sonogashira 偶联过程，能以优良的产率和非对映选择性生成各种不同取代的脱芳构炔基化吲哚啉产物[54]。相较于质子极性溶剂，反应更适合于在非质子极性溶剂中进行。

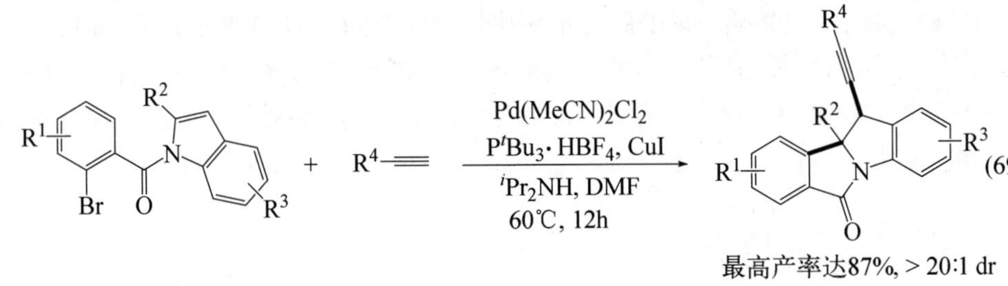

反应条件示意:

反应物 + R⁴−≡ 经 Pd(MeCN)₂Cl₂, P^{t}Bu₃·HBF₄, CuI, ^{i}Pr₂NH, DMF, 60℃, 12h 条件反应 (69)

最高产率达87%, > 20:1 dr

产物示例收率：82%, 82%, 47%, 54%, 87%, 62%, 52%, 60%

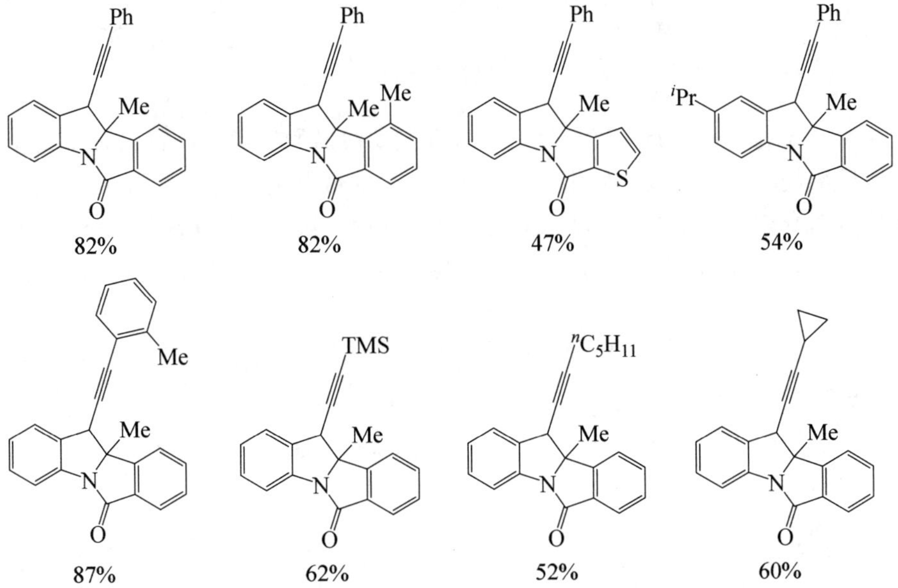

光学活性化合物在有机和药物合成领域应用广泛，因此钯催化的带有碳碳双键结构片段的芳基卤化物对映选择性 Heck/Sonogashira 串联反应受到不少关注。以 Pd(dba)₂ 作为催化剂前体，(R)-BINAP 作为手性膦配体，可以实现酰基吲哚不对称脱芳构化反应，以良好的收率、对映选择性和非对映选择性得到光学活性吲哚啉衍生物[55]。

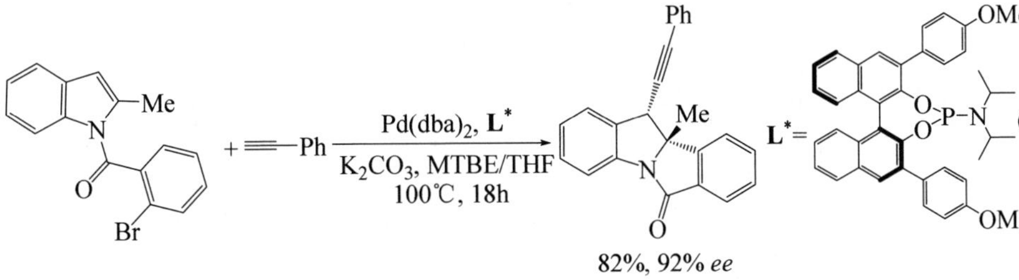

反应式 (70): 酰基吲哚 + Ph−≡ 经 Pd(dba)₂, L*, K₂CO₃, MTBE/THF, 100℃, 18h 条件得到产物，82%, 92% ee

L* 为手性膦配体结构

第 9 章 过渡金属催化串联反应及其在药物合成上的应用　　295

Pd/Cu 协同催化邻碘丙烯酰胺与末端炔烃对映选择性 Heck/Sonogashira 反应，通过使用光学活性的二茂铁化合物作为配体，可以合成具有光学活性的含三氟甲基季碳中心的吲哚酮衍生物[56]。值得注意的是，添加剂碳酸银对反应的产率和对映选择性至关重要。

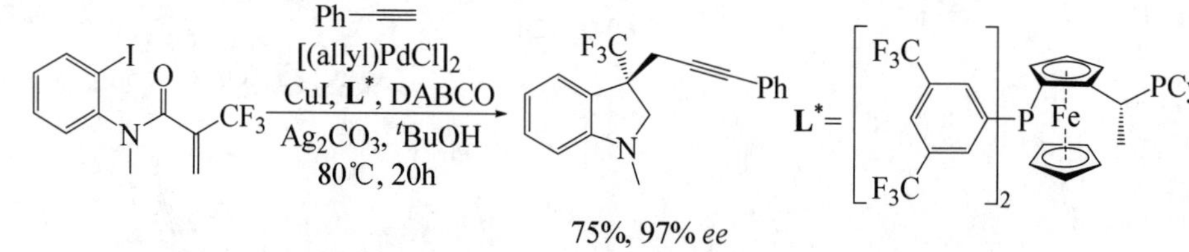

$$(71)$$

钯催化邻位连有乙烯基的芳基卤化物和末端炔烃的不对称 Heck/Sonogashira 反应，实现烯烃的不对称双功能化反应，合成一系列带有炔丙基季碳中心的光学活性的苯并呋喃化合物[57]。此外，以具有药物活性的甲基炔诺酮和炔雌醇甲醚作为偶联组分，分别以 81% 和 83% 的产率、高于 20∶1 的非对映选择性得到相应的产物。

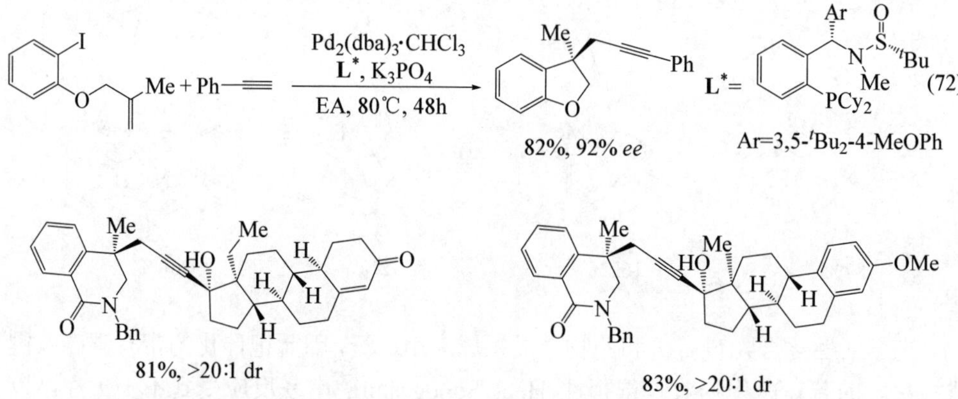

$$(72)$$

三取代烯烃通过酰胺键与邻溴苯基相连的底物在 $Pd(OAc)_2$/CuI 与 BINAP-型手性膦酰胺配体构成的催化体系催化下与端炔烃进行不对称 Heck/Sonogashira 串联反应，以高对映选择性和非对映选择性得到一系列 1,2-芳基炔基化的含两个相邻季碳中心的光学活性的螺环吲哚啉酮衍生物[58]。

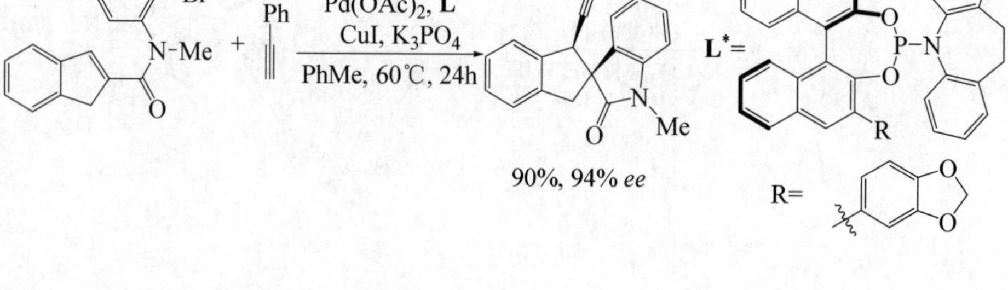

$$(73)$$

N-（邻乙烯基苯基）氨基甲酰氯衍生物与端炔烃在 Pd(OAc)$_2$/PPh$_3$ 催化下进行 Heck/Sonogashira 串联环化反应，得到炔丙基取代季碳中心的吲哚啉酮衍生物，反应无须共催化剂铜盐即可顺利进行[59]。

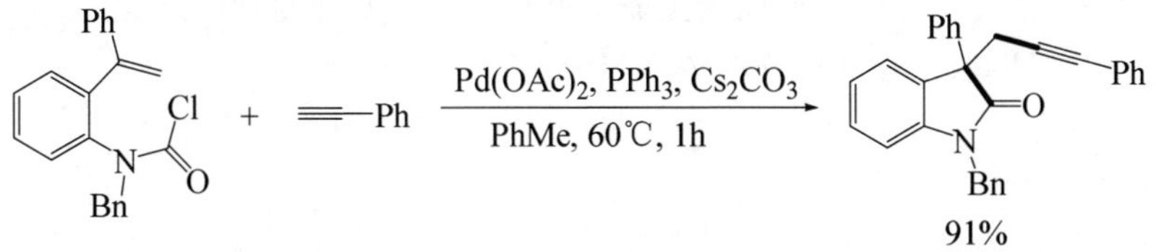

$$\text{(74)}$$

1-碘代-4-氮杂-1,6-二烯类化合物也能在钯催化下与端炔烃进行 Heck/Sonogashira 串联环化反应，以良好的收率得到一系列 5-炔丙基-1,2,5,6-四氢吡啶衍生物，该反应也无须铜盐存在[60]。

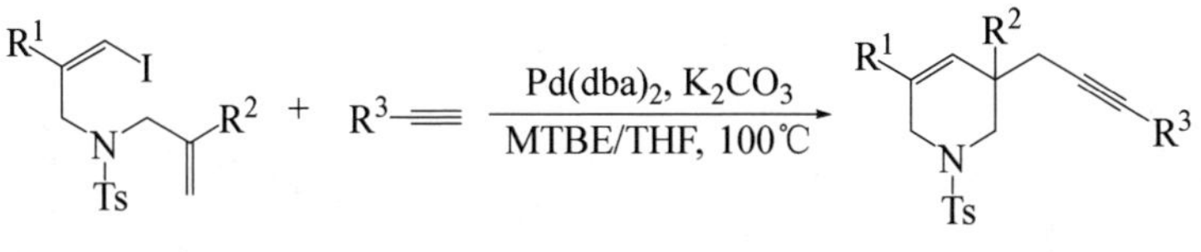

$$\text{(75)}$$

R^1, R^2=烷基，环烷基；R^3=芳基

除了端炔烃，丙炔酸也可作为偶联组分参与 Heck/Sonogashira 串联反应。例如，N-(2-溴苯甲酰基)吲哚与丙炔酸在催化体系 Pd(MeCN)$_2$Cl$_2$/XPhos 催化下发生脱芳构-脱羧环化偶联反应，以良好的收率和化学选择性生成一系列稠合二氢吲哚衍生物。

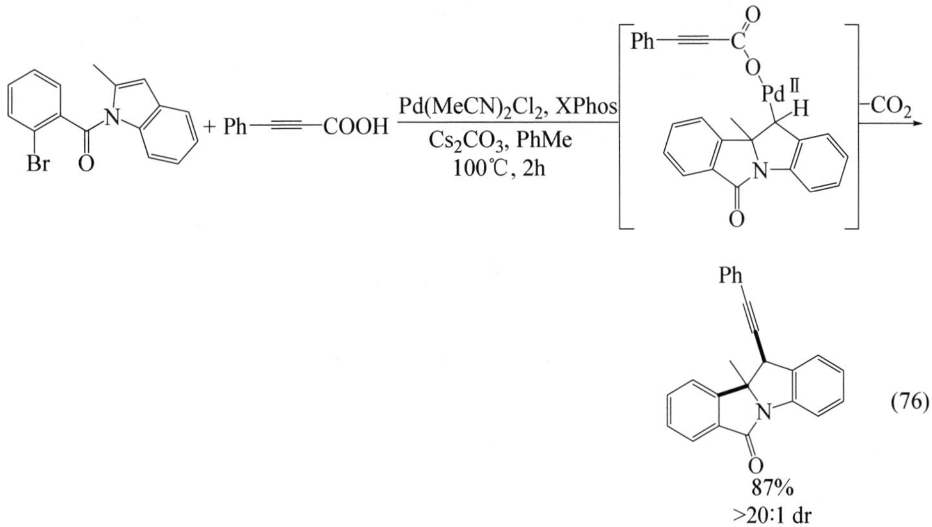

$$\text{(76)}$$

87%
>20:1 dr

反应机理如图 9-13 所示：首先，芳基溴化物氧化加成到 Pd^0L_n 生成苄基钯中间体 **A**，随后分子内的吲哚双键和苄基钯配位生成中间体 **B**，进而中间体 **B** 发生迁移插入得到五元环钯中间体 **C**；此时，苯炔酸通过羧基单氧与钯配位生成中间体 **D**，随后中间体 **D** 脱去一分子二氧化碳生成中间体 **E**，最后 **E** 通过还原消除生成偶联环合产物，并再生催化剂活性组分 Pd^0L_n[61]。

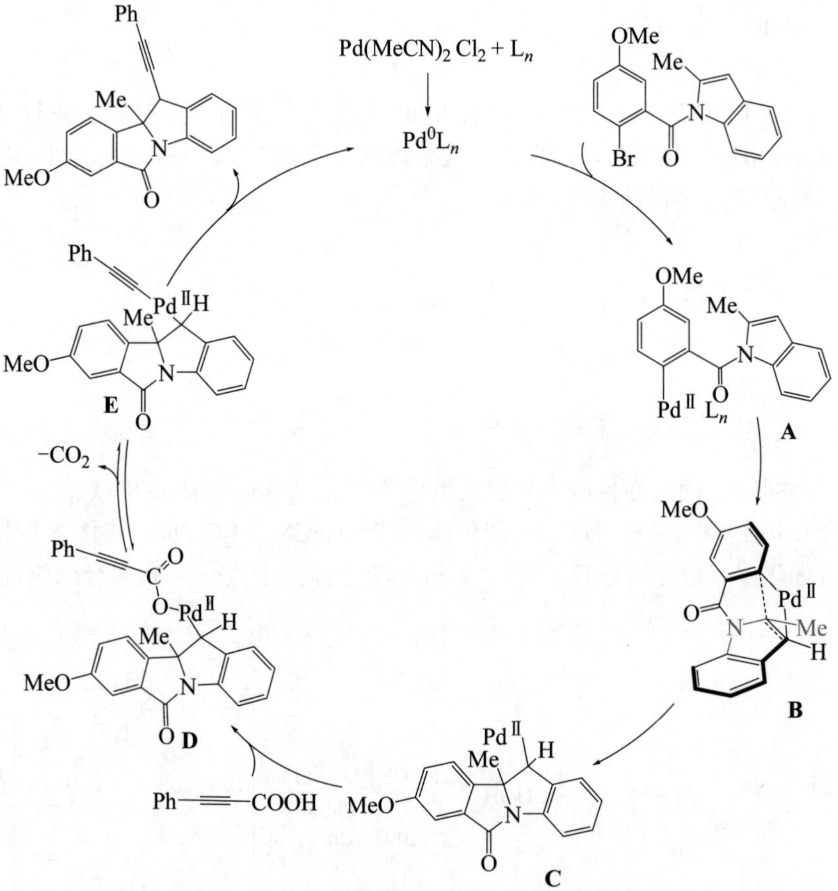

图 9-13　Pd 催化脱芳构-脱羧环化偶联反应机理

N-(2-碘苯甲酰基)吲哚与丙炔酸在钯/手性磷酰胺配体催化体系催化下发生对映选择性 Heck/Sonogashira 串联环化反应，以良好的产率、优异的对映体选择性和非对映体选择性生成一系列含 C2、C3-季碳中心的光学活性的四环吲哚啉酮衍生物[62]。

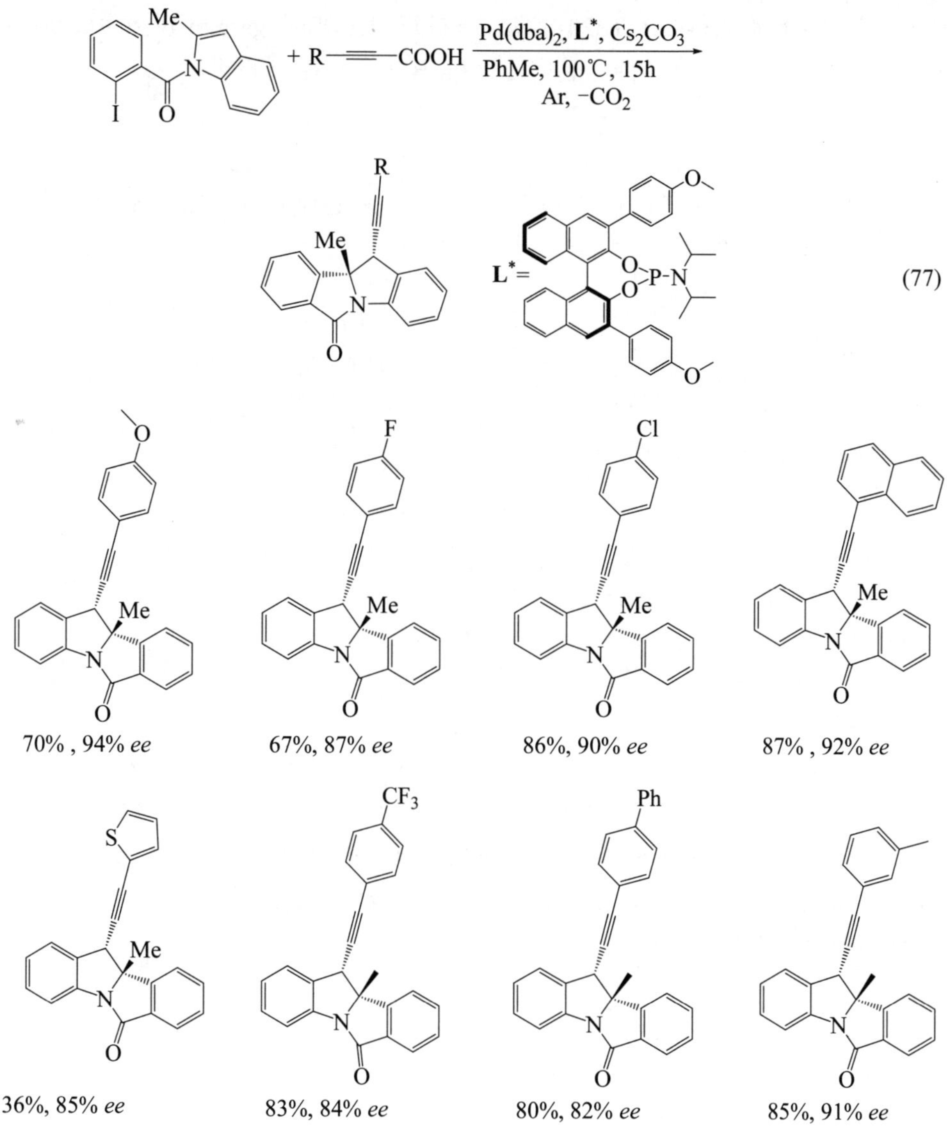

Me
I
O
N
+ R
COOH
Pd(dba)2, L*, Cs2CO3
PhMe, 100℃, 15h
Ar, −CO2
R
Me
N
O
L*=
O
P
N
70% , 94% ee
F
67%, 87% ee
Cl
86%, 90% ee
87% , 92% ee
S
36%, 85% ee
CF3
83%, 84% ee
Ph
80%, 82% ee
85%, 91% ee

(77)

9.3.3 Heck/Sonogashira 串联反应在药物和天然产物合成上的应用

Heck/Sonogashira 串联反应避免了中间体的纯化分离、基团保护等烦琐过程，实现复杂化合物的 “一锅” 法高效构建，具有原子经济性和步骤经济性等优点，在药物分子和天然产物合成上应用广泛。

二氢吲哚酮分子骨架 **A** 是一类重要的杂环结构单元，具有重要的药物活性，用作 5-HT_7 受体和神经激肽受体的拮抗剂。以 *N*-（邻碘苯基）-2-甲基丙烯酰胺和炔丙胺化合物为起始原料在 Pd(PPh$_3$)$_4$ 催化下进行 Heck/Sonogashira 反应生成环化中间体 TT-16；TT-16 经 Pt/C 催化氢化、脱 PMB 生成目标分子[63]。

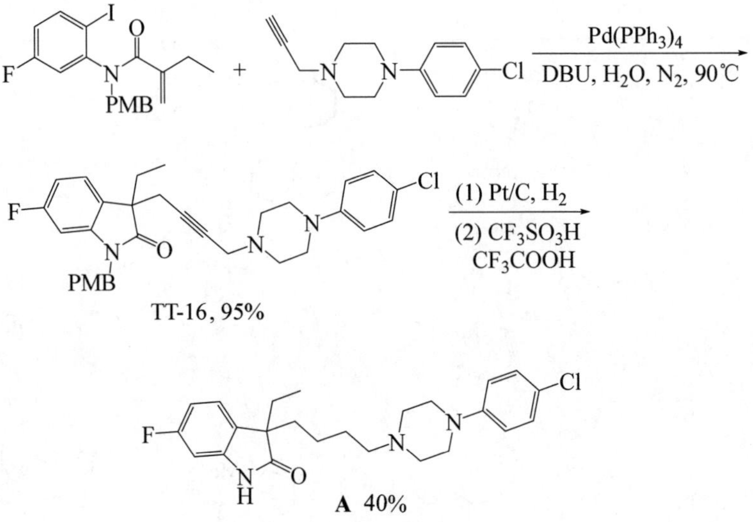

$$(78)$$

选择性5-HT$_7$受体拮抗剂

EGIS-12,233 化合物为 5-HT$_{6/7}$ 受体的一种高效选择性拮抗剂，广泛用作新型抗焦虑药和抗抑郁药。以邻碘丙烯酰胺衍生物为起始原料，炔丙胺化合物为偶联组分，在手性膦配体存在下通过 Pd/Cu 协同催化进行对映选择性 Heck/Sonogashira 反应，生成光学活性的关键中间体 TT-17，该中间体再经催化氢化、脱甲氧基甲基最后得到目标分子 EGIS-12,233[64]。

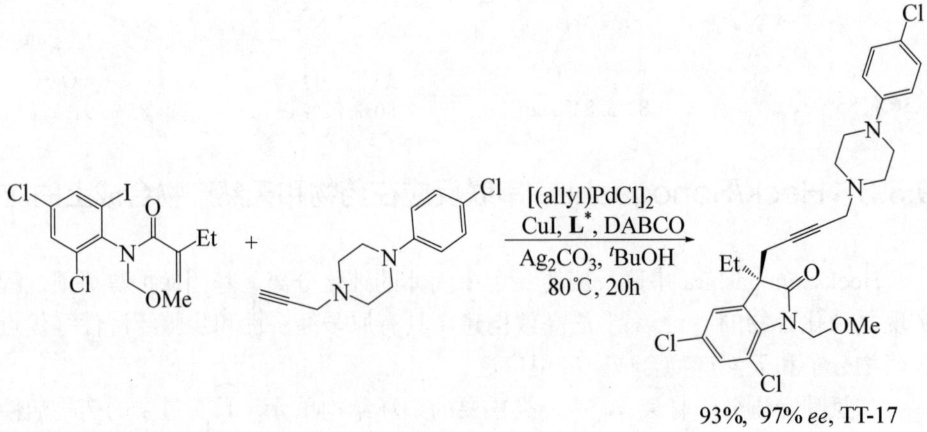

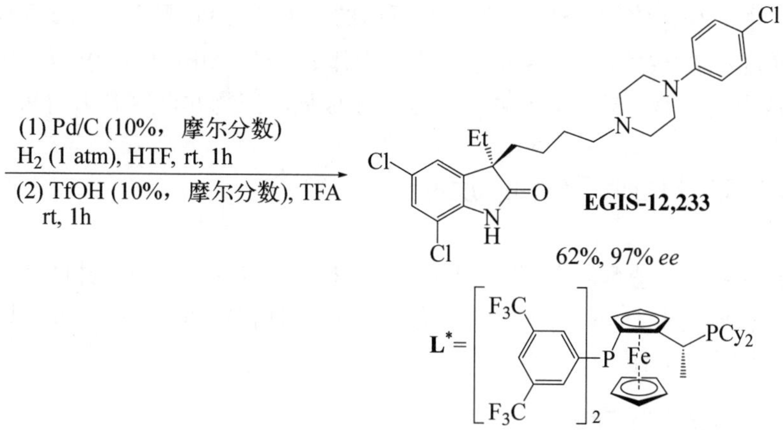

$$\text{(1) Pd/C (10\%, 摩尔分数)} \atop \text{H}_2 \text{ (1 atm), HTF, rt, 1h}$$
$$\text{(2) TfOH (10\%, 摩尔分数), TFA, rt, 1h}$$

EGIS-12,233
62%, 97% ee

$$L^* = \left[(F_3C)_2C_6H_3)_2P \right]_2 Fe \text{—PCy}_2$$

(79)

9.4 Heck/Suzuki 串联反应

9.4.1 Heck/Suzuki 串联反应及反应机理

与 Heck 反应引发的其他串联反应类似，芳基或烯基卤化物或类卤化物与零价钯氧化加成、烯烃插入生成烷基钯中间体不发生 β-消除，而是与加入的硼试剂（苯硼酸、联硼酸酯、乙烯基三氟硼酸盐、乙烯基硼酸酯、苯基三氟硼酸盐、苯基硼酸酯等）进行偶联反应生成相应的偶联产物，为 Heck/Suzuki 串联反应。

以 N-（邻卤苯基）丙烯酰胺与硼试剂的 Heck/Suzuki 串联反应为例，反应机理如图 9-14 所示：首先，N-（邻卤苯基）丙烯酰胺的卤代芳基与钯催化剂 Pd^0L_2 氧化

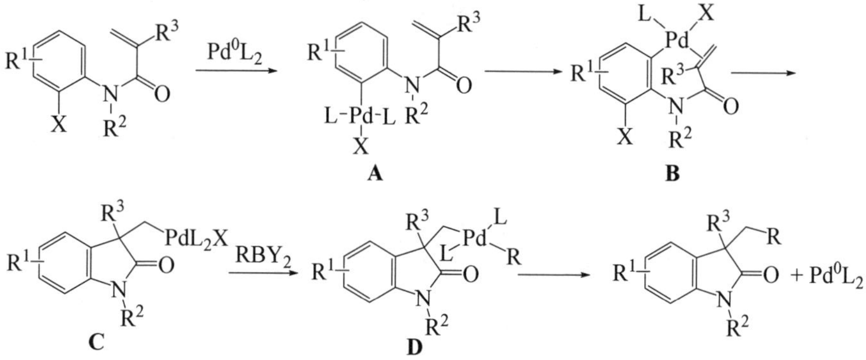

X=Cl, Br, I；R^1, R^2, R^3=各种不同取代的给电子基、吸电子基、卤素等

图 9-14 钯催化的 Heck/Suzuki 串联反应机理

加成生成芳基钯卤化物中间体 **A**，然后分子内的乙烯基与 Pd^{II} 配位形成 π-配合物中间 **B**，**B** 通过乙烯基插入转化为 σ-配合物中间体 **C**，**C** 与硼试剂 RBY_2 发生转移金属反应生成中间体 **D**，最后 **D** 通过还原消除生成偶联产物并再生催化剂 Pd^0L_2。

9.4.2 基本的 Heck/Suzuki 反应

芳基卤化物、烯基溴化物、芳基四氟重氮硼酸盐和各种不同的有机硼化合物在钯催化下都能够发生 Heck/Suzuki 串联反应，得到结构多样的偶联产物，以下按反应底物的不同进行介绍。

9.4.2.1 芳基碘化物为底物

结构中同时含邻碘芳基和丙烯酰基结构单元的丙烯酰胺或丙烯酸酯类化合物在钯催化剂催化下与芳基硼酸发生 Heck/Suzuki 串联反应，化学选择性地生成苯并吡咯或苯并呋喃衍生物[65]。此外，N,N-(2-丙烯基)(2-甲基烯丙基)-2-碘苯甲酰胺在钯催化剂催化下与四苯基硼酸钠在无外加碱存在的条件下进行偶联反应，一次构建两个五元杂环，得到具有两个季碳中心的 6/5/5-稠环化合物。该反应实际上属于 Heck/Heck/Suzuki 串联反应。

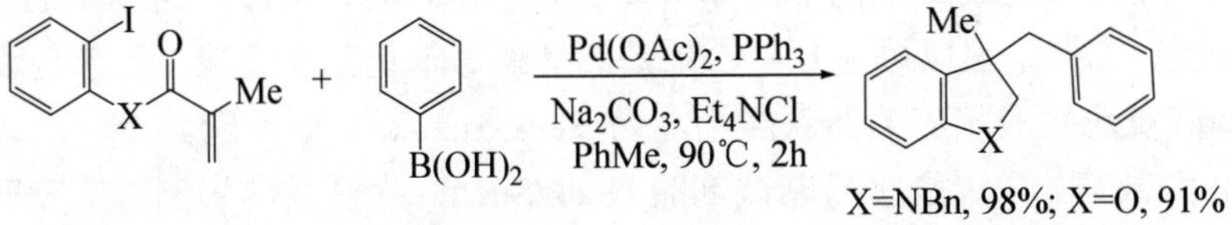

(80) X=NBn, 98%; X=O, 91%

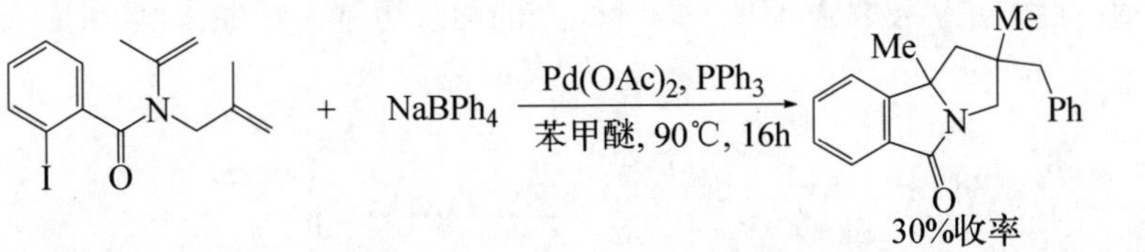

(81) 30%收率

N-邻碘芳甲基-2-丁炔酰胺或 2-丁炔酸邻碘芳甲酯与芳基硼酸在钯催化剂催化下顺利进行偶联反应，以高区域选择性和立体选择性生成单一异构体的苯并稠合杂环化合物[66]。除了芳基硼酸，以反式苯乙烯硼酸为偶联试剂，也能顺利进行反应并以中等产率得到反式构型的异色满酮或异喹啉酮衍生物。

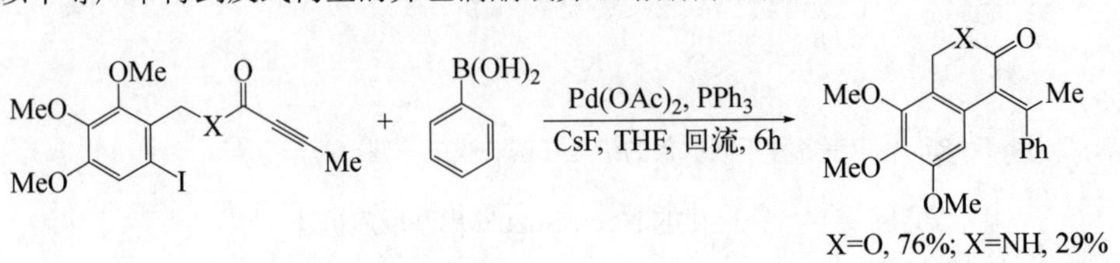

(82) X=O, 76%; X=NH, 29%

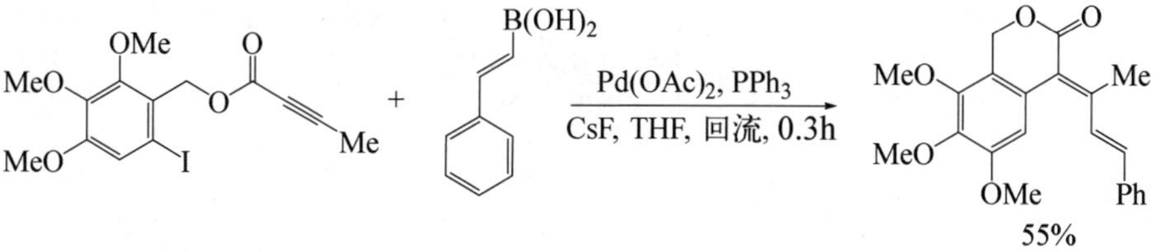

N-邻碘芳基-*β*-芳基丙烯酰胺类化合物与有机硼化合物在 Pd(PPh$_3$)$_4$ 催化下一步反应构建两个碳碳键,实现了 3,3-二取代二氢吲哚酮衍生物的便捷合成[67]。反应过程如下:首先在钯催化剂作用下,芳基碘化物发生分子内 Heck 反应生成五元环烷基钯中间体,该中间体进一步与有机硼试剂发生分子间 Suzuki 偶联得到吲哚啉骨架结构。苯硼酸、乙烯基三氟硼酸盐、乙烯基硼酸酯、苯基三氟硼酸盐、苯基硼酸酯都能作为偶联试剂参与此类 Heck/Suzuki 串联反应。

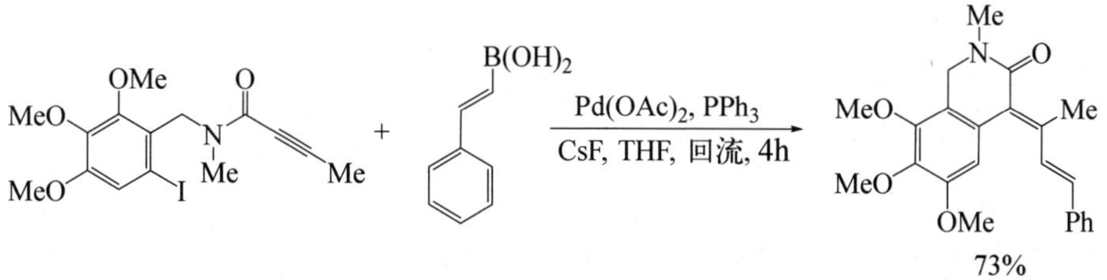

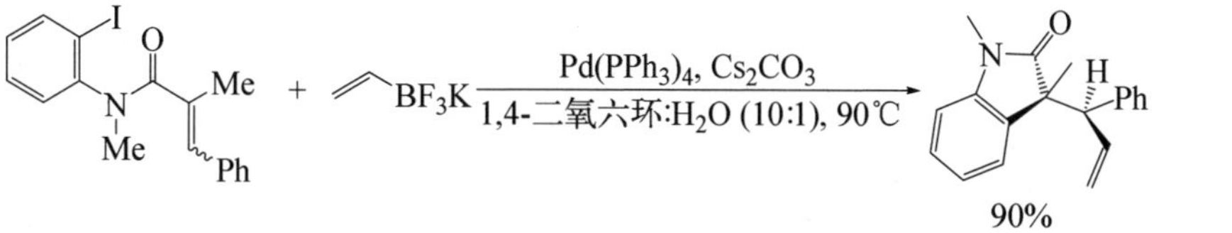

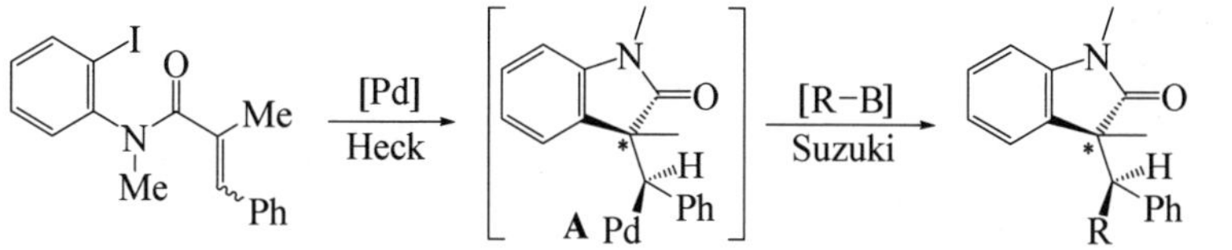

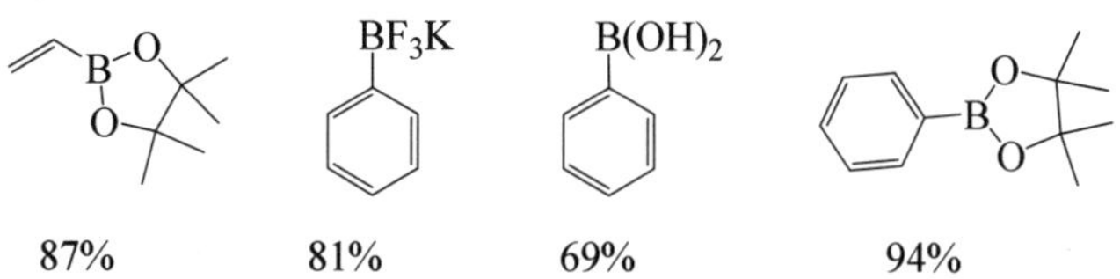

联硼酸酯(B$_2$Pin$_2$)也可以作为有机硼试剂参与 Heck/Suzuki 反应。以 *N*-邻碘芳基丙烯酰胺与 B$_2$Pin$_2$ 有机硼试剂为底物,在钯催化剂和微波辐射的条件下能以高化学选择性得到二氢吲哚硼酸酯衍生物[68]。此外,邻碘苯胺、异烯丙酸、苯甲醛与异氰化物能发生四组分 Ugi 多米诺串联反应,生成三级胺偶联产物,进一步通过 Heck/Suzuki 反应得到二氢吲哚硼酸酯产物。

$$\text{[structure]} + B_2Pin_2 \xrightarrow[\text{MW, 120°C 40min}]{\text{Pd(PPh}_3)_4, \text{Na}_2\text{CO}_3}{\text{ACN:H}_2\text{O (9:1)}} \text{[product]} \quad 98\% \qquad (86)$$

$$\text{[aryl iodide amine]} + \text{[methacrylic acid]} + \text{[aldehyde]} + R^3-N{\equiv} \xrightarrow[\text{MeOH, rt}]{\text{U-4CR}} \text{[intermediate]}$$

$$+ B_2Pin_2 \xrightarrow[\text{MW, 120°C 40min}]{\substack{\text{Pd(PPh}_3)_4 \ (2\%, \text{摩尔分数}) \\ \text{Na}_2\text{CO}_3 \ (3.0 \ \text{equiv.}) \\ \text{ACN:H}_2\text{O (9:1)}}} \text{[product]} \quad \begin{array}{l} R^1\text{=H, } R^2\text{=Me, } R^3={}^t\text{Bu, 82\%} \\ R^1\text{=Me, } R^2\text{=Cl, } R^3\text{=TMB, 64\%} \end{array} \qquad (87)$$

N-炔丙基邻碘苯甲酰胺与芳基硼酸或杂芳基硼酸在钯催化剂催化下进行 Heck/Suzuki 串联反应,高立体选择性地生成 3,4-二氢异喹啉酮衍生物;而且,在有机碱 DBU 作用下,反应产物 3,4-二氢异喹啉酮的环外双键很容易地转化为环内双键[69]。

$$\text{[substrate]} \xrightarrow[\substack{\text{K}_3\text{PO}_4, \text{DMF} \\ \text{rt, 3~14h}}]{\substack{R^1-B(OH)_2 \\ \text{Pd(OAc)}_2, \text{PPh}_3}} \text{[dihydroisoquinolinone]} \xrightarrow[\text{异构化}]{\text{DBU}} \text{[isoquinolinone]} \qquad (88)$$

91%   54%   82%   75%

在手性钯催化体系催化下,邻碘苯基与非活泼烯烃通过适当长度碳链或杂碳链相连的化合物与有机硼酸进行不对称 Heck/Suzuki 串联环化反应,以良好的收率和高对映选择性生成含有季碳中心的光学活性的苯并呋喃、苯并吡啶和苯并吡喃等化合物[70]。有机硼酸可以是芳基硼酸、杂芳基硼酸、烷基硼酸以及烯基硼酸,从而实现非活泼烯烃的双芳基化、芳基烷基化和芳基烯基化反应。

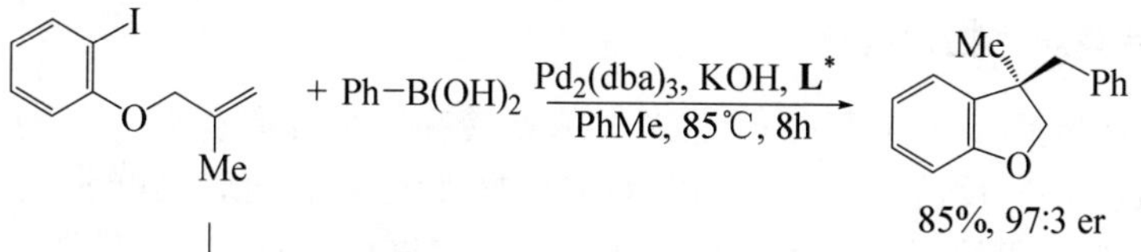

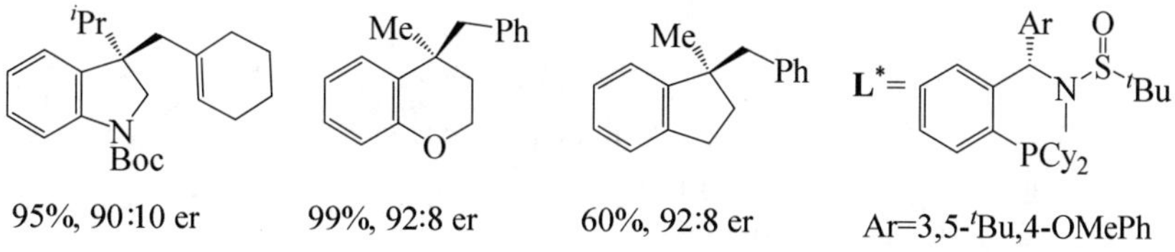

我国科学家在本领域内也做出了杰出的贡献，复旦大学张俊良教授和长春工业大学刘宇教授课题组共同发现了 N-烯丙基邻碘苯甲酰胺或邻碘苯非活泼烯烃基醚与片呐醇硼酸酯的不对称 Heck/Suzuki 串联反应，以良好的收率和对映选择性生成光学活性的二取代异喹啉酮衍生物[71, 72]。

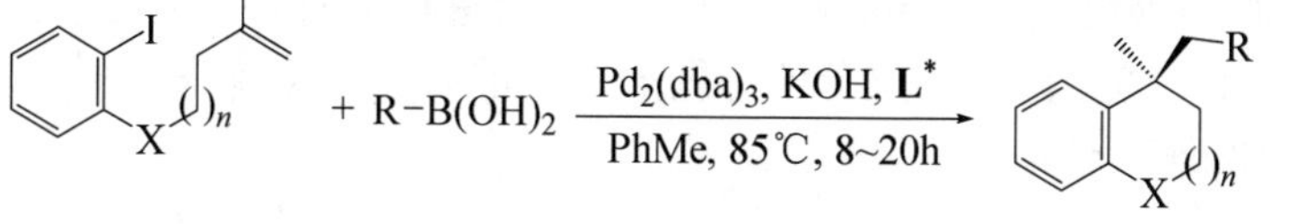

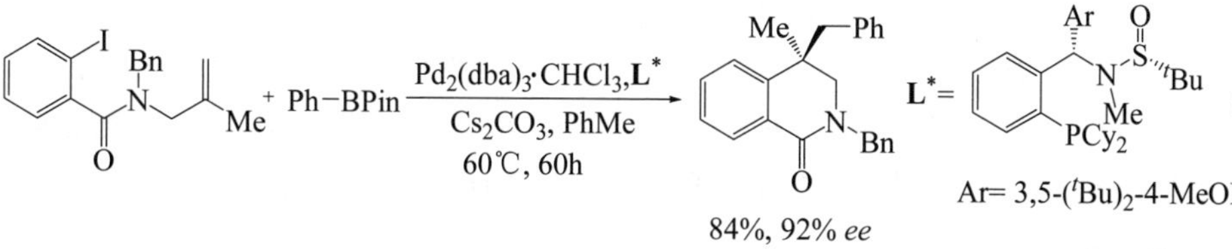

N-邻碘苯甲酰基吲哚与芳基硼酸在 $Pd_2(dba)_3$/BINOL-型手性磷酰胺配体/Cu_2O 共同催化下进行吲哚不对称双芳基化反应，以优良的产率、优异的对映选择性和非对映选择性生成光学活性的四环吲哚啉衍生物[73]。

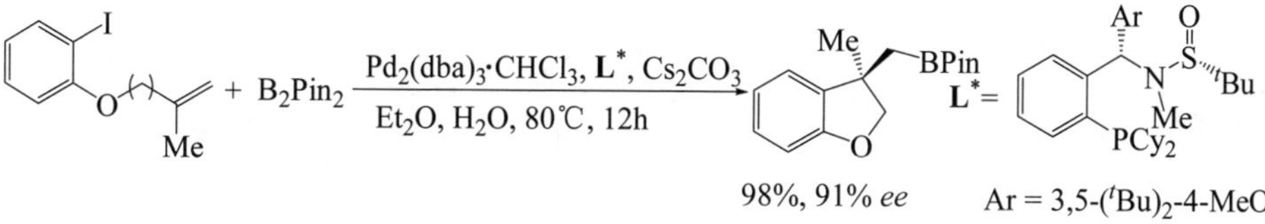

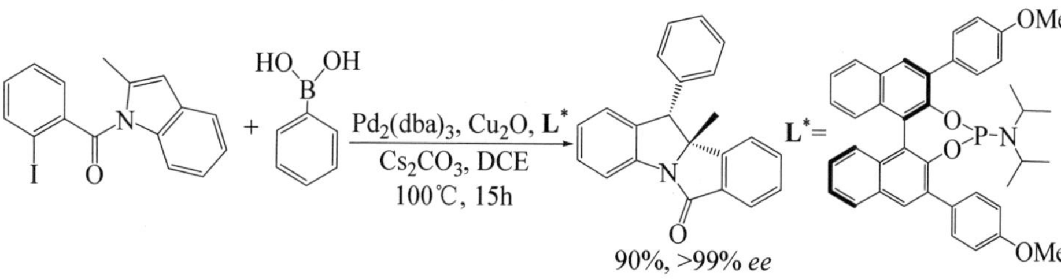

9.4.2.2 芳基溴化物为底物

以邻溴苯基-3-烯丁基胺衍生物与苯硼酸在钯催化剂催化下进行 Heck 环化/Suzuki 交叉偶联反应,以良好的非对映选择性生成含三级和四级季碳中心的四氢喹啉衍生物;当以反式苯乙烯硼酸替代芳基硼酸进行反应时,不能进行 Heck/Suzuki 串联反应,而是芳基溴与反式苯乙烯硼酸直接进行 Suzuki 偶联反应,以 53%的产率得到相应的偶联产物[74]。

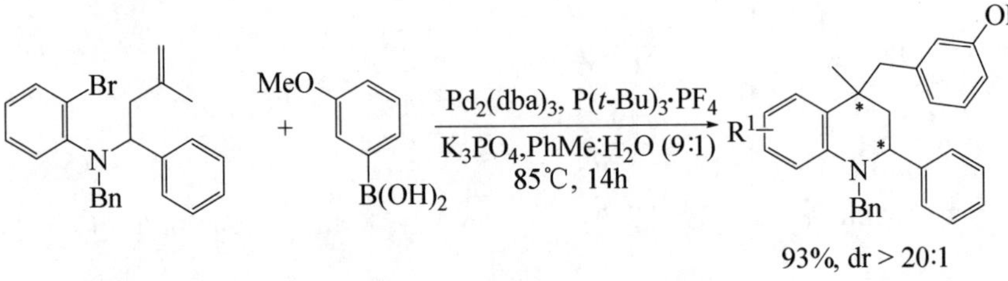

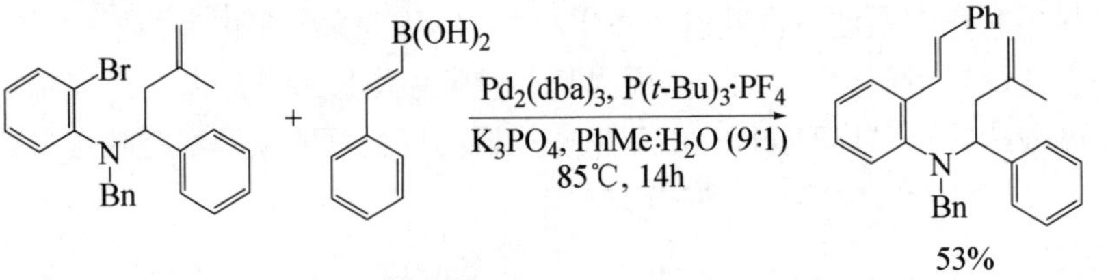

在微波辅助的钯催化下,N-(邻溴苯乙基)炔丙酰胺与硼酸进行 Heck/Suzuki 串联反应,实现中环产物苯并吖庚因的高效合成[75]。除了苯硼酸,其他芳基硼酸、苯基三氟硼酸盐、苯硼酸酯、氮硼酸酯也可以作为有机硼试剂进行此类偶联反应。

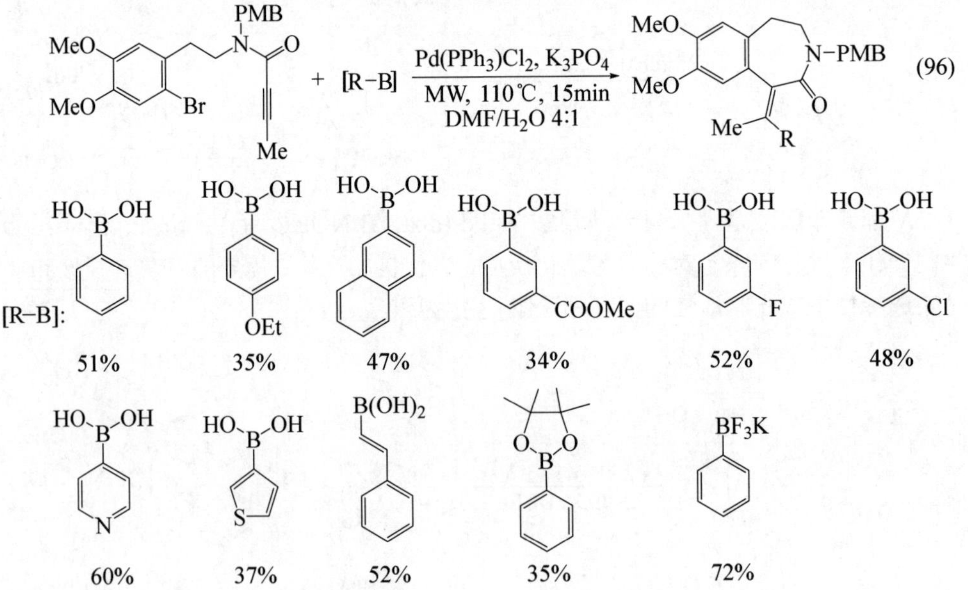

N-吲哚邻溴苯甲酰胺衍生物与苯基硼酸酯在钯催化下进行 Heck/Suzuki 串联反应，实现吲哚脱芳构反式 1,2-双芳基化，高效合成一系列四环吲哚衍生物，P^tBu_3 作为配体可以有效地抑制直接 Suzuki 偶联副反应的发生[76]。

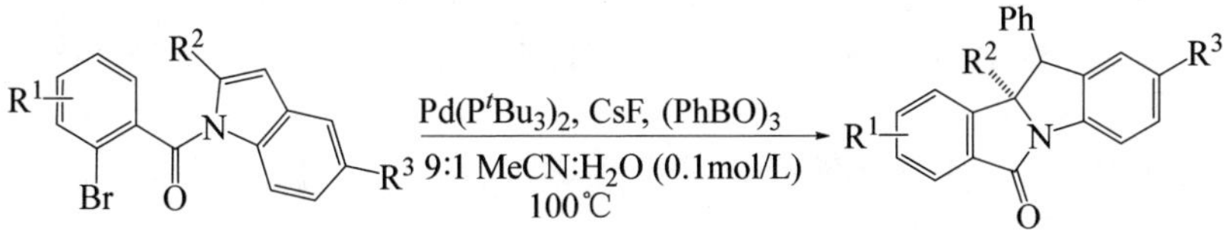

$Pd(P^tBu_3)_2$, CsF, $(PhBO)_3$
9:1 MeCN:H_2O (0.1mol/L)
100℃

(97)

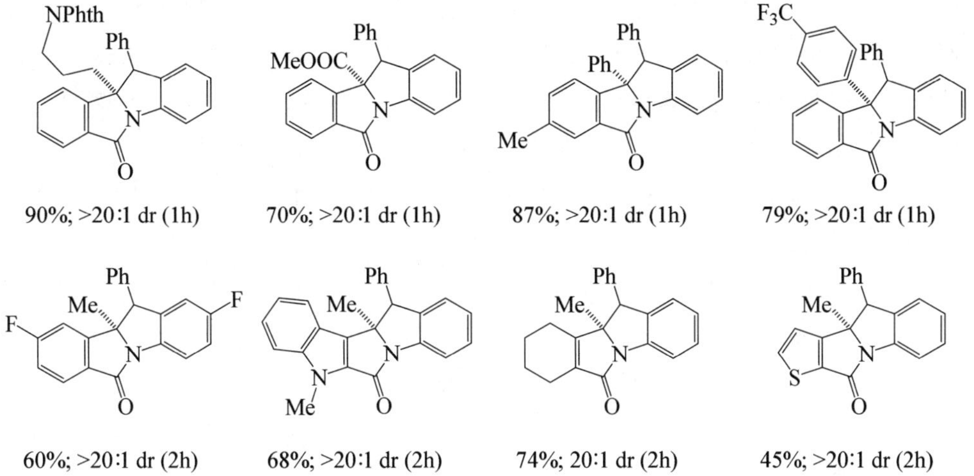

90%; >20:1 dr (1h)
70%; >20:1 dr (1h)
87%; >20:1 dr (1h)
79%; >20:1 dr (1h)
60%; >20:1 dr (2h)
68%; >20:1 dr (2h)
74%; 20:1 dr (2h)
45%; >20:1 dr (2h)

除了邻溴苯甲酰胺衍生物，类似的邻溴杂芳甲酰胺衍生物也能发生 Heck/Suzuki 串联环化反应。例如，*N*-（邻溴吡啶）烯丙胺衍生物与苯硼酸在钯催化剂催化下生成不同取代的氮嘌呤类化合物。此外，一些溴代吡唑基烯丙胺类化合物也能和苯硼酸在相同的催化反应条件下发生 Heck/Suzuki 串联反应，生成二氢吡咯并吡唑衍生物，但收率较低[77]。

$Pd_2(dba)_3$, Na_2CO_3
DMF, 100℃, 18h
67%收率

(98)

$Pd_2(dba)_3$ (5%, 摩尔分数)
K_2CO_3 (2.0 equiv)
DMF, 100℃, 18h

(99)

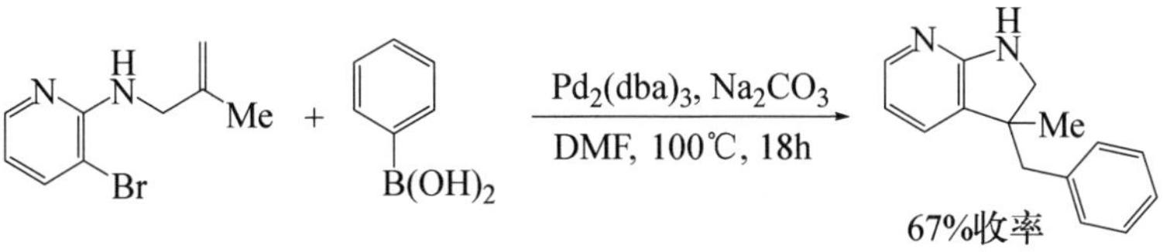

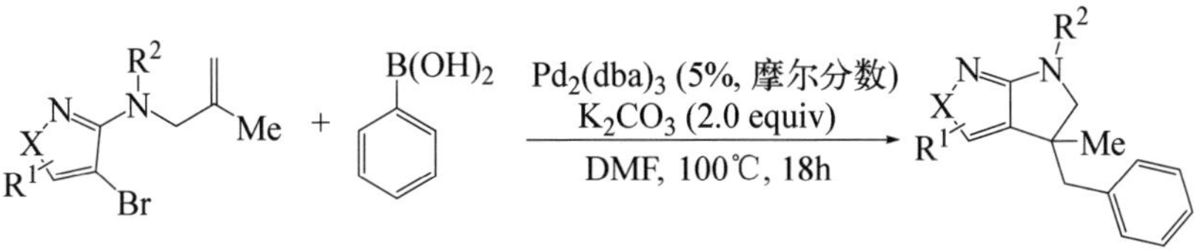

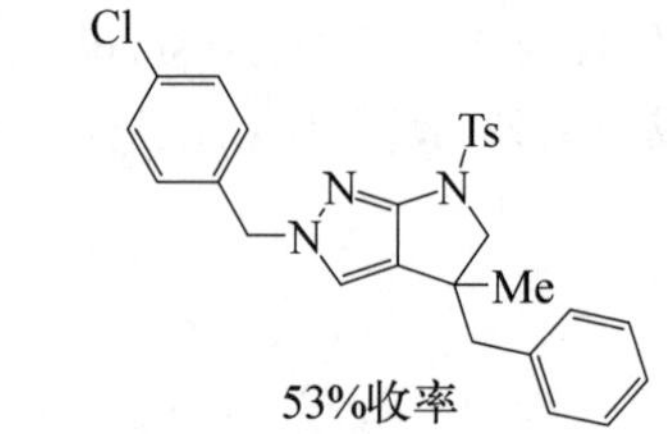

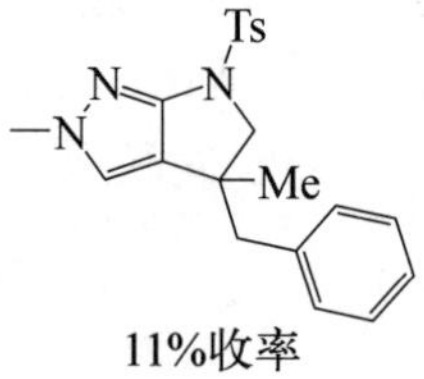

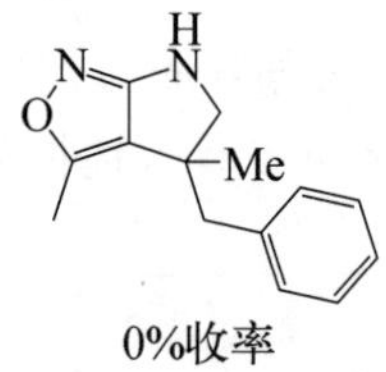

53%收率　　　　　　11%收率　　　　　　0%收率

  N-（邻溴芳基）-α-萘甲酰胺类化合物以萘环作为共轭二烯单元与四苯硼酸钠在手性钯催化剂催化下进行不对称 Heck/Suzuki 串联反应，完成 1,4-双芳基脱芳构化反应，以良好的区域选择性和非对映选择性生成 1,4-二氢萘-苯并吡咯烷酮螺环化合物。萘环双键脱芳插入能够抑制直接 Suzuki 偶联，从而得到预期的 Heck/Suzuki 偶联产物。α-萘甲-邻溴苯基醚同样能够进行此类反应得到1,4-双芳基化产物。此外，该催化反应体系还能扩展到 α-萘联邻溴苯、炔烃和四芳基硼酸盐三组分的 Heck/Suzuki 串联反应，以良好的收率和优异的非对映选择性生成苯并环己烯-苯并环戊烯螺环产物[78]。

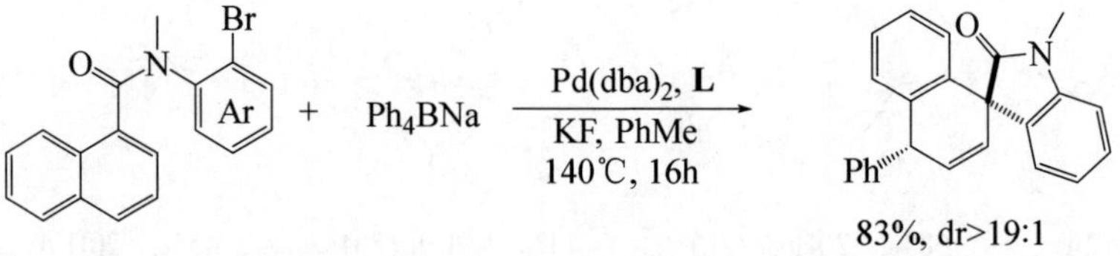

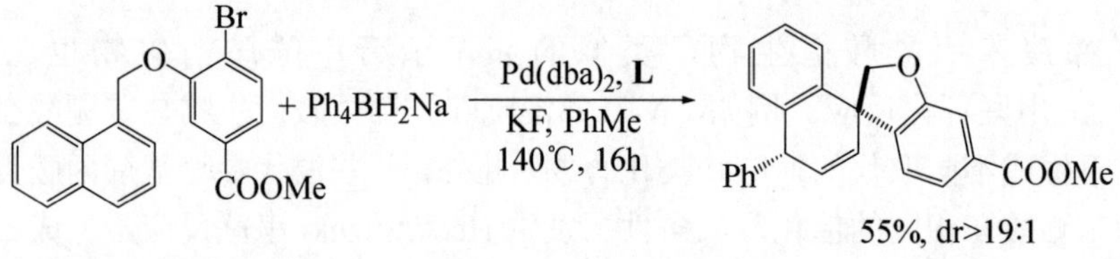

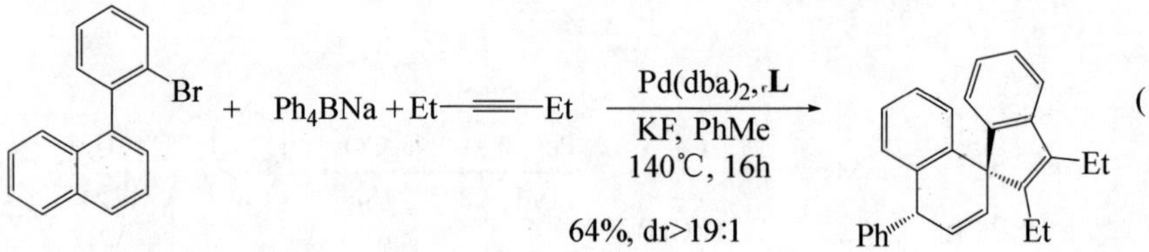

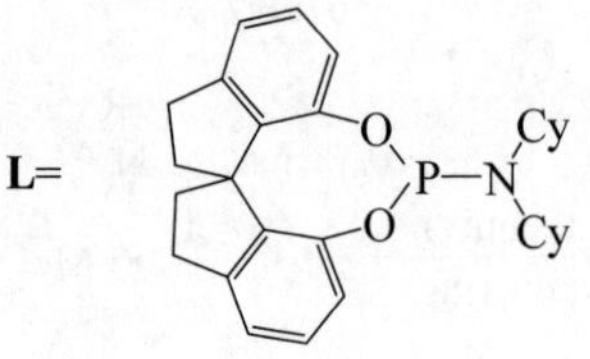

以 *N*-吲哚邻溴苯甲酰胺为底物，芳基硼酸作为偶联剂，在 Pd(OAc)$_2$/Cu$_2$O 共催化剂体系催化下能够顺利进行 Heck/Suzuki 串联反应合成结构多样的二氢吲哚酮类化合物。

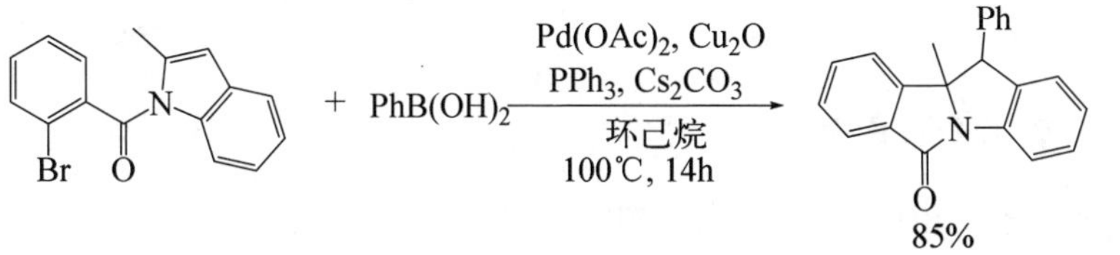

$$\text{(103)}$$

反应机理如图 9-15：首先芳基溴化物氧化加成到 Pd^0 生成二价钯中间体 **A**，随后吲哚双键与二价钯配位得到中间体 **B**，进一步发生迁移插入生成五元环钯中间体 **C**，苯硼酸作为偶联试剂对五元环钯中间体 **C** 亲核进攻得到中间体 **D**，最后经过还原消除得到二氢吲哚酮产物[79]。

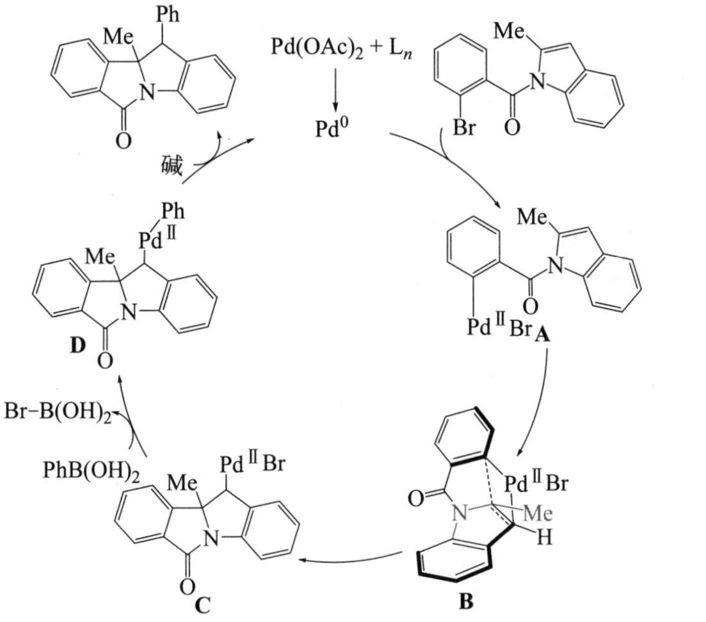

图 9-15　钯催化 *N*-吲哚邻溴苯甲酰胺 Heck/Suzuki 串联反应机理

仍以 *N*-吲哚邻溴苯甲酰胺为底物，将偶联试剂芳基硼酸代之以 1,1-二甲基 1-苯基硅烷硼酸酯，在钯催化剂催化下立体选择性地进行分子内吲哚脱芳构化-芳基硅化反应，以良好的收率得到含有硅基的四环吲哚衍生物[80]。

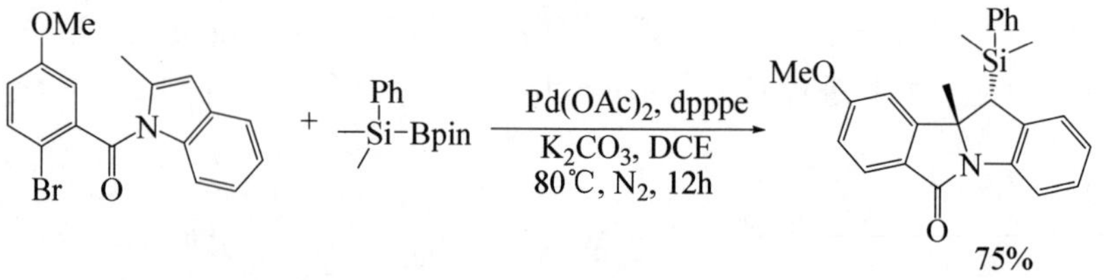

[反应式 (104)]

9.4.2.3 芳基氯化物为底物

以 N-(邻氯苯基)丙烯酰胺类化合物为底物，芳基硼酸作为偶联试剂，在镍催化剂催化下发生 Heck/Suzuki 串联反应，生成含有季碳中心的 3,3-二取代吲哚啉酮衍生物。反应得以进行的关键是 N-(邻氯苯基)丙烯酰胺与镍催化剂形成五元环甲基镍中间体 A，一旦 A 生成，其进一步与硼酸发生 Suzuki 偶联生成最终产物[81]。

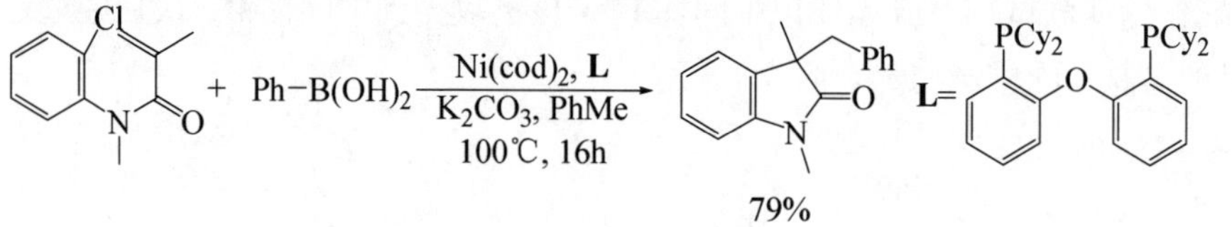

[反应式 (105)]

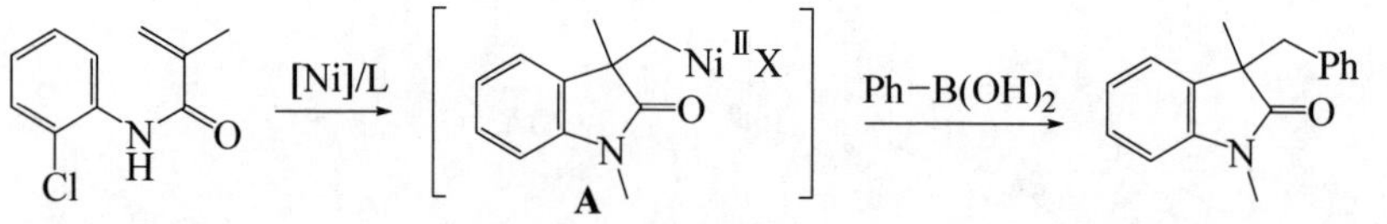

在 Pd(OAc)$_2$/BINOL-型手性磷酰胺配体催化下，上述底物与偶联试剂联硼酸酯进行不对称 Heck/Suzuki 串联反应，实现光学活性的吲哚啉酮类化合物的高效合成[82]。

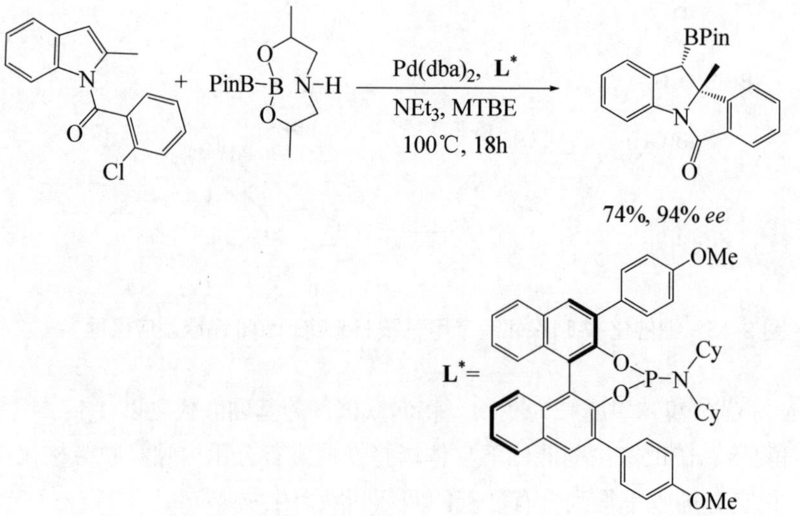

[反应式 (106)]

9.4.2.4 烯基溴化物为底物

钯催化 N-烯丙基-N-(2-溴烯丙基)化合物与芳基硼酸发生 Heck 环化/Suzuki 偶联反应,为合成 3-亚甲基-4-苄基吡咯烷提供了一种简单的方法。然而,反应过程中会发生直接的 Heck 反应,经由 β-氢消除生成二氢吡啶和 3,4-二取代吡咯两种副产物。当取代基 R^1 为磺酰基时,中间体 **A** 中的磺酰基氧可以稳定烷基钯中间体,从而抑制其发生 β-氢消除生成直接 Heck 偶联副产物[83]。

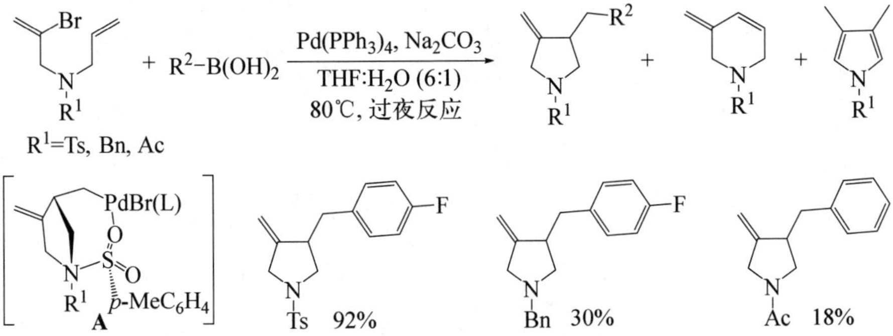

3-氮杂-1-溴-1,5-二烯类化合物在钯催化剂存在下先通过氧化加成、阴离子交换、烯烃插入完成分子内 Heck 环化生成二氢吡咯甲基钯中间体,该中间体与偶联试剂芳基硼酸进行转金属化、还原消除完成分子间 Suzuki 偶联反应,最终得到多取代二氢吡咯衍生物[84]。

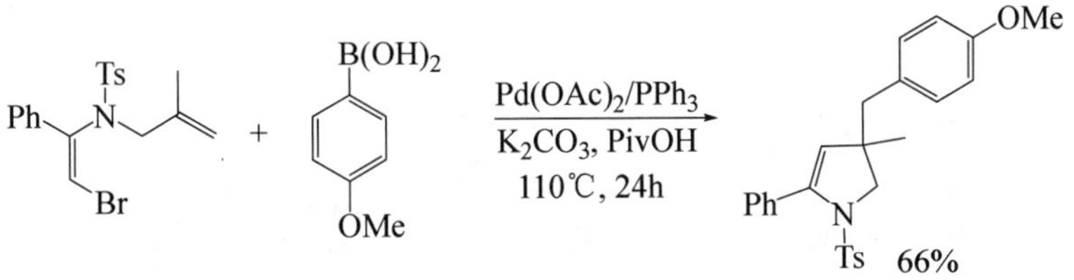

9.4.2.5 芳基四氟重氮硼酸盐为底物

邻烯丙氧基苯四氟硼酸重氮盐与芳基硼酸在手性钯催化剂催化下发生不对称 Heck/Suzuki 串联反应,完成烯烃不对称双芳基化,以良好的产率和对映选择性得到光学活性的苯并二氢呋喃衍生物。添加剂 Ag_2CO_3 不可或缺,其既作为碱促进反应的进行,又起到氯离子清除剂的作用[85]。

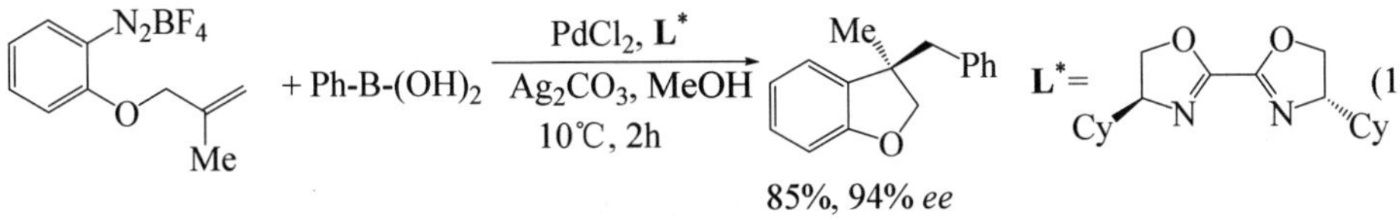

9.4.2.6 其他化合物为底物

邻位含烯丙基的碳酸二芳甲酯与芳基硼酸酯在钯催化剂 [PdCl(η^3-C_3H_5)]$_2$/DTBMP 催化下进行 Heck/Suzuki 串联反应，以良好的产率和非对映选择性生成 1,2-二取代茚满衍生物。具有远程位阻的三[3,5-二(叔丁基)-4-甲氧基苯基]膦 (DTBMP)配体是实现高化学选择性和立体选择性的关键[86]。

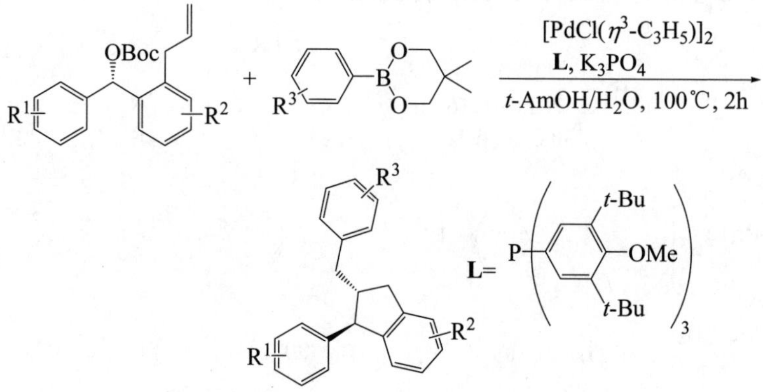

$$(110)$$

最高产率达99%，trans/cis >20:1

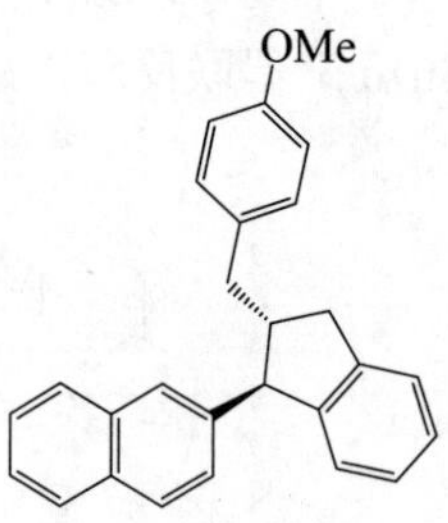

89%, 97:3 er, >99 es

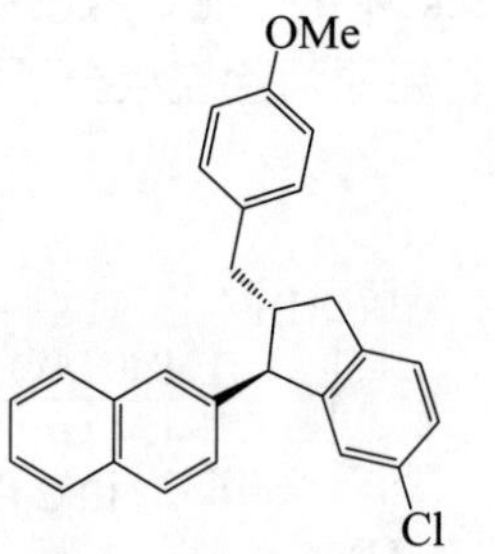

80%, 95:5 er, >99 es

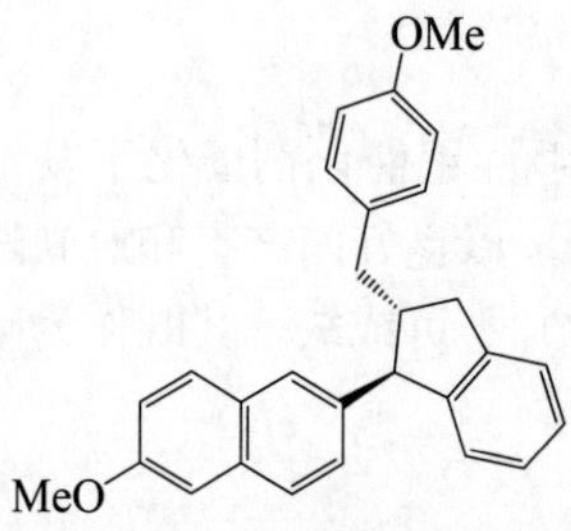

75%, 93:7 er, 98% es

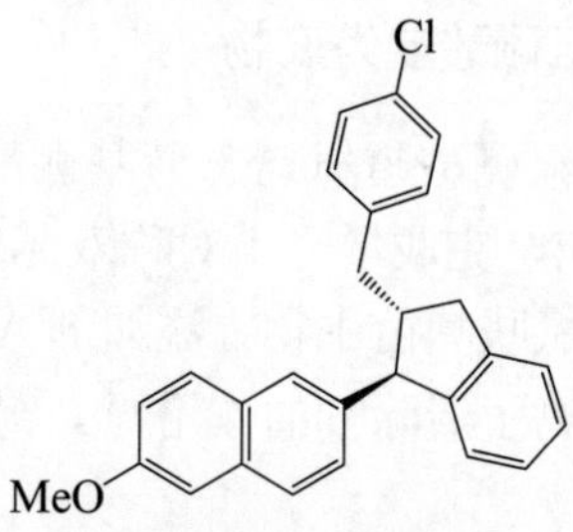

70%, 95:5 er, >99 es

Pd(PPh$_3$)$_4$/CuCl 协同催化邻烯基-N-氯甲酰苯胺与1,1-二硼基甲烷的串联环化反应，高效合成一系列含 Bpin 的二氢吲哚酮衍生物[87]。

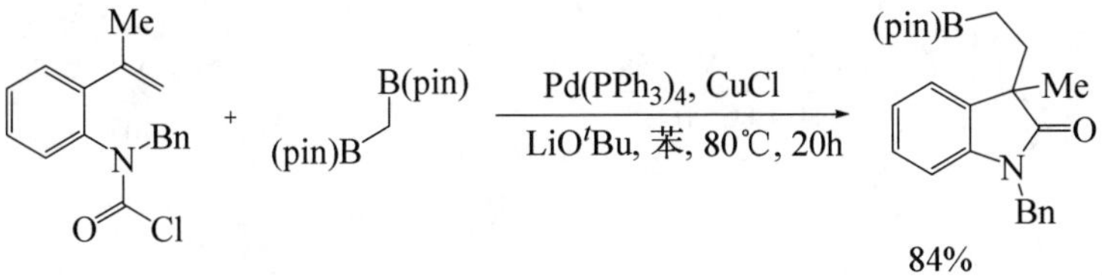

(111)

9.4.3 Heck/Suzuki 串联反应在药物和天然产物合成上的应用

异喹诺酮 TT-18 是一种高效的微管蛋白聚合抑制剂，可与秋水仙碱位点结合从而抑制微管蛋白的聚合，能够选择性治疗血管肿瘤。其结构和合成路线如下所示：以三甲氧基苯腈为起始原料，经过碘化、氰基还原、酰胺化反应可得关键的中间体 N-（邻碘芳甲基）炔丙酰胺 TT-18'，TT-18'在钯催化剂催化下与芳基硼酸发生 Heck/Suzuki 偶联即可得到目标产物异喹诺酮 TT-18 [88]。

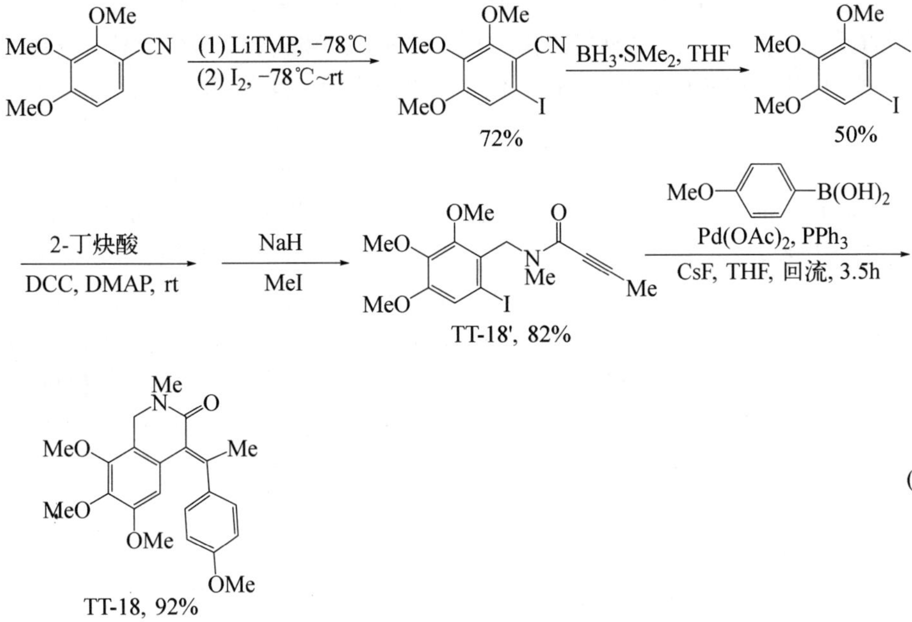

(112)

反应式（113）目标产物是一种 CB2 受体激动剂（CB2 receptor agonist），能够显著增加 HL-60 细胞中 P-ERK 的表达。在其合成中，邻烯丙氧基芳基四氟硼酸重氮盐与苯硼酸在手性钯催化剂催化下进行不对称 Heck/Suzuki 串联反应得到关键光学活性中间体 TT-19，TT-19 与 Mo(CO)$_6$ 在微波协助和钯催化剂催化下发生羰基化反应生成目标分子[89]。

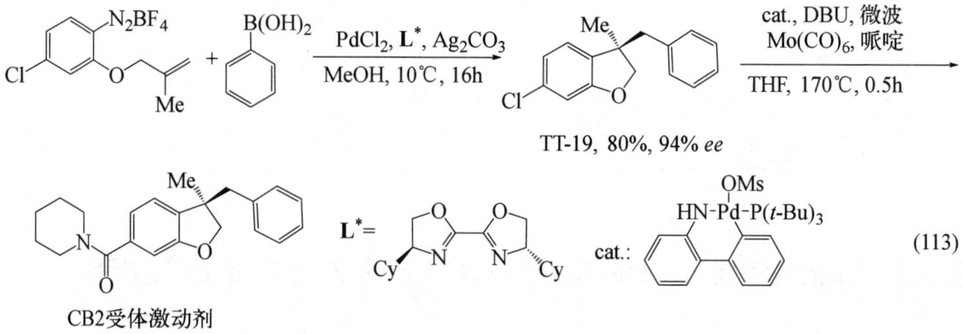

(113)

CB2受体激动剂

9.5 Heck/C–H 官能团化串联反应

9.5.1 Heck/C–H 官能团化串联反应及机理

过渡金属催化 C-H 键官能团化是通过 C-H 键活化直接在有机化合物骨架引入官能团的反应。通过 C-H 键官能团化构筑新的碳碳键和碳杂键是合成具有生物活性和药理活性化合物的重要手段，在有机和药物合成领域具有广泛应用。由于电负性（C=2.5，H=2.2）和空间位阻等因素的客观存在，复杂有机化合物中不同碳氢键的键能相当，而 C-H 键又是极性较小的共价键，键能较高，因此温和条件下实现 C-H 键选择性官能团化具有非常大的挑战性。目前，已发展出两个主要的 C-H 键选择性官能团化策略：①在底物分子中预先引入导向基团如 2-吡啶基、酰氨基实现选择性 C-H 键官能团化。此类反应一般是导向基与中心金属配位以实现导向基邻位选择性活化，然后再发生氧化加成、还原消除等步骤实现 C-H 键选择性官能团化。其缺点是需要预先引入导向基团，反应结束后再移除，增加了反应步骤的同时降低了反应的原子经济性和步骤经济性。②利用 Heck 反应生成的 σ-钯中间体代替导向基团，通过合理的底物分子设计，用于活化分子内或分子间的 C-H 键，实现 C-H 键选择性官能团化。

以钯催化的结构中同时含有邻碘苯基、乙烯基和吡咯环结构底物的分子内 Heck/C-H 键官能团化串联反应为例，介绍此类反应的机理[90]。如图 9-16 所示，底物的碘苯基团与 Pd⁰ 氧化加成、乙烯基插入、环化生成第一个 σ-Pd^{II} 中间体 **A**，随后吡咯环中的碳碳双键插入 **A** 中的 Pd-C 之间生成第二个 σ-Pd^{II} 中间体 **B**，最后 **B** 进行 β-氢消除生成含有一个季碳中心的吡咯螺环化合物。

以烯基碘苯与 2-(三甲基硅烷基)苯基三氟甲磺酸酯的串联反应为例介绍 σ-钯中间体与亲核试剂反应实现 C-H 官能团化反应的机理[91]。如图 9-17 所示，首先碘苯基与 Pd⁰ 氧化加成生成二价钯中间体 **A**，随后经过 5/6-exo-trig 环化形成新戊/己基甲基

σ-钯中间体 **B**；**B** 的 C(sp³)–Pd^II 键亲电加成得到由 2-(三甲基硅烷基)苯基三氟甲磺酸酯消除生成的苯炔中间体 **E**，生成芳基钯中间体 **C**，**C** 随即通过碱诱导 C-H 键官能团化生成七元环钯中间体 **D**，最后 **D** 通过还原消除得到 9,10-二氢菲衍生物。

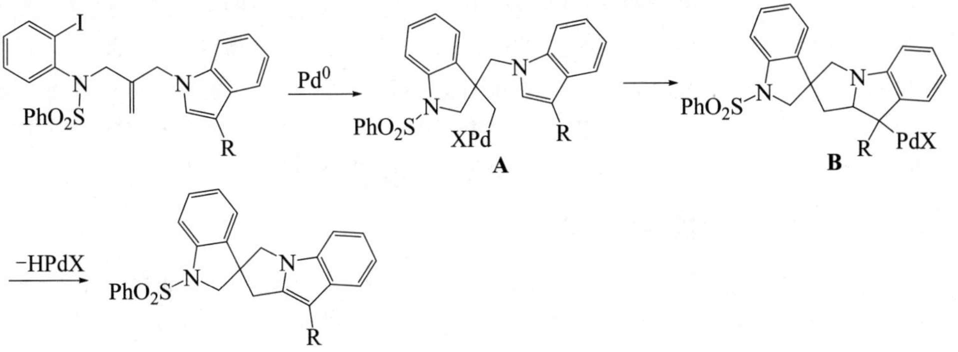

图 9-16　钯催化 σ-钯中间体直接 β-氢消除反应机理

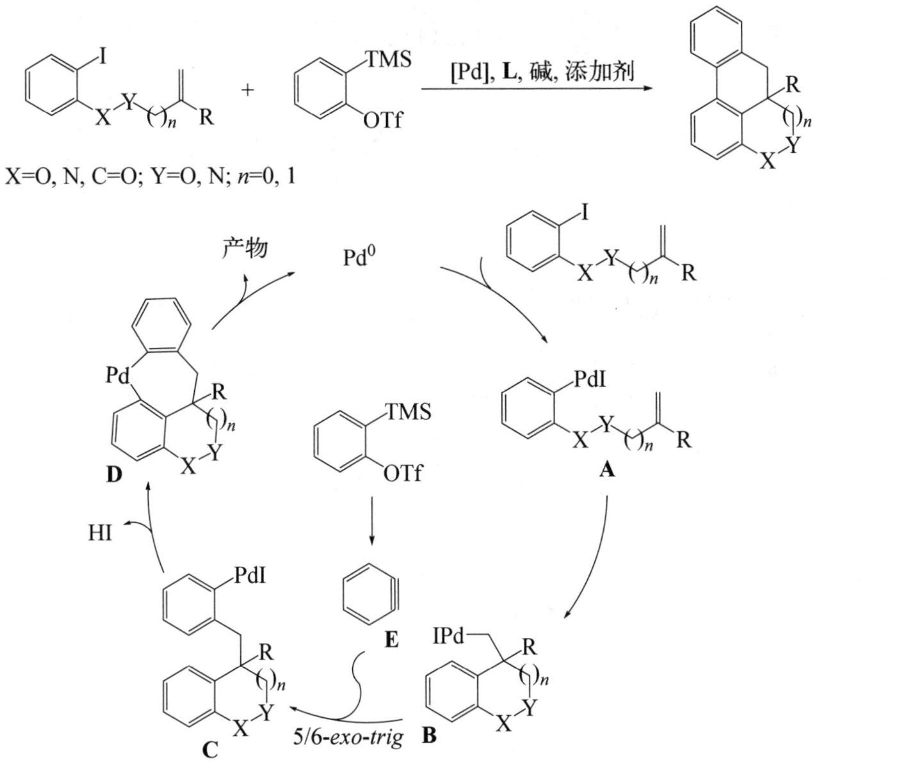

图 9-17　σ-钯中间体与亲核试剂反应机理

第 9 章　过渡金属催化串联反应及其在药物合成上的应用

9.5.2 基本的 Heck/C-H 官能团化串联反应

通过 Heck/C-H 键官能团化串联反应，利用 Heck 反应生成的 σ-钯中间体代替导向基团，可以避免导向基的引入和移除，一锅反应实现多个碳碳键的构建，极大地提高了反应的效率及原子经济性。根据 σ-钯中间体反应类型的不同，此类反应主要有两种：①Heck 反应生成的 σ-钯中间体直接发生 β-氢消除得到相应的碳环、螺环、稠环化合物；②Heck 反应生成的 σ-钯中间体被反应体系亲核试剂捕获发生分子内或分子间亲核加成，然后发生还原消除得到对应产物。

9.5.2.1　σ-钯中间体直接 β-氢消除

在钯催化剂催化下，含 β-溴苯乙烯和炔基结构单元的化合物通过分子内碳钯化、苯环 C-H 键功能化串联反应，一步构建两个碳-碳键，同时形成一个五元环和六元环，得到稠合三环衍生物[92]。

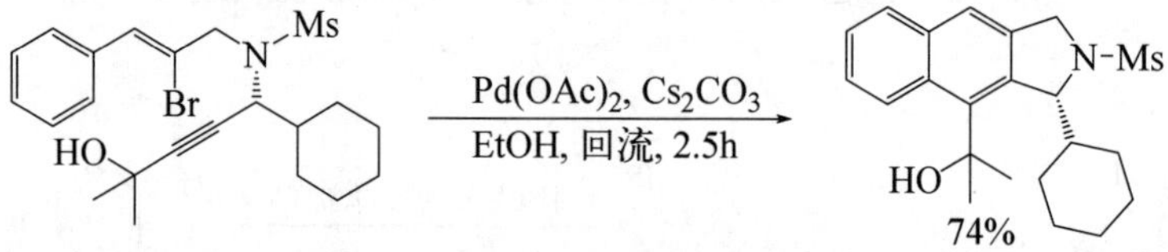

(114)

以累积多烯和芳基碘化物为起始原料，在钯催化下发生 Heck/C-H 官能团化串联反应，通过 σ-钯中间体直接 β-氢消除可以得到含有芳基的四取代烯烃茚衍生物[93]。

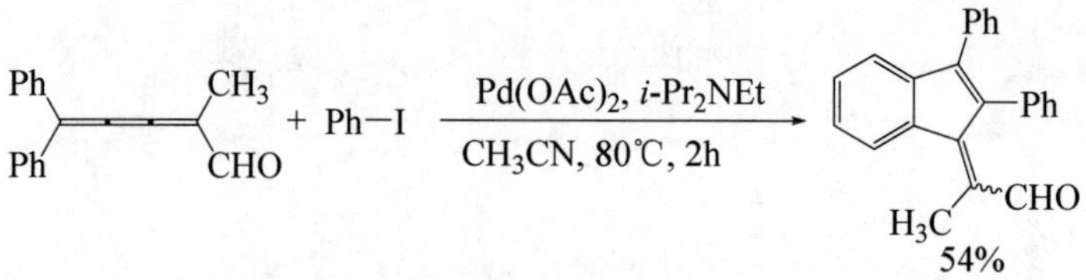

(115)

钯催化 N-芳基丙炔酰胺、碘苯、邻硝基碘苯三组分串联反应，经过 Sonagashira 偶联/碳钯化/C-H 键活化/Heck C-C 键偶联等反应，生成不同取代的二氢吲哚酮衍生物[94]。

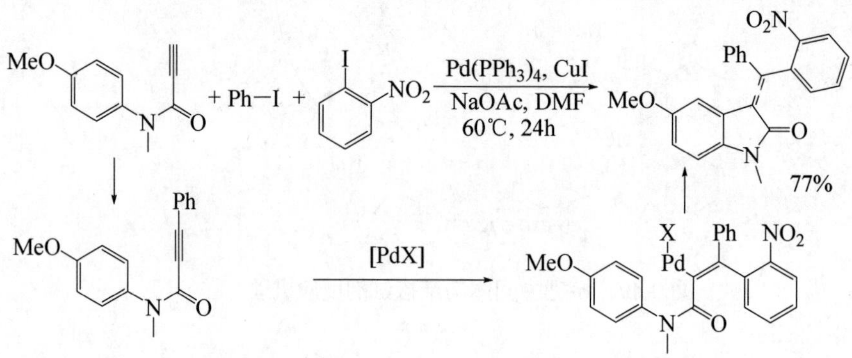

(116)

以邻溴代苯基叔丁基醚为底物,在钯催化剂作用下发生 Heck/C-H 官能团化串联反应,通过选择性活化甲基 C(sp^3)-H 键,能够合成不同取代的苯并呋喃衍生物[95]。

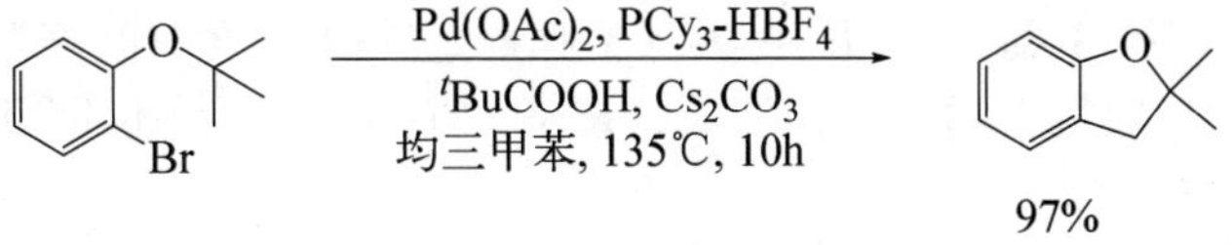

$$\text{邻溴代苯基叔丁基醚} \xrightarrow[\text{均三甲苯, 135℃, 10h}]{\text{Pd(OAc)}_2, \text{PCy}_3\text{-HBF}_4, \ ^t\text{BuCOOH, Cs}_2\text{CO}_3} \text{2,2-二甲基苯并二氢呋喃} \quad 97\% \tag{117}$$

N-(邻溴苯基)-2-亚甲基-3-苯基丙酰胺衍生物在钯催化剂催化下进行 Heck/C-H 键官能团化串联反应,生成二氢吲哚酮螺环产物[96]。

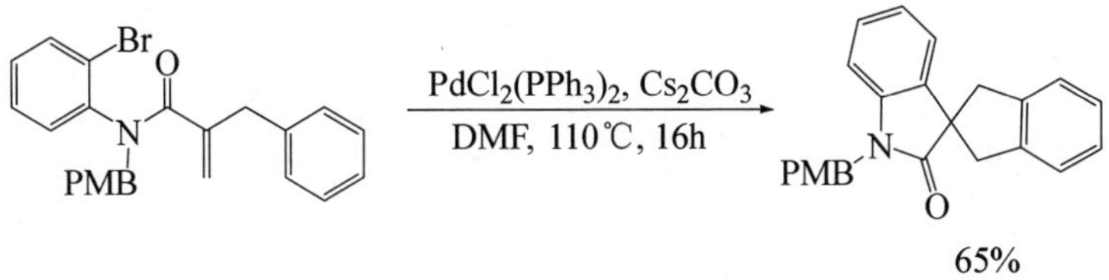

$$\xrightarrow[\text{DMF, 110℃, 16h}]{\text{PdCl}_2(\text{PPh}_3)_2, \text{Cs}_2\text{CO}_3} \quad 65\% \tag{118}$$

含邻溴苯基丙基、苄基的哌啶烯酮类化合物在钯催化下进行 Heck/C-H 键官能团化反应,生成新型螺旋五环化合物,其中的 6/6 螺环结构经由溴苯氧化加成、双键迁移插入、苯基 C-H 键活化、还原消除等基元步骤构成[97]。

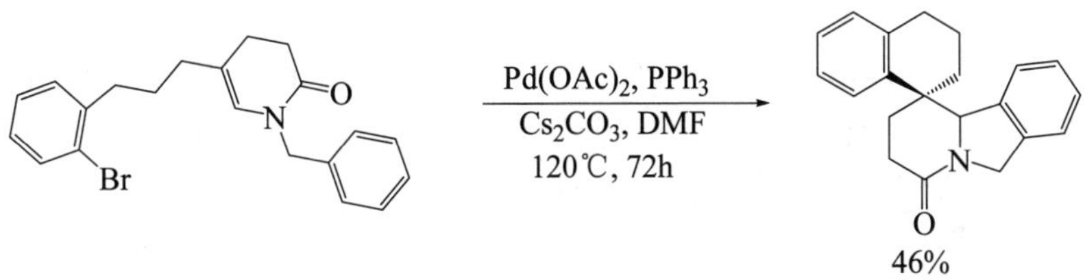

$$\xrightarrow[\text{Cs}_2\text{CO}_3, \text{DMF} \\ 120℃, 72h]{\text{Pd(OAc)}_2, \text{PPh}_3} \quad 46\% \tag{119}$$

N-苯基丙烯酰胺 α-位与卤代苯卤原子邻位通过二碳或碳杂链相连的底物分子在 Pd(OAc)$_2$/XPhos/K$_2$CO$_3$ 催化体系催化下能够顺利地进行 Heck/C-H 键官能团化串联反应,高效合成苯并二氢呋喃、二氢吲哚、二氢茚骨架螺 3,4-二氢喹啉酮骨架的螺环化合物[98]。

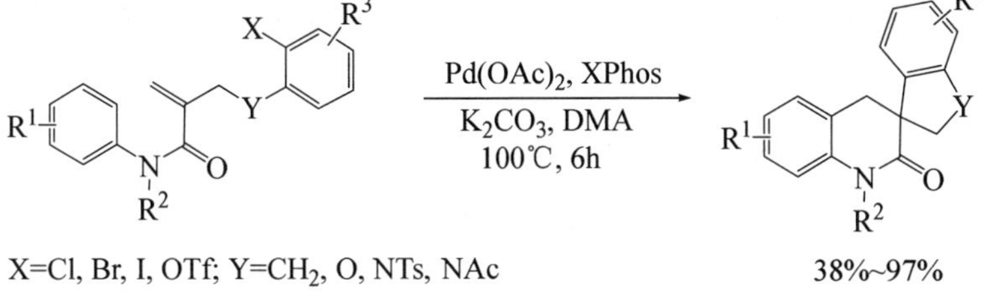

$$\xrightarrow[\text{K}_2\text{CO}_3, \text{DMA} \\ 100℃, 6h]{\text{Pd(OAc)}_2, \text{XPhos}} \quad 38\%\sim97\% \tag{120}$$

X=Cl, Br, I, OTf; Y=CH$_2$, O, NTs, NAc

邻氯乙烯基吡啶与噻吩硼酸在铑/钯双金属催化下发生串联环化反应，"一锅"合成二氢喹啉并噻吩环衍生物。该反应也可以分为两个独立步骤进行，例如，3-氯-5-三氟甲-2-乙烯基吡啶与 3-噻吩硼酸在[Rh(C_2H_4)Cl]$_2$ 催化下进行 Heck 偶联反应生成相应的偶联产物 3-氯-5-三氟甲-2-（3-噻吩乙基）；该中间产物进一步在 Pd(OAc)$_2$/PCy$_3$-HBF$_4$ 催化下通过氧化加成、苯基 C-H 键活化、还原消除等基元反应生成二氢喹啉并噻吩[99]。

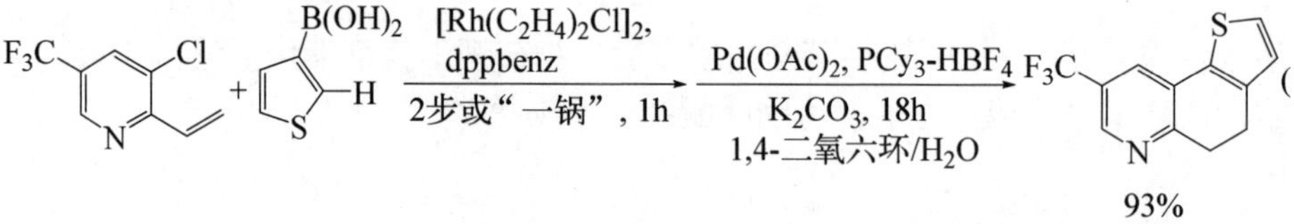

$$\text{(121)}$$

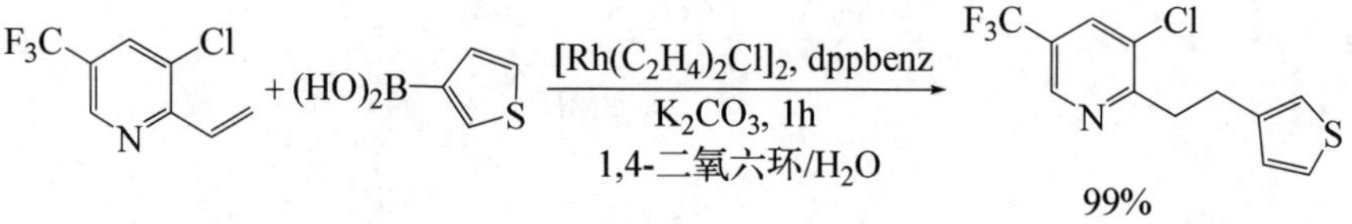

$$\text{(122)}$$

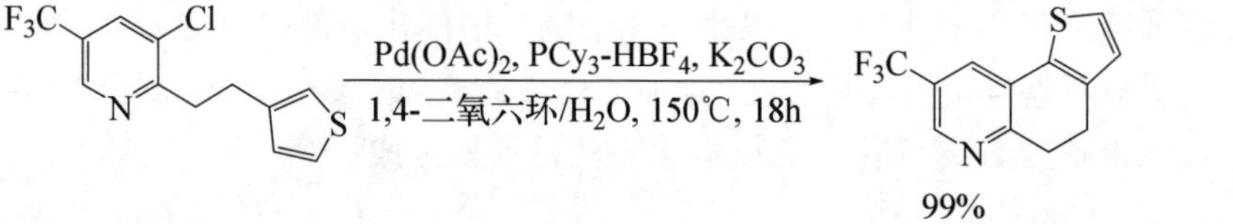

$$\text{(123)}$$

邻碘芳基（2-芳基烯丙基）胺类化合物与 2-重氮-2-芳基乙酸酯在钯催化剂催化下，经过分子内 Heck 螺环化、远程 C—H 键活化、重氮羰基卡宾插入等基元步骤，构建两个季碳中心，最终以良好的化学选择性生成螺二氢吲哚衍生物[100]。

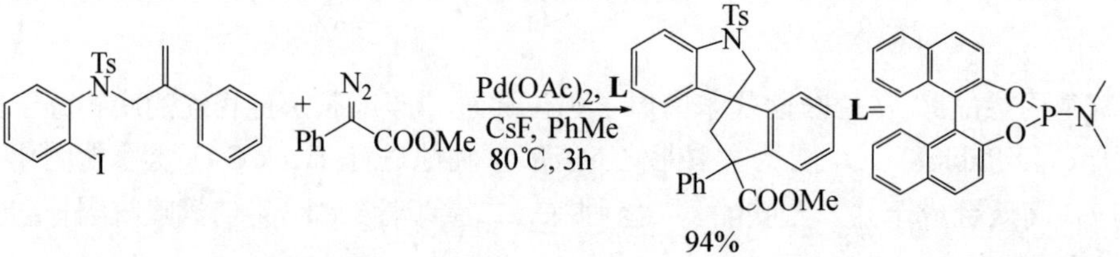

$$\text{(124)}$$

邻碘芳基烯丙基醚与 1-甲基-4-苯基三唑在钯催化剂催化下进行 Heck/分子间芳杂环 C-H 键活化串联反应，得到连有 1,2,3-三唑骨架结构的苯并二氢呋喃[101]。

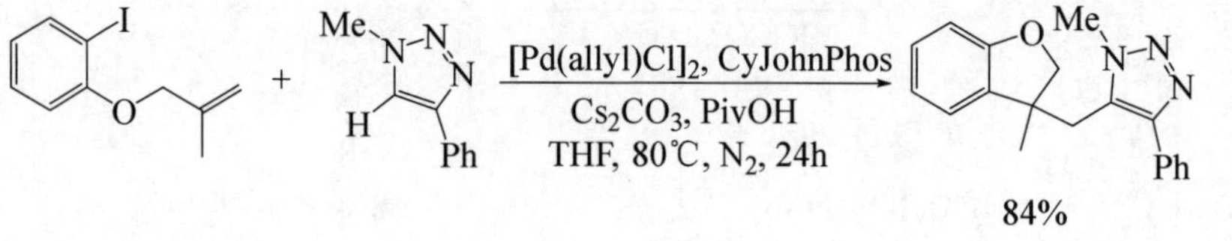

$$\text{(125)}$$

9.5.2.2 σ-钯中间体与亲核试剂反应

在钯催化剂催化下 N-邻碘苯基-2-胺乙基丙烯酰胺进行分子内 Heck/C-H 官能团化串联反应,生成具有生物活性的螺吡咯烷-3,3'-吲哚酮衍生物。反应经历了氧化加成、双键迁移插入、氨基对烷基钯中间体亲核加成、还原消除等步骤[102]。

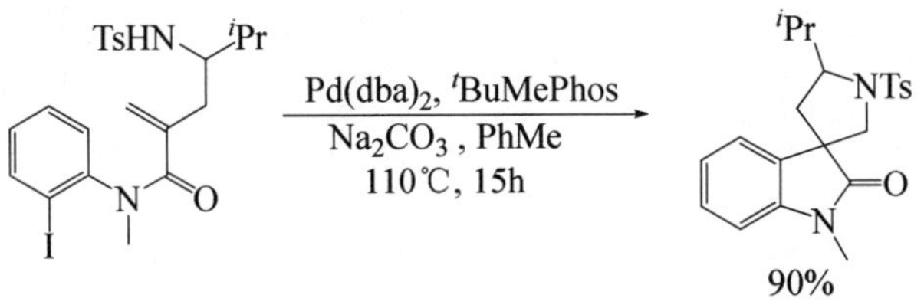

$$(126)$$

邻碘苯基与 1,5-二烯单元通过酰胺键相连的化合物与对甲基苯磺酰脒在钯催化剂催化下进行 Heck 环化/分子间亲核串联反应,生成芳乙烯基环戊烷螺二氢吲哚酮衍生物[103]。

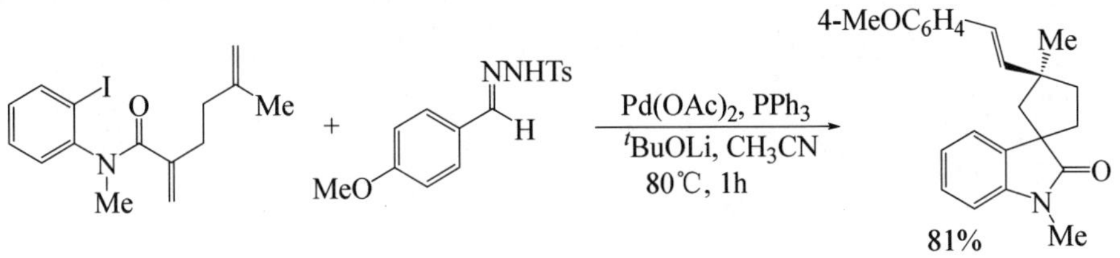

$$(127)$$

邻碘苯甲基烯丙基醚类化合物在钯催化剂催化下与二氮甲基酮发生 Heck 反应/C-H 键活化/胺化串联反应,能够以良好的收率得到稠环和螺环化合物。当底物烯基碘化物的双键上连有甲基时,反应首先发生氧化加成、双键迁移插入得到五元环二价钯中间体 **A**,其进一步与二氮嘧啶酮发生氧化加成生成四价钯中间体 **B**,最后通过还原消除一分子异氰酸叔丁酯得到 6/6/5 稠环化合物;当底物烯基碘化物的双键上连有苯基时,经过氧化加成、双键迁移插入、苯基 C-H 键活化得到五元螺环二价钯中间体 **C**,然后与二氮嘧啶酮发生氧化加成和还原消除生成 6/6/5/6 螺环化合物[104]。

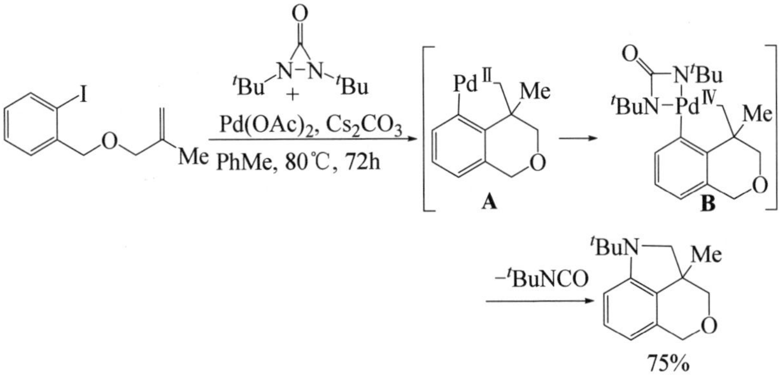

$$(128)$$

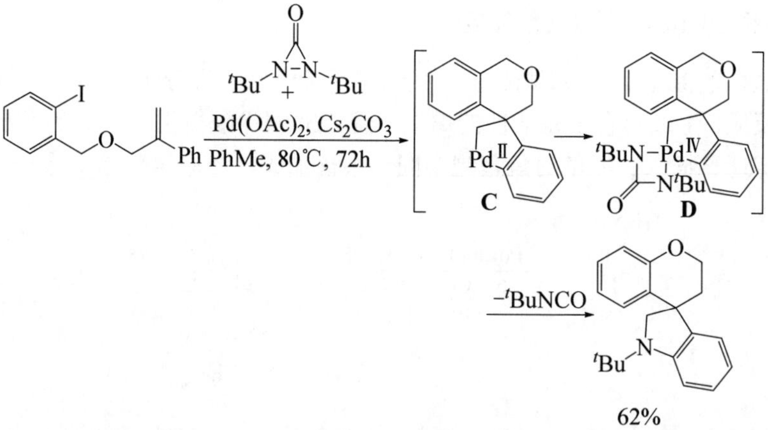

$$（129）\quad 62\%$$

邻碘苯基与苯乙烯通过酰氨基、醚键连接的化合物与邻三甲硅基苯基三氟甲磺酸酯在钯催化剂催化下发生串联环化反应，生成不同取代的二氢吲哚酮和苯并呋喃酮衍生物，该反应经历碳钯化、C-H键活化、苯炔迁移插入等步骤[105]。

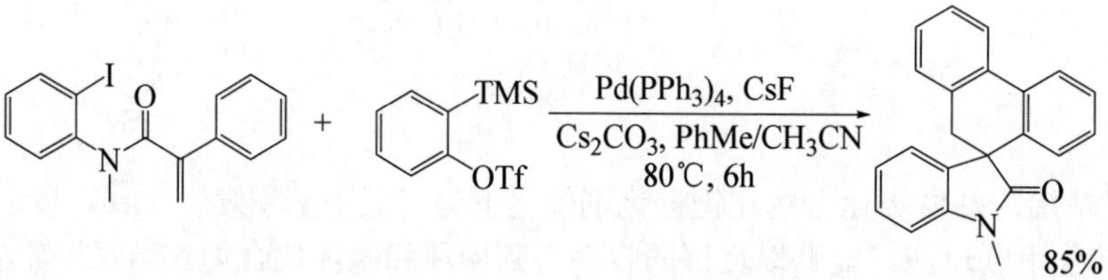

$$（130）\quad 85\%$$

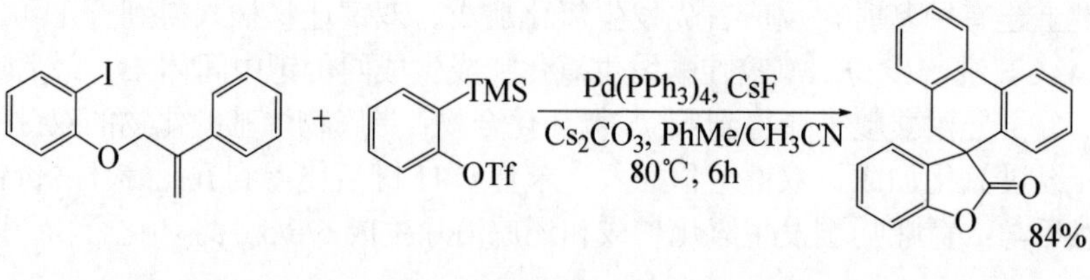

$$（131）\quad 84\%$$

N-(2-碘苯基)-N-甲基-2-苯基丙烯酰胺与二溴甲烷在 Pd(OAc)$_2$/P(o-tol)$_3$ 催化下可实现远程 C-H 键活化和 CH$_2$Br$_2$ 双烷基化串联反应，以优良的产率合成复杂结构的二氢吲哚酮衍生物[106]。

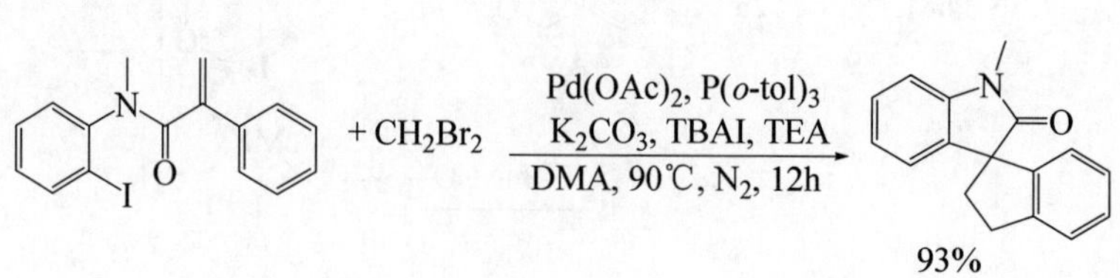

$$（132）\quad 93\%$$

以烯丙酰邻碘苯甲酰混合亚胺为起始原料,邻溴苯甲酸作为偶联试剂,在钯催化剂催化下经历氧化加成、分子内碳钯化、C-H 键活化、氧化加成、分子内脱羧、还原消除等步骤,通过 C(sp^2)-Br 键的裂解和脱羧过程,实现芳环对五元环钯中间体的插入,以高度区域选择性生成异喹啉二酮衍生物[107]。

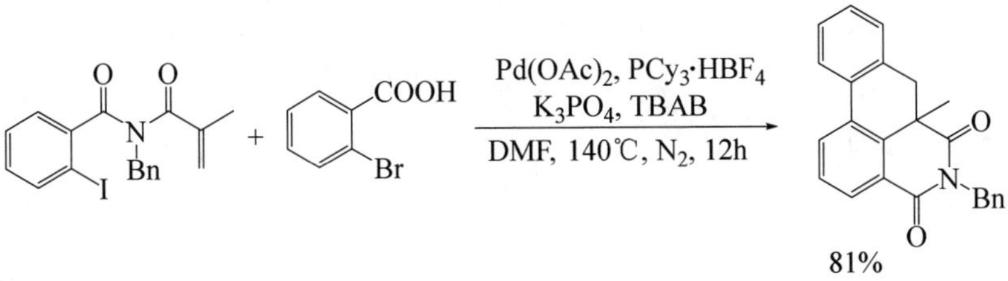

$$\text{邻碘苯甲酰烯丙酰胺} + \text{邻溴苯甲酸} \xrightarrow[\text{DMF, 140℃, N}_2\text{, 12h}]{\text{Pd(OAc)}_2\text{, PCy}_3\cdot\text{HBF}_4 \\ \text{K}_3\text{PO}_4\text{, TBAB}} \text{产物} \quad 81\% \tag{133}$$

钯催化亲核试剂(Z)-3-碘烯丙醇或胺和联烯胺类化合物[4+2]串联环化反应,高区域选择性地生成二氢吡喃或二氢吡啶衍生物[108]。其中以烯丙醇作为亲核试剂时,在不加配体的条件下反应即能顺利地进行。

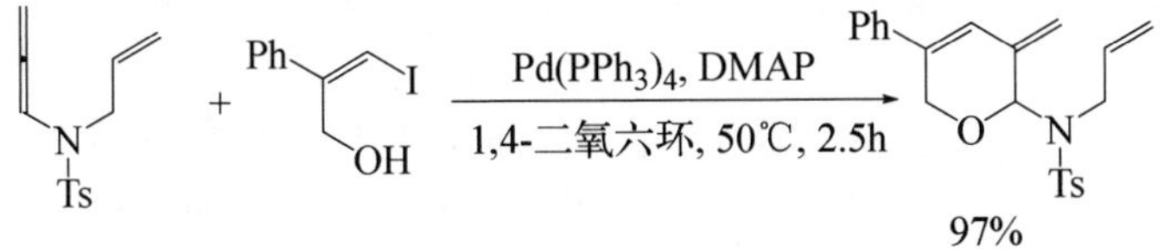

$$\xrightarrow[\text{1,4-二氧六环, 50℃, 2.5h}]{\text{Pd(PPh}_3)_4\text{, DMAP}} \quad 97\% \tag{134}$$

$$\xrightarrow[\text{1,4-二氧六环, 80℃, 5h}]{\text{Pd}_2(\text{dba})_3\text{, PPh}_3\text{, TEA}} \quad 71\% \tag{135}$$

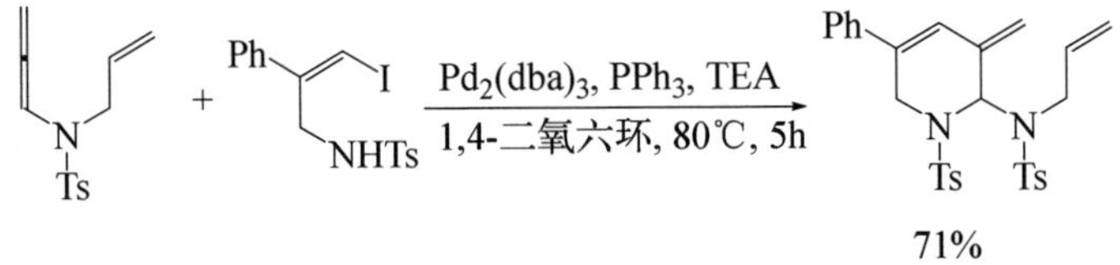

我国科学家在本领域内有杰出的研究成果,西北大学关正辉教授课题组以 N-邻溴苯基丙烯酰胺为底物与不同的亲核试剂在手性单齿膦酰胺配体辅助的钯催化剂催化下,进行一氧化碳插入,对映选择性地生成光学活性的含有季碳中心 β-羰基取代的二氢吲哚化合物[109]。

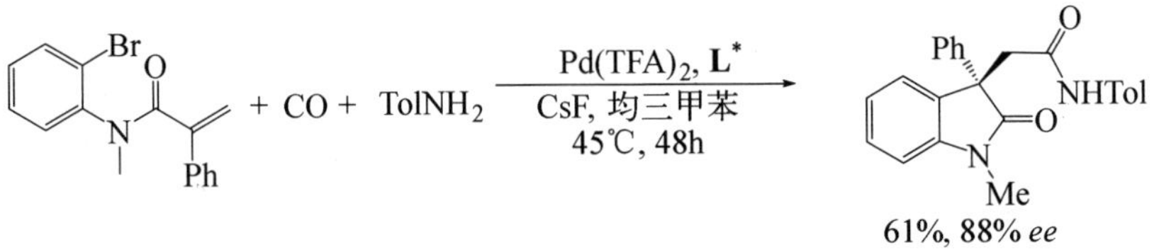

$$+ \text{CO} + \text{TolNH}_2 \xrightarrow[\text{45℃, 48h}]{\text{Pd(TFA)}_2\text{, L}^* \\ \text{CsF, 均三甲苯}} \quad 61\%,\ 88\%\ ee \tag{136}$$

$$+ \text{CO} + \text{MeOH} \xrightarrow[\text{45℃, 48h}]{\text{Pd(TFA)}_2\text{, L}^* \\ \text{CsF, 均三甲苯}} \quad 89\%,\ 92\%\ ee \tag{137}$$

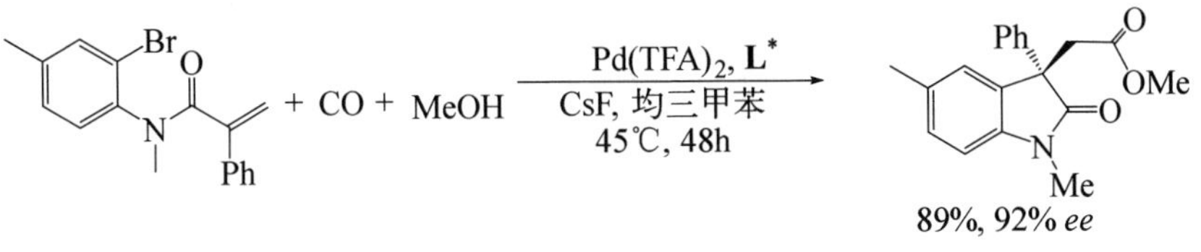

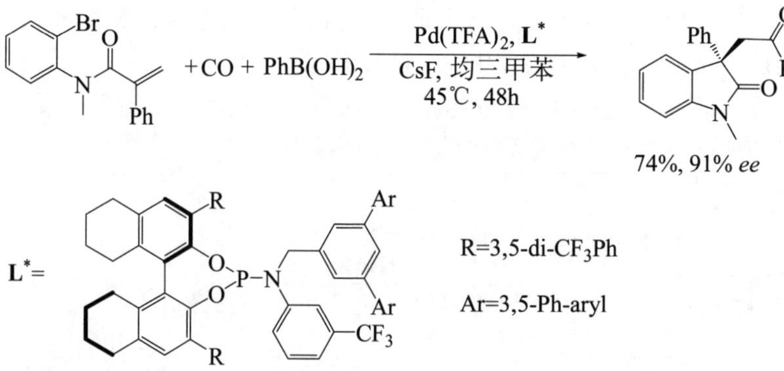

$$\text{(138)}$$

74%, 91% ee

L* = [phosphoramidite ligand structure]

R=3,5-di-CF$_3$Ph
Ar=3,5-Ph-aryl

9.5.3 Heck/C-H 官能团化串联反应在药物和天然产物合成上的应用

含氮杂环、稠环是许多具有生物活性的药物分子和天然产物的核心骨架结构，过渡金属催化 Heck/C-H 官能团化反应为构建特定含氮杂环、稠环提供了新的合成途径。

spiropentacyclics 具有独特的螺环结构，含此结构的三尖杉碱表现出显著的抗白血病活性。其合成路线如下所示：该反应以简单易得的邻溴碘苯和 4-戊烯醇为原料进行分子间 Heck 偶联、进一步氧化得到相应的醛，然后醛与四氢吡咯缩合成烯胺，经过 Michael 加成、水解脱去四氢吡咯，最后与苄胺发生分子间缩合反应得到烯基环内酰胺化合物 TT-20，进一步在钯催化剂作用下发生 Heck/C-H 键官能团化得到目标分子[110]。

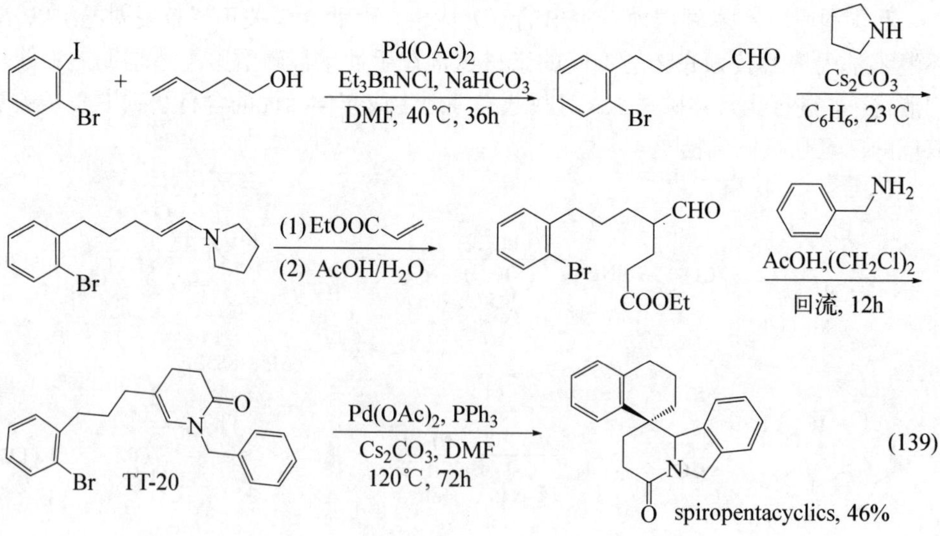

$$\text{(139)}$$

spiropentacyclics, 46%

spiro[pyrrolidine-3,3'-oxindole]是许多生物碱的核心结构单元,此类化合物具有显著的生物和药物活性。其结构和合成路线如下:以简单易得的邻溴苯胺为原料,与茴香醛发生还原胺化得到 PMB 保护的苯胺,随后与单甲基丙二酸钾盐反应生成酰胺化合物,其进一步在碱的作用下与苄溴发生亲核取代反应得到烷基化中间体 TT-21。中间体 TT-21 在碱性条件下水解、酸化可得丙烯酰胺中间体 TT-22,最后在钯催化剂催化下发生 Heck/C-H 键官能团化串联反应得到目标分子[111]。

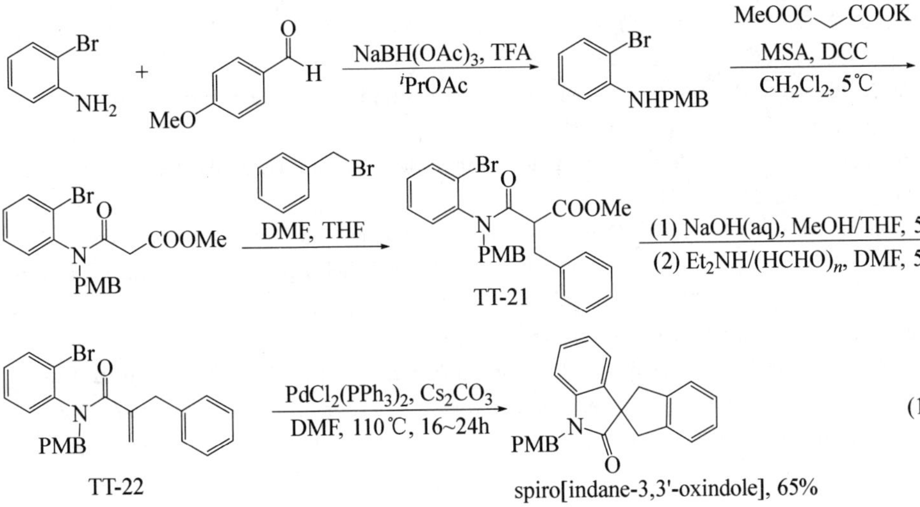

生物活性六氢吡咯吲哚及其二聚生物碱(+)-毒扁豆碱[(+)-physostigmine]广泛存在于天然产物和药物分子。以芳基丙烯酰胺和苄醇为底物,在钯催化剂作用下通过一氧化碳插入发生 Heck/C-H 官能团化反应,得到关键的光学活性二氢吲哚酮中间体,再经多步转化可生成对应的目标分子[109]。

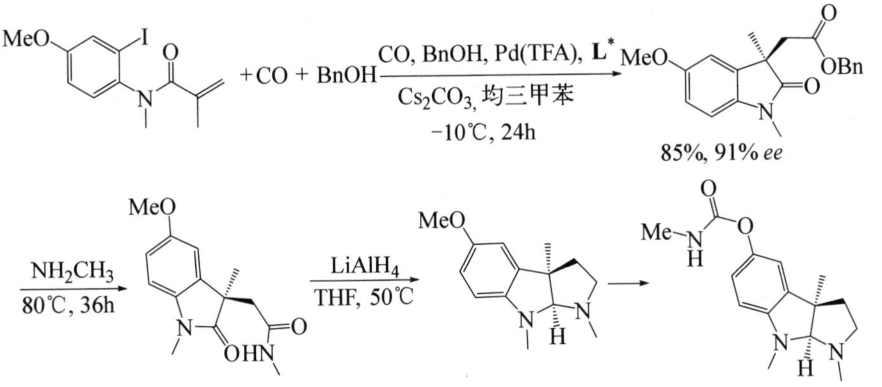

L* = [structure]   R=3,5-di-CF₃Ph
Ar = 3,5-Ph-aryl

参考文献

[1] Trost B M, Van Vranken D L. Asymmetric transition metal-catalyzed allylic alkylations [J]. Chemical Reviews, 1996, 96(1): 395-422.

[2] Wu X, Gong L Z. Palladium(0)-catalyzed difunctionalization of 1,3-dienes: from racemic to enantioselective [J]. Synthesis, 2018, 51(1): 122-134.

[3] Patel B A, Dickerson J E, Heck R F. Palladium-catalyzed arylation of conjugated dienes [J]. The Journal of Organic Chemistry, 2002, 43(26): 5018-5020.

[4] O'Connor J M, Stallman B J, Clark W G, et al. Some aspects of palladium-catalyzed reactions of aryl and vinylic halides with conjugated dienes in the presence of mild nucleophiles [J]. The Journal of Organic Chemistry, 2002, 48(6): 807-809.

[5] Chen S S, Meng J, Li Y H, et al. Palladium-catalyzed enantioselective heteroannulation of 1,3-dienes by functionally substituted aryl iodides [J]. The Journal of Organic Chemistry, 2016, 81(19): 9402-9408.

[6] Cheng X, Ma S. [Pd(PPh₃)₄]-catalyzed diastereoselective synthesis of trans-1,2-diazetidines from 2,3-allenyl hydrazines and aryl halides [J]. Angewandte Chemie International Edition, 2008, 47(24): 4581-4583.

[7] Larock R C, Zenner J M. Enantioselective, palladium-catalyzed hetero- and carboannulation of allenes using functionally-substituted aryl and vinylic iodides [J]. The Journal of Organic Chemistry, 2002, 60(3): 482-483.

[8] Shu W, Yu Q, Ma S. Development of a new spiro-box ligand and its application in highly enantioselective palladium-catalyzed cyclization of 2-Iodoanilines with allenes [J]. Advanced Synthesis & Catalysis, 2009, 351(17): 2807-2810.

[9] Larock R C, Berrios-Pena N G, Fried C A, et al. Palladium-catalyzed annulation of 1,4-dienes using ortho-functionally-substituted aryl halides [J]. The Journal of Organic Chemistry, 2002, 58(17): 4509-4510.

[10] Larock R C, Wang Y, Lu Y, et al. Synthesis of aryl-substituted allylic amines via palladium-catalyzed coupling of aryl iodides, nonconjugated dienes, and amines [J]. The Journal of Organic Chemistry, 2002, 59(26): 8107-8114.

[11] Han X, Larock R C. Synthesis of highly functionalized polycyclics via Pd-catalyzed intramolecular coupling of aryl/vinylic halides, non-conjugated dienes and nucleophiles [J]. Synlett, 1998(7): 748-750.

[12] Ahmar M, Barieux J J, Cazes B, et al. Carbopalladation of allenic hydrocarbons [J]. Tetrahedron, 1987, 43(3): 513-526.

[13] Larock R C, Fried C A. Palladium-catalyzed carboannulation of 1,3-dienes by aryl halides [J]. Journal of The American Chemical Society, 2002, 112(15): 5882-5884.

[14] Wu X, Lin H C, Li M L, et al. Enantioselective 1,2-difunctionalization of dienes enabled by chiral palladium complex-catalyzed cascade arylation/allylic alkylation reaction [J]. Journal of The American Chemical Society, 2015, 137(42): 13476-13479.

[15] Liao L, Jana R, Urkalan K B, et al. A palladium-catalyzed three-component cross-coupling of conjugated dienes or terminal alkenes with vinyl triflates and boronic acids [J]. Journal of The American Chemical Society, 2011, 133(15): 5784-5787.

[16] Huang T H, Chang H M, Wu M Y, et al. Palladium-catalyzed three-component assembling of allenes, organic halides, and arylboronic acids [J]. The Journal of Organic Chemistry, 2002, 67(1): 99-105.

[17] Liao L, Jana R, Urkalan K B, et al. A palladium-catalyzed three-component cross-coupling of conjugated dienes or terminal alkenes with vinyl triflates and boronic acids [J]. Journal of The American Chemical Society, 2011, 133(15): 5784-5787.

[18] Stokes B J, Liao L, Andrade A. M, et al. A palladium-catalyzed three-component-coupling strategy for the differential vicinal diarylation of terminal 1,3-dienes [J]. Organic Letters, 2014, 16(17): 4666-4669.

[19] Luo L, Zheng H, Liu J, et al. Highly chemo- and regioselective construction of spirocarbocycles by a Pd(0)-catalyzed dearomatization of phenol-based biaryls with 1,3-dienes [J]. Organic Letters, 2016, 18(9): 2082-2085.

[20] Tao Z L, Adili A, Shen H C, et al. Catalytic enantioselective assembly of homoallylic alcohols from dienes, aryldiazonium salts, and aldehydes [J]. Angewandte Chemie International Edition, 2016, 55(13): 4322-4326.

[21] Yoshida M, Sugimoto K, Ihara M. Palladium-catalyzed ring expansion reaction of (Z)-1-(1,3-butadienyl)cyclobutanols with aryl iodides. stereospecific synthesis of (Z)-2-(3-aryl-1-propenyl) cyclopentanones [J]. Organic Letters, 2004, 6(12): 1979-1982.

[22] Zhang Y, Shen H C, Li Y Y, et al. Access to chiral tetrahydrofluorenes through a palladium-catalyzed enantioselective tandem intramolecular Heck/Tsuji-Trost reaction [J]. Chenmical Communication, 2019, 55(26): 3769-3772.

[23] Shen H C, Wang P S, Tao Z L, et al. An enantioselective multicomponent carbonyl allylation of aldehydes with dienes and alkynyl bromides enabled by chiral palladium phosphate [J]. Advanced Synthesis & Catalysis, 2017, 359(14): 2383-2389.

[24] Wang J, Liu Y, Xiong Z, et al. Palladium-catalysed dearomative aryl/cycloimidoylation of indoles [J]. Chenmical Communication, 2020, 56(22): 3249-3252.

[25] Larock R C, Berrios-Pena N G, Fried C A. Regioselective, palladium-catalyzed hetero- and carboannulation of 1,2-dienes using functionally substituted aryl halides [J]. The Journal of Organic Chemistry, 2002, 56(8): 2615-2617.

[26] Ma S, Shi Z, Wu S. Enantioselective synthesis of β-arylbutenolides via palladium(0) catalysed asymmetric coupling cyclisation reaction of racemic allenic carboxylic acids with aryl iodides [J]. Tetrahedron: Asymmetry, 2001, 12(2): 193-195.

[27] Xie X, Ma S. Palladium-catalyzed asymmetric coupling cyclization of terminal gamma-allenols with aryl iodides [J]. Chenmical Communication, 2013, 49(50): 5693-5695.

[28] Larock R C, Berrios-Pena N G, Fried C A. Regioselective, palladium-catalyzed hetero- and carboannulation of 1,2-dienes using functionally substituted aryl halides [J]. The Journal of Organic Chemistry, 2002, 56(8): 2615-2617.

[29] Inuki S, Oishi S, Fujii N, et al. Total synthesis of (+/−)-lysergic acid, lysergol, and isolysergol by palladium-catalyzed domino cyclization of amino allenes bearing a bromoindolyl group [J]. Organic Letters, 2008, 10(22): 5239-5242.

[30] Overman L E, Rosen M. Total synthesis of (−)-spirotryprostatin B and three stereoisomers [J].

Angewandte Chemie International Edition, 2000, 39(24): 4596-4599.

[31] Hoffmann-Röder A, Krause N. Synthesis and properties of allenic natural products and pharmaceuticals [J]. Angewandte Chemie International Edition, 2004, 43(10): 1196-1216.

[32] Azizollahi H, Mehta V P, García-López J A. Pd-catalyzed cascade reactions involving skipped dienes: from double carbopalladation to remote C-C cleavage [J]. Chenmical Communication, 2019, 55(69): 10281-10284.

[33] Abelman M M, Overman L E. Palladium-catalyzed polyene cyclizations of dienyl aryl iodides [J]. Journal of The American Chemical Society, 2002, 110(7): 2328-2329.

[34] Grigg R, Sridharan V, Sukirthalingam S. Alkylpalladium(II) species. Reactive intermediates in a bis-cyclisation route to strained polyfused ring systems [J]. Tetrahedron Letters, 1991, 32(31): 3855-3858.

[35] Grigg R, Fretwell P, Meerholtz C, et al. Palladium catalysed synthesis of spiroindolines [J]. Tetrahedron, 1994, 50(2): 359-370.

[36] Coya E, Sotomayor N, Lete E. Enantioselective palladium-catalyzed heck-heck cascade reactions: ready access to the tetracyclic core of lycorane alkaloids [J]. Advanced Synthesis & Catalysis, 2015, 357(14-15): 3206-3214.

[37] Trost B M, Ryan M C, Rao M, et al. Construction of enantioenriched [3.1.0] bicycles via a ruthenium-catalyzed asymmetric redox bicycloisomerization reaction [J]. Journal of The American Chemical Society, 2014, 136(50): 17422-17425.

[38] Huang X, Nguyen M H, Pu M, et al. Asymmetric reductive and alkynylative Heck bicyclization of enynes to access conformationally restricted aza[3.1.0]bicycles [J]. Angewandte Chemie International Edition, 2020, 59(27): 10814-10818.

[39] Hu Y M, Zhou J, Long X T, et al. Palladium-catalyzed cascade reactions of benzyl halides with N-allyl-N-(2-butenyl)-p-toluenesulfonamide [J]. Tetrahedron Letters, 2003, 44(27): 5009-5010.

[40] Hu J, Hirao H, Li Y, et al. Palladium-catalyzed asymmetric intermolecular cyclization [J]. Angewandte Chemie International Edition, 2013, 52(33): 8676-8680.

[41] Szlosek-Pinaud M, Diaz P, Martinez J, et al. Efficient synthetic approach to heterocycles possessing the 3,3-disubstituted-2,3-dihydrobenzofuran skeleton via diverse palladium-catalyzed tandem reactions [J]. Tetrahedron, 2007, 63(16): 3340-3349.

[42] Li Y, Ding Z, Lei A, et al. Ni-Catalyzed enantioselective reductive aryl-alkenylation of alkenes: application to the synthesis of (+)-physovenine and (+)-physostigmine [J]. Organic Chemistry Frontiers, 2019, 6(18): 3305-3309.

[43] Ma T, Chen Y, Li Y, et al. Nickel-catalyzed enantioselective reductive aryl fluoroalkenylation of alkenes [J]. ACS Catalysis, 2019, 9(10): 9127-9133.

[44] Ju B, Chen S, Kong W. Pd-catalyzed enantioselective double heck reaction [J]. Organic Letters, 2019, 21(23): 9343-9347.

[45] Overman L E, Ricca D J, Tran V D. First total synthesis of scopadulcic acid B [J]. Journal of the American Chemical Society, 2002, 115(5): 2042-2044.

[46] Fox M E, Li C, Marino J P, et al. Enantiodivergent total syntheses of (+)- and (−)-scopadulcic acid A [J]. Journal of the American Chemical Society, 1999, 121(23): 5467-5480.

[47] Maddaford S P, Andersen N G, Cristofoli W A, et al. Total synthesis of (+)-xestoquinone using an asymmetric palladium-catalyzed polyene cyclization [J]. Journal of the American Chemical Society, 1996, 118(44): 10766-10773.

[48] Ju B, Chen S, Kong W. Pd-catalyzed enantioselective double heck reaction [J]. Organic Letters, 2019, 21(23): 9343-9347.

[49] Hu J, Hirao H, Li Y, et al. Palladium-catalyzed asymmetric intermolecular cyclization [J]. Angewandte Chemie International Edition, 2013, 52(33): 8676-8680.

[50] Danodia A K, Saunthwal R K, Patel M, et al. Pd-catalyzed one-pot sequential unsymmetrical cross-coupling reactions of aryl/heteroaryl 1,2-dihalides [J]. Organic & Biomolecular Chemistry, 2016, 14(27): 6487-6496.

[51] Zhou M B, Huang X C, Liu Y Y, et al. Alkylation of terminal alkynes with transient sigma-alkylpalladium(II) complexes: a carboalkynylation route to alkyl-substituted alkynes [J]. Chemistry A European Journal, 2014, 20(7): 1843-1846.

[52] Zhang X, Liu A, Chen W. Pd-catalyzed efficient one-pot sequential cross-coupling reactions of aryl dihalides [J]. Organic Letters, 2008, 10(17): 3849-3852.

[53] Wang D C, Wang H X, Hao E J, et al. Synthesis of 3,3-disubstituted oxindoles containing a 3-(4-aminobut-2-ynyl) unit via domino Heck-Sonogashira reaction in water [J]. Advanced Synthesis & Catalysis, 2016, 358(3): 494-499.

[54] Liu R R, Xu T F, Wang Y G, et al. Palladium-catalyzed dearomative arylalkynylation of indoles [J]. Chemical Communication, 2016, 52(94): 13664-13667.

[55] Liu R R, Wang Y G, Li Y L, et al. Enantioselective dearomative difunctionalization of indoles by palladium-catalyzed heck/sonogashira sequence [J]. Angewandte Chemie International Edition, 2017, 56(26): 7475-7478.

[56] Bai X, Wu C, Ge S, et al. Pd/Cu-catalyzed enantioselective sequential Heck/Sonogashira coupling: asymmetric synthesis of oxindoles containing trifluoromethylated quaternary stereogenic centers[J]. Angewandte Chemie International Edition, 2020, 59(7): 2764-2768.

[57] Zhou L, Li S, Xu B, et al. Enantioselective difunctionalization of alkenes by a palladium-catalyzed Heck/Sonogashira sequence[J]. Angewandte Chemie International Edition, 2020, 59(7): 2769-2775.

[58] Zhu J W, Zhou B, Cao Z Y, et al. Stereoselective 1,2-dicarbofunctionalization of trisubstituted alkenes by palladium-catalyzed Heck/Suzuki or Heck/Sonogashira domino sequence[J]. CCS Chemistry, 2021, 3(9): 2340-2349.

[59] Sun W, Shi X, Chen C, et al. Palladium-catalyzed cascade cyclization/alkynylation of alkene-tethered carbamoyl chlorides with terminal alkynes: synthesis of alkyne-functionalized oxindoles[J]. Asian Journal of Organic Chemistry, 2020, 9(4): 575-578.

[60] Chi X, Xia T, Yang Y, et al. Highly diastereoselective synthesis of an octahydro-1H-cyclpenta[c]pyridine skeleton via a Pd/Au-relay catalyzed reaction of (Z)-1-iodo-1,6-diene and alkyne[J]. Organic Chemistry Frontiers, 2022, 9(12): 3186-3191.

[61] Chen S, Wu X X, Wang J, et al. Palladium-catalyzed intramolecular dearomatization of indoles via decarboxylative alkynyl termination[J]. Organic Letters, 2016, 18(16): 4016-4019.

[62] Li Y, Yue-Su M S, Zhang H Y, et al. Synthesis of tetracyclic indolines through palladium-catalyzed asymmetric dearomative reaction of aryl iodides[J]. ChemistrySelect, 2021, 6(19): 4719-4724.

[63] Wang D C, Wang H X, Hao E J, et al. Synthesis of 3,3-disubstituted oxindoles containing a 3-(4-aminobut-2-ynyl) unit via domino Heck-Sonogashira reaction in water[J]. Advanced Synthesis & Catalysis, 2016, 358(3): 494-499.

[64] Bai X, Wu C, Ge S, et al. Pd/Cu-catalyzed enantioselective sequential Heck/Sonogashira coupling:

asymmetric synthesis of oxindoles containing trifluoromethylated quaternary stereogenic centers[J]. Angewandte Chemie International Edition, 2020, 59(7): 2764-2768.

[65] Grigg R, Sansano J M, Santhakumar V, et al. Palladium catalysed tandem cyclisation-anion capture processes. Part 3. Organoboron anion transfer agents[J]. Tetrahedron, 1997, 53(34): 11803-11826.

[66] Arthuis M, Pontikis R, Florent J C. Stereoselective synthesis of novel highly substituted isochromanone and isoquinolinone-containing exocyclic tetrasubstituted alkenes[J]. The Journal of Organic Chemistry, 2009, 74(5): 2234-2237.

[67] Seashore-Ludlow B, Somfai P. Domino carbopalladation–cross-coupling for the synthesis of 3, 3-disubstituted oxindoles[J]. Organic letters, 2012, 14(15): 3858-3861.

[68] Vachhani D D, Butani H H, Sharma N, et al. Domino Heck/borylation sequence towards indolinone-3-methyl boronic esters: trapping of the σ-alkylpalladium intermediate with boron[J]. Chemical Communications, 2015, 51(80): 14862-14865.

[69] Mondal A, Kundu P, Jash M, et al. Palladium-catalysed stereoselective synthesis of 4-(diarylmethylidene)-3, 4-dihydroisoquinolin-1(2H)-ones: expedient access to 4-substituted isoquinolin-1(2H)-ones and isoquinolines[J]. Organic & Biomolecular Chemistry, 2018, 16(6): 963-980.

[70] Zhang Z M, Xu B, Wu L, et al. Enantioselective dicarbofunctionalization of unactivated alkenes by palladium-catalyzed tandem Heck/Suzuki coupling reaction[J]. Angewandte Chemie International Edition, 2019, 58(41): 14653-14659.

[71] Chen Q, Li S, Xie X, et al. Pd-catalyzed enantioselective dicarbofunctionalization of alkene to access disubstituted dihydroisoquinolinone[J]. Organic Letters, 2021, 23(11): 4099-4103.

[72] Wu Y, Wu L, Zhang Z M, et al. Enantioselective difunctionalization of alkenes by a palladium-catalyzed Heck/borylation sequence[J]. Chemical Science, 2022, 13(7): 2021-2025.

[73] Li Y, Zhang H Y, Zhang Y, et al. Palladium-catalyzed asymmetric intramolecular dearomative Heck annulation of aryl halides to furnish indolines[J]. The Journal of Organic Chemistry, 2021, 86(21): 14640-14651.

[74] Jonathan E Wilson. Diastereoselective synthesis of tetrahydroquinolines via a palladium-catalyzed Heck-Suzuki cascade reaction[J]. Tetrahedron Letters, 2012, 53(18): 2308-2311.

[75] Peshkov A A, Peshkov V A, Pereshivko O P, et al. Heck-Suzuki tandem reaction for the synthesis of 3-benzazepines[J]. The Journal of Organic Chemistry, 2015, 80(13): 6598-6608.

[76] Petrone D A, Kondo M, Zeidan N, et al. Pd(0)-catalyzed dearomative diarylation of indoles[J]. Chemistry: A European Journal, 2016, 22(16): 5684-5691.

[77] Schempp T T, Daniels B E, Staben S T, et al. A general strategy for the construction of functionalized azaindolines via domino palladium-catalyzed Heck cyclization/Suzuki coupling[J]. Organic Letters, 2017, 19(13): 3616-3619.

[78] Zhou B, Wang H, Cao Z Y, et al. Dearomative 1, 4-difunctionalization of naphthalenes via palladium-catalyzed tandem Heck/Suzuki coupling reaction[J]. Nature Communications, 2020, 11(1): 4380.

[79] Duan S B, Gao X J, Zhang H Y, et al. Palladium-catalyzed intramolecular tandem dearomatization of indoles for the synthesis of tetracyclic indolines[J]. Arabian Journal of Chemistry, 2021, 14(6):103155.

[80] Wang G, Wei M, Liu T, et al. Palladium-catalyzed stereoselective intramolecular Heck dearomative silylation of indoles[J]. Advanced Synthesis & Catalysis, 2022, 364(4): 909-913.

[81] Li Y, Wang K, Ping Y, et al. Nickel-catalyzed domino Heck cyclization/Suzuki coupling for the synthesis of 3, 3-disubstituted oxindoles [J]. Organic Letters, 2018, 20(4): 921-924.

[82] Shen C, Zeidan N, Wu Q, et al. Pd-catalyzed dearomative arylborylation of indoles [J]. Chemical Science, 2019, 10(10): 3118-3122.

[83] Lee C W, Oh K S, Kim K S, et al. Suppressed β-hydride elimination in palladium-catalyzed cascade cyclization-coupling reactions: an efficient synthesis of 3-arylmethylpyrrolidines [J]. Organic Letters, 2000, 2(9): 1213-1216.

[84] Peng J, Zhao Y, Zhou J, et al. Synthesis of dihydropyrrole derivatives by a palladium-catalyzed Heck and Suzuki cross-coupling cascade reaction[J]. Synthesis, 2014, 46(15): 2051-2056.

[85] Ju B, Chen S, Kong W. Enantioselective palladium-catalyzed diarylation of unactivated alkenes[J]. Chemical Communications, 2019, 55(95): 14311-14314.

[86] Matsude A, Hirano K, Miura M. Highly stereoselective synthesis of 1, 2-disubstituted indanes by Pd-catalyzed Heck/Suzuki sequence of diarylmethyl carbonates[J]. Organic Letters, 2020, 22(8): 3190-3194.

[87] Zhang C, Wu X, Wang C, et al. Pd/Cu-catalyzed domino cyclization/deborylation of alkene-tethered carbamoyl chloride and 1, 1-diborylmethane[J]. Organic Letters, 2020, 22(16): 6376-6381.

[88] Arthuis M, Pontikis R, Florent J C. Stereoselective synthesis of novel highly substituted isochromanone and isoquinolinone-containing exocyclic tetrasubstituted alkenes[J]. The Journal of Organic Chemistry, 2009, 74(5): 2234-2237.

[89] Ju B, Chen S, Kong W. Enantioselective palladium-catalyzed diarylation of unactivated alkenes[J]. Chemical Communications, 2019, 55(95): 14311-14314.

[90] Grigg R, Fretwell P, Meerholtz C, et al. Palladium catalysed synthesis of spiroindolines[J]. Tetrahedron, 1994, 50(2): 359-370.

[91] Yao T, He D. Palladium-catalyzed domino Heck/aryne carbopalladation/C-H functionalization: synthesis of heterocycle-fused 9, 10-dihydrophenanthrenes[J]. Organic Letters, 2017, 19(4): 842-845.

[92] Ohno H, Yamamoto M, Iuchi M, et al. Palladium-catalyzed tandem cyclization of bromoenynes through aromatic C-H bond functionalization[J]. Angewandte Chemie, 2005, 117(32): 5233-5236.

[93] Furuta T, Asakawa T, Iinuma M, et al. Domino Heck-C-H activation reaction of unsymmetrically substituted[3]cumulene[J]. Chemical communications, 2006, (34): 3648-3650.

[94] Pinto A, Neuville L, Zhu J. Palladium-catalyzed three-component synthesis of 3-(diarylmethylene) oxindoles through a domino Sonagashira/carbopalladation/C-H activation/C-C bond-forming sequence[J]. Angewandte Chemie, 2007, 119(18): 3355-3359.

[95] Lafrance M, Gorelsky S I, Fagnou K. High-yielding palladium-catalyzed intramolecular alkane arylation: reaction development and mechanistic studies[J]. Journal of the American Chemical Society, 2007, 129(47): 14570-14571.

[96] Ruck R T, Huffman M A, Kim M M, et al. Palladium-catalyzed trandem Heck reaction/C-H functionalization-preparation of spiro-indane-oxindoles[J]. Angewandte Chemie International Edition, 2008, 47(25): 4711-4714.

[97] Satyanarayana G, Maichle-Mössmer C, Maier M E. Formation of pentacyclic structures by a domino sequence on cyclic enamides[J]. Chemical Communications, 2009, (12): 1571-1573.

[98] Piou T, Neuville L, Zhu J. Spirocyclization by palladium-catalyzed domino Heck-direct C-H arylation reactions: synthesis of spirodihydroquinolin-2-ones[J]. Organic Letters, 2012, 14(14): 3760-3763.

[99] Lied F, Brodnik Žugelj H, Kress S, et al. Employing Pd-catalyzed C-H arylation in multicomponent-multicatalyst reactions (MC)2R: one-pot synthesis of dihydrobenzoquinolines[J]. ACS Catalysis, 2017, 7(2): 1378-1382.

[100] Liu J G, Chen W W, Gu C X, et al. Access to spiroindolines and spirodihydrobenzofurans via Pd-catalyzed domino Heck spiroyclization through C-H activation and carbene insertion[J]. Organic Letters, 2018, 20(9): 2728-2732.

[101] Ishu K, Kumar D, Maurya N K, et al. Dicarbofunctionalization of unactivated alkenes by palladium-catalyzed domino Heck/intermolecular direct hetero arylation with heteroarenes[J]. Organic & Biomolecular Chemistry, 2021, 19(10): 2243-2253.

[102] Jaegli S, Erb W, Retailleau P, et al. Palladium-catalyzed domino process to spirooxindoles: ligand effect on aminopalladation versus carbopalladation[J]. Chemistry: A European Journal, 2010, 16(20): 5863-5867.

[103] Liu X, Ma X, Huang Y, et al. Pd-catalyzed heck-type cascade reactions with *N*-tosyl hydrazones: an efficient way to alkenes via in situ generated alkylpalladium[J]. Organic Letters, 2013, 15(18): 4814-4817.

[104] Zheng H, Zhu Y, Shi Y. Palladium(0)-catalyzed Heck reaction/C-H activation/amination sequence with diaziridinone: a facile approach to indolines[J]. Angewandte Chemie, 2014, 126(42): 11462-11466.

[105] Yoon H, Lossouarn A, Landau F, et al. Pd-catalyzed spirocyclization via C-H activation and benzyne insertion[J]. Organic Letters, 2016, 18(24): 6324-6327.

[106] Shao C, Wu Z, Ji X, et al. An approach to spirooxindoles via palladium-catalyzed remote C-H activation and dual alkylation with CH_2Br_2[J]. Chemical Communications, 2017, 53(75): 10429-10432.

[107] Luo X, Zhou L, Lu H, et al. Palladium-catalyzed domino Heck/C-H activation/decarboxylation: a rapid entry to fused isoquinolinediones and isoquinolinones[J]. Organic Letters, 2019, 21(24): 9960-9964.

[108] Yan F, Liang H, Ai B, et al. Palladium-catalyzed intermolecular [4+2] formal cycloaddition with (*Z*)-3-iodo allylic nucleophiles and allenamides[J]. Organic & Biomolecular Chemistry, 2019, 17(10): 2651-2656.

[109] Chen M, Wang X, Yang P, et al. Palladium-catalyzed enantioselective Heck carbonylation with a monodentate phosphoramidite ligand: asymmetric synthesis of (+)-physostigmine, (+)-physovenine and (+)-folicanthine[J]. Angewandte Chemie International Edition, 2020, 59(29): 12199-12205.

[110] Satyanarayana G, Maichle-Mössmer C, Maier M E. Formation of pentacyclic structures by a domino sequence on cyclic enamides[J]. Chemical Communications, 2009, (12): 1571-1573.

[111] Ruck R T, Huffman M A, Kim M M, et al. Palladium-catalyzed tandem Heck reaction/C-H functionalization-preparation of spiro-indane-oxindoles[J]. Angewandte Chemie International Edition, 2008, 47(25): 4711-4714.